Filtering in the Time and Frequency Domains

Filtering in the Time and Frequency Domains

Herman J. Blinchikoff
& Anatol I. Zverev

Westinghouse Electric Corporation

A Wiley-Interscience Publication

JOHN WILEY & SONS New York · London · Sydney · Toronto

Copyright © 1976 by John Wiley & Sons, Inc.

All rights reserved. Published simultaneously in Canada.

No part of this book may be reproduced by any means, nor transmitted, nor translated into a machine language without the written permission of the publisher.

Library of Congress Cataloging in Publication Data:

Blinchikoff, Herman J
   Filtering in the time and frequency domains.

   "A Wiley-Interscience publication."
   Bibliography: p.
   Includes index.
   1. Electric filters. I. Zverev, Anatol I., joint author. II. Title.

TK7872.F5B56   621.3815'32   76-120
ISBN 0-471-98679-8

Printed in the United States of America

10 9 8 7 6 5 4 3 2 1

To Nannette, Marlene, Laurie Jo, Cari, and my parents

H. J. B.

To my son Alex

A.I.Z.

Preface

Filtering, a fundamental signal process in electronics, includes such diverse operations as channeling, demodulating, equalizing, detecting, decoding, phase-splitting, integrating, and differentiating. Consequently, it is increasingly important for system-oriented engineers as well as for equipment designers to know the capabilities and limitations of filters in order to ensure effective system design and synthesis.

Books on filtering are written primarily for the filter designer and are therefore too detailed for the broader audience. The nonspecialist wants a treatment of filtering that deals with pertinent questions related to insertion loss, group delay, delay equalization, trade-offs between attenuation and group delay, trade-offs between attenuation and impulse response overshoots, and the feasibility of specified responses in both time and frequency domains.

Accordingly, the first goal of this book is to provide helpful information to the filter designer and to the electronics engineer who does not specialize in filter design, so that each can perform his task more efficiently and knowledgeably. The second goal is to show that the area of filtering includes devices and concepts that historically have not been treated within this discipline.

Much of the material presented here was given in course form to our colleagues during the years 1969–1975. Most participants were engineers interested in the general aspects of filtering. That this course was well received encouraged us to write this book, a text for electrical engineering students as well as for practicing engineers in industrial applied science schools. Our intent is to establish a relationship between industrial thinking and university teaching.

Thus in this book the systems engineer can find what the filter can do for him. Knowledge of the response trade-offs and the required effort for filter design and synthesis is very important in the decision-making process associated with electronic systems. To aid him in this task we provide the necessary denormalization equations for the design curves in *Handbook of Filter Synthesis* by A. I. Zverev (Wiley, 1967). This allows the significant responses of the various filter types to be easily determined. The two books complement each other and offer a more complete picture of the filtering possibilities.

To establish the appropriate mathematical relationships we use four basic transforms—Fourier, Laplace, Hilbert, and z. The first two relate the frequency-domain and time-domain characterizations while the Hilbert transform relates attenuation and phase (group delay) for minimum-phase networks. The z-transform is the tool for analyzing discrete-time systems.

Although the word "filter" usually suggests a device designed to exhibit a specified magnitude response, the widespread use of pulsed systems (such as radar, television, and communication) and the newest signal processing schemes have greatly expanded the concept of a filter.

The pulsed systems verified that the frequency responses were not powerful enough to adequately describe the important system behavior that occurred in the time domain. As a result, the impulse response emerged as the more important system characterization. However, the frequency-domain approach to analysis and design has such justifiably deep roots that it is still difficult to appreciate that in many cases optimum system performance can be obtained only through the time domain.

The time-domain approach and the extension of the filtering concept are exemplified by the matched filter, whose output signal-to-noise ratio is optimized when its impulse response is the mirror image of the transmitted pulse. This is a true filter in every sense of the word, yet frequency need not be mentioned to classify and design it. The filter is signal selective rather than frequency selective. No attempt is made to preserve the shape of the input signal at the output; rather, effort is directed toward signal detection.

Filtering aspects are discussed in terms of lumped-constant networks because these components and their analysis are familiar. We generally avoid explicit discussion of the hardware and concentrate instead on the function of the hardware, while also including many applications and implementations. The solid state engineer can use this material for designing and analyzing active filters, whose basic element is the operational amplifier. By using lumped equivalent circuits for his cavities and distributed elements, the microwave engineer can transfer the data presented here to his own expertise. The designer of digital circuits, surface wave devices, and optical systems will also find this information beneficial.

In Chapter 1, using the differential equation as the fundamental system description, we show how to obtain the filtering functions associated with physical systems; namely, the impulse response, step response, weighting function, and convolution integral. Chapter 2 introduces the Fourier and Laplace transforms, which lead to the frequency-domain system descriptions including the transfer function, magnitude response, phase response, and group-delay response. An introduction to the Hilbert transform, which is useful for relating specific network functions, concludes the chapter. These first two chapters should be helpful to those who have not specifically studied the time- and frequency-domain relationships.

Theoretical and realizable lowpass responses, including limitations in the time and frequency domains, are discussed in Chapter 3. The important attributes of each response are described, including a special section on rise time and noise bandwidth, two system design parameters.

In Chapter 4 we concentrate on the transformation of the normalized lowpass prototype into other filter types. The narrow-band bandpass filter is discussed in detail because its analysis is applicable to crystal, helical, coaxial cavity, stripline, interdigital, and waveguide filters.

We consider the all-pass function—a function that is useful for phase and group delay equalization and for the simulation of specified delays—in Chapter 5. Practical filter networks always include losses, and in Chapter 6 we discuss their effect on the theoretical responses. Useful design curves depict the attenuation response, insertion loss, and bandwidth change due to finite-Q elements.

In Chapter 7 we switch the focus from classical filter treatment to a consideration of the filtering of signals in a noisy environment, in particular, the matched filter. Since most communication and radar systems incorporate a form of

Preface

matched filtering, we feel that this subject should become a part of the filter engineer's repertoire. Following an in-depth study of matched filtering, we investigate a practical example of it—pulse compression using linear FM and then the Barker codes. We do not emphasize the hardware implementation although optical, surface wave, digital, and charge transfer devices are briefly discussed.

Chapter 8 treats two methods of time-domain synthesis, the quasi-stationary approach to the analysis of linear systems excited by modulated inputs, and the subject of average time delay. Chapter 9 is devoted to digital filtering and includes a discussion of the z-transform, which is then used to derive the important filter characterizations. Although introductory in nature, this chapter provides the specifics of digital filtering and relates them to familiar analog filter descriptions.

Problems and answers are supplied for each chapter, and numerous examples are given throughout. Suggested prerequisites for a course using this book include Laplace transform theory, circuit theory, and basic differential equations.

We express our gratitude to our many colleagues, in particular to Roy Anderson, Howard Fitzhugh II, Earnest Harrison, James Klekotka, Dr. Raymond Martin, Richard Morrison, and Samuel Zimmerman for their helpful criticisms, suggestions, and comments. We also thank Mayer Savetman who programmed the computer for many of the calculations, and Ms. Dawn Martinsen and Mrs. Kathleen D. Gibbons who accurately typed several drafts of the manuscript.

THE AUTHORS

Baltimore, Maryland
December 1975

Contents

Chapter 1. Time–Domain Analysis   1

1.1 Definitions   2
1.2 The Differential Equation   3
1.3 Solution of the Homogeneous Equation   5
1.4 Solution of the Nonhomogeneous Equation   8
     1.4.1 Specific Case, 9
     1.4.2 More General Case, 11
1.5 Convolution Integral   12
1.6 Impulse Function   15
1.7 Step Function   17
1.8 Impulse Response, Step Response, and Weighting Function   18
1.9 Summary   23
References   24
Problems   24

Chapter 2. Frequency–Domain Analysis   27

2.1 The Fourier Transform   27
     2.1.1 Real Time Functions, 29
     2.1.2 Causal Time Functions, 30
     2.1.3 Symmetric Time Functions, 31
     2.1.4 Parseval's Theorem, 32
2.2 The Laplace Transform   37
2.3 The Inverse Transform   43
2.4 Solution of Differential Equations by Laplace Transforms   49
2.5 The Transfer Function   53
     2.5.1 Derivation, 53
     2.5.2 Poles and Zeros, 56
     2.5.3 Steady-State Responses, 58
     2.5.4 s-Plane Geometry, 64
2.6 Group Delay and Phase Delay   66
     2.6.1 Definitions, 66
     2.6.2 System Delay and Signal Distortion, 70
     2.6.3 Phase-Intercept Distortion, 72
     2.6.4 Modulated-Signal Delay, 73
2.7 The Hilbert Transform   75
References   76
Problems   77

Chapter 3. Linear System Responses — 79

3.1 Ideal Low–Pass Responses — 81
   3.1.1 Rectangular Magnitude with Linear Phase, 81
   3.1.2 Impulse Response, 82
   3.1.3 Step Response, 83
   3.1.4 Gaussian Magnitude with Linear Phase, 84
   3.1.5 Impulse Response, 85
   3.1.6 Step Response, 86
   3.1.7 Summary of Results, 87
   3.1.8 Paley-Wiener Condition, 88
   3.1.9 Minimum-Phase Functions, 89
   3.1.10 Realizable Networks, 91
   3.1.11 Minimum-Phase Network with Rectangular Magnitude Response, 93
   3.1.12 Minimum-Phase Network with Linear Phase Response, 94

3.2 Mathematical Approximations — 94
   3.2.1 Taylor Approximation, 96
   3.2.2 Chebyshev Approximation, 100
   3.2.3 Least-Squares Approximation, 103

3.3 Realizable Low–Pass Response Characterization — 106
   3.3.1 The Normalized Low–Pass Response, 107
   3.3.2 All-Pole Networks, 107
   3.3.3 Group Delay Function, 108

3.4 Rectangular Magnitude Response Approximations — 109
   3.4.1 Maximally Flat (Butterworth), 109
   3.4.2 Equiripple (Chebyshev), 117
   3.4.3 Legendre, 123
   3.4.4 Least Squares, 123

3.5 Constant Group Delay Approximations — 124
   3.5.1 Maximally Flat (Bessel) Delay, 124
   3.5.2 Equiripple (Chebyshev) Delay, 128
   3.5.3 Least-Squares Delay, 128

3.6 Ideal Time–Domain Approximations — 129
   3.6.1 Gaussian Response, 130
   3.6.2 Equiripple Transient Response Overshoots, 132
   3.6.3 Monotonic Step Response with Minimum Rise Time, 134

3.7 Special Cases — 135
   3.7.1 Transitional Responses, 135
   3.7.2 Specific Pole Locations, 136
   3.7.3 Synchronous Response, 136
   3.7.4 Cauer Filters, 137

3.8 Rise Time and Noise Bandwidth — 139
   3.8.1 Rise Time, 139
   3.8.2 Noise Bandwidth, 146
   3.8.3 Relationship Between Noise Bandwidth and Impulse Response Energy, 149

References — 149
Problems — 151

Chapter 4. Frequency Transformations — 153

4.1 Normalized Parameters — 153
4.2 Low–Pass Filter — 154
    4.2.1 Transient Responses, 155
    4.2.2 Element Values, 155
    4.2.3 Example of Low–Pass Calculation, 156
4.3 High–Pass Filter — 157
    4.3.1 Conventional Transformation, 157
    4.3.2 Transient Responses, 158
    4.3.3 Relationship Between LP and HP Transient Responses, 158
    4.3.4 Element Values, 163
    4.3.5 Example of High–Pass Calculation, 163
    4.3.6 Preservation of LP Transient Characteristics, 165
    4.3.7 Application of New High–Pass Transformation, 166
4.4 Bandpass Filter — 167
    4.4.1 Conventional Transformation, 167
    4.4.2 Group Delay Behavior, 171
    4.4.3 Pole-Zero Locations, 172
    4.4.4 Element Values, 173
    4.4.5 Example of Bandpass Calculation, 174
4.5 Narrow–Band Bandpass Filter — 178
    4.5.1 Basic Definitions, 178
    4.5.2 Example of Attenuation Calculation, 178
    4.5.3 Pole-Zero Locations, 179
    4.5.4 Group Delay Behavior, 179
    4.5.5 Example of Group Delay Calculation, 183
    4.5.6 Example of Wide-Band Group Delay Calculation, 183
    4.5.7 Transient Responses, 184
    4.5.8 Realizations with Nodal and Mesh Networks, 186
    4.5.9 Example of Network Design, 191
    4.5.10 Features of Nodal and Mesh Networks, 194
4.6 Wide-Band Constant-Delay Bandpass Filter — 196
    4.6.1 Transfer Function and Least-Squares Approximation, 196
    4.6.2 Tables of Optimum Parameters, 197
    4.6.3 Normalized Attenuation and Group Delay Curves, 199
    4.6.4 Element Values, 202
4.7 Bandstop Filter — 205
    4.7.1 Conventional Transformation, 205
    4.7.2 Example of Bandstop Calculation, 206
    4.7.3 Narrow-Band Transformation, 207
References — 209
Problems — 210

Chapter 5. All–Pass Functions — 212

5.1 Applications of All–Pass Filters — 212
    5.1.1 Delay Equalization, 212

      5.1.2 Synthesis of Linear Delay for Pulse Compression, 213
      5.1.3 Digital System Delay Equalization, 213
      5.1.4 Phase-Splitting Networks, 213
      5.1.5 Approximation Techniques, 214

5.2 All–Pass Transfer Function   215
      5.2.1 First-Order All–Pass Function, 215
      5.2.2 Example of First-Order All–Pass Delay Calculation, 217
      5.2.3 Butterworth Delay Equalization with a First-Order All–Pass Function, 217
      5.2.4 Second-Order All–Pass Function, 220
      5.2.5 Example of Second-Order All–Pass Delay Calculation, 222
      5.2.6 Transient Responses of a Second-Order All–Pass System, 224
      5.2.7 Butterworth Delay Equalization with a Second-Order All–Pass Function, 225
      5.2.8 Effect of Delay Equalization on Transient Responses, 228
      5.2.9 Transient Responses of Butterworth Filters with Linear Phase, 230
      5.2.10 Narrow-Band Filter Equalization, 232

5.3 Least-Squares Approximation   233
      5.3.1 Numerical Integration, 233
      5.3.2 Equalization of Filter Group Delay, 234
      5.3.3 Starting Values for Computation, 235
      5.3.4 Butterworth Delay Equalization with a Second-Order All–Pass Function, 235

5.4 Network Realizations   237
      5.4.1 First-Order Lattice Section, 238
      5.4.2 Second-Order Lattice Section, 238
      5.4.3 First-Order Bridge Networks, 240
      5.4.4 Second-Order Bridge Networks, 240
      5.4.5 Effect of Losses in a Second-Order Network, 242
      5.4.6 Example of Bandpass Delay Equalization, 244

References   246
Problems   246

Chapter 6. Finite-Q Elements and Predistortion   248

6.1 The Quality Factor   248
6.2 Lossy-Filter Responses   249
      6.2.1 Transfer Function, 249
      6.2.2 Transient Responses, 250
      6.2.3 Butterworth Filter Analysis, 251
      6.2.4 Chebyshev and Gaussian Filter Analysis, 270
      6.2.5 Q Requirement for Bandpass Filters, 272
      6.2.6 Illustrative Example, 272

6.3 Insertion Loss   274
      6.3.1 Mismatch Loss, 275
      6.3.2 Resistive Loss, 276

Contents

      6.3.3 Insertion Loss Curves, 276
      6.3.4 Approximate Expression for Insertion Loss, 278
      6.3.5 Minimum Insertion Loss Filter, 281
      6.3.6 Insertion Loss Comparison of Various Filters, 282
6.4 Predistortion   286
      6.4.1 Minimum Value of Quality Factor for Various Responses, 286
      6.4.2 Insertion Loss, 287
      6.4.3 Example of Insertion Loss Calculation, 288
      6.4.4 Example of Predistorted Filter Design, 288
      6.4.5 Summary, 289
References   289
Problems   289

Chapter 7. Optimum Linear Filtering   292

7.1 Autocorrelation, Cross–Correlation, and Power Density Functions   293
      7.1.1 Periodic Function Analysis, 294
      7.1.2 Aperiodic (Transient) Function Analysis, 297
      7.1.3 Example of Aperiodic Function Calculation, 298
      7.1.4 Random Function Analysis, 301
      7.1.5 Properties of the Autocorrelation Function, 304
      7.1.6 Cross–Correlation Function, 304
      7.1.7 Relationship Between Convolution and Cross–Correlation, 305
      7.1.8 Output Autocorrelation Function of a Linear System, 307
      7.1.9 Output Power Density Spectrum of a Linear System, 308
7.2 Linear Mean–Square Estimation   309
      7.2.1 Wiener-Kolmogoroff Filtering, 309
      7.2.2 Kalman Filtering, 311
      7.2.3 Summary, 312
7.3 Matched Filtering   312
      7.3.1 Matched Filter Derivation, 313
      7.3.2 Filtering the Rectangular Pulse, 316
      7.3.3 Frequency-Domain Characterization, 319
      7.3.4 Matched Filter Synthesis, 321
      7.3.5 Synthesis of the Rectangular Impulse Response, 322
      7.3.6 Synthesis of the Trapezoidal Impulse Response, 323
      7.3.7 Relationship of Matched Filtering to Cross–Correlation, 326
      7.3.8 Nonoptimal Conditions, 329
      7.3.9 Nonwhite Input Noise Spectrum, 333
7.4 Pulse Compression Using Linear Frequency Modulation   335
      7.4.1 The Linear FM Signal, 335
      7.4.2 The Linear FM Matched Filter, 338
      7.4.3 Compression Mechanism, 341
      7.4.4 Sidelobe Reduction, 345
      7.4.5 The Tapped Delay Line, 347
      7.4.6 Optimum Filter Realizations, 348

7.5	Phase–Coded Waveforms	350
	7.5.1 The Barker Codes, 353	
	7.5.2 Amplitude Spectra of Pulse Sequences, 355	
	7.5.3 Sidelobe Reduction, 357	
7.6	Matched Filter Technology	365
	7.6.1 Optical Signal Processor, 365	
	7.6.2 Surface Acoustic Wave Device, 366	
	7.6.3 Digital Filter, 366	
	7.6.4 Charge Transfer Device, 367	
References		368
Problems		369

Chapter 8. Time–Domain Operations — 371

8.1	Approximations to a Prescribed Function	372
	8.1.1 Method of Moments, 372	
	8.1.2 Approximation of the Rectangular Pulse by the Method of Moments, 374	
	8.1.3 Approximation of the Triangular Function by the Method of Moments, 376	
	8.1.4 Method of Least Squares, 377	
	8.1.5 Least-Squares Approximation for Specified Exponentials, 378	
	8.1.6 Approximation of the Rectangular Pulse by Two Specified Exponentials, 379	
	8.1.7 Least-Squares Approximation for Optimum Amplitudes and Exponents, 380	
	8.1.8 Approximation of the Rectangular Pulse by One Exponential, 381	
	8.1.9 Approximation of the Rectangular Pulse by Two Exponentials, 382	
	8.1.10 Orthogonal Filter, 382	
8.2	Response of Linear, Time-Invariant Systems to Modulated Waveforms	383
	8.2.1 Spectral Approach, 383	
	8.2.2 FM Waveform Analysis by the Spectral Approach, 384	
	8.2.3 Dynamic Approach, 385	
	8.2.4 Quasi-Stationary Analysis, 387	
	8.2.5 Response to an Amplitude-Modulated Signal, 391	
	8.2.6 Response to a Frequency-Modulated Signal, 394	
8.3	Response of Linear, Time-Variant Systems to Modulated Waveforms	395
	8.3.1 Quasi-Stationary Response, 396	
	8.3.2 Quasi-Stationary Analysis of a First-Order System, 397	
	8.3.3 Separable Systems, 399	
	8.3.4 Quasi-Stationary Analysis of Separable Systems, 401	
	8.3.5 Approximation of a Nonseparable System by a Separable System, 402	
	8.3.6 Summary, 403	

8.4	Average Delay Through a Time-Invariant System	403
	8.4.1 Definition of Average Delay, 403	
	8.4.2 Average Delay of a Sinusoid, 404	
	8.4.3 Average Delay of the Impulse Function, 405	
	8.4.4 Average Delay of the Step Function, 406	
	8.4.5 Average Delay of the Rectangular Pulse, 408	
	8.4.6 Summary of Delay Definitions, 410	
References		410
Problems		412

Chapter 9. Digital Filtering   414

9.1	The Uniform Sampling Theorem	415
	9.1.1 Theorem and Proof, 416	
	9.1.2 Reconstruction of Time Function, 417	
	9.1.3 Physical Interpretation of the Sampling Theorem, 418	
	9.1.4 Interpolation Functions, 420	
9.2	Definition of a Digital Filter	421
9.3	The Difference Equation	422
	9.3.1 Example of a Difference Equation Computation, 423	
	9.3.2 Digital Filter Simulation of a First-Order Difference Equation, 424	
9.4	The z-Transform	424
	9.4.1 Definitions and Properties, 426	
	9.4.2 Mapping the s-Plane into the z-Plane, 431	
	9.4.3 Frequency-Domain Characteristics, 432	
	9.4.4 The Inverse z-Transform, 434	
9.5	Application of z-Transforms to Difference Equations	436
	9.5.1 Classical Solution of Difference Equations, 436	
	9.5.2 Solution by z-Transforms, 437	
	9.5.3 System Function and Unit-Sample Response, 438	
	9.5.4 Example of System Response Calculation, 439	
	9.5.5 Frequency-Domain Functions, 440	
9.6	Introduction to Design Techniques	442
	9.6.1 The Impulse-Invariant Method, 442	
	9.6.2 Example Illustrating Impulse-Invariant Method, 445	
	9.6.3 The Bilinear z-Transform, 447	
	9.6.4 Example of Bilinear z-Transform Design, 449	
	9.6.5 Nonrecursive Filters, 450	
	9.6.6 Example of Nonrecursive Filter Computation, 452	
	9.6.7 Fourier Series Approach to Nonrecursive Filtering, 454	
	9.6.8 Example of the Fourier Series Design Technique, 456	
	9.6.9 Example of Nonrecursive Filter Realizations, 458	
	9.6.10 Window (Weighting) Functions, 458	
	9.6.11 Example of MTI Filter Design, 460	
9.7	Digital Networks, Error Sources, and the FFT	462
	9.7.1 Digital Networks, 463	

9.7.2 Example of Filter Realization in Parallel and Cascade Forms, 466
 9.7.3 Errors Due to Practical Hardware, 467
 9.7.4 The Fast Fourier Transform, 469
 9.7.5 Summary, 471

References — 472
Problems — 472

Answers to Problems — 475

Index — 487

Chapter One

Time-Domain Analysis

In most scientific areas the establishment of a useful system model is necessary to describe the system behavior. An effective model need not duplicate the system's physical mechanism, but it is essential that its analysis agree with system measurements to within an acceptable tolerance. For example, the molecules in a lattice structure are to a first approximation represented by masses connected by springs. Obviously this is not the physical arrangement in the solid, but this model has provided useful insights into the system performance, in addition to important results that have been verified by system measurements. Once a suitable model has been established, analyses can replace measurements to determine system behavior.

The linear system has proved to be a useful model for many physical processes. This is fortunate because the principle of superposition applies to these systems, thus allowing general response formulations. The nonlinear system model, however, does not admit of this general response formulation, and analysis is usually confined to specific systems.

Many filtering devices, independent of the hardware realization, are accurately modeled by a linear system. An important member of this class is the system described by an ordinary linear differential equation. So many models are characterized by this differential equation that we may consider this equation as a new model—more precisely, as a mathematical model—with its application in the physical, biological, and social sciences. Consequently its analysis is of theoretic and practical interest.

Here we connect a physical system and the differential equation describing it. We introduce the homogeneous equation and show that once its solutions are known, an explicit solution of the general equation is obtained. We then relate this solution to the impulse response, step response, and convolution integral, which are necessary concepts for the understanding and analysis of filtering systems. Finally, the impulse response is shown to characterize a system that is either physically realizable or physically unrealizable, depending on the time-scale ordering.

Most studies of differential equations do not treat this precise problem. Some are too concerned with the mathematical aspects of the equation, while others include unnecessary material, thereby obscuring the transition from the homogeneous solutions to the impulse response of a physical system. This self-contained discussion is intended to clarify this topic.

The important class of filters composed of conventional lumped elements is governed by constant-coefficient differential equations, where the independent variable is time. Thus our approach can be identified as the time-domain analysis of these lumped-constant filters. The concepts introduced, however, are also useful for understanding filtering systems that are described by partial differential equations (such as microwave transmission line filters) and by difference equations (such as digital filters).

The time-domain approach is useful when the problem is stated in terms of a differential equation, or alternatively, when the system impulse response is defined. For the nonconstant coefficient equation, analysis in the time domain is usually the only feasible approach.

Time is the primary variable, thus the physical description of the filtering system should be expressed in the time domain. After all, the input and output signals are defined in this domain. However, most filter specifications are expressed in terms of steady-state frequency concepts such as the magnitude, phase, and group delay responses. The common usage of such responses arises from the mathematical treatment of the problem.

When the system is time-invariant in addition to being linear, the coefficients of the differential equation describing it are constants. These systems are useful as models for many portions of real-life systems. Extensive use is made of the Fourier and Laplace transforms for analysis and design of these systems, leading to frequency as the independent variable rather than time. From a practical consideration, these transformations often lead to mathematical simplifications, thus justifying their use. However, when no definite advantage is gained by the use of these transformations, the time variable should be retained, since then a physical interpretation is attached to the mathematics. Even experienced engineers have become so preoccupied with the notion of frequency that they fail to appreciate the limitations, disadvantages, and subtleties of this concept. Optimum system performance is often only achievable by time-domain approaches.

The advent of television, radar, and other systems using pulses has emphasized that the response to an impulse function is the more important characteristic of the system and that the frequency concept is just a useful tool for achieving a desired time response. Today it is often more convenient and beneficial to think about and analyze filtering systems in the time domain.

1.1 DEFINITIONS

A system is a device, real or abstract, that interrelates the excitation and the response. Further classification of the system is necessary if it is to be analyzed. The classifications considered here pertain to linearity, causality, and time variance. The necessary terms are now defined.

A system is linear if the input $c_1 f_1(t) + c_2 f_2(t)$ produces an output $c_1 g_1(t) + c_2 g_2(t)$ for all $f_1(t)$ and $f_2(t)$, when it is known that an input $f_1(t)$ produces an output $g_1(t)$ and an input $f_2(t)$ produces an output $g_2(t)$. The c_1 and c_2 are arbitrary constants but may be complex numbers. This property of superposition is characteristic of linear systems. Two examples illustrate this idea.

Example 1-1. Consider a system that multiplies the input by 2. Then $g_1(t) = 2f_1(t)$ and $g_2(t) = 2f_2(t)$. If the input is then $c_1 f_1(t) + c_2 f_2(t)$, the output is $2(c_1 f_1(t) + c_2 f_2(t)) = c_1(2f_1(t)) + c_2(2f_2(t)) = c_1 g_1(t) + c_2 g_2(t)$, and the system is linear.

The Differential Equation

Example 1-2. Consider a system whose response is the square of the input. Then $g_1(t) = f_1^2(t)$ and $g_2(t) = f_2^2(t)$. Clearly, if the input is $c_1 f_1(t) + c_2 f_2(t)$, the output $(c_1 f_1(t) + c_2 f_2(t))^2$ is not $c_1 f_1^2(t) + c_2 f_2^2(t)$; hence this system is nonlinear.

If the system parameters remain constant with time, then the system is time-invariant. An ordinary differential equation with constant coefficients describes a linear time-invariant system. For this system, an input delay of T seconds leads to an output delay of T seconds. Thus if an input $f(t)$ produces an output $u(t)$, then $f(t - T)$ produces an output $u(t - T)$.

A system is time-variant if the system parameters vary with time in a prescribed manner. An ordinary differential equation with variable coefficients describes a linear time-variant system. This system has the property that an input delay of T seconds does not generally lead to an output delay of T seconds. In fact the system response depends on the instant that the input is applied. An example of a time-variant system is an inductor and varactor diode in parallel. For small signal operation, the capacitance of the varactor is a function of the bias voltage across it. Since the voltage may be a function of time, the capacitance, hence the resonant frequency are both time-dependent.

Except for a few sections, this book is concerned with time-invariant systems, but since the following theory on differential equations likewise applies to time-variant systems, we consider both until it is advantageous to work with the former only.

A system is nonanticipative if the output value at any instant is *independent* of the input values at all later instants. This system, also referred to as a causal or physically realizable system, has the property that if the input is zero for $t < t_0$, then the output is also zero for $t < t_0$.

A system is anticipative if the output value at any instant is *not independent* of the input value at a later instant. This system is also referred to as a noncausal or physically unrealizable system. An anticipative system responds before an input is applied; that is, it anticipates the input.

We occasionally consider anticipatory systems that exhibit optimum, unachievable responses that are useful as standards against which to compare realizable approximations. Furthermore, these anticipative responses often yield accurate rules of thumb that are useful in practice.

1.2 THE DIFFERENTIAL EQUATION

The differential equation implicitly relates the system input and output functions, for in the general case derivatives of both functions appear in the equation. However, it fails to define a unique response for each input function unless additional information about the system is given. This includes the initial system state (initial conditions), the anticipatory nature of the system (ordering of the time scale), and the domain for which the differential equation description is valid (the time interval).

We now consider a single-input, single-output, linear system whose input $f(t)$ and output $u(t)$ are implicitly related on the interval $t_1 \leq t \leq t_2$ by the nth-order ordinary differential equation

$$a_0(t)\frac{d^n u}{dt^n} + a_1(t)\frac{d^{n-1} u}{dt^{n-1}} + \cdots + a_n(t)u = f(t) \quad (1.2\text{-}1)$$

or

$$L(p, t)u = f \quad (1.2\text{-}2)$$

where

$$p^n = \frac{d^n}{dt^n}$$

$L(p, t) = a_0(t)p^n + a_1(t)p^{n-1} + \cdots + a_n(t)$, the differential operator

The $a_i(t)$ are assumed to be defined and continuous throughout the interval $t_1 \leq t \leq t_2$;† and we assume $a_0(t) \neq 0$ on this interval. The initial conditions for this system are expressed by the first $(n-1)$ derivatives of $u(t)$:

$$u(t_0) = 0$$

$$\left.\frac{du}{dt}\right|_{t=t_0} = u'(t_0) = 0$$

$$\vdots \quad (1.2\text{-}3)$$

$$\left.\frac{d^{n-1} u}{dt^{n-1}}\right|_{t=t_0} = u^{(n-1)}(t_0) = 0$$

where t_0 is a particular instant of time within the interval $(t_1 \leq t_0 \leq t_2)$. The right side of (1.2-1) may also include higher-order derivatives of $f(t)$, and this situation is discussed in Section 1.4.2. At present we confine our attention to systems characterized by (1.2-1).

Example 1-3. As an example of a linear system and the differential equation describing it, consider the RL network in Fig. 1-1, where the resistance and inductance values are each a prescribed function of time. The first-order differential equation relating the output current $i(t)$ and the input voltage $v(t)$ is

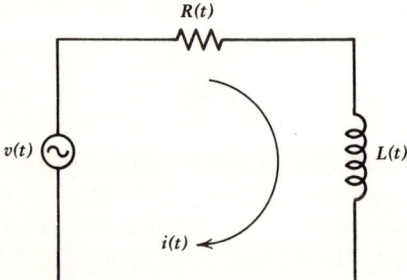

Figure 1-1. *RL* network.

---

†For the closed interval $t_1 \leq t \leq t_2$, derivatives at $t = t_1$ are evaluated as derivatives to the right of t_1 and those at $t = t_2$ are evaluated to the left of t_2.

Solution of the Homogeneous Equation

obtained by Kirchhoff's voltage law, which states that the sum of the voltage drops across the components equals the input voltage, that is,

$$L(t)\frac{di}{dt} + R(t)i = v(t) \qquad (1.2\text{-}4)$$

Comparing (1.2-4) with (1.2-1), we find that $n = 1$, $a_0(t) = L(t)$, and $a_1(t) = R(t)$.

It is generally unrecognized that the differential equation of (1.2-1) defines $(n + 1)$ nonequivalent systems, of which one is nonanticipative, one is anticipative, and the rest operate on both the past and future of the input [7]. Most treatments of linear systems consider only the nonanticipative case, since this is the one that relates to physical processes. However, an anticipative and nonanticipative system are defined once the time-scale ordering is specified, and we show this in Section 1.8. Then the words "past" and "future" take on conventional meanings and provide the means for distinguishing between these two systems [6, 7].

Henceforth, the system described by (1.2-1) is assumed nonanticipative until we discuss the time-scale ordering in Section 1.8. With this understanding, the initial conditions in (1.2-3), referred to as the *zero initial conditions*, imply that the system is in a relaxed state immediately *prior* to $t = t_0$, that is, there is no output if there is no input. For example, an RL series network is not in a relaxed state if there is an initial current in the inductor.

In general, the system initial conditions are not the zero initial conditions. However, from the theory of differential equations we can form u, the general solution of the differential equation, as

$$u = u_p + u_h \qquad (1.2\text{-}5)$$

where u_p is *any* particular solution and u_h is the sum of n linearly independent homogeneous solutions [2]. For the particular solution, we choose the one that satisfies the zero initial conditions; thus all initial conditions are satisfied by appropriately scaling the homogeneous solutions.

1.3 SOLUTION OF THE HOMOGENEOUS EQUATION

The existence theorem for a linear differential equation guarantees that there is a unique function $u(t)$ that identically satisfies (1.2-1) and the specified initial conditions on the interval $t_1 \leq t \leq t_2$ [1, 3]. This is a powerful theorem, for it tells us that a unique solution does exist, and our search for it is not in vain. Furthermore, no matter how we find this solution, it is the correct solution.

The particular solution to (1.2-1) with zero initial conditions can be explicitly expressed as an integral, whose integrand is the product of a kernel function and the input function $f(t)$. The significant aspect of this kernel function is that it is completely determined once the solutions of the homogeneous equation associated with (1.2-1) are known [2, 4, 8]. Because the homogeneous solutions are so necessary for the general solution, their important properties are now presented, together with explicit solutions of the constant-coefficient homogeneous equation.

The homogeneous differential equation related to (1.2-1) is obtained, in the notation of (1.2-2), as

$$L(p, t)u = 0 \tag{1.3-1}$$

and arises when the input function $f(t) = 0$. A solution of this equation, indicated by $u_i(t)$ and abbreviated u_i, is called a homogeneous solution.

Since $L(p, t)$ is an nth-order differential operator, there are n linearly independent (L.I.) functions, designated $u_1, u_2, \ldots, u_n$, that satisfy (1.3-1). The most general solution of the homogeneous equation u_h is composed of any linear combination of these n L.I. solutions [3], which can always be reduced to the form

$$u_h = c_1 u_1 + c_2 u_2 + \cdots + c_n u_n \tag{1.3-2}$$

where $c_1, c_2, \ldots, c_n$ are constants. It is important to note that the n L.I. solutions are not unique. Example 1-4 illustrates this fact for $n = 2$.

The linear independence of $u_1, \ldots, u_n$ implies that the equality $k_1 u_1 + \cdots + k_n u_n = 0$ can hold at every instant within the given interval only if the constants k_i are all zero. This condition is equivalent to the statement that $u_1, \ldots, u_n$ are linearly independent on the interval if and only if $W(t)$, the Wronskian determinant in (1.3-3), is not equal to zero for any value of t on the interval [1].

$$W(t) = \begin{vmatrix} u_1(t) & u_2(t) & \ldots & u_n(t) \\ u_1'(t) & u_2'(t) & \ldots & u_n'(t) \\ \vdots & \vdots & & \vdots \\ u_1^{(n-1)}(t) & u_2^{(n-1)}(t) & \ldots & u_n^{(n-1)}(t) \end{vmatrix} \tag{1.3-3}$$

Example 1-4. To show that the two L.I. solutions of a second-order homogeneous differential equation are not unique, we assume these solutions are $u_1 = e^{-t}$ and $u_2 = e^{-2t}$. The linear independence of u_1 and u_2 is verified by demonstrating that $W(t) \neq 0$ for any t on the finite interval.

$$W(t) = \begin{vmatrix} e^{-t} & e^{-2t} \\ -e^{-t} & -2e^{-2t} \end{vmatrix} = -e^{-3t} \tag{1.3-4}$$

Then u_h, from (1.3-2), is

$$u_h = c_1 e^{-t} + c_2 e^{-2t} \tag{1.3-5}$$

However, another equally valid set of L.I. solutions is $u_1 = e^{-t}$ and $u_2 = e^{-t} + e^{-2t}$, for which the Wronskian determinant is

$$W(t) = \begin{vmatrix} e^{-t} & e^{-t} + e^{-2t} \\ -e^{-t} & -e^{-t} - 2e^{-2t} \end{vmatrix} = -e^{-3t} \tag{1.3-6}$$

assuring their linear independence. Then, from (1.3-2), the general homogeneous solution is

$$u_h = c_0 e^{-t} + c_2 (e^{-t} + e^{-2t}) = (c_0 + c_2) e^{-t} + c_2 e^{-2t} \tag{1.3-7}$$

Equations 1.3-5 and 1.3-7 become identical if $(c_0 + c_2)$ is re-defined as c_1. Thus the general solution of the homogeneous equation is the same, even though a different

Solution of the Homogeneous Equation

set of L.I. solutions is initially selected. This conclusion is also valid for the nth-order equation.

The non-constant-coefficient homogeneous differential equation allows no simple ways of determining its solutions which are applicable to general equations of order n. Only for $n = 1$ can a general formulation of the solution be stated. However, these solutions can be obtained numerically on a digital computer. Thus we can assume that the homogeneous solutions are known.

For the constant-coefficient equation, the general solution u_h can always be represented as a sum of products of polynomials and exponentials; that is, each L.I. solution is of the form $t^k e^{mt}$ [3]. For this case, the problem reduces to a standard algebra problem and can be considered completely solved.

We now give the prescription for solving the homogeneous equation with constant coefficients, and follow it with an example illustrating the method.

1. Assume a solution of the form $u = e^{mt}$, where m is a constant, possibly complex.
2. Substitute this expression for u into (1.3-1) to give

$$a_0 m^n e^{mt} + a_1 m^{n-1} e^{mt} + \cdots + a_n e^{mt} = p(m) e^{mt} = 0 \qquad (1.3\text{-}8)$$

where

$$p(m) = a_0 m^n + a_1 m^{n-1} + \cdots + a_n \qquad (1.3\text{-}9)$$

and $p(m)$ is the *characteristic polynomial* associated with the given differential equation; e^{mt} is a solution of the differential equation if m is chosen as a root of

$$p(m) = 0 \qquad (1.3\text{-}10)$$

the *characteristic equation* associated with the differential equation.

3. Find the n roots of the characteristic equation.
4. Assign
   (a) to each simple real root m the function e^{mt}.
   (b) to each simple complex-root pair $\alpha \pm j\beta$, the functions $e^{\alpha t} \cos \beta t$, $e^{\alpha t} \sin \beta t$.
   (c) to each real root m of multiplicity k, the functions $e^{mt}, te^{mt}, \ldots, t^{k-1} e^{mt}$.
   (d) to each complex-root pair $\alpha \pm j\beta$, of multiplicity k, the functions

   $$e^{\alpha t} \cos \beta t, te^{\alpha t} \cos \beta t, \ldots, t^{k-1} e^{\alpha t} \cos \beta t$$

   $$e^{\alpha t} \sin \beta t, te^{\alpha t} \sin \beta t, \ldots, t^{k-1} e^{\alpha t} \sin \beta t$$

5. The n functions $u_1, \ldots, u_n$ thus obtained are L.I. solutions of (1.3-1), and (1.3-2) is then the general solution of the constant-coefficient homogeneous equation.

We previously mentioned that all solutions were of the form $t^k e^{mt}$, where t^k is included for multiple roots, yet the solutions just given contain sinusoidal functions. The apparent contradiction is resolved by noting that sine and cosine functions are linear combinations of exponentials with imaginary exponents, but they offer the advantage of being real functions. This is clarified by Example 1-5.

Example 1-5. The complex roots of the characteristic equation associated with a differential equation with real coefficients must occur in complex conjugate pairs.

Let these roots be $\alpha + j\beta$ and $\alpha - j\beta$. For this case $k = 0$, and the two L.I. solutions form a part of the general homogeneous solution as

$$c_3 e^{(\alpha+j\beta)t} + c_4 e^{(\alpha-j\beta)t} \tag{1.3-11}$$

We now convert the partial solution in (1.3-11) to a sum of real functions. First use Euler's identity

$$e^{\pm j\theta} = \cos\theta \pm j\sin\theta \tag{1.3-12}$$

in (1.3-11) to yield

$$e^{\alpha t}[c_3(\cos\beta t + j\sin\beta t) + c_4(\cos\beta t - j\sin\beta t)]$$
$$= e^{\alpha t}[(c_3 + c_4)\cos\beta t + j(c_3 - c_4)\sin\beta t] \tag{1.3-13}$$

Then identify $c_3 + c_4$ as the new constant c_1 and $j(c_3 - c_4)$ as c_2 to give

$$c_1 e^{\alpha t}\cos\beta t + c_2 e^{\alpha t}\sin\beta t \tag{1.3-14}$$

The functions $e^{\alpha t}\cos\beta t$ and $e^{\alpha t}\sin\beta t$ are indeed L.I., and the form of each is now the same as that prescribed in part (b) of item 4.

Example 1-6. We now determine the general solution of the constant-coefficient homogeneous equation

$$\frac{d^5 u}{dt^5} + 2\frac{d^3 u}{dt^3} + \frac{du}{dt} = 0 \tag{1.3-15}$$

Substitute $u = e^{mt}$ into (1.3-15) to obtain the characteristic equation

$$p(m) = m^5 + 2m^3 + m = 0 \tag{1.3-16}$$

The five roots of this polynomial are

$$m = 0 \qquad m = \pm j \qquad m = \pm j \tag{1.3-17}$$

and the function associated with each is obtained from item 4 in the prescription. For $m = 0$, from (a), we have

$$u_1 = e^{0 \cdot t} = 1 \tag{1.3-18}$$

For the double root at $\pm j$, from (d), $\alpha = 0$, $\beta = 1$, $k = 2$, and

$$\begin{aligned} u_2 &= e^{0 \cdot t}\cos t = \cos t \\ u_3 &= te^{0 \cdot t}\cos t = t\cos t \\ u_4 &= e^{0 \cdot t}\sin t = \sin t \\ u_5 &= te^{0 \cdot t}\sin t = t\sin t \end{aligned} \tag{1.3-19}$$

The general solution, from (1.3-2), is then

$$u_h = c_1 + c_2\cos t + c_3 t\cos t + c_4\sin t + c_5 t\sin t \tag{1.3-20}$$

1.4 SOLUTION OF THE NONHOMOGENEOUS EQUATION

Here we present the solution of the nonhomogeneous equation for two cases. The first case assumes that the differential equation does not include derivatives of the input function $f(t)$, and the second assumes that derivatives of $f(t)$ appear in the

Solution of the Nonhomogeneous Equation

differential equation. Each solution is presented in the form of an integral whose integrand is known once the n L.I. homogeneous solutions are specified.

1.4.1 Specific Case

The differential equation in (1.2-1) contains no derivatives of $f(t)$, and its solution, subject to zero initial conditions, is given by the following theorem [2, 4].

Theorem. Let $u_1, \ldots, u_n$ be any L.I. solutions of the homogeneous equation, and let $W(t)$ be the corresponding Wronskian. We define the function $k(t, \tau)$ as

$$k(t, \tau) = \frac{1}{a_0(\tau)W(\tau)} \begin{vmatrix} u_1(\tau) & u_2(\tau) & \cdots & u_n(\tau) \\ u_1'(\tau) & u_2'(\tau) & \cdots & u_n'(\tau) \\ \vdots & \vdots & & \vdots \\ u_1^{(n-2)}(\tau) & u_2^{(n-2)}(\tau) & \cdots & u_n^{(n-2)}(\tau) \\ u_1(t) & u_2(t) & \cdots & u_n(t) \end{vmatrix} \quad (1.4\text{-}1)$$

Here $k(t, \tau)$ is known as the kernel function associated with (1.2-1) and satisfies the following conditions at the instant $t = \tau$,

$$\left.\frac{\partial^i k(t, \tau)}{\partial t^i}\right|_{t=\tau} = 0 \quad i = 0, 1, \ldots, n-2$$

$$\left.\frac{\partial^{n-1} k(t, \tau)}{\partial t^{n-1}}\right|_{t=\tau} = \frac{1}{a_0(\tau)} \quad (1.4\text{-}2)$$

For the particular case $n = 1$,

$$k(t, \tau) = \frac{1}{a_0(\tau)} \frac{u_1(t)}{u_1(\tau)} \quad (1.4\text{-}3)$$

The solution of (1.2-1) with zero initial conditions is then

$$u_p(t) = \int_{t_0}^{t} k(t, \tau) f(\tau) \, d\tau \quad (1.4\text{-}4)$$

Proof. Use Leibnitz's rule for differentiating integrals [2] and the conditions in (1.4-2) to obtain

$$u_p'(t) = \int_{t_0}^{t} \frac{\partial k}{\partial t} f(\tau) \, d\tau + k(t, t) f(t) = \int_{t_0}^{t} \frac{\partial k}{\partial t} f(\tau) \, d\tau$$

$$\vdots$$

$$u_p^{(n-1)}(t) = \int_{t_0}^{t} \frac{\partial^{n-1} k}{\partial t^{n-1}} f(\tau) \, d\tau \quad (1.4\text{-}5)$$

$$u_p^{(n)}(t) = \int_{t_0}^{t} \frac{\partial^n k}{\partial t^n} f(\tau) \, d\tau + \frac{f(t)}{a_0(t)}$$

Substituting from (1.4-5) into (1.2-1) gives

$$\int_{t_0}^{t} \left[a_0(t) \frac{\partial^n k}{\partial t^n} + a_1(t) \frac{\partial^{n-1} k}{\partial t^{n-1}} + \cdots + a_n(t) k \right] f(\tau) \, d\tau + f(t) = f(t) \quad (1.4\text{-}6)$$

The sum of terms in the brackets is zero since $k(t, \tau)$ is a linear combination of $u_1, \ldots, u_n$, and is therefore a solution of the homogeneous equation. Hence (1.4-6) is satisfied and the theorem is proved.

The kernel function is unique, independent of the choice of the u_i's; and, from its definition (1.4-1), it is a continuous function of t and τ. Expansion of the determinant in (1.4-1) shows that $k(t, \tau)$ is separable in t and τ, that is, it may always be written in the bilinear form

$$k(t, \tau) = \sum_{i=1}^{n} u_i(t) v_i(\tau) \tag{1.4-7}$$

where $v_i(\tau)$ is a complicated function of the n L.I. homogeneous solutions and their first $(n-1)$ derivatives. As a sidenote, Kaplan [2] shows that the v_i's are the L.I. solutions of the adjoint equation corresponding to (1.3-1).

For a constant-coefficient equation, $k(t, \tau)$ simplifies to a function of only the difference in t and τ, taking the form

$$k(t-\tau) = \sum_{i=1}^{n} A_i u_i(t-\tau) \tag{1.4-8}$$

where A_i is a constant.

We have thus established that knowledge of the L.I. solutions of the homogeneous equation is sufficient to uniquely determine the kernel function.

Equation 1.4-4 represents the solution to (1.2-1) that satisfies the zero initial conditions. For other initial conditions, the general solution, from (1.2-5), is expressible as

$$u(t) = \int_{t_0}^{t} k(t, \tau) f(\tau) \, d\tau + c_1 u_1(t) + \cdots + c_n u_n(t) \tag{1.4-9}$$

where the constants $c_1, \ldots, c_n$ are determined to ensure that the specified conditions at $t = t_0$ are satisfied.

We have now established that the general solution can be represented as an integral that takes on zero value at $t = t_0$, and a group of terms that establish the specified conditions at $t = t_0$. The integral in (1.4-9) is known as the *superposition integral*. For zero initial conditions, $c_1 = c_2 = \cdots = c_n = 0$, and the total solution is then given by the superposition integral.

Example 1-7. Consider the differential equation

$$\frac{d^2u}{dt^2} + 3\frac{du}{dt} + 2u = f(t) \tag{1.4-10}$$

with $t_0 = 0$ and the initial conditions $u(0) = u_0$ and $u'(0) = u_0'$. The characteristic equation, from (1.3-9) and (1.3-10), is

$$p(m) = m^2 + 3m + 2 = 0 \tag{1.4-11}$$

the roots are $m = -1$ and $m = -2$, and two L.I. homogeneous solutions are

$$u_1(t) = e^{-t} \qquad u_2(t) = e^{-2t} \tag{1.4-12}$$

From (1.3-4) the Wronskian determinant $W(t) = -e^{-3t}$, and from (1.4-1) the kernel

Solution of the Nonhomogeneous Equation

function is

$$k(t, \tau) = \frac{1}{1 \cdot W(\tau)} \begin{vmatrix} e^{-\tau} & e^{-2\tau} \\ e^{-t} & e^{-2t} \end{vmatrix} = e^{-(t-\tau)} - e^{-2(t-\tau)} \quad (1.4\text{-}13)$$

Comparing (1.4-13) with (1.4-7) we find that $v_1(\tau) = e^{\tau}$ and $v_2(\tau) = -e^{2\tau}$. Because the differential equation has constant coefficients, $k(t, \tau)$ also has the form shown in (1.4-8), where $A_1 = 1$ and $A_2 = -1$. The general solution, from (1.4-9), is

$$u(t) = \int_0^t [e^{-(t-\tau)} - e^{-2(t-\tau)}] f(\tau) \, d\tau + c_1 e^{-t} + c_2 e^{-2t} \quad (1.4\text{-}14)$$

Inserting the two initial conditions in (1.4-14), we get the two simultaneous equations

$$\begin{aligned} u(0) &= u_0 = c_1 + c_2 \\ u'(0) &= u_0' = -c_1 - 2c_2 \end{aligned} \quad (1.4\text{-}15)$$

whose solutions are $c_1 = 2u_0 + u_0'$ and $c_2 = -u_0 - u_0'$. Then the total solution is

$$u(t) = \int_0^t [e^{-(t-\tau)} - e^{-2(t-\tau)}] f(\tau) \, d\tau + (2u_0 + u_0') e^{-t} - (u_0 + u_0') e^{-2t} \quad (1.4\text{-}16)$$

1.4.2 More General Case

We now consider what happens when the differential equation includes derivatives of the input $f(t)$, for many physical systems are so characterized. The right-hand side of (1.2-1) then has the form

$$b_0(t) \frac{d^m f}{dt^m} + \cdots + b_m(t) f \quad m < n \quad (1.4\text{-}17)$$

which may be interpreted as a more general input. With this interpretation, the solution of the differential equation with zero initial conditions, from (1.4-4), is

$$u_p(t) = \int_{t_0}^t k(t, \tau) \left[b_0(\tau) \frac{d^m f(\tau)}{d\tau^m} + \cdots + b_m(\tau) f(\tau) \right] d\tau \quad (1.4\text{-}18)$$

It is desirable to express $u_p(t)$ as

$$u_p(t) = \int_{t_0}^t k_1(t, \tau) f(\tau) \, d\tau \quad (1.4\text{-}19)$$

where $k_1(t, \tau)$ is a modified kernel function. In this form each new input function to a given system can be inserted into (1.4-19) and the solution can be determined, in contrast to (1.4-18), where derivatives of each new input must first be obtained. The relationship between $k_1(t, \tau)$ and $k(t, \tau)$ is [8]

$$\begin{aligned} k_1(t, \tau) = (-1)^m \frac{\partial^m}{\partial \tau^m} [b_0(\tau) k(t, \tau)] \\ + (-1)^{m-1} \frac{\partial^{m-1}}{\partial \tau^{m-1}} [b_1(\tau) k(t, \tau)] + \cdots + b_m(\tau) k(t, \tau) \end{aligned} \quad (1.4\text{-}20)$$

When $m = 0$, $k_1(t, \tau) = b_0(\tau) k(t, \tau)$.

Example 1-8. We now determine the solution of the differential equation

$$\frac{d^2u}{dt^2} + 3\frac{du}{dt} + 2u = 4\frac{df}{dt} + f \qquad f(t) = \begin{cases} 0 & t < 0 \\ e^{-t} & t \geq 0 \end{cases} \qquad (1.4\text{-}21)$$

by the technique just described. Zero initial conditions ($t_0 = 0$) are assumed. The homogeneous portion of (1.4-21) is the same as the homogeneous equation in Ex. 1-7, hence

$$k(t - \tau) = e^{-(t-\tau)} - e^{-2(t-\tau)} \qquad (1.4\text{-}22)$$

and $k_1(t - \tau)$, from (1.4-20) where $b_0(t) = 4$, $b_1(t) = 1$, and $m = 1$, is

$$k_1(t - \tau) = -\frac{\partial}{\partial \tau}[4k(t - \tau)] + k(t - \tau)$$

$$= -4[e^{-(t-\tau)} - 2e^{-2(t-\tau)}] + e^{-(t-\tau)} - e^{-2(t-\tau)}$$

$$= 7e^{-2(t-\tau)} - 3e^{-(t-\tau)} \qquad (1.4\text{-}23)$$

The general solution from (1.4-19) is then

$$u(t) = \int_0^t [7e^{-2(t-\tau)} - 3e^{-(t-\tau)}]f(\tau)\,d\tau \qquad (1.4\text{-}24)$$

With the exponential input given in (1.4-21)

$$u(t) = \begin{cases} 0 & t < 0 \\ 7e^{-t} - 7e^{-2t} - 3te^{-t} & t \geq 0 \end{cases} \qquad (1.4\text{-}25)$$

which can be verified as the solution to (1.4-21).

1.5 CONVOLUTION INTEGRAL

The convolution integral is of fundamental importance in the analysis and design of a linear time-invariant system, for it is an explicit representation of the system response. Its relationship to physical systems is considered in Section 1.8.

Let $p(t)$ and $q(t)$ each be a piecewise continuous function for $-\infty < t < \infty$. The convolution of p and q, denoted by $p*q$, is defined to be the third function

$$v(t) = p*q = \int_{-\infty}^{\infty} p(t - \tau)q(\tau)\,d\tau \qquad (1.5\text{-}1)$$

and the integral on the right is called the *convolution integral*. Furthermore, the convolution of q and p, $q*p$, is identical to $p*q$, that is,

$$q*p = p*q \qquad (1.5\text{-}2)$$

This is easily shown by letting $x = t - \tau$, $dx = -d\tau$ in (1.5-1), yielding

$$p*q = \int_{-\infty}^{\infty} p(x)q(t - x)\,dx = q*p \qquad (1.5\text{-}3)$$

An important special case, useful in later work, occurs when both $p(t)$ and $q(t)$

Convolution Integral

are zero for $t < 0$. Then (1.5-1) becomes

$$v(t) = \int_0^t p(t - \tau)q(\tau)\, d\tau \qquad (1.5\text{-}4)$$

From our discussion on constant-coefficient differential equations, the kernel function $k_1(t, \tau)$ can be expressed as $k_1(t - \tau)$, and the superposition integral in (1.4-19) then becomes

$$u_p(t) = \int_{t_0}^t k_1(t - \tau)f(\tau)\, d\tau \qquad (1.5\text{-}5)$$

Thus the solution of the differential equation can be expressed as the convolution of its kernel function and the driving function $f(t)$. In many practical situations $f(t) = 0$ for $t < 0$, hence $t_0 = 0$, and (1.5-5) reduces to the form of (1.5-4); namely,

$$u_p(t) = \int_0^t k_1(t - \tau)f(\tau)\, d\tau \qquad (1.5\text{-}6)$$

A graphical interpretation of the convolution integral is now given. It is emphasized that the following description is not the physical process of the system; rather, it is a graphical computation of the integral in (1.5-6). Assume that $f(\tau)$ and $k_1(\tau)$ are the functions appearing in Fig. 1-2. By its definition, (1.4-1) and (1.4-20), $k_1(\tau)$ is a continuous function and is so indicated in Fig. 1-2(b). However, only the solid-line portion of $k_1(\tau)$ enters into the convolution integral computation. This will become apparent as we proceed. Then $k_1(\tau - t)$, in Fig. 1-3a, represents $k_1(\tau)$ delayed by t. A sign change in the argument of $k_1(\tau - t)$ causes a reflection of the function about the point $\tau = t$, as in Fig. 1-3b.

For each value of t, the integral in (1.5-6) is equal to the area under the function $f(\tau)k_1(t - \tau)$ from $\tau = 0$ to $\tau = t$. The relative positions of $f(\tau)$ and $k_1(t - \tau)$ for three values of t and the corresponding values of the integral are given in Figs. 1-4 and 1-5, respectively. The area is indicated by the shading in Fig. 1-4. Note that values of $k_1(t - \tau)$ for which τ is greater than t do not enter into the computations. This is the hatched portion of $k_1(t - \tau)$.

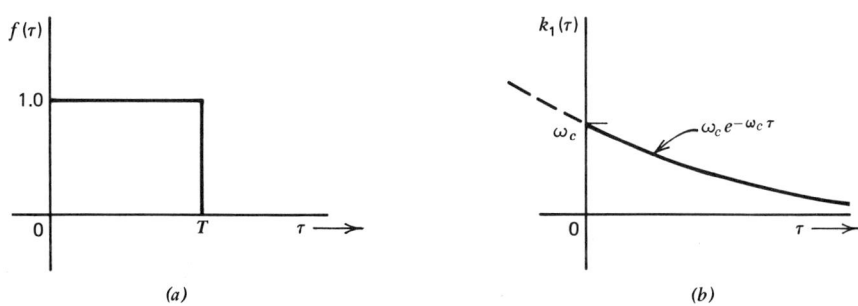

Figure 1-2. Plot of $f(\tau)$ and $k_1(\tau)$.

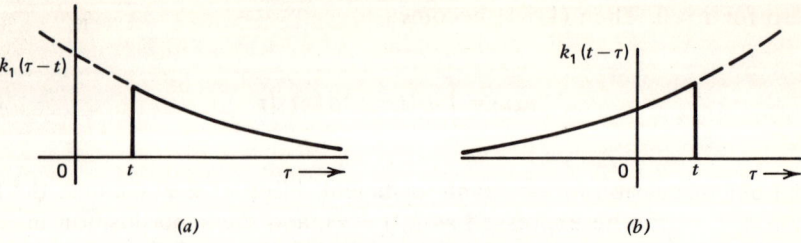

Figure 1-3. Plots of (a) $k_1(\tau - t)$ and (b) $k_1(t - \tau)$.

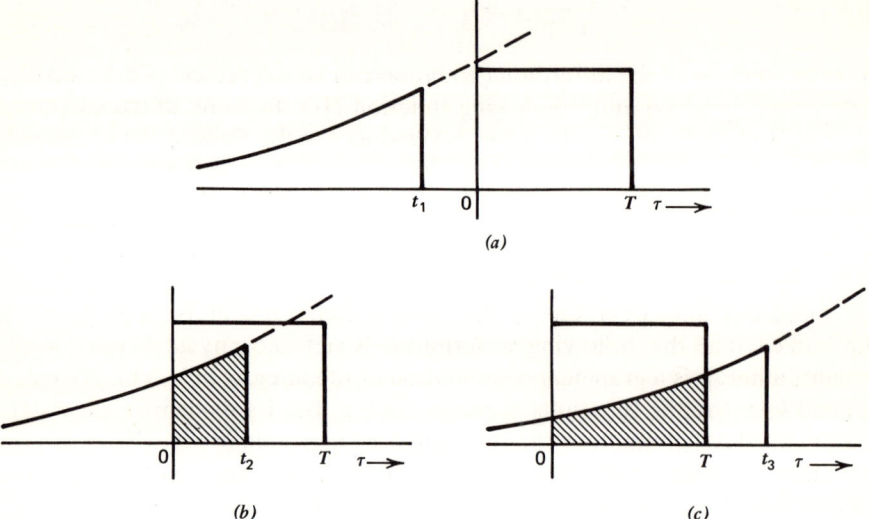

Figure 1-4. Convolution integral evaluation.

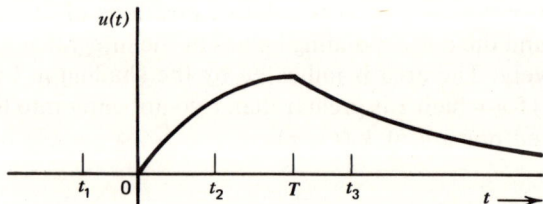

Figure 1-5. The convolution $u(t)$ of $f(t)$ and $k_1(t)$.

This graphical technique is also useful for estimating the integral value, which usually requires less time than computation of the integral. The general shape of $u(t)$ in Fig. 1-5 is easily determined from a pictorial sequence similar to that in Fig. 1-4. The graphical interpretation is additionally useful for obtaining the proper integral limits. Without this technique, determination of the limits can be difficult when one or both functions are described by different functions over various time intervals. This fact is illustrated in Example 1-9.

Impulse Function

Example 1-9. We now determine $u(t)$, the convolution of $k_1(t)$ and $f(t)$ in Fig. 1-2. Because $f(t)$ is of finite duration, the integral limits are more easily obtained by examination of Fig. 1-4. When $t < 0$ (Fig. 1-4a), there is no function overlap, and $u(t) = 0$. For $0 < t < T$ (Fig. 1-4b), the overlap occurs between 0 and t (t_2 in the figure); thus the upper integral limit is t. However, when $t \geq T$ (Fig. 1-4c), the overlap occurs only between 0 and T; thus the upper limit is now T. From (1.5-6), $u(t)$ is then

$$u(t) = \begin{cases} 0 & t \leq 0 \\ \omega_c \int_0^t e^{-\omega_c(t-\tau)} \cdot 1 \, d\tau = (1 - e^{-\omega_c t}) & 0 \leq t \leq T \\ \omega_c \int_0^T e^{-\omega_c(t-\tau)} \cdot 1 \, d\tau = (e^{\omega_c T} - 1)e^{-\omega_c t} & t \geq T \end{cases} \quad (1.5\text{-}7)$$

and is shown in Fig. 1-5.

1.6 IMPULSE FUNCTION

For use in analyzing quantum mechanical problems, the physicist Paul Dirac defined the function $\delta(t - a)$, shown in Fig. 1-6, to have the properties

$$\delta(t - a) = \begin{cases} 0 & t \neq a \\ \infty & t = a \end{cases} \quad (1.6\text{-}1)$$

$$\int_{a-\epsilon}^{a+\epsilon} \delta(t - a) \, dt = 1 \quad \epsilon > 0 \quad (1.6\text{-}2)$$

Equation 1.6-1 implies that the value of $\delta(t - a)$ is zero except at $t = a$, where it is infinite, and (1.6-2) implies that the area under $\delta(t - a)$ is unity. This function has several names, the most common being unit-impulse function, impulse function, delta function, and Dirac delta function. The word "unit" in the term "unit-impulse" is associated with the area under $\delta(t)$ as described by (1.6-2). Throughout this text we refer to $\delta(t)$ as the impulse function, with the understanding that its area is unity.

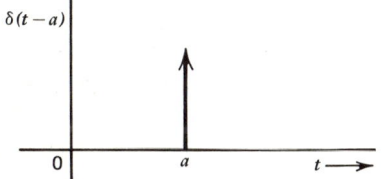

Figure 1-6. The unit impulse function.

The impulse function has proved to be extremely valuable in applied mathematics and system analysis, but clearly it does not fall into the class of ordinary functions. In this text, however, the impulse function is treated as an ordinary function because of the useful results that are obtained. For mathematical purists, its existence is legitimized by a branch of mathematics known as the theory of distributions, described in Appendix I of Ref. 5 and the references listed there.

One should not be overly concerned that no ordinary function can have the properties given in (1.6-1) and (1.6-2). The situation is similar to that encountered in solving the equation $x^2 = -1$. Since no real number satisfies this equation, we invent an imaginary number j that has the property $j^2 = -1$. Likewise, we invent an "ideal" function $\delta(t)$ to have the properties just described.

The impulse function can be generated as the limit of a number of functions, one being the rectangular pulse in Fig. 1-7. This pulse has unit area, regardless of

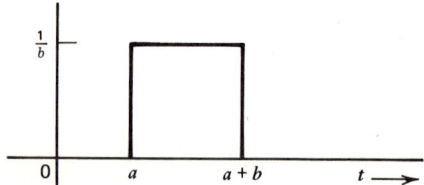

Figure 1-7. Rectangular pulse of unit area.

the value of b; and in the limit as $b \to 0$, it approaches the impulse function. Incidentally, a very narrow pulse is a common way of simulating the impulse function in the laboratory.

Another useful property of the impulse function is now given. If the impulse function appears as a factor in the integrand and $f(t)$ is continuous at $t = b$, the following simple result is obtained.

$$\int_a^c f(t)\,\delta(t-b)\,dt = \begin{cases} f(b) & a < b < c \\ 0 & b < a, b > c \end{cases} \qquad (1.6\text{-}3)$$

This is proved by noting that $\delta(t-b)$ is zero everywhere except at $t = b$, so we can change limits and then make ϵ as small as we like. Thus

$$\int_a^c f(t)\,\delta(t-b)\,dt = \int_{b-\epsilon}^{b+\epsilon} f(t)\,\delta(t-b)\,dt \qquad \epsilon > 0 \qquad (1.6\text{-}4)$$

Since $f(t)$ is assumed to be continuous at $t = b$, its change in value becomes very small as $\epsilon \to 0$; in fact, by decreasing ϵ we can make $f(t)$ as near to $f(b)$ over the range of integration as we like. It can then be removed from the integral, and aided by (1.6-2), we obtain the desired result.

$$\int_{b-\epsilon}^{b+\epsilon} f(t)\,\delta(t-b)\,dt = f(b) \int_{b-\epsilon}^{b+\epsilon} \delta(t-b)\,dt = f(b) \cdot 1 = f(b) \qquad (1.6\text{-}5)$$

1.7 STEP FUNCTION

Another function that we find useful in the study of filtering systems is the unit step function, defined as

$$u_{-1}(t-a) = \begin{cases} 0 & t < a \\ 1 & t \geq a \end{cases} \quad (1.7\text{-}1)$$

and plotted in Fig. 1-8. We shall refer to this function as the step function, although the unit amplitude is implied. This function has two salient properties: its value is zero when its argument is negative, and its value is unity when its argument is positive. When the argument is zero, the value of $u_{-1}(t)$ is undefined; therefore we assign it the value of unity.

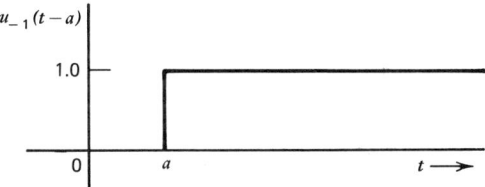

Figure 1-8. The unit step function.

To show the relationship between the step function and the impulse function, we integrate $\delta(t-a)$ from $-\infty$ to t. Then,

$$\int_{-\infty}^{t} \delta(x-a)\, dx = \begin{cases} 0 & t < a \\ 1 & t \geq a \end{cases} \quad (1.7\text{-}2)$$

But this is just $u_{-1}(t-a)$! We can therefore write

$$u_{-1}(t-a) = \int_{-\infty}^{t} \delta(x-a)\, dx \quad (1.7\text{-}3)$$

or

$$\delta(t-a) = \frac{d}{dt} u_{-1}(t-a) \quad (1.7\text{-}4)$$

Here, the d/dt symbol serves in the formal sense but of course has no meaning at the discontinuity. However, its use is justified in the same manner as we justify the use of the impulse function, namely, as an aid to understanding physical systems.

Step functions can also be used to synthesize waveforms composed of pulses. The unit-amplitude pulse in Fig. 1-9a can be expressed as the linear combination of two step functions as

$$p(t-a) = u_{-1}(t-a) - u_{-1}[t-(a+T)] \quad (1.7\text{-}5)$$

The first step function starts at $t = a$ and the second step function starts at

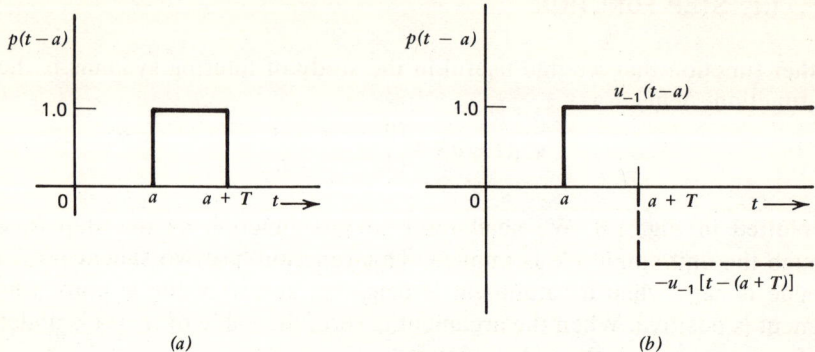

Figure 1-9. (a) Basic pulse, (b) synthesis of $p(t-a)$ from two step functions.

$t = a + T$, as in Fig. 1-9b. Subtraction of the second from the first gives the pulse in Fig. 1-9a. A waveform composed of a linear combination of pulses can then be constructed in terms of this basic pulse.

1.8 IMPULSE RESPONSE, STEP RESPONSE, AND WEIGHTING FUNCTION

The system impulse response $h(t, \tau)$ is defined to be the value of the response at the instant t to an impulse function applied at the instant τ, when the system is initially at rest (in a relaxed state). This response is the important characteristic of a linear system, for knowledge of $h(t, \tau)$ enables the system response to an arbitrary input function to be explicitly determined.

An impulse function at $t = \tau$ is represented as $\delta(t - \tau)$. Accordingly the impulse response is obtained by solving the differential equation in (1.2-1), with $f(t) = \delta(t - \tau)$, subject to the zero initial conditions. We have already shown, however, that the superposition integral in (1.4-19) is the solution of this differential equation for an arbitrary driving function $f(t)$. Therefore, after changing the integration variable from τ to η, the impulse response $h(t, \tau)$, is expressible as

$$h(t, \tau) = \int_{t_0}^{t} k_1(t, \eta) \delta(\eta - \tau) \, d\eta \tag{1.8-1}$$

It is generally assumed that the time-scale ordering requires that $t > t_0$, but this is not necessarily true. We now show that two different solutions are obtained depending on whether t is assumed greater than t_0 or t is assumed less than t_0.

If t is assumed greater than t_0, use of (1.6-3) reduces (1.8-1) to

$$h(t, \tau) = \begin{cases} k_1(t, \tau) & t_0 < \tau < t \\ 0 & t < \tau \end{cases} \tag{1.8-2}$$

At $t = \tau$ we set $h(\tau, \tau) = k_1(\tau, \tau)$. For the special case $k_1(t, \tau) = k(t, \tau)$ of (1.4-1), (1.4-2) states that $k(\tau, \tau) = 0$ for $n > 1$. Figure 1-10 shows the relative position of

Impulse Response, Step Response, and Weighting Function

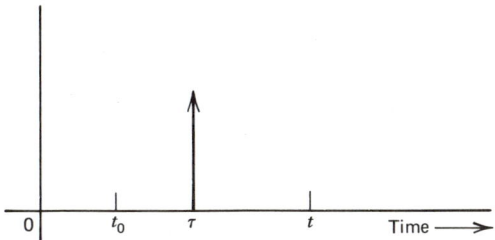

Figure 1-10. Relative position of the impulse function for nonzero output ($t > t_0$).

the impulse function for a nonzero output. Then $h(t, \tau)$ can be expressed as

$$h(t, \tau) = u_{-1}(t - \tau)k_1(t, \tau) \qquad (1.8\text{-}3)$$

This very important result states that the impulse response is zero for $t < \tau$ and equal to $k_1(t, \tau)$ for $t \geq \tau$. Thus the impulse response is nonanticipative (causal), since there is no response before the impulse function is applied.

If t is assumed less than t_0, (1.8-1) becomes

$$h(t, \tau) = -\int_t^{t_0} k_1(t, \eta)\delta(\eta - \tau)\, d\eta \qquad (1.8\text{-}4)$$

Application of (1.6-3) then yields

$$h(t, \tau) = \begin{cases} -k_1(t, \tau) & t < \tau < t_0 \\ 0 & t > \tau \end{cases} \qquad (1.8\text{-}5)$$

Figure 1-11 shows the relative position of the impulse function for a nonzero output. For this case the impulse function must be applied prior to $t = t_0$. Then

$$h(t, \tau) = -u_{-1}(\tau - t)k_1(t, \tau) \qquad (1.8\text{-}6)$$

Thus $h(t, \tau)$ is zero for $t > \tau$ and is $-k_1(t, \tau)$ for $t \leq \tau$. We see that $h(t, \tau)$ is strictly anticipatory, being nonzero only for the time before the impulse function is applied and zero thereafter.

The results of (1.8-3) and (1.8-6) substantiate the earlier statements that the differential equation can characterize either an anticipative or a nonanticipative system, depending on the ordering of the time scale.

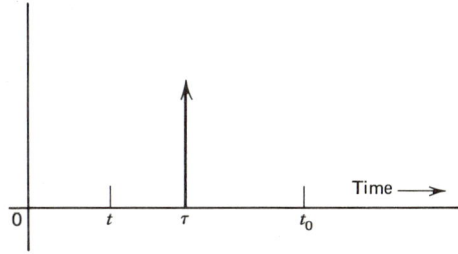

Figure 1-11. Relative position of the impulse function for nonzero output ($t < t_0$).

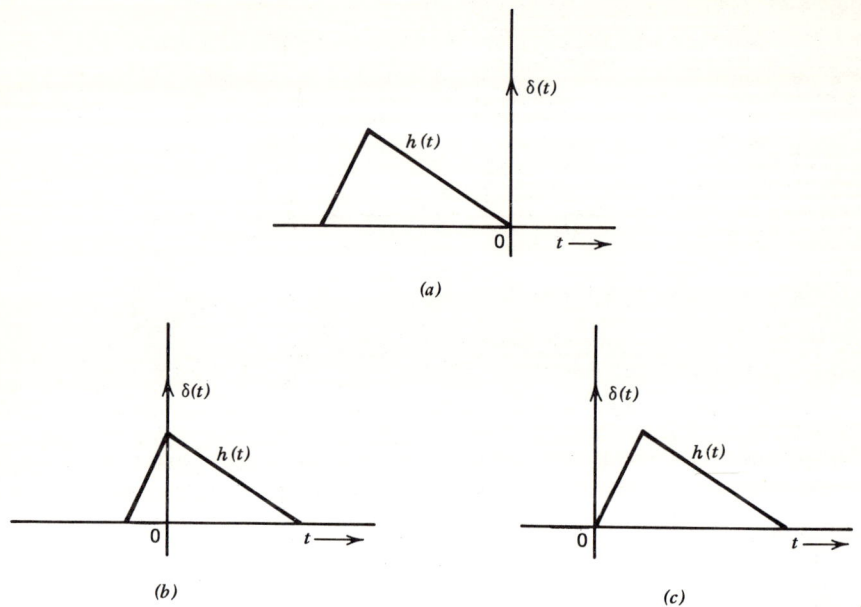

Figure 1-12. Three system responses to $\delta(t)$. (a) and (b) Anticipatory, (c) nonanticipatory.

We can also determine the anticipatory nature of the system by examining the impulse response. Let us examine three system responses to an impulse function applied at $t = 0$. The responses in Figs. 1-12a and 1-12b are anticipatory because each response begins before $\delta(t)$ is applied. The response in Fig. 1-12c is nonanticipatory because the response is zero before $\delta(t)$ is applied.

Since $k_1(t, \tau)$ is separable in t and τ, so also is $h(t, \tau)$. If the system is time-invariant, $h(t, \tau)$ can further be reduced to a function of $(t - \tau)$. Thus the impulse response of a time-variant system depends on the application time of the impulse function, whereas the impulse response of a time-invariant system depends only on the difference between the observation time and the impulse function application time. This time difference is often called the age variable.

Because we are primarily concerned with real physical systems, the systems are necessarily causal. For this condition, the proper ordering of the time scale is the more popular $t > t_0$, and $h(t, \tau)$ is given by (1.8-3). Then $k_1(t, \tau)$ in the superposition integral can be replaced by $h(t, \tau)$, since $h(t, \tau) = k_1(t, \tau)$ over the integration interval. The input-output relationship for a system at rest, from (1.4-19), is then

$$u(t) = \int_{t_0}^{t} h(t, \tau) f(\tau) \, d\tau \tag{1.8-7}$$

and for a time-invariant system, $u(t)$ can be written

$$u(t) = \int_{t_0}^{t} h(t - \tau) f(\tau) \, d\tau \tag{1.8-8}$$

Since $h(t - \tau)$ is zero outside the integration interval, the integral limits can be

Impulse Response, Step Response, and Weighting Function

changed to $-\infty$ and $+\infty$ without changing the integral value. Then $u(t)$ is expressed as

$$u(t) = \int_{-\infty}^{\infty} h(t-\tau)f(\tau)\, d\tau \qquad (1.8\text{-}9)$$

which we recognize as the general convolution integral of (1.5-1). It is very common in practice for the input $f(t)$ to be zero for $t < 0$. Then $t_0 = 0$ and (1.8-8) becomes

$$u(t) = \int_{0}^{t} h(t-\tau)f(\tau)\, d\tau \qquad (1.8\text{-}10)$$

which is the convolution integral of (1.5-4). Thus the convolution of the input signal and the system impulse response yields the system response.

We now have reached the amazing conclusion that a linear system with constant parameters is uniquely characterized by the single function $h(t)$, which is the system response to an impulse function applied at $t = 0$. It is this function that is termed the system's impulse response. Knowledge of $h(t)$ allows the determination of the system response to an arbitrary input by way of the convolution integral.

Now that we have established that convolving the system input and impulse response yields the system output, let us return to Section 1.5 and attach physical meaning to the folding operation shown there for the graphical solution of the integral. Refer to Fig. 1-2, where $k_1(\tau)$ is now interpreted as the impulse response. Therefore the hatched portion of $k_1(\tau)$ is now zero because $h(\tau) = 0$ for $\tau < 0$. If we do not reflect $k_1(\tau - t)$ about the point $\tau = t$, then $k_1(\tau - t)$ overlaps $f(\tau)$ for values of t between $-\infty$ and T, and a response is obtained before the input function is applied. This is the response of an anticipative system and is not the response of the causal system under discussion. The reflection of $k_1(\tau - t)$ shown in Fig. 1-3 remedies this, resulting in the situation represented in Fig. 1-4. There the impulse response first overlaps $f(\tau)$ at $t = 0$, consistent with physical reasoning. As t increases, the overlap of $f(\tau)$ increases in the proper sequence, yielding the causal system response in Fig. 1-5.

The impulse response is often referred to as the *weighting function*. The latter terminology arises because the output function from (1.8-10), at one instant t, depends on the weighted average of the input function over the interval 0 to t. At each instant t, $h(t - \tau)$ tells how well the system remembers past values of $f(t)$ and weights them accordingly. Although described by the same function of t and τ, the weighting function and the impulse response have different physical meanings. The impulse response is a function of t with τ as a fixed parameter, namely, the application time of the impulse function. The weighting function, however, is a function of τ with t as a fixed parameter.

The response of a linear system to a step function is of frequent interest in practice and is obtained by integrating the impulse response. This is verified by letting $f(\tau) = u_{-1}(\tau)$ in (1.8-7). Then

$$g(t) = \int_{0}^{t} h(t, \tau)\, d\tau \qquad (1.8\text{-}11)$$

For a time-invariant system, $g(t)$ is termed the step response and is given by

$$g(t) = \int_0^t h(t-\tau)\, d\tau = -\int_t^0 h(\eta)\, d\eta = \int_0^t h(\eta)\, d\eta \qquad (1.8\text{-}12)$$

The response to the delayed step function $u_{-1}(t-\tau)$ is obtained by replacing t by $t - \tau$,

$$g(t-\tau) = \int_0^{t-\tau} h(\eta)\, d\eta \qquad (1.8\text{-}13)$$

Example 1-10. To determine the impulse response and the step response of the causal system described by the differential equation in Example 1-8, we use the relation

$$\frac{d^2u}{dt^2} + 3\frac{du}{dt} + 2u = 4\frac{df}{dt} + f \qquad (1.8\text{-}14)$$

Since $k_1(t-\tau)$ is given by (1.4-23), from (1.8-3) the impulse response ($\tau = 0$) is

$$h(t) = u_{-1}(t)[7e^{-2t} - 3e^{-t}] \qquad (1.8\text{-}15)$$

The step response, obtained from (1.8-12), is then

$$g(t) = u_{-1}(t)\left[\frac{1}{2} - \frac{7}{2}e^{-2t} + 3e^{-t}\right] \qquad (1.8\text{-}16)$$

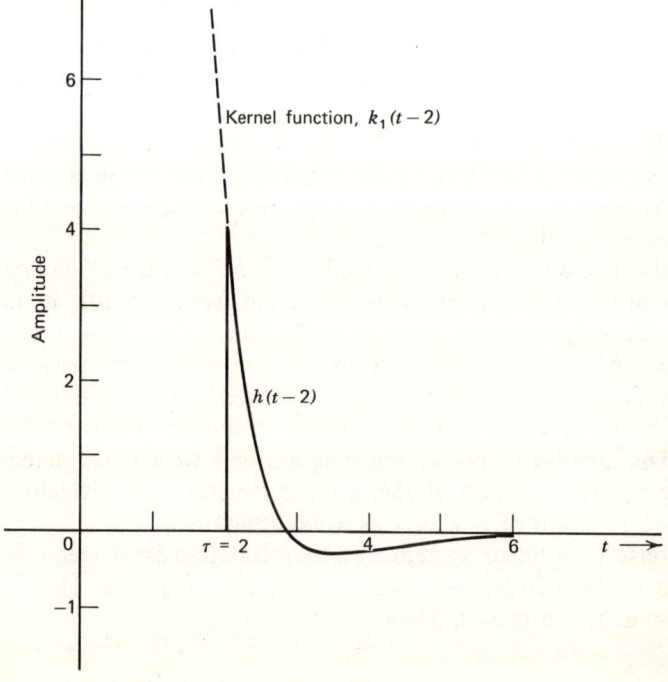

Figure 1-13. Kernel function ($\tau = 2$) and $h(t-2)$ for the system in Example 1-10.

Summary

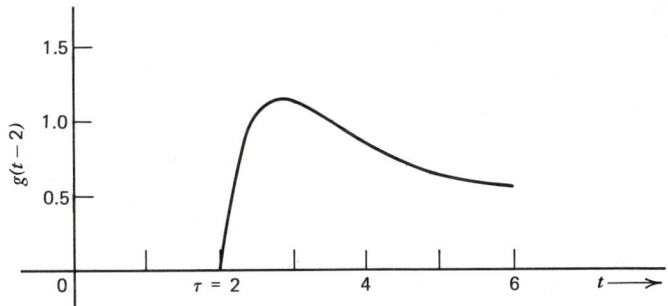

Figure 1-14. The $g(t-2)$ for the system in Example 1-10.

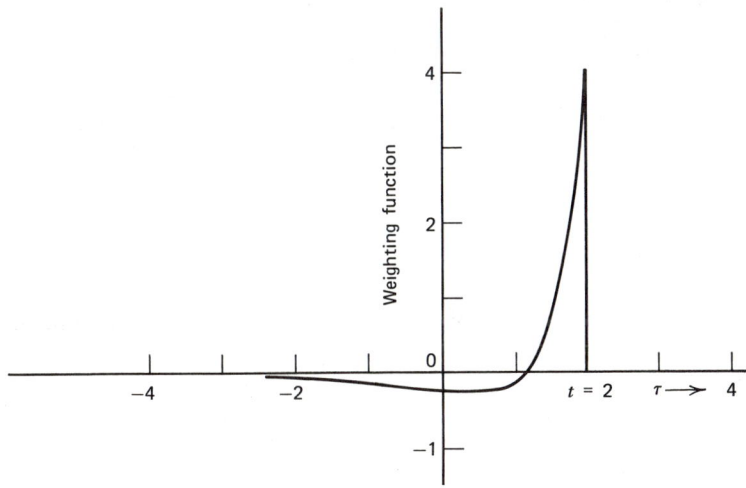

Figure 1-15. Weighting function ($t=2$) for the system in Example 1-10.

To illustrate the introduction of a delay $\tau = 2$, we show $k_1(t-2)$ and $h(t-2)$ in Fig. 1-13 and $g(t-2)$ in Fig. 1-14. The kernel function (dashed line) is nonzero for $t < 2$, whereas $h(t-2)$ and $g(t-2)$ are both zero for $t < 2$ because the system is nonanticipatory. For $t \geq 2$, $h(t-2)$ and $k_1(t-2)$ are identical. The weighting function $h(t-\tau)$ appears in Fig. 1-15 as a function of τ, with $t = 2$. It is the mirror image of the impulse response and weights the input function from the time of excitation to $t = 2$. The integral of this weighted function then yields the system output at $t = 2$. This process is the familiar convolution, depicted in Fig. 1-4.

1.9 SUMMARY

In this chapter we have consolidated the mathematical developments necessary to relate the physical system and the differential equation describing it. This included the derivation of the impulse response, step response, and convolution integral. These quantities, which are important to the theory of filtering, were shown to be a function of the L.I. solutions of the homogeneous equation. It is hoped that this time-domain approach has clarified the transition from the basic differential equation concepts to the everyday tools of the linear system analyst.

REFERENCES

1. Ince, E. L., *Ordinary Differential Equations*, Dover, New York, 1926.
2. Kaplan, W., *Operational Methods for Linear Systems*, Addison-Wesley, Reading, Mass., 1962.
3. Kaplan, W., *Ordinary Differential Equations*, Addison-Wesley, Reading, Mass., 1960.
4. Miller, K. S., "Properties of Impulsive Responses and Green's Functions," *IRE Trans. Circuit Theory*, Vol. CT-2, pp. 26–31, March 1955.
5. Papoulis, A., *The Fourier Integral and Its Applications*, McGraw-Hill, New York, 1962.
6. Weiss, L., "Dual Dynamical Systems and Their Representation by System Functions," *Int. J. Control*, Vol. 1, No. 5, pp. 475–485, May 1965.
7. Zadeh, L. A., "Time-Varying Networks, I", *Proc. IRE*, Vol. 49, pp. 1488–1503, October 1961.
8. Zadeh, L. A., "The Determination of the Impulsive Response of Variable Networks," *J. Appl. Phys.*, Vol. 21, pp. 642–645, July 1950.

PROBLEMS

1-1. Are the functions in each of the sets below linearly dependent or linearly independent?

(a) t, t^2, t^3 (e) $e^{-t}, \frac{1}{2}(t-1)$
(b) $\sin t, \cos t$ (f) $t^3 + 1, 4t^3 + 4$
(c) $\sin t, \sin 2t$ (g) $x^2 + 8, x^2 + 2$
(d) $\sin^2 t, \cos^2 t, \cos 2t$ (h) $e^{jt}, \sin t$

1-2. Find the general solution of each of the following differential equations.

(a) $\dfrac{d^2 u}{dt^2} - 16u = 0$ (d) $\dfrac{d^3 u}{dt^3} - 8u = 0$

(b) $\dfrac{d^3 u}{dt^3} - 3\dfrac{d^2 u}{dt^2} + 3\dfrac{du}{dt} - u = 0$ (e) $\dfrac{d^2 u}{dt^2} + 6\dfrac{du}{dt} + 25u = 0$

(c) $\dfrac{d^4 u}{dt^4} = 0$

1-3. For the differential equation

$$\frac{d^2 u}{dt^2} + (a+b)\frac{du}{dt} + abu = f(t)$$

(a) Determine the kernel function $k(t, \tau)$ when $a \neq b$ and when $a = b$. Since this is a constant-coefficient differential equation, $k(t, \tau)$ is a function of the difference $(t - \tau)$.

(b) Find the general solution when $f(t) = u_{-1}(t)$, assuming zero initial conditions, $t_0 = 0$, and $a \neq b$.

1-4. For a second-order, non-constant-coefficient differential equation, $k(t, \tau)$ is

Problems

expressible as

$$k(t, \tau) = \sum_{i=1}^{2} u_i(t) v_i(\tau)$$

where the $u_i(t)$ are the homogeneous solutions. Express $v_1(\tau)$ and $v_2(\tau)$ as a function of $u_1(\tau)$, $u_2(\tau)$, $u_1'(\tau)$, $u_2'(\tau)$, and $a_0(\tau)$.

1-5. For the differential equation

$$(t^2 + 9t + 20)\frac{d^2 u}{dt^2} + (t^2 + 10t + 23)\frac{du}{dt} + (t+3)u = f(t) \qquad (-4 < t)$$

two linearly independent solutions are $u_1(t) = e^{-t}$ and $u_2(t) = 1/(t+5)$.

(a) Find $k(t, \tau)$. Verify (1.4-2).
(b) With $t_0 = 0$ and initial conditions $u(0) = \frac{3}{2}$, $u'(0) = -1$, express the general solution as the sum of an integral and the homogeneous solutions.
(c) For zero initial conditions, determine $u(t)$ when $f(t) = \delta(t-2)$.

1-6. Evaluate the following integrals

(a) $\displaystyle\int_{5}^{100} e^{-2t} \delta(t-24)\, dt$

(b) $\displaystyle\int_{0}^{\infty} e^x \delta(x - \pi)\, dx$

(c) $\displaystyle\int_{0}^{\infty} (t^2 + 1)\delta(t+1)\, dt$

1-7. Plot
(a) $u_{-1}(t-2)$
(b) $u_{-1}(t) - u_{-1}(t-2)$
(c) $tu_{-1}(t) - 2(t-1)u_{-1}(t-1) + (t-2)u_{-1}(t-2)$

1-8. The differential equation relating the input voltage $f(t)$ and the output voltage $u(t)$ for the RL network below is

$$L\frac{du}{dt} + Ru = Rf(t)$$

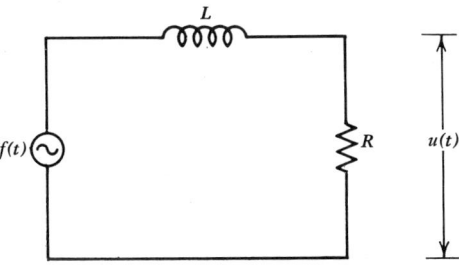

Figure P.1-8.

(a) Find the kernel function $k(t - \tau)$ and the impulse response $h(t - \tau)$.
(b) Plot the weighting function at the instant $t = 0$.
(c) Determine the response to the rectangular pulse of unit amplitude and duration T (Fig. 1-2a). There are two regions to consider: $t < T$ and $t > T$. Plot the general shape of the response.
(d) Compute $u(t)$ for the following inputs.
 (i) $u_{-1}(t)$ (ii) $\delta(t)$ (iii) $\delta(t - 5)$

1-9. The response of a linear time-invariant system to $f(t) = \delta(t - 5)$ is

$$u(t) = e^{5\alpha}\frac{(\alpha^2 + \beta^2)}{\beta} e^{-\alpha t}[\sin \beta t \cos 5\beta - \cos \beta t \sin 5\beta]u_{-1}(t - 5)$$

(a) What are the values of α and β so that the differential equation describing the filter is

$$\frac{d^2u}{dt^2} + 4\frac{du}{dt} + 13u = 13f(t)$$

(b) Write the expression for the kernel function.

Chapter Two
Frequency-Domain Analysis

Although the differential equation is a basic system description, obtaining this equation can be tedious and time-consuming. Consequently for a time-invariant system this approach is avoided in practice, except in special cases. The Fourier and Laplace transforms offer an alternative approach for characterizing and analyzing these systems. Insight into system behavior is often obtained by the transform method.

These transforms change a function of one variable into a function of another variable, and, when applied to problems in the physical sciences, the transform pairs and variables may correspond to physical quantities. We assign time and frequency as the transform variables because these are the variables associated with the filtering devices considered in this text. Accordingly, one of this chapter's goals is to relate the frequency-domain concepts introduced here to their time-domain counterparts introduced in Chapter 1.

We first discuss the mathematical aspects of these transforms and then apply the Laplace transform method to a differential equation to obtain its solution. This leads to the introduction and discussion of the transfer function, poles and zeros, and the steady-state responses—magnitude, phase, and group delay. The final section introduces a third transform, the Hilbert transform, which is useful for relating various functions of a specified system.

The basic premise underlying the Fourier and Laplace transforms is that a time function may be decomposed into a linear combination of exponentials. Furthermore, when the exponents are imaginary, the composite signal is representable as a spectrum of sinusoidal components.

Today, so much test equipment is built to test networks in terms of their steady-state responses to sinusoidal inputs that many of the quantities associated with sinusoidal behavior have acquired a very real "physical" meaning. This identification can often lead to conceptual difficulties. We must remember that time is the primary variable. In fact, we showed in Chapter 1 that the idea of frequency never needs to be introduced to describe the system and then determine its input-output relationship. Nevertheless, the steady-state concept is a valuable tool for designing a network with a specified time response. A great many computational and practical benefits can be had by using the steady-state response, and this choice in practice is certainly justified.

2.1 THE FOURIER TRANSFORM

The Fourier transform of an arbitrary time function $f(t)$ is indicated by $\mathscr{F}[f(t)]$ (spoken "Fourier transform of $f(t)$") and given by

$$\mathscr{F}[f(t)] = F(\omega) = \int_{-\infty}^{\infty} f(t)e^{-j\omega t} \, dt \qquad (2.1\text{-}1)$$

the frequency-domain representation of $f(t)$. Here ω is the radian frequency and is related to f, the frequency in hertz, by

$$\omega = 2\pi f \text{ rads/sec} \qquad (2.1\text{-}2)$$

Lowercase letters are reserved for the time functions; capital letters indicate their transforms.

The integral on the right in (2.1-1) is known as the Fourier integral, and it is useful for analyzing an aperiodic function, sometimes called a transient function. We use the Fourier series to obtain the frequency representation of a periodic function. This subject is briefly discussed in Section 7.1.1 and is covered thoroughly in Refs. 4 and 5. The Fourier integral is often derived from the Fourier series of a periodic function by allowing the period to approach infinity, thereby making the periodic function approach an aperiodic one.

The function $F(\omega)$ is generally complex, represented as

$$F(\omega) = R(\omega) + jX(\omega) = A(\omega)e^{j\theta(\omega)} \qquad (2.1\text{-}3)$$

where R, X, A, and θ are real functions. Here, $R(\omega)$ is the real part of $F(\omega)$, and $X(\omega)$ is the imaginary part of $F(\omega)$. The amplitude-density spectrum of $f(t)$, which is sometimes called the Fourier spectrum, is the absolute value of $F(\omega)$. This quantity $A(\omega)$ is given by

$$A(\omega) = |F(\omega)| = \sqrt{R^2(\omega) + X^2(\omega)} \qquad (2.1\text{-}4)$$

and the phase angle $\theta(\omega)$ is given by

$$\theta(\omega) = \tan^{-1} \frac{X(\omega)}{R(\omega)} \qquad (2.1\text{-}5)$$

The time function $f(t)$ can be recovered by the inversion formula

$$f(t) = \frac{1}{2\pi} \int_{-\infty}^{\infty} F(\omega)e^{j\omega t} \, d\omega \qquad (2.1\text{-}6)$$

that is, $f(t)$, which might consist of a number of completely different analytic pieces in the various parts of the t axis, is now represented by a single expression. Rewriting (2.1-6) as

$$f(t) = \int_{-\infty}^{\infty} \left[\frac{F(\omega) \, d\omega}{2\pi} \right] e^{j\omega t} \qquad (2.1\text{-}7)$$

reveals that $f(t)$ has been decomposed into an infinite number of sinusoids, represented by $e^{j\omega t}$, whose amplitudes are infinitesimal, each containing the factor $d\omega$. Thus the amplitude-density spectrum $A(\omega)$ is not the actual amplitude characteristic of $f(t)$, rather, it is a characteristic that shows relative magnitude only, hence, is called a density function. The phase angle of the infinitesimal sinusoids is given by $\theta(\omega)$.

Equations 2.1-1 and 2.1-6 are the classical Fourier transform pair, allowing

The Fourier Transform

passage between the time domain and the frequency domain, and vice versa. Not all functions are Fourier transformable in the classical sense. Suppose, however, that the function $f(t)$ is of bounded variation; that is, it can be represented by a curve of finite length in any finite interval of time. Then a sufficient but not necessary condition for the existence of the Fourier integral is that $f(t)$ be absolutely integrable, that is,

$$\int_{-\infty}^{\infty} |f(t)|\, dt < \infty \qquad (2.1\text{-}8)$$

Often the Fourier transform of functions not satisfying (2.1-8), such as the impulse function, can be obtained by introducing mathematical processes allowable only within the theory of distributions [5]. Then the Fourier theory can be extended to a more general class of functions.

2.1.1 Real Time Functions

The recovery of $f(t)$ from $F(\omega)$ depends on the frequency information over the infinite spectrum. We now show that if $f(t)$ is real, we need only examine $F(\omega)$ over the positive frequencies. This is a useful result, since most time functions encountered in practice are real.

Initially assume that $f(t)$ is complex as

$$f(t) = f_1(t) + jf_2(t) \qquad (2.1\text{-}9)$$

and its transform is given by (2.1-3). Then from (2.1-6) we recover $f(t)$ as

$$f(t) = \frac{1}{2\pi} \int_{-\infty}^{\infty} [R(\omega) + jX(\omega)][\cos \omega t + j \sin \omega t]\, d\omega \qquad (2.1\text{-}10)$$

using Euler's identity from (1.3-12)

$$e^{j\omega t} = \cos \omega t + j \sin \omega t \qquad (2.1\text{-}11)$$

Separating into real and imaginary parts, we obtain the general result

$$f_1(t) = \frac{1}{2\pi} \int_{-\infty}^{\infty} [R(\omega) \cos \omega t - X(\omega) \sin \omega t]\, d\omega$$

$$f_2(t) = \frac{1}{2\pi} \int_{-\infty}^{\infty} [R(\omega) \sin \omega t + X(\omega) \cos \omega t]\, d\omega \qquad (2.1\text{-}12)$$

If we now restrict $f(t)$ to be real, $f_2(t)$ must be zero. This will occur if $R(\omega)$ is an even function and $X(\omega)$ is an odd function defined as

$$\begin{aligned} R(\omega) &= R(-\omega) \quad &\text{even function} \\ X(\omega) &= -X(-\omega) \quad &\text{odd function} \end{aligned} \qquad (2.1\text{-}13)$$

Under these conditions, the products $R(\omega) \sin \omega t$ and $X(\omega) \cos \omega t$ are each odd functions and integrate to zero over the infinite interval. Likewise $f_1(t)$ need only be integrated from zero to infinity and then doubled, because $R(\omega)\cos \omega t$ and $X(\omega)\sin \omega t$ are each even functions. Then

$$f(t) = f_1(t) = \frac{1}{\pi} \int_0^\infty [R(\omega)\cos \omega t - X(\omega)\sin \omega t]\, d\omega \qquad (2.1\text{-}14)$$

Therefore, if $f(t)$ is real, the information at positive frequencies is sufficient to reconstruct $f(t)$.

2.1.2 Causal Time Functions

If $f(t)$, in addition to being real, is further restricted to be causal ($f(t) = 0$ for $t < 0$), either the real or imaginary part of $F(\omega)$ is sufficient to reconstruct $f(t)$, that is,

$$f(t) = \frac{2}{\pi} \int_0^\infty R(\omega)\cos \omega t\, d\omega = -\frac{2}{\pi} \int_0^\infty X(\omega)\sin \omega t\, d\omega \qquad t \geq 0 \qquad (2.1\text{-}15)$$

These relationships are obtained by expressing (2.1-14) as

$$f(t) = f_e(t) + f_o(t) \qquad (2.1\text{-}16)$$

where $f_e(t)$ is the even portion of $f(t)$, given by

$$f_e(t) = \frac{1}{\pi} \int_0^\infty R(\omega)\cos \omega t\, d\omega \qquad (2.1\text{-}17)$$

and $f_o(t)$ is the odd portion of $f(t)$, given by

$$f_o(t) = -\frac{1}{\pi} \int_0^\infty X(\omega)\sin \omega t\, d\omega \qquad (2.1\text{-}18)$$

However for $t < 0$, $f_e(-t) = f_e(t)$, $f_o(-t) = -f_o(t)$, and (2.1-16) becomes

$$f(t) = f_e(t) - f_o(t) \qquad t < 0 \qquad (2.1\text{-}19)$$

Since the system is causal, $f_e(t) = f_o(t)$, and $f(t)$, from (2.1-16), is then given by either

$$f(t) = \begin{cases} 0 & t < 0 \\ 2f_e(t) & t \geq 0 \end{cases} \qquad (2.1\text{-}20)$$

or

$$f(t) = \begin{cases} 0 & t < 0 \\ 2f_o(t) & t \geq 0 \end{cases} \qquad (2.1\text{-}21)$$

yielding the relationships of (2.1-15).

It should be clear from (2.1-15) that $R(\omega)$ and $X(\omega)$ are related, and Papoulis [5] shows that the elimination of $f(t)$ results in the following explicit equations:

$$X(\omega) = -\frac{2}{\pi} \int_0^\infty \int_0^\infty R(\lambda)\cos \lambda t \sin \omega t\, d\lambda\, dt \qquad (2.1\text{-}22)$$

The Fourier Transform

$$R(\omega) = -\frac{2}{\pi} \int_0^\infty \int_0^\infty X(\lambda) \sin \lambda t \cos \omega t \, d\lambda \, dt \qquad (2.1\text{-}23)$$

Since double integrals are not particularly appealing, in Section 2.7 we introduce single integral representations known as the Hilbert transforms.

2.1.3 Symmetric Time Functions

If the time function $f(t)$ is symmetric about $t = t_0$, that is, $f(t) = f(2t_0 - t)$, the phase function $\theta(\omega)$ is a linear function of the frequency ω. This result is obtained by considering the symmetric time function in Fig. 2-1a. Then from (2.1-1), the Fourier transform of $f(t)$ is

$$F(\omega) = \int_{t_a}^{2t_0 - t_a} f(t) e^{-j\omega t} \, dt \qquad (2.1\text{-}24)$$

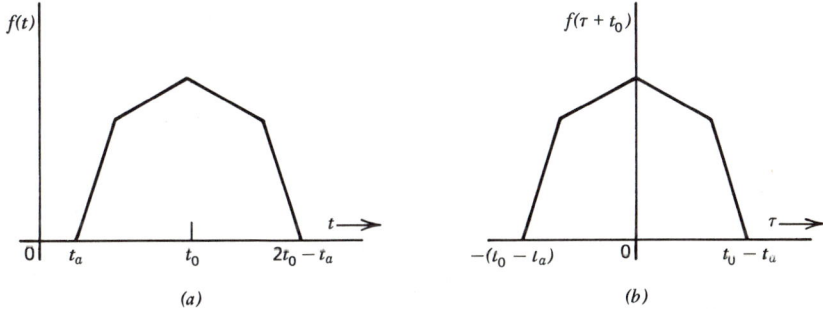

Figure 2-1. Time function symmetric (a) about $t = t_0$, (b) about $\tau = 0$.

Now introduce the new variable

$$\tau = t - t_0 \qquad (2.1\text{-}25)$$

to position $f(t)$ symmetrically about $\tau = 0$ as in Fig. 2-1b. Then

$$F(\omega) = \int_{-(t_0 - t_a)}^{t_0 - t_a} f(\tau + t_0) e^{-j\omega(\tau + t_0)} \, d\tau = e^{-j\omega t_0} \int_{-(t_0 - t_a)}^{t_0 - t_a} f(\tau + t_0) e^{-j\omega \tau} \, d\tau \qquad (2.1\text{-}26)$$

and after use of (1.3-12), $F(\omega)$ becomes

$$F(\omega) = e^{-j\omega t_0} \left[\int_{-(t_0 - t_a)}^{t_0 - t_a} f(\tau + t_0) \cos \omega \tau \, d\tau - j \int_{-(t_0 - t_a)}^{t_0 - t_a} f(\tau + t_0) \sin \omega \tau \, d\tau \right] \qquad (2.1\text{-}27)$$

Since $f(\tau + t_0)$ is an even function, as Fig. 2-1b clearly shows, and $\sin \omega \tau$ is an odd function, their product is an odd function, and the second integral is zero. The integrand of the first integral is an even function, and there we need only integrate from $\tau = 0$ to $\tau = t_0 - t_a$ and double the result to give

$$F(\omega) = 2e^{-j\omega t_0} \int_0^{t_0-t_a} f(\tau + t_0) \cos \omega\tau \, d\tau \qquad (2.1\text{-}28)$$

The integral is real, hence $F(\omega)$ is now in the form of (2.1-3), where

$$A(\omega) = 2 \left| \int_0^{t_0-t_a} f(\tau + t_0) \cos \omega\tau \, d\tau \right| \qquad (2.1\text{-}29)$$

and

$$\theta(\omega) = -\omega t_0 \qquad (2.1\text{-}30)$$

Thus the phase shift is linear with frequency, and the axis of symmetry occurs at t_0, the proportionality constant in (2.1-30).

This property has important applications in filter theory, as we shall see later, for the symmetry of the filter's impulse response is a measure of the filter's phase linearity. Furthermore the linear phase characteristic is a highly desirable filter phase response because then the group delay is constant.

2.1.4 Parseval's Theorem

The following basic result

$$\int_{-\infty}^{\infty} |f(t)|^2 \, dt = \frac{1}{2\pi} \int_{-\infty}^{\infty} |F(\omega)|^2 \, d\omega = \frac{1}{2\pi} \int_{-\infty}^{\infty} A^2(\omega) \, d\omega \qquad (2.1\text{-}31)$$

is known as Parseval's theorem; it is very useful in network analysis. If $f(t)$ is the voltage across a one-ohm resistor, then

$$E = \int_{-\infty}^{\infty} |f(t)|^2 \, dt \qquad (2.1\text{-}32)$$

is the total energy delivered to the resistor. If $A(\omega)$ is the amplitude-density function of $f(t)$, then $A^2(\omega)$ is the energy-density function, and the total energy delivered to the resistor, from (2.1-31), is the area under the $A^2(\omega)/2\pi$ curve,

$$E = \frac{1}{2\pi} \int_{-\infty}^{\infty} A^2(\omega) \, d\omega \qquad (2.1\text{-}33)$$

We can therefore determine E in either the time or the frequency domain from (2.1-31).

Example 2-1. We determine the transform of the exponential $f(t) = u_{-1}(t)e^{-\alpha t}$ from (2.1-1) as

$$F(\omega) = \int_0^{\infty} e^{-\alpha t} e^{-j\omega t} \, dt = \frac{1}{\alpha + j\omega} = \frac{\alpha}{\alpha^2 + \omega^2} - j\frac{\omega}{\alpha^2 + \omega^2} \qquad (2.1\text{-}34)$$

The Fourier Transform

From (2.1-3), (2.1-4), and (2.1-5)

$$R(\omega) = \frac{\alpha}{\alpha^2 + \omega^2} \qquad X(\omega) = -\frac{\omega}{\alpha^2 + \omega^2}$$
$$A(\omega) = \frac{1}{\sqrt{\alpha^2 + \omega^2}} \qquad \theta(\omega) = -\tan^{-1}\left(\frac{\omega}{\alpha}\right) \tag{2.1-35}$$

The functions $A(\omega)$ and $\theta(\omega)$ appear in Fig. 2-2. Note that because $f(t)$ is real, $R(\omega)$ is an even function and $X(\omega)$ is an odd function.

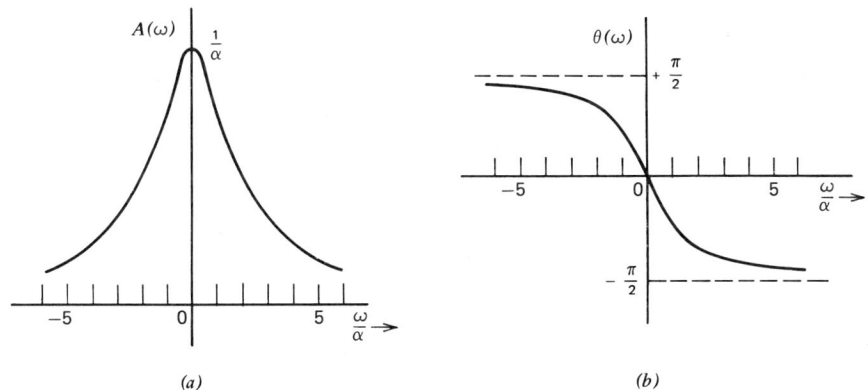

(a) (b)

Figure 2-2. (a) Fourier spectrum, (b) phase angle for $u_{-1}(t)e^{-\alpha t}$.

The time function $f(t)$ is recovered from the inversion integral (2.1-14) as

$$f(t) = \frac{1}{\pi}\int_0^\infty \frac{\alpha \cos \omega t}{\omega^2 + \alpha^2}\, d\omega + \frac{1}{\pi}\int_0^\infty \frac{\omega \sin \omega t}{\omega^2 + \alpha^2}\, d\omega \tag{2.1-36}$$

These integrals are evaluated from a book of definite integrals [2]. Then

$$f(t) = \frac{1}{2}e^{-\alpha t} + \frac{1}{2}\begin{cases} -e^{-\alpha t} & t < 0 \\ e^{-\alpha t} & t \geq 0 \end{cases} \tag{2.1-37}$$

or

$$f(t) = u_{-1}(t)e^{-\alpha t} \tag{2.1-38}$$

which is the correct time function.

Alternatively, knowledge of either $R(\omega)$ or $X(\omega)$ completely determines $f(t)$, because $f(t)$ is causal. Its even part $f_e(t)$, given by (2.1-17), is

$$f_e(t) = \frac{1}{\pi}\int_0^\infty \frac{\alpha \cos \omega t}{\omega^2 + \alpha^2}\, d\omega \tag{2.1-39}$$

The integral value, from (2.1-37), is $(\pi/2)e^{-\alpha t}$ so

$$f_e(t) = \frac{1}{\pi} \times \frac{\pi}{2} e^{-\alpha t} = \frac{1}{2} e^{-\alpha t} \tag{2.1-40}$$

From (2.1-20)

$$f(t) = \begin{cases} 2f_e(t) = e^{-\alpha t} & t \geq 0 \\ 0 & t < 0 \end{cases} \quad (2.1\text{-}41)$$

This is the correct time function and can also be obtained by considering only $X(\omega)$, using (2.1-18) and (2.1-21).

Consider $f(t)$ as the voltage across a one-ohm resistor. The energy delivered to the resistor is computed in the time domain from (2.1-32) as

$$E = \int_0^\infty e^{-2\alpha t}\, dt = \frac{1}{2\alpha} \quad (2.1\text{-}42)$$

and in the frequency domain, from (2.1-33), as

$$E = \frac{1}{2\pi} \int_{-\infty}^{\infty} \frac{d\omega}{\omega^2 + \alpha^2} = \frac{1}{2\pi\alpha} \tan^{-1}\left(\frac{\omega}{\alpha}\right)\bigg|_{-\infty}^{\infty} = \frac{1}{2\alpha} \quad (2.1\text{-}43)$$

As expected from Parseval's theorem, the energy is the same whether computed in the time or frequency domain.

Example 2-2. The impulse function $\delta(t - t_0)$ is not Fourier transformable in the classical sense, but we proceed as if it were transformable. Substitute $\delta(t - t_0)$ into (2.1-1) and use the result in (1.6-3) to obtain

$$F(\omega) = \int_{-\infty}^{\infty} \delta(t - t_0) e^{-j\omega t}\, dt = e^{-j\omega t_0} \quad (2.1\text{-}44)$$

The amplitude-density spectrum is unity for all frequencies, and the phase angle is the linear function $-\omega t_0$. Since we experienced no difficulty in applying (2.1-1), the difficulty must occur in the use of the inversion integral. From (2.1-6)

$$f(t) = \frac{1}{2\pi} \int_{-\infty}^{\infty} e^{-j\omega t_0} e^{j\omega t}\, d\omega = \frac{1}{2\pi} \int_{-\infty}^{\infty} \cos \omega(t - t_0)\, d\omega + \frac{j}{2\pi} \int_{-\infty}^{\infty} \sin \omega(t - t_0)\, d\omega$$
$$(2.1\text{-}45)$$

The value of the second integral is zero because $\sin \omega(t - t_0)$ is an odd function, while the first integral is meaningless unless it is interpreted within the theory of distributions. Therefore, $f(t)$ must equal $\delta(t - t_0)$, establishing the relationships

$$\delta(t - t_0) = \frac{1}{2\pi} \int_{-\infty}^{\infty} e^{j\omega(t - t_0)}\, d\omega = \frac{1}{2\pi} \int_{-\infty}^{\infty} \cos \omega(t - t_0)\, d\omega \quad (2.1\text{-}46)$$

Example 2-3. The transform of the exponential $e^{j\omega_0 t}$ is obtained using the result in (2.1-46). From (2.1-1)

$$F(\omega) = \int_{-\infty}^{\infty} e^{-j(\omega - \omega_0)t}\, dt = \int_{-\infty}^{\infty} \cos(\omega - \omega_0)t\, dt = 2\pi\delta(\omega - \omega_0) \quad (2.1\text{-}47)$$

The Fourier Transform

yielding the transform pair

$$e^{j\omega_0 t} \leftrightarrow 2\pi\delta(\omega - \omega_0) \qquad (2.1\text{-}48)$$

Thus the transform of $e^{j\omega_0 t}$ is an impulse function at $\omega = \omega_0$ of area 2π. For $\omega_0 = 0$, $e^{j\omega_0 t}$ is unity and its Fourier transform is an impulse function at $\omega = 0$, as in Fig. 2-3. This result agrees with the physical situation. The frequency content of a direct current wave is located at zero frequency.

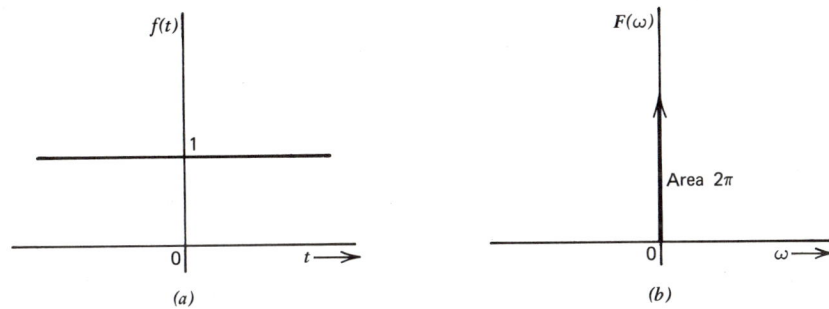

Figure 2-3. (a) The unit constant and (b) its Fourier transform.

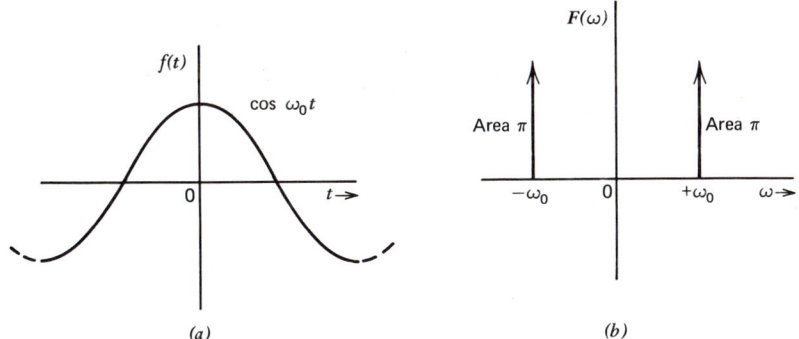

Figure 2-4. (a) The function $\cos \omega_0 t$ and (b) its Fourier transform.

Example 2-4. The transform of

$$\cos \omega_0 t = \frac{e^{j\omega_0 t}}{2} + \frac{e^{-j\omega_0 t}}{2} \qquad (2.1\text{-}49)$$

from (2.1-48) is

$$F(\omega) = \pi\delta(\omega - \omega_0) + \pi\delta(\omega + \omega_0) \qquad (2.1\text{-}50)$$

where $F(\omega)$ consists of two impulse functions, each of area π, located at $+\omega_0$ and $-\omega_0$, as in Fig. 2-4b.

Example 2-5. The transform of sgn t in Fig. 2-5 is now shown to be

$$F(\omega) = \frac{2}{j\omega} \qquad (2.1\text{-}51)$$

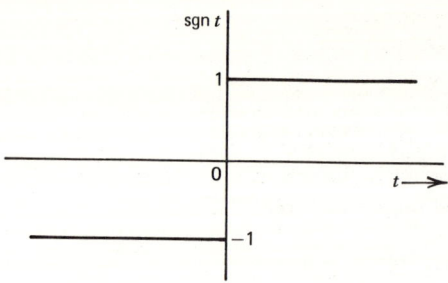

Figure 2-5. The function sgn t.

even though (2.1-8) is not satisfied. Substitute $F(\omega)$ into (2.1-6) to get

$$f(t) = \frac{1}{2\pi} \int_{-\infty}^{\infty} \frac{2e^{j\omega t}}{j\omega} d\omega = \frac{1}{\pi j} \int_{-\infty}^{\infty} \frac{\cos \omega t}{\omega} d\omega + \frac{1}{\pi} \int_{-\infty}^{\infty} \frac{\sin \omega t}{\omega} d\omega \quad (2.1\text{-}52)$$

The first integral is zero because the integrand is an odd function, and the second integral, from a table of definite integrals such as Ref. 2, is $-\pi$, 0, or π, as t is negative, zero, or positive. Then

$$f(t) = \operatorname{sgn} t = \begin{cases} -1 & t < 0 \\ 0 & t = 0 \\ +1 & t > 0 \end{cases} \quad (2.1\text{-}53)$$

Having shown that $F(\omega)$ is indeed the transform of $f(t)$, we now attempt to determine this transform directly from (2.1-1) as

$$F(\omega) = \int_{-\infty}^{\infty} \operatorname{sgn} t (\cos \omega t - j \sin \omega t) \, dt \quad (2.1\text{-}54)$$

Because sgn t is an odd function, the integral reduces to

$$F(\omega) = -j2 \int_{0}^{\infty} \sin \omega t \, dt \quad (2.1\text{-}55)$$

Therefore, the transform pair is valid if

$$\int_{0}^{\infty} \sin \omega t \, dt = \frac{1}{\omega} \quad (2.1\text{-}56)$$

which can be verified by interpreting the quantities in (2.1-56) as distributions.

Example 2-6. The step function $u_{-1}(t)$ in Fig. 1-8 is also not absolutely integrable in the classical sense, but from (2.1-1),

$$F(\omega) = \int_{0}^{\infty} e^{-j\omega t} dt = \int_{0}^{\infty} \cos \omega t \, dt - j \int_{0}^{\infty} \sin \omega t \, dt \quad (2.1\text{-}57)$$

Using the relationships in (2.1-46) and (2.1-56), we establish the transform of $f(t)$

The Laplace Transform

as

$$F(\omega) = \frac{1}{2} \times 2\pi\delta(\omega) - j \times \frac{1}{\omega} = \pi\delta(\omega) + \frac{1}{j\omega} \qquad (2.1\text{-}58)$$

Comparison of (2.1-51) and (2.1-58) reveals that except for the impulse function at $\omega = 0$, the amplitude-density spectrum of $u_{-1}(t)$ is one-half the amplitude-density spectrum of sgn t. There is no DC component in sgn t, for its average value is zero, whereas the average value of $u_{-1}(t)$ is $\frac{1}{2}$, represented by $\pi\delta(\omega)$. The factor of $\frac{1}{2}$ in the amplitude-density spectrum is alternatively justified by considering the function $\frac{1}{2}$ sgn t in Fig. 2-6. Then $u_{-1}(t)$ is obtained by adding the constant $\frac{1}{2}$ to $\frac{1}{2}$ sgn t as

$$u_{-1}(t) = \frac{1}{2} + \frac{1}{2} \text{ sgn } t \qquad (2.1\text{-}59)$$

Clearly $u_{-1}(t)$ and sgn t differ only by the factor $\frac{1}{2}$ and the constant $\frac{1}{2}$. The latter is represented in the frequency domain by an impulse function of area π at $\omega = 0$ (see Fig. 2-3), while the factor of $\frac{1}{2}$ carries over to the frequency domain.

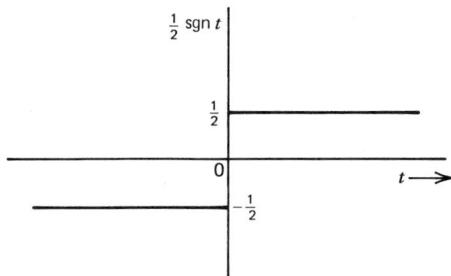

Figure 2-6. The function $\frac{1}{2}$ sgn t.

2.2 THE LAPLACE TRANSFORM

Recent years have seen the Laplace transform compete with the Fourier transform as the primary analysis tool for time-invariant systems. This is not surprising because the Lapace transform is applicable to a more general class of functions, and it is not necessary to resort to the theory of distributions to justify the existence of certain transforms. We consider the one-sided Laplace transform, which applies to time functions that are zero for values of time less than zero, a common situation in practice. However functions that are not zero for $t < 0$ are not rejected by the theory, for these functions are transformable by the two-sided Laplace transform [6].

The Laplace transformation is a mathematical operation indicated symbolically by $\mathscr{L}[f(t)]$ (spoken "Laplace transform of $f(t)$"), and defined for functions $f(t)$ that are zero for $t < 0$ as

$$\mathscr{L}[f(t)] = F(s) = \int_0^\infty f(t) e^{-st} \, dt \qquad (2.2\text{-}1)$$

The complex frequency

$$s = \sigma + j\omega \tag{2.2-2}$$

is a generalization of the sinusoidal frequency and is discussed in Section 2.5.2. As with the Fourier transform, lowercase letters indicate the time functions and capital letters are reserved for their transforms.

Many time functions that are not absolutely integrable, hence are not Fourier transformable in the classical sense, do have Laplace transforms. This is achieved by the convergence factor $e^{-\sigma t}$ appearing in the integrand, allowing the integral to exist for most functions encountered in physical systems. This is seen by rewriting (2.2-1) as

$$F(s) = \int_0^\infty (f(t)e^{-\sigma t})e^{-j\omega t}\, dt \tag{2.2-3}$$

which can be interpreted as the Fourier transform of $f(t)e^{-\sigma t}u_{-1}(t)$. The function $e^{-\sigma t}$ approaches zero fast enough that for most functions the product $e^{-\sigma t}f(t)$ approaches zero as t approaches infinity. Thus a function $f(t)$ is transformable if the product $|f(t)|e^{-\sigma t}$ is absolutely integrable for some real value of σ, that is, the following integral must exist:

$$\int_0^\infty |f(t)|e^{-\sigma t}\, dt < \infty \tag{2.2-4}$$

Functions such as t^t and e^{t^2} are not Laplace transformable, but these functions have little importance in engineering. However if t^t were the system input Laplace transforms could still be used for analysis. This input would surely saturate a physical system at some time $t = T$ and thereby create a nonlinear system. But over the linear portion, the integral in (2.2-4) from $t = 0$ to $t = T$ does exist for the input t^t.

In the following examples we determine the transforms of some useful functions and operations, which are summarized in Table 2-1. For additional transforms, the reader is referred to the many books on the subject (*e.g.*, Refs. 1 and 4).

Example 2-7. For $f(t) = \delta(t - t_0)$, an impulse function occurring at $t = t_0$, (2.2-1), aided by (1.6-3), gives

$$F(s) = \int_0^\infty \delta(t - t_0)e^{-st}\, dt = e^{-st_0} \tag{2.2-5}$$

For $t_0 = 0$, the transform reduces to unity. To keep the mathematics consistent, we consider the impulse to occur slightly to the right of $t = 0$, indicated by $t = 0_+$. Then the impulse function in (2.2-5) occurs within the limits of the integral, allowing (1.6-3) to be valid. Henceforth this fact is understood. Thus we drop the + subscript and

$$\mathscr{L}[\delta(t)] = 1 \tag{2.2-6}$$

Example 2-8. For $f(t) = e^{-\alpha t}$,

$$F(s) = \int_0^\infty e^{-(s+\alpha)t}\, dt = -\left.\frac{e^{-(s+\alpha)t}}{s+\alpha}\right|_0^\infty = \frac{1}{s+\alpha} \tag{2.2-7}$$

The Laplace Transform

Table 2-1 Laplace Transform Pairs

$f(t)$	$F(s)$
1. $\delta(t-t_0)$	e^{-st_0}
2. $u_{-1}(t)$	$\dfrac{1}{s}$
3. $e^{-\alpha t}$	$\dfrac{1}{s+\alpha}$
4. $e^{-\alpha t}\sin\beta t$	$\dfrac{\beta}{(s+\alpha)^2+\beta^2}$
5. $e^{-\alpha t}\cos\beta t$	$\dfrac{s+\alpha}{(s+\alpha)^2+\beta^2}$
6. $\cos\beta t$	$\dfrac{s}{s^2+\beta^2}$
7. $\sin\beta t$	$\dfrac{\beta}{s^2+\beta^2}$
8. $t^n f(t)$	$(-1)^n \dfrac{d^n F(s)}{ds^n}$
9. t^n	$\dfrac{n!}{s^{n+1}}$
10. $\dfrac{d^n f(t)}{dt^n}$	$s^n F(s) - s^{n-1}f(0) - \cdots - f^{(n-1)}(0)$
11. $\int_0^t f(x)\,dx$	$\dfrac{1}{s}F(s)$
12. $u_{-1}(t-t_0)f(t-t_0)$	$e^{-t_0 s}F(s),\ t_0 \geqslant 0$
13. $h(t)*f(t) = \int_0^t h(t-\tau)f(\tau)\,d\tau$	$H(s)F(s)$
14. $2e^{-\alpha t}(a\cos\beta t - b\sin\beta t)$	$\dfrac{a+jb}{s+\alpha-j\beta} + \dfrac{a-jb}{s+\alpha+j\beta}$
15. $f_1(t)f_2(t)$	$F_1(s)*F_2(s)$

The evaluation of the upper limit is zero, provided the real part of s is greater than $-\alpha$. This defines the region of convergence for the transform to exist; by analytic continuation [4], however, the transform is valid for other values of s. For the remaining examples it is understood that the value of σ is large enough to ensure that the integral exists.

As α approaches zero, the exponential function approaches the step function, and the transform approaches $1/s$. Thus

$$\mathscr{L}[u_{-1}(t)] = \frac{1}{s} \tag{2.2-8}$$

It is often convenient to tabulate the transforms of relatively complicated functions and then let one or more of the parameters take on limiting values to obtain the transforms of simpler functions. This is now illustrated by examples.

Example 2-9. The Laplace transform of $e^{-\alpha t}\sin\beta t$ and $e^{-\alpha t}\cos\beta t$ is obtained by first using Euler's identity in (1.3-12) to form the function

$$e^{-\alpha t}e^{-j\beta t} = e^{-\alpha t}(\cos\beta t - j\sin\beta t) \tag{2.2-9}$$

Then from (2.2-1)

$$\mathcal{L}[e^{-(\alpha+j\beta)t}] = \int_0^\infty e^{-(\alpha+j\beta)t} e^{-st} dt = \int_0^\infty e^{-\alpha t} \cos \beta t \, e^{-st} dt - j \int_0^\infty e^{-\alpha t} \sin \beta t \, e^{-st} dt \qquad (2.2\text{-}10)$$

This partitioning shows that the real part of the first integral is the transform of $e^{-\alpha t} \cos \beta t$, while the imaginary part of the first integral is the transform of $-e^{-\alpha t} \sin \beta t$. From (2.2-7)

$$\mathcal{L}[e^{-(\alpha+j\beta)t}] = \frac{1}{s+\alpha+j\beta} = \frac{s+\alpha}{(s+\alpha)^2+\beta^2} - j\frac{\beta}{(s+\alpha)^2+\beta^2} \qquad (2.2\text{-}11)$$

and we obtain the transforms

$$\mathcal{L}[e^{-\alpha t} \cos \beta t] = \frac{s+\alpha}{(s+\alpha)^2+\beta^2} \qquad (2.2\text{-}12)$$

$$\mathcal{L}[e^{-\alpha t} \sin \beta t] = \frac{\beta}{(s+\alpha)^2+\beta^2} \qquad (2.2\text{-}13)$$

The transforms of $\cos \beta t$ and $\sin \beta t$, letting $\alpha = 0$ in (2.2-12) and (2.2-13), are

$$\mathcal{L}[\cos \beta t] = \frac{s}{s^2+\beta^2} \qquad (2.2\text{-}14)$$

$$\mathcal{L}[\sin \beta t] = \frac{\beta}{s^2+\beta^2} \qquad (2.2\text{-}15)$$

Example 2-10. Consider the function $t^n f(t)$, where n is an integer. Then $F(s)$ is the transform of $f(t)$ given by

$$F(s) = \int_0^\infty f(t) e^{-st} dt \qquad (2.2\text{-}16)$$

Differentiate both sides with respect to s, which yields

$$\frac{dF(s)}{ds} = -\int_0^\infty t f(t) e^{-st} dt \qquad (2.2\text{-}17)$$

Repeated differentiation leads to the result

$$\mathcal{L}[t^n f(t)] = \int_0^\infty t^n f(t) e^{-st} dt = (-1)^n \frac{d^n F(s)}{ds^n} \qquad (2.2\text{-}18)$$

For the special case $f(t) = u_{-1}(t)$, $F(s) = 1/s$ and

$$\frac{d^n F(s)}{ds^n} = \frac{(-1)^n n!}{s^{n+1}} \qquad (2.2\text{-}19)$$

From (2.2-18), the transform of t^n is then

$$\mathcal{L}[t^n] = (-1)^n \times \frac{(-1)^n n!}{s^{n+1}} = \frac{n!}{s^{n+1}} \qquad (2.2\text{-}20)$$

The Laplace Transform

Example 2-11. The Laplace transform of df/dt, the derivative of $f(t)$, is

$$\mathcal{L}\left[\frac{df}{dt}\right] = \int_0^\infty \frac{df}{dt} e^{-st}\, dt \qquad (2.2\text{-}21)$$

We now use the formula for integration by parts,

$$\int_a^b u\, dv = uv\Big|_a^b - \int_a^b v\, du \qquad (2.2\text{-}22)$$

by letting $u = e^{-st}$, $du = -se^{-st}\, dt$, $dv = df$, and $v = f$. Then

$$\mathcal{L}\left[\frac{df}{dt}\right] = f(t)e^{-st}\Big|_0^\infty + s\int_0^\infty f(t)e^{-st}\, dt = sF(s) - f(0) \qquad (2.2\text{-}23)$$

where $\lim_{t \to \infty} f(t)e^{-st} = 0$ for some value of σ, and $F(s)$ is the transform of $f(t)$. This is the first transform we have encountered that includes an initial condition, namely, the value of $f(t)$ at $t = 0$. Assuming $f(0) = 0$, the derivative of $f(t)$ with respect to time corresponds to multiplying $F(s)$ by s in the frequency domain.

Repeated use of this technique yields the transform for $d^n f/dt^n$,

$$\mathcal{L}\left[\frac{d^n f}{dt^n}\right] = s^n F(s) - s^{n-1} f(0) - s^{n-2} f'(0) - \cdots - f^{(n-1)}(0) \qquad (2.2\text{-}24)$$

where $f^{(i)}(0)$ is the ith derivative of f with respect to t, evaluated at $t = 0$. Note that n initial conditions are included in this transform.

Example 2-12. The transform of the integral of $f(t)$, $\int_0^t f(x)\, dx$, from (2.2-1), is

$$\mathcal{L}\left[\int_0^t f(x)\, dx\right] = \int_0^\infty \left[\int_0^t f(x)\, dx\right] e^{-st}\, dt \qquad (2.2\text{-}25)$$

Again, use (2.2-22) with

$$u = \int_0^t f(x)\, dx \qquad dv = e^{-st}\, dt$$
$$du = f(t)\, dt \qquad v = -\frac{1}{s} e^{-st} \qquad (2.2\text{-}26)$$

giving

$$\mathcal{L}\left[\int_0^t f(x)\, dx\right] = -\frac{1}{s} e^{-st} \int_0^t f(x)\, dx\Big|_0^\infty + \frac{1}{s}\int_0^\infty f(t)e^{-st}\, dt \qquad (2.2\text{-}27)$$

The first term is zero at $t = 0$ and $t = \infty$; thus

$$\mathcal{L}\left[\int_0^t f(x)\, dx\right] = \frac{1}{s} F(s) \qquad (2.2\text{-}28)$$

where $F(s)$ is the Laplace transform of $f(t)$. Integration of $f(t)$ in the time domain corresponds to dividing $F(s)$ by s in the frequency domain.

Example 2-13. The transform of the delayed function $u_{-1}(t-t_0)f(t-t_0)$ is

$$\mathcal{L}[u_{-1}(t-t_0)f(t-t_0)] = \int_0^\infty u_{-1}(t-t_0)f(t-t_0)e^{-st}\,dt \qquad (2.2\text{-}29)$$

The integral is zero from $t=0$ to $t=t_0$, which means that we can change the lower limit to t_0. Then by the change of variable $x = t - t_0$,

$$\mathcal{L}[u_{-1}(t-t_0)f(t-t_0)] = \int_0^\infty f(x)e^{-s(x+t_0)}\,dx = e^{-st_0}F(s) \qquad (2.2\text{-}30)$$

Thus the transform of a function $f(t)$ delayed by t_0 is simply the transform of $f(t)$ multiplied by e^{-st_0}.

Example 2-14. The transform of the convolution integral (1.8-10) is given by

$$\mathcal{L}[h*f] = \int_0^\infty \left[\int_0^t h(t-\tau)f(\tau)\,d\tau\right]e^{-st}\,dt \qquad (2.2\text{-}31)$$

Since $h(t-\tau)=0$ for $\tau > t$, the upper limit of the convolution integral can be changed to ∞ without affecting the result. Then rearranging the integrals,

$$\mathcal{L}[h*f] = \int_0^\infty f(\tau)\,d\tau \int_0^\infty e^{-st}h(t-\tau)\,dt \qquad (2.2\text{-}32)$$

Since $h(t-\tau) = 0$ for $t < \tau$, the lower limit of the second integral can be changed to τ, yielding

$$\mathcal{L}[h*f] = \int_0^\infty f(\tau)\,d\tau \int_\tau^\infty e^{-st}h(t-\tau)\,dt \qquad (2.2\text{-}33)$$

Introduce the new variable $x = t - \tau$. Then

$$\mathcal{L}[h*f] = \int_0^\infty f(\tau)\,d\tau \int_0^\infty e^{-s(x+\tau)}h(x)\,dx = \int_0^\infty f(\tau)e^{-s\tau}\,d\tau \int_0^\infty h(x)e^{-sx}\,dx$$

$$= F(s)H(s) \qquad (2.2\text{-}34)$$

Therefore convolution of two time-domain functions is equivalent to multiplication of their transforms in the frequency domain, a very important result for the analysis of time-invariant systems.

Example 2-15. Consider the product of the two time functions $f_1(t)$ and $f_2(t)$, with respective transforms $F_1(s)$ and $F_2(s)$. This product has practical significance because it is the result of amplitude modulating the signal $f_2(t)$ by $f_1(t)$. Its transform is

$$\mathcal{L}[f_1(t)f_2(t)] = \int_0^\infty f_1(t)f_2(t)e^{-st}\,dt \qquad (2.2\text{-}35)$$

To further simplify this expression, we use the inverse transform relationship

The Inverse Transform

(2.3-1), although it has not yet been discussed. However the significance of the results here justifies its use now as an intermediate step in the analysis. Therefore substitute this integral expression for $f_1(t)$ into (2.2-35) to yield

$$\mathscr{L}[f_1(t)f_2(t)] = \int_0^\infty f_2(t)e^{-st}\left[\frac{1}{2\pi j}\int_c F_1(y)e^{yt}\,dy\right]dt \qquad (2.2\text{-}36)$$

Interchange the order of integration and adjust the contour c to c_1 to give

$$\mathscr{L}[f_1(t)f_2(t)] = \frac{1}{2\pi j}\int_{c_1} F_1(y)\left[\int_0^\infty f_2(t)e^{-(s-y)t}\,dt\right]dy$$

$$= \frac{1}{2\pi j}\int_{c_1} F_1(y)F_2(s-y)\,dy \qquad (2.2\text{-}37)$$

This integral should look familiar, for it is the convolution integral in the complex frequency domain. Thus the transform of a product of two time functions is the convolution of their transforms. The results of this example and Example 2-14 show that convolution in one domain implies multiplication in the other domain. A rigorous derivation of (2.2-37) is given in Ref. 1.

A listing of important transform pairs is supplied in Table 2-1.

2.3 THE INVERSE TRANSFORM

We have shown how to transform from the time domain to the frequency domain; now we consider the reverse operation. One method of recovering the time function $f(t)$ from the frequency function $F(s)$ is through the inversion integral

$$f(t) = \mathscr{L}^{-1}[F(s)] = \frac{1}{2\pi j}\int_c F(s)e^{st}\,ds \qquad (2.3\text{-}1)$$

where c is a suitable contour in the complex s-plane. The symbol $\mathscr{L}^{-1}$ is spoken "the inverse Laplace transform of." Evaluation of this integral requires complex-variable theory, in particular the method of residues. An equally valid technique, requiring about the same amount of work but not calling for complex-variable theory, is the method of partial fractions. Two definitions are presented before we proceed with this technique.

A *rational function* (or rational fraction) is the ratio of two polynomials, that is, the rational function $H(s)$ has the form

$$H(s) = \frac{b_0 s^m + b_1 s^{m-1} + \cdots + b_m}{a_0 s^n + a_1 s^{n-1} + \cdots + a_n} \qquad (2.3\text{-}2)$$

If the denominator polynomial is a constant, $H(s)$ is a polynomial. Thus a polynomial is also a rational function.

A *proper fraction* is a rational function for which the degree of the numerator polynomial is less than the degree of the denominator polynomial. Otherwise the rational function is an improper fraction. For example, $H_1(s)$ in (2.3-3) is a proper fraction, and $H_2(s)$ and $H_3(s)$ in (2.3-4) are improper fractions.

$$H_1(s) = \frac{4s^3 + 8s^2 + 3s + 2}{10s^4 + 6s^3 + s^2 + 4s + 1} \tag{2.3-3}$$

$$H_2(s) = \frac{4s^2 + 9s + 3}{s + 2} \quad \text{and} \quad H_3(s) = \frac{s^2 + 2s + 3}{s^2 + s + 1} \tag{2.3-4}$$

An improper fraction can be decomposed into a proper fraction plus other terms by long division. This is illustrated for $H_2(s)$.

$$\begin{array}{r} 4s + 1 \\ s+2 \overline{\smash{)}4s^2 + 9s + 3} \\ \underline{4s^2 + 8s} \\ s + 3 \\ \underline{s + 2} \\ 1 \end{array} \tag{2.3-5}$$

Now $H_2(s)$ is expressed as

$$H_2(s) = 4s + 1 + \frac{1}{s + 2} \tag{2.3-6}$$

where $1/(s + 2)$ is a proper fraction. Similarly

$$H_3(s) = 1 + \frac{s + 2}{s^2 + s + 1} \tag{2.3-7}$$

Let us begin to describe the method of partial fractions by noting that most transforms that arise in the analysis of systems described by constant-coefficient differential equations are rational functions of s. For determining the inverse of these transforms we use a table of available transform pairs, similar to Table 2-1. When the function of s appears in the transform table our problem is solved. It is, however, unlikely that a given problem will yield transforms that are already tabulated. Our task then is to set forth a procedure for reducing an arbitrary rational function into a sum of elementary transforms that are already tabulated. This technique is similar to that employed in finding the integral of a function. The function is decomposed into a sum of elementary functions whose integrals have already been tabulated.

An example now shows the partial fraction expansion of a rational function $F(s)$, after which its inverse transform is determined.

Example 2-16. Let $F(s)$ be the proper fraction

$$F(s) = \frac{2s^2 + 3s + 3}{s^3 + 6s^2 + 11s + 6} \tag{2.3-8}$$

First we determine the roots of the denominator polynomial to be $-1, -2, -3$. Then we express $F(s)$ as a sum of partial fractions.

$$F(s) = \frac{2s^2 + 3s + 3}{(s+1)(s+2)(s+3)} = \frac{A_1}{s+1} + \frac{A_2}{s+2} + \frac{A_3}{s+3} \tag{2.3-9}$$

The constant A_1 is determined by multiplying both sides of (2.3-9) by $(s + 1)$, yielding

$$A_1 + \frac{A_2(s+1)}{s+2} + \frac{A_3(s+1)}{s+3} = \frac{2s^2 + 3s + 3}{(s+2)(s+3)} \tag{2.3-10}$$

The Inverse Transform

and then letting $s = -1$. The value of A_1 is

$$A_1 = \left.\frac{2s^2 + 3s + 3}{(s+2)(s+3)}\right|_{s=-1} = 1 \qquad (2.3\text{-}11)$$

Similarly

$$A_2 = \left.\frac{2s^2 + 3s + 3}{(s+1)(s+3)}\right|_{s=-2} = -5 \qquad (2.3\text{-}12)$$

$$A_3 = \left.\frac{2s^2 + 3s + 3}{(s+1)(s+2)}\right|_{s=-3} = 6 \qquad (2.3\text{-}13)$$

Then

$$F(s) = \frac{1}{s+1} - \frac{5}{s+2} + \frac{6}{s+3} \qquad (2.3\text{-}14)$$

and its inverse transform is obtained from Table 2-1, entry 3, which gives the transform pair

$$\frac{1}{s+\alpha} \leftrightarrow e^{-\alpha t} \qquad (2.3\text{-}15)$$

Therefore

$$f(t) = \mathcal{L}^{-1}[F(s)] = [e^{-t} - 5e^{-2t} + 6e^{-3t}]u_{-1}(t) \qquad (2.3\text{-}16)$$

This technique of expanding the rational function into partial fractions is valid so long as all the denominator polynomial roots are simple (no root occurs more than once). Then the rational function $F(s)$ can be expressed as the sum of partial fractions

$$F(s) = \frac{b_0 s^m + \cdots + b_m}{(s-s_1)(s-s_2)\cdots(s-s_n)} = \frac{A_1}{s-s_1} + \cdots + \frac{A_n}{s-s_n} \quad (n > m) \quad (2.3\text{-}17)$$

where the A_i's are given by

$$A_i = (s - s_i)F(s)\bigg|_{s=s_i} \qquad (2.3\text{-}18)$$

Complex roots always occur in conjugate pairs because the polynomial has real coefficients. The A's corresponding to these roots also occur in conjugate pairs. Furthermore, these two partial fractions, consisting of complex numbers, can always be combined to yield a real function of s.

Example 2-17. Let the roots be $-\alpha \pm j\beta$, $A_1 = a + jb$, and $A_2 = a - jb$. The two fractions are now combined to form a real function of s.

$$F(s) = \frac{a+jb}{s+\alpha-j\beta} + \frac{a-jb}{s+\alpha+j\beta} = 2a\frac{s+\alpha}{(s+\alpha)^2+\beta^2} - 2b\frac{\beta}{(s+\alpha)^2+\beta^2} \qquad (2.3\text{-}19)$$

Now $F(s)$ is expressed as a real function of s. From Table 2-1, entries 4 and 5, the inverse transform is

$$f(t) = 2e^{-\alpha t}(a\cos\beta t - b\sin\beta t) \qquad (2.3\text{-}20)$$

Therefore when complex roots occur, we need only know a, b, α, and β to obtain the inverse transform in (2.3-20). This transform pair is entry 14 in Table 2-1.

When multiple roots are present (one root occurs more than once), the preceding scheme is slightly modified. Suppose we have the rational function

$$F(s) = \frac{3}{(s+1)^2(s+3)} \qquad (2.3\text{-}21)$$

where the root at $s = -1$ occurs twice. Then $F(s)$ is expressed as

$$F(s) = \frac{A_1}{s+1} + \frac{A_2}{(s+1)^2} + \frac{A_3}{s+3} \qquad (2.3\text{-}22)$$

From (2.3-18) A_3 is found to be $A_3 = \frac{3}{4}$. However A_1 and A_2 cannot be determined as previously described. To demonstrate this, we multiply both sides in (2.3-22) by $(s+1)$ yielding

$$(s+1)F(s) = A_1 + \frac{A_2}{s+1} + \frac{(s+1)A_3}{s+3} \qquad (2.3\text{-}23)$$

If we now let $s = -1$ as before, the term $A_2/(s+1)$ approaches infinity and neither A_1 nor A_2 is determinable. Instead, multiply both sides of (2.3-22) by $(s+1)^2$ as

$$(s+1)^2 F(s) = (s+1)A_1 + A_2 + \frac{(s+1)^2}{s+3} A_3 \qquad (2.3\text{-}24)$$

Evaluation of (2.3-24) at $s = -1$ now gives

$$A_2 = (s+1)^2 F(s) \Big|_{s=-1} = \frac{3}{2} \qquad (2.3\text{-}25)$$

The constant A_1 is found by differentiating (2.3-24) with respect to s and evaluating both sides at $s = -1$ as

$$\frac{d}{ds}[(s+1)^2 F(s)]_{s=-1} = A_1 + \frac{A_3(s+1)(s+5)}{(s+3)^2} \Big|_{s=-1} \qquad (2.3\text{-}26)$$

The second term on the right is zero, and the value of A_1 is then

$$A_1 = \frac{d}{ds}\left[\frac{3}{s+3}\right]_{s=-1} = \frac{-3}{(s+3)^2}\Big|_{s=-1} = -\frac{3}{4} \qquad (2.3\text{-}27)$$

Substituting the A_i's into (2.3-22) gives

$$F(s) = \frac{-\frac{3}{4}}{s+1} + \frac{\frac{3}{2}}{(s+1)^2} + \frac{\frac{3}{4}}{s+3} \qquad (2.3\text{-}28)$$

The inverse transform of the second term is given in Table 2-1, entry 8, with $f(t) = e^{-\alpha t}$

$$te^{-\alpha t} \leftrightarrow \frac{1}{(s+\alpha)^2} \qquad (2.3\text{-}29)$$

Then the inverse transform of $F(s)$ is

$$f(t) = \frac{3}{4}(e^{-3t} + 2te^{-t} - e^{-t})u_{-1}(t) \qquad (2.3\text{-}30)$$

From this example we observe that repeated differentiation allows us to find the constants of the partial fraction expansion when there are multiple roots. In

The Inverse Transform

summary, the inverse transform of a rational function with simple or multiple roots can be found as follows.

1. By long division, decompose the rational function into a proper fraction $F(s)$ plus other terms.
2. Factor the denominator polynomial of $F(s)$. Then $F(s)$ is of the form

$$F(s) = \frac{G(s)}{(s-s_1)^{m_1}(s-s_2)^{m_2}\cdots(s-s_k)^{m_k}} \tag{2.3-31}$$

where m_i is the multiplicity of the root s_i.

3. Express $F(s)$ as the partial fraction expansion

$$F(s) = \frac{A_{m_1}}{(s-s_1)^{m_1}} + \frac{A_{m_1-1}}{(s-s_1)^{m_1-1}} + \cdots + \frac{A_1}{(s-s_1)} + \frac{B_{m_2}}{(s-s_2)^{m_2}} + \cdots$$
$$+ \frac{B_1}{(s-s_2)} + \cdots + \frac{K_{m_k}}{(s-s_k)^{m_k}} + \cdots + \frac{K_1}{(s-s_k)} \tag{2.3-32}$$

The coefficients are given by

$$A_{m_1} = (s-s_1)^{m_1} F(s)\Big|_{s=s_1}$$
$$A_{m_1-1} = \frac{d}{ds}[(s-s_1)^{m_1} F(s)]\Big|_{s=s_1}$$
$$\vdots$$
$$A_1 = \frac{1}{(m_1-1)!}\frac{d^{m_1-1}}{ds^{m_1-1}}[(s-s_1)^{m_1} F(s)]\Big|_{s=s_1}$$
$$B_{m_2} = (s-s_2)^{m_2} F(s)\Big|_{s=s_2}$$
$$B_{m_2-1} = \frac{d}{ds}[(s-s_2)^{m_2} F(s)]\Big|_{s=s_2} \tag{2.3-33}$$
$$\vdots$$
$$B_1 = \frac{1}{(m_2-1)!}\frac{d^{m_2-1}}{ds^{m_2-1}}[(s-s_2)^{m_2} F(s)]\Big|_{s=s_2}$$
$$\vdots$$
$$K_{m_k} = (s-s_k)^{m_k} F(s)\Big|_{s=s_k}$$
$$K_{m_k-1} = \frac{d}{ds}[(s-s_k)^{m_k} F(s)]\Big|_{s=s_k}$$
$$\vdots$$
$$K_1 = \frac{1}{(m_k-1)!}\frac{d^{m_k-1}}{ds^{m_k-1}}[(s-s_k)^{m_k} F(s)]\Big|_{s=s_k}$$

4. Refer to a table of Laplace transforms, such as Table 2-1, to determine the time functions corresponding to each partial fraction in (2.3-32).

Example 2-18. The inverse transform of

$$H(s) = \frac{s^4 + 6s^3 + 22s^2 + 30s + 14}{s^4 + 6s^3 + 22s^2 + 30s + 13} \quad (2.3\text{-}34)$$

is now obtained.

1. $H(s)$ is an improper fraction because the degree of the numerator polynomial is the same as the degree of the denominator polynomial. By long division reduce $H(s)$ to the proper fraction

$$H(s) = 1 + \frac{1}{s^4 + 6s^3 + 22s^2 + 30s + 13} = 1 + F(s) \quad (2.3\text{-}35)$$

2. The roots of the denominator polynomial of $F(s)$ are $-1, -1, -2 \pm j3$. Then $F(s)$ in factored form is

$$F(s) = \frac{1}{(s+1)^2(s+2+j3)(s+2-j3)} \quad (2.3\text{-}36)$$

3. Now $F(s)$ is written as a partial fraction expansion according to (2.3-32)

$$F(s) = \frac{A_2}{(s+1)^2} + \frac{A_1}{s+1} + \frac{B_1}{(s+2+j3)} + \frac{C_1}{(s+2-j3)} \quad (2.3\text{-}37)$$

Here $m_1 = 2$, $m_2 = 1$, $m_3 = 1$. The constants are found from (2.3-33) to be

$$A_2 = (s+1)^2 F(s)\big|_{s=-1} = \frac{1}{(s+2)^2 + 9}\bigg|_{s=-1} = \frac{1}{10}$$

$$A_1 = \frac{d}{ds}\left[(s+1)^2 F(s)\right]\bigg|_{s=-1} = \frac{-2(s+2)}{[(s+2)^2+9]^2}\bigg|_{s=-1} = -\frac{1}{50} \quad (2.3\text{-}38)$$

$$B_1 = (s+2+j3)F(s)\big|_{s=-2-j3} = \frac{1}{(s+1)^2(s+2-j3)}\bigg|_{s=-2-j3} = \frac{1}{100} - \frac{j}{75}$$

$$C_1 = B_1^* = \frac{1}{100} + \frac{j}{75}$$

where the superscript symbol (*) indicates conjugation. Thus

$$(\alpha + j\beta)^* = \alpha - j\beta$$

The partial fraction expansion of $H(s)$ is

$$H(s) = 1 + \frac{1/10}{(s+1)^2} - \frac{1/50}{s+1} + \frac{1/100 - j/75}{s+2+j3} + \frac{1/100 + j/75}{s+2-j3} \quad (2.3\text{-}39)$$

4. The time function $h(t)$ is obtained by referring to Table 2-1. The last two terms in (2.3-39) are transformable by entry 14, with $\alpha = 2, \beta = 3, a = 1/100,$ and $b = 1/75$. Then

$$h(t) = \delta(t) + \left[\frac{1}{10} t e^{-t} - \frac{1}{50} e^{-t} + 2e^{-2t}\left(\frac{1}{100}\cos 3t - \frac{1}{75}\sin 3t\right)\right] u_{-1}(t) \quad (2.3\text{-}40)$$

We have briefly outlined a method for finding inverse transforms without resorting to Laplace transform theory. We express the transform as a partial fraction expansion and refer to a table of transform pairs to obtain the time

Solution of Differential Equations

function corresponding to each fraction. The Laplace transform theory has already been used in constructing the table! Proficiency in finding inverse transforms increases as one works more and more problems.

Manual inversion of high-order rational functions is very tedious and time-consuming. Consequently this task is usually performed on a digital computer by one of the many algorithms that are available.

2.4 SOLUTION OF DIFFERENTIAL EQUATIONS BY LAPLACE TRANSFORMS

Here we use the Laplace transform to solve the constant-coefficient differential equation. However other important results arise. We obtain the expression for the transfer function, which characterizes the system in the frequency domain. Also we establish the link between the system's time-domain and frequency-domain representations, namely, that the transfer function is the Laplace transform of the system's response to an impulse function applied at $t = 0$.

The constant-coefficient differential equation

$$a_0 \frac{d^n u}{dt^n} + \cdots + a_n u = b_0 \frac{d^m f}{dt^m} + \cdots + b_m f \tag{2.4-1}$$

with zero initial conditions is converted to the algebraic equation

$$a_0 s^n U(s) + \cdots + a_n U(s) = b_0 s^m F(s) + \cdots + b_m F(s) \tag{2.4-2}$$

by using the transform in (2.2-24). The $U(s)$ is then

$$U(s) = \frac{b_0 s^m + \cdots + b_m}{a_0 s^n + \cdots + a_n} F(s) \tag{2.4-3}$$

hence the solution to the differential equation is the inverse transform of $U(s)$

$$u(t) = \mathcal{L}^{-1}[U(s)] \tag{2.4-4}$$

Example 2-19. The differential equation

$$\frac{d^2 u}{dt^2} + 5 \frac{du}{dt} + 4u = 3 \frac{df}{dt} \quad (f(t) = u_{-1}(t) e^{-2t}) \tag{2.4-5}$$

is now solved using Laplace transforms. A comparison of (2.4-1) and (2.4-5) shows that $n = 2$, $a_0 = 1$, $a_1 = 5$, $a_2 = 4$, $m = 1$, $b_0 = 3$, and $b_1 = 0$. Substitution of these values into (2.4-3) gives

$$U(s) = \frac{3s}{s^2 + 5s + 4} \times \frac{1}{s+2} = \frac{3s}{(s+1)(s+2)(s+4)} \tag{2.4-6}$$

where $\mathcal{L}[e^{-2t}] = 1/(s+2)$. Expand $U(s)$ into the partial fractions

$$U(s) = \frac{3}{s+2} - \frac{1}{s+1} - \frac{2}{s+4} \tag{2.4-7}$$

and the inverse transform is

$$u(t) = [3e^{-2t} - e^{-t} - 2e^{-4t}] u_{-1}(t) \tag{2.4-8}$$

which is verified as the solution to (2.4-5). Higher-order differential equations are similarly solved. Furthermore, initial conditions can be introduced in the derivative transform as shown in (2.2-24).

A very important fact may have been overlooked in obtaining $u(t)$ in Example 2-19. In (2.4-6) we determined the roots of the denominator polynomial before we expanded $U(s)$ into partial fractions. But this polynomial is the characteristic polynomial associated with the homogeneous differential equation. Regardless of whether we use Laplace transforms or the prescription in Section 1.3 to solve the differential equation, we must always find the roots of the same polynomial.

The result in (2.4-3) has an important interpretation with respect to the physical system described by the differential equation. Divide both sides by $F(s)$ to give

$$\frac{U(s)}{F(s)} = \frac{b_0 s^m + \cdots + b_m}{a_0 s^n + \cdots + a_n} \qquad (2.4\text{-}9)$$

The right-hand side of (2.4-9) is independent of the input transform $F(s)$ and output transform $U(s)$. Each differential equation coefficient is a function of only the system parameters. This quantity designated $H(s)$ is defined to be the system transfer function,

$$H(s) = \frac{U(s)}{F(s)} \qquad (2.4\text{-}10)$$

For a system described by the differential equation in (2.4-1), (2.4-9) reveals that $H(s)$ is always a rational function in s.

Equation 2.4-10 states that $H(s)$ is the Laplace transform of the output divided by the Laplace transform of the input when a single input is applied, all other inputs being set to zero. This definition, although derived for the system described by (2.4-1), is applicable to a more general class of linear, time-invariant systems. For example, consider the system whose input $f(t)$ and output $u(t)$ are related by

$$u(t) = a_0 f(t) + a_1 f(t - T) \qquad (2.4\text{-}11)$$

where a_0 and a_1 are constants and T is a constant delay. Then the system transfer function, according to (2.4-10), is

$$H(s) = \frac{U(s)}{F(s)} = \frac{a_0 F(s) + a_1 e^{-sT} F(s)}{F(s)} = a_0 + a_1 e^{-sT} \qquad (2.4\text{-}12)$$

which is not a rational function in s.

Alternately, $H(s)$ can be defined as

$$H(s) = \frac{\text{system response to an exponential } (e^{st})}{e^{st}} \qquad (2.4\text{-}13)$$

For the system described by (2.4-11), the response to an exponential is

$$u(t) = a_0 e^{st} + a_1 e^{s(t-T)} \qquad (2.4\text{-}14)$$

and from (2.4-13)

$$H(s) = \frac{a_0 e^{st} + a_1 e^{s(t-T)}}{e^{st}} = a_0 + a_1 e^{-sT} \qquad (2.4\text{-}15)$$

which agrees with (2.4-12).

Solution of Differential Equations

Example 2-20. We now determine $H(s)$ for the system described by (2.4-5), first computing the system's response to an exponential (e^{st}) and then dividing the response by e^{st} according to (2.4-13). The response is given by the convolution integral. First, however, we determine the impulse response from the kernel function.

The solutions to the homogeneous portion of (2.4-5) are $u_1(t) = e^{-4t}$ and $u_2(t) = e^{-t}$. From (1.4-8) the kernel function is

$$k(t - \tau) = A_1 e^{-4(t-\tau)} + A_2 e^{-(t-\tau)} \tag{2.4-16}$$

and from the conditions in (1.4-2)

$$\begin{aligned} A_1 + A_2 &= 0 \\ -4A_1 - A_2 &= 1 \end{aligned} \tag{2.4-17}$$

yielding $A_2 = \frac{1}{3}$, $A_1 = -\frac{1}{3}$. Thus

$$k(t - \tau) = \tfrac{1}{3} e^{-(t-\tau)} - \tfrac{1}{3} e^{-4(t-\tau)} \tag{2.4-18}$$

and $k_1(t - \tau)$, with $m = 1$, $b_0 = 3$, and $b_1 = 0$, is obtained from (1.4-20) as

$$k_1(t - \tau) = -3 \frac{\partial}{\partial \tau} k(t - \tau) = 4e^{-4(t-\tau)} - e^{-(t-\tau)} \tag{2.4-19}$$

The system response is given by (1.4-19) with $f(\tau) = e^{s\tau}$. The exponential begins at $t = -\infty$; thus $t_0 = -\infty$. Then

$$u(t) = \int_{-\infty}^{t} [4e^{-4(t-\tau)} - e^{-(t-\tau)}] e^{s\tau} \, d\tau$$

$$= 4e^{-4t} \int_{-\infty}^{t} e^{(s+4)\tau} \, d\tau - e^{-t} \int_{-\infty}^{t} e^{(s+1)\tau} \, d\tau$$

$$= \frac{3s}{s^2 + 5s + 4} e^{st} \tag{2.4-20}$$

Divide the response by e^{st} to find the transfer function

$$H(s) = \frac{3s}{s^2 + 5s + 4} \tag{2.4-21}$$

which agrees with (2.4-6) after dividing $U(s)$ by $F(s) = 1/(s+2)$. Thus both definitions of $H(s)$ are equivalent.

We have shown that $U(s)$, the Laplace transform of the output function, is related to $F(s)$, the Laplace transform of the input function, by (2.4-10), as

$$U(s) = H(s)F(s) \tag{2.4-22}$$

This is a very useful result because we can use it to find the system response for any Laplace transformable input.

We are now able to find system outputs by solving differential equations in the time domain or by first going to the frequency domain for computation and then returning to the time domain. A very skilled solver of differential equations may argue that there is no particular advantage derived from the frequency-domain approach, but for most, the Laplace transform approach offers the following advantages:

1. Initial conditions are easily introduced.
2. The solution can be looked up in available tables, avoiding the integration necessary in evaluating the convolution integral or the various special techniques for solving differential equations.
3. Insight into system behavior is often obtained. Section 2.5 elaborates on this.

The connection between time- and frequency-domain system characterizations is now derived. Consider the system response when the input $f(t)$ is an impulse function. From (2.2-6) its transform is unity, and (2.4-22) therefore reduces to

$$U(s) = H(s) \cdot 1 \qquad (2.4\text{-}23)$$

The inverse transform of $U(s)$ is the system response to an impulse, which by definition is the impulse response $h(t)$. Hence

$$h(t) = \mathscr{L}^{-1}[H(s)] \qquad (2.4\text{-}24)$$

or

$$H(s) = \mathscr{L}[h(t)] = \int_0^\infty h(t)e^{-st}\, dt \qquad (2.4\text{-}25)$$

We have thus established the important connection between the time-domain and frequency-domain descriptions of a linear time-invariant system, namely, $H(s)$ is the Laplace transform of $h(t)$, the response of the system to an impulse at $t = 0$ and, of course, the inverse transform of $H(s)$ is $h(t)$.

This result can also be obtained by remembering that the system output is the convolution of the system impulse response $h(t)$ and the input $f(t)$,

$$u(t) = h(t) * f(t) \qquad [\text{see } (1.8\text{-}9)] \qquad (2.4\text{-}26)$$

But since convolution in the time domain corresponds to multiplication in the frequency domain, from (2.2-34)

$$U(s) = H(s)F(s) \qquad (2.4\text{-}27)$$

where $H(s)$ is the transform of $h(t)$, and $F(s)$ is the transform of $f(t)$. This equation, however, is the same as (2.4-22), where $H(s)$ is the transfer function. Thus $H(s)$ must be the Laplace transform of $h(t)$, and we arrive at the result given by (2.4-25).

The system step response $g(t)$ is obtained by substituting $F(s) = 1/s$ (the transform of $u_{-1}(t)$) into (2.4-22) and taking the inverse of both sides, yielding

$$u(t) = g(t) = \mathscr{L}^{-1}\left[\frac{H(s)}{s}\right] \qquad (2.4\text{-}28)$$

We showed in Section 1.8 that the step response is the integral of the impulse response. In the frequency domain, this corresponds to $(1/s)H(s)$ [see (2.2-28)], which agrees with (2.4-28).

The Transfer Function

2.5 THE TRANSFER FUNCTION

The transfer function $H(s)$ is the frequency-domain description of a linear time-invariant system and is a necessary function for analysis and synthesis in this domain. Hence we introduce the transfer function of systems (filters) composed of lumped constants (those described by ordinary constant-coefficient differential equations) and give a method for determining it. Because these lumped constants have analogies in other systems, the analysis here is broadly applicable. Then the system response to sinusoidal inputs is examined, yielding the steady-state magnitude, phase, and group delay responses, important quantities in everyday engineering practice. Finally, the poles and zeros of $H(s)$ are introduced and we explain their geometric significance.

2.5.1 Derivation

Before the transfer function can be determined, the input and output quantities must be specified. The voltage transfer function $T(s)$ is the ratio of the output voltage to the input voltage, while the current transfer function $\alpha(s)$ is the ratio of the output current to the input current. Both are dimensionless. The transfer impedance $Z_{12}(s)$ relates the output voltage to the input current and has the dimension of ohms, while the transfer admittance $Y_{12}(s)$ relates the output current to the input voltage and has the dimension of mhos. However, we shall continue to use the term "transfer function" represented by $H(s)$ unless it is more meaningful to use one of the other descriptions.

Example 2-21. We now obtain the transfer function of the network in Fig. 2-7 using the definition in (2.4-13). The differential equation describing this system, for the input e^{st}, is

$$LC\frac{d^2v_0}{dt^2} + RC\frac{dv_0}{dt} + v_0 = e^{st} \tag{2.5-1}$$

From (2.4-13), the output voltage is

$$v_0 = H(s)e^{st} \tag{2.5-2}$$

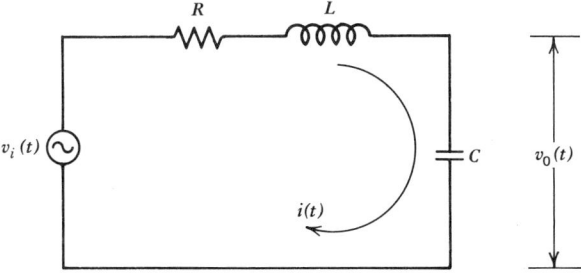

Figure 2-7. RLC series network.

and the first two derivatives of v_0 are

$$\frac{dv_0}{dt} = sH(s)e^{st}$$

$$\frac{d^2v_0}{dt^2} = s^2H(s)e^{st}$$

(2.5-3)

Substitute from (2.5-2) and (2.5-3) into (2.5-1) to give

$$(LCs^2 + RCs + 1)H(s) = 1 \qquad (2.5\text{-}4)$$

The transfer function is then

$$H(s) = \frac{1}{LCs^2 + RCs + 1} \qquad (2.5\text{-}5)$$

The important features of $H(s)$ in (2.5-5) are revealed by rearranging $H(s)$ as

$$H(s) = \frac{1/sC}{sL + R + 1/sC} \qquad (2.5\text{-}6)$$

First note that each term in (2.5-6) has the dimension of ohms, where the complex frequency s replaces the usual $j\omega$. Thus sL is the generalized impedance of an inductor L, and $1/sC$ is the generalized impedance of a capacitor C. Also, although not shown here, sM is the generalized impedance of a mutual inductor M. Next note that the voltage ratio $H(s)$ in (2.5-6) is the same as the one obtained by replacing each element by its generalized impedance and applying the basic laws of network analysis, such as Kirchhoff's voltage and current laws. The latter method is not a coincidence but a valid procedure for obtaining the transfer function. It eliminates the intermediate step of obtaining the differential equation. Example 2-22 illustrates this method.

Example 2-22. The transfer function of the network in Fig. 2-8 is obtained by first replacing each element by its generalized impedance as in Fig. 2-9. The impedance of the RC parallel circuit is

$$Z_1 = \frac{R}{1 + RCs} \qquad (2.5\text{-}7)$$

and the output voltage $V_0(s)$ is obtained by the voltage division of the input as

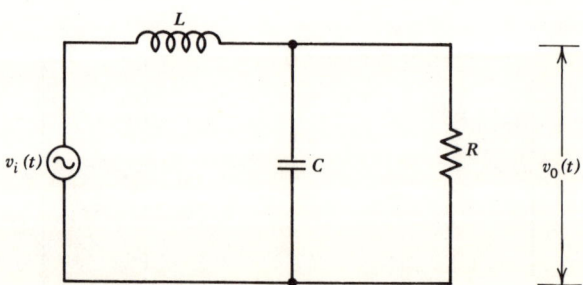

Figure 2-8. Singly terminated second-order network.

The Transfer Function

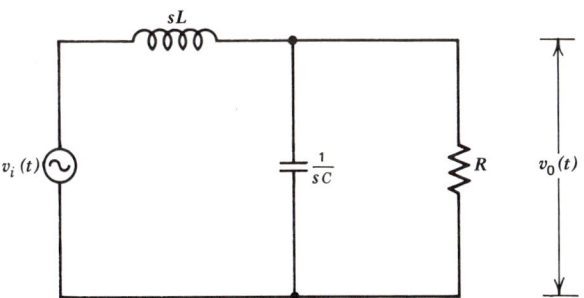

Figure 2-9. Network replaced by generalized impedances.

$$V_o(s) = \frac{Z_1}{sL + Z_1} V_i(s) = \frac{1}{LCs^2 + (L/R)s + 1} V_i(s) \qquad (2.5\text{-}8)$$

Then the transfer function $H(s)$ is

$$H(s) = \frac{V_o(s)}{V_i(s)} = \frac{1}{LCs^2 + (L/R)s + 1} \qquad (2.5\text{-}9)$$

For this example, $H(s)$ represents $T(s)$, the voltage transfer function.

We have established in principle how to find the transfer function of any linear time-invariant network composed of lumped components. We need only replace each element by its generalized impedance (resistors remain the same) and use any of the basic techniques (e.g., node or mesh analysis) to determine the desired input-output ratio. An example illustrates the determination of $Z_{12}(s)$.

Example 2-23. The transfer impedance $Z_{12}(s)$ of the RLC network in Fig. 2–10 is now determined. By Kirchhoff's current law,

$$I(s) = \frac{V_o(s)}{sL} + \frac{V_o(s)}{1/sC} + \frac{V_o(s)}{R} \qquad (2.5\text{-}10)$$

yielding the transfer impedance

$$Z_{12}(s) = \frac{V_o(s)}{I(s)} = \frac{Ls}{LCs^2 + (L/R)s + 1} \qquad (2.5\text{-}11)$$

Because $T(s)$, $\alpha(s)$, $Z_{12}(s)$, and $Y_{12}(s)$ are always rational functions of s, the technique presented in Section 2.3 can be used to recover the time function.

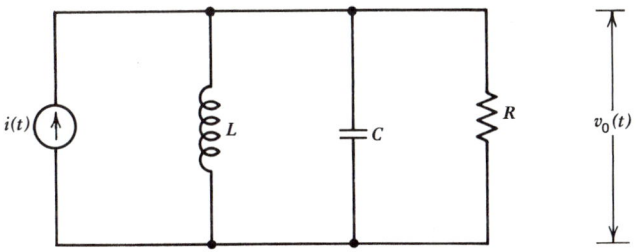

Figure 2-10. RLC parallel network.

Example 2-24. We now determine the response of the network in Example 2-22 to a voltage impulse at $t = 0$. Let $R = 1$ ohm, $L = 1$ henry (H), and $C = \sqrt{2}/2$ farad (F). The output voltage is then the network impulse response; $v_0(t)$ is the inverse transform of $H(s)$. From (2.5-9)

$$v_0(t) = h(t) = \mathcal{L}^{-1}[H(s)] = \mathcal{L}^{-1}\left[\frac{1}{\left(s + \frac{\sqrt{2}}{2}\right)^2 + \left(\frac{\sqrt{2}}{2}\right)^2}\right] \tag{2.5-12}$$

From (2.2-13), with $\alpha = \sqrt{2}/2$ and $\beta = \sqrt{2}/2$,

$$h(t) = \sqrt{2}\, e^{-(\sqrt{2}/2)t} \sin\frac{\sqrt{2}}{2}t\, u_{-1}(t) \tag{2.5-13}$$

We shall later see that $h(t)$ is the impulse response of the second-order Butterworth filter.

2.5.2 Poles and Zeros

We now introduce the poles and zeros of $H(s)$ and their pictorial representation. From (2.4-9), $H(s)$ is the rational function in s,

$$H(s) = \frac{b_0 s^m + \cdots + b_m}{a_0 s^n + \cdots + a_n} = \frac{N(s)}{D(s)} \tag{2.5-14}$$

where $N(s)$ is the numerator polynomial and $D(s)$ is the denominator polynomial. The roots of $N(s)$ and $D(s)$ are given special names because of their importance in characterizing $H(s)$; the former are called the zeros of $H(s)$ and the latter are called the poles of $H(s)$. Specifications of the poles, zeros, and a gain constant H_0 completely characterize the system, for then the transfer function is known. Equation 2.5-14 can be rewritten as

$$H(s) = H_0 \frac{s^m + (b_1/b_0)s^{m-1} + \cdots + b_m/b_0}{s^n + (a_1/a_0)s^{n-1} + \cdots + a_n/a_0} \tag{2.5-15}$$

where the gain constant $H_0 = b_0/a_0$.

The zeros are complex frequencies at which the transfer function has the value zero, while the poles are those complex frequencies at which the value of the transfer function is infinite. The poles have further significance, for they are the natural frequencies of the system, thus determine the general system behavior. Refer to (2.3-31) and note that before the inverse transform is analytically determined, the roots of $D(s)$ must be found. These roots determine whether the time-domain response increases with time, oscillates, or decreases with time.

We have defined s as the complex frequency $\sigma + j\omega$. More precisely s is a complex variable whose real part is σ (neper frequency) and whose imaginary part is ω (radian frequency). Consequently any function of s is a function of a complex variable and has a real and imaginary part. One of these, of course, may be zero. The complex frequency s is a generalization of the familiar sinusoidal frequency $j\omega$ ($\sigma = 0$).

With $\sigma = 0$, and with the aid of (2.1-11), the exponential representation

$$e^{st} = e^{(\sigma + j\omega)t} = e^{\sigma t} e^{j\omega t} \tag{2.5-16}$$

reduces to

$$e^{j\omega t} = \cos \omega t + j \sin \omega t \tag{2.5-17}$$

The Transfer Function

This form reveals the sinusoidal nature of the function with $\sigma = 0$. In many practical problems, for ease of analysis, the function $e^{j\omega t}$ is used as an input, with the understanding that the real part of the response corresponds to the input $\cos \omega t$ and the imaginary part of the response corresponds to the input $\sin \omega t$. Alternatively, $\sin \omega t$ and $\cos \omega t$ can be simulated by a superposition of $e^{j\omega t}$ and $e^{-j\omega t}$.

The pictorial representation of poles and zeros furnishes insight into the system behavior (see Section 2.5.4). The plane that contains the two numbers defining a pole or zero (σ and ω) is known as the "s-plane." When plotted, a pole is indicated by a cross and a zero is indicated by a circle. The order "q" of the pole or zero is indicated by (q) close to the pole or zero. No number is given if the order is unity. The gain constant H_0 sometimes appears in the diagram. Figure 2-11 is a typical s-plane plot, and it is emphasized that a plot completely characterizes the system (to within a gain constant if H_0 is not given).

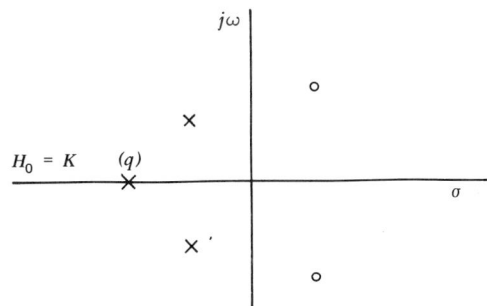

Figure 2-11. An s-plane pole-zero diagram.

A transfer function has an equal number of poles and zeros, although some of the zeros may occur at infinity. Only finite poles and zeros, however, are plotted on the pole-zero diagram. Three regions of the s-plane deserve special mention because of responses peculiar to poles lying in these regions. The region to the left of the $j\omega$ axis is known as the left-half plane, and the region to the right of the $j\omega$ axis is called the right-half plane. The third region is the infinitesimal strip known as the $j\omega$ axis. From our discussion on inverse transforms, we know that poles in the left-half plane give rise to exponentially decreasing time functions, and poles in the right-half plane give rise to exponentially increasing time functions. Poles on the $j\omega$ axis result in oscillating time functions.

Example 2-25. The poles and zeros of the transfer function

$$H(s) = 5 \frac{s^3 - 3s^2 + 9s + 13}{s^4 + 14s^3 + 89s^2 + 296s + 400} \tag{2.5-18}$$

and the corresponding pole-zero diagram are now obtained. The numerator polynomial $N(s)$ is factored as

$$N(s) = (s+1)(s^2 - 4s + 13) \tag{2.5-19}$$

and its roots, the finite zeros of $H(s)$, are -1 and $2 \pm j3$. The denominator polynomial $D(s)$ is factored as

$$D(s) = (s+4)^2(s^2+6s+25) \tag{2.5-20}$$

and its roots, the poles of $H(s)$, are $-4, -4, -3 \pm j4$. There is also a zero at infinity resulting in an equal number of poles and zeros. Figure 2-12 is the pole-zero plot in the s-plane.

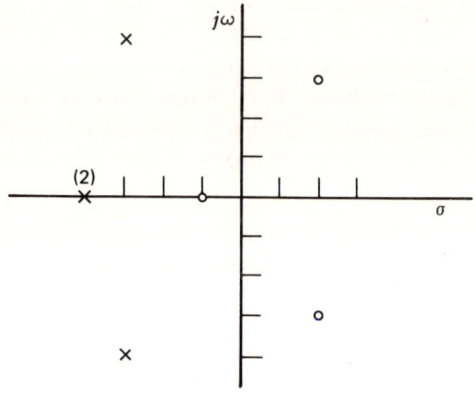

Figure 2-12. Pole-zero diagram for Example 2-25.

2.5.3 Steady-State Responses

We now consider the system behavior when the input is a sinusoidal voltage or current that has been on for all time. That the sinusoid is extremely important in engineering is evidenced by the great amount of test equipment available for measurement of these frequencies. The functional form of this wave (a voltage in this case) is

$$v(t) = V_M \cos \omega t \tag{2.5-21}$$

where V_M is the peak voltage and ω is the radian frequency of the waveform. Remember that $\cos \omega t$ and $\sin \omega t$ are components of the more general function $e^{j\omega t}$. From (2.1-11)

$$\cos \omega t = \text{Re } e^{j\omega t}$$
$$\sin \omega t = \text{Im } e^{j\omega t} \tag{2.5-22}$$

where Re and Im designate the real and imaginary parts, respectively, of the quantity following.

Let $e^{j\omega_0 t}$ be the input to a system characterized by the transfer function in (2.5-14), assuming its poles lie in the left-half plane. The output transform, from (2.4-22), is then

$$U(s) = H(s) \cdot \frac{1}{s - j\omega_0} \tag{2.5-23}$$

Now $U(s)$ is separated into the two partial fractions

$$U(s) = \frac{N(s)}{D(s)(s-j\omega_0)} = \frac{G(s)}{D(s)} + \frac{N(j\omega_0)}{D(j\omega_0)} \cdot \frac{1}{s-j\omega_0} \tag{2.5-24}$$

The Transfer Function

where $G(s)$ is a polynomial in s. The coefficient of $1/(s - j\omega_0)$, obtained by the technique in Section 2.3, is $H(j\omega_0)$. Because the zeros of $D(s)$ are in the left-half plane, the time function corresponding to $G(s)/D(s)$ approaches zero for very large values of time. Then for $t \gg 0$,

$$u(t) \approx \mathscr{L}^{-1}\left[\frac{H(j\omega_0)}{s - j\omega_0}\right] = H(j\omega_0)e^{j\omega_0 t} \qquad (2.5\text{-}25)$$

that is, the response is the input exponential multiplied by the complex number $H(j\omega_0)$. The response at large values of time is the same as the response obtained for the case of the input being on for all time; all transients have disappeared.

The input frequency is ω_0, but since ω_0 can be any frequency, the subscript is removed. The $H(j\omega)$ is a complex function that can be arranged as the sum of a real part $R(\omega)$ and an imaginary part $X(\omega)$, just as we did for the Fourier development in Section 2.1. Then

$$H(j\omega) = R(\omega) + jX(\omega) = |H(j\omega)|e^{j\theta(\omega)} \qquad (2.5\text{-}26)$$

where

$$|H(j\omega)| = \sqrt{R^2(\omega) + X^2(\omega)} \qquad (2.5\text{-}27)$$

and

$$\theta(\omega) = \tan^{-1}\frac{X(\omega)}{R(\omega)} \qquad (2.5\text{-}28)$$

Then (2.5-25) can be expressed as

$$u(t) = |H(j\omega)|e^{j[\omega t + \theta(\omega)]} \qquad (2.5\text{-}29)$$

The real part and the imaginary part of $u(t)$ are the respective responses to $\cos \omega t$ and $\sin \omega t$. This is shown from Euler's identity of (2.1-11) and system linearity.

$$u(t)(\text{for } \cos \omega t \text{ input}) = |H(j\omega)| \cos[\omega t + \theta(\omega)]$$
$$u(t)(\text{for } \sin \omega t \text{ input}) = |H(j\omega)| \sin[\omega t + \theta(\omega)] \qquad (2.5\text{-}30)$$

Thus the magnitude of the sinusoid with frequency ω after passing through the system is $|H(j\omega)|$, and the sinusoid experiences a phase shift $\theta(\omega)$. We define $|H(j\omega)|$ as the magnitude or steady-state response of the system, and $\theta(\omega)$ is the system phase response. Furthermore, the magnitude response is related to the transfer function by

$$|H(j\omega)|^2 = H(s)H(-s)|_{s=j\omega} \qquad (2.5\text{-}31)$$

Note that we retain $j\omega$ as the argument when $H(s)$ is evaluated at $s = j\omega$, whereas we use ω as the argument for the Fourier transform. This practice reminds us that the particular function, in this case $H(s)$, can be evaluated at any frequency in the complex s-plane, whereas the Fourier transform is always evaluated along the imaginary axis.

This raises the following question. Given a function $f(t)$ which is zero for $t < 0$, when is the Fourier transform of $f(t)$ identical to the Laplace transform of $f(t)$ evaluated at $s = j\omega$? They are the same if the $j\omega$ axis is within the region of convergence of the Laplace transform integral in (2.2-1). If this is true, the Fourier inversion integral in (2.1-6) can be used to recover the time function corresponding to $F(s)|_{s=j\omega}$ because $F(s)|_{s=j\omega}$ is also the Fourier transform of $f(t)$.

We briefly discussed the region of convergence in Example 2-8. There the $j\omega$ axis lies in the region of convergence if $\alpha > 0$, hence the Fourier transform of $e^{-\alpha t}$ exists. However, as $\alpha \to 0$, $f(t)$ approaches the step function, for which the classical Fourier transform does not exist. The integral for the Laplace transform exists if $\sigma > 0$, which clearly excludes the $j\omega$ axis. Thus the Fourier transform of $u_{-1}(t)$ cannot be obtained by evaluating $F(s)$ at $s = j\omega$.

The $|H(j\omega)|$ is also the ratio of the output sinusoid amplitude to the input sinusoid amplitude. Consider $u(t)$ in (2.5-29), which is the system response to a unit-amplitude sinusoid. Then

$$\frac{\text{output amplitude}}{\text{input amplitude}} = \frac{|H(j\omega)|}{1} = |H(j\omega)| \qquad (2.5\text{-}32)$$

If the input is multiplied by a constant, system linearity requires that the output also be multiplied by the same factor. Thus the output-input amplitude ratio remains at $|H(j\omega)|$.

Because $|H(j\omega)|$ can take on an enormous range of values, the logarithm of $H(j\omega)$ is often used to obtain more convenient values. If the natural logarithm is used

$$-\ln H(j\omega) = -\ln[|H(j\omega)|e^{j\theta(\omega)}] = -\ln|H(j\omega)| - j\theta(\omega) = \alpha(\omega) + j\beta(\omega) \qquad (2.5\text{-}33)$$

where $\alpha(\omega)$ is the attenuation in nepers,

$$\alpha(\omega) = -\ln|H(j\omega)| \text{ nepers} \qquad (2.5\text{-}34)$$

while $\beta(\omega)$ is the phase-lag angle in radians and is the negative of the phase angle $\theta(\omega)$,

$$\beta(\omega) = -\theta(\omega) \text{ rads} \qquad (2.5\text{-}35)$$

Many books consider $\beta(\omega)$ as the phase of filtering networks, but this has the opposite slope of $\theta(\omega)$. Confusion can be avoided by noting which phase angle is being considered.

A common definition of attenuation is obtained using the base 10 logarithm,

$$A = -20 \log_{10}|H(j\omega)| \text{ decibels} \qquad (2.5\text{-}36)$$

The units of this attenuation are decibels (dB). The number of decibels per neper is found from (2.5-34) and (2.5-36) as

$$\frac{A}{\alpha(\omega)} = \frac{-20\log_{10}|H(j\omega)|}{-\ln|H(j\omega)|} = \frac{20\log_{10}|H(j\omega)|}{\ln 10 \log_{10}|H(j\omega)|} = \frac{20}{\ln 10} = 8.686 \qquad (2.5\text{-}37)$$

or 1 neper = 8.686 dB. By considering the attenuation rather than the magnitude, we no longer run into inconveniently large or small numbers. Remember also that 1 rad = $180°/\pi$ = 57.3°.

Example 2-26. The magnitude and phase response of the network in Fig. 2-13 are now determined. The transfer function $H(s)$ is

$$H(s) = \frac{V_o(s)}{V_i(s)} = \frac{\frac{1}{2}}{s^3 + 2s^2 + 2s + 1} \qquad (2.5\text{-}38)$$

For this case $H(s)$ is $T(s)$, the output-input voltage ratio. The poles of $H(s)$ lie on

The Transfer Function

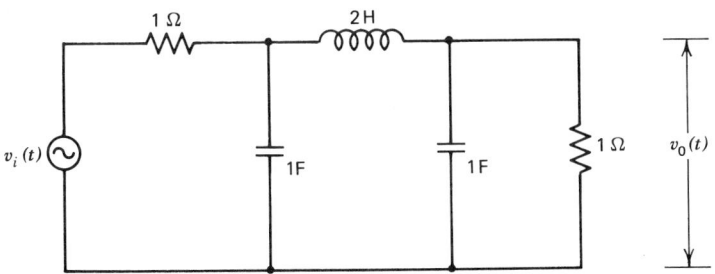

Figure 2-13. A third-order network.

the unit circle at $s = -1$ and $s = -1/2 \pm j\sqrt{3}/2$. Evaluate $H(s)$ at $s = j\omega$ to give

$$H(j\omega) = \frac{\frac{1}{2}}{1 - 2\omega^2 + j\omega(2 - \omega^2)} \quad (2.5\text{-}39)$$

The magnitude response is the absolute value of $H(j\omega)$, given by

$$|H(j\omega)| = \frac{\frac{1}{2}}{\sqrt{(1 - 2\omega^2)^2 + \omega^2(2 - \omega^2)^2}} = \frac{\frac{1}{2}}{\sqrt{1 + \omega^6}} \quad (2.5\text{-}40)$$

which is also obtainable from (2.5-31). This response, known as the third-order Butterworth or maximally flat low-pass response, is discussed in more detail in Chapter 3. At $\omega = 0$, $|H(j0)| = \frac{1}{2}$, the voltage division due to the two resistors. At $\omega = 1$, $|H(j1)| = 1/(2\sqrt{2})$, and for large values of ω, $|H(j\omega)| \approx \frac{1}{2}\omega^3$. Relative to the gain at $\omega = 0$, the attenuation from (2.5-36) is

$$A = -20 \log_{10} \frac{1}{\sqrt{1 + \omega^6}} = 10 \log_{10}(1 + \omega^6) \text{ decibels} \quad (2.5\text{-}41)$$

At $\omega = 1$, $A = 10 \log_{10} 2 = 3.01$ dB. In practice this value of attenuation is referred to as 3 dB. The frequency at which the response is $1/\sqrt{2}$ times the maximum response is called the 3-dB frequency.

The phase response, from (2.5-28), is

$$\theta(\omega) = \tan^{-1} \frac{\omega(\omega^2 - 2)}{1 - 2\omega^2} \quad (2.5\text{-}42)$$

which is $0°$ at $\omega = 0$ and $-135°$ at $\omega = 1$. For large values of ω, $\theta(\omega)$ approaches $-270°$. The phase-lag angle β is $+135°$ at $\omega = 1$. The $|H(j\omega)|$ expressed in decibels is shown in Fig. 2-14a; $\theta(\omega)$ appears in Fig. 2-14b.

The transfer function can be recovered from the magnitude response by reversing the procedure given by (2.5-31). For a given $|H(j\omega)|^2$:

1. Replace ω^2 by $-s^2$.
2. Factor numerator and denominator polynomial.
3. Identify left-half plane poles with $H(s)$ and right-half plane poles with $H(-s)$. Placement of zeros is arbitrary; however, Section 3.1.9 shows that the minimum-phase function is obtained when only the left-half plane zeros are associated with $H(s)$.

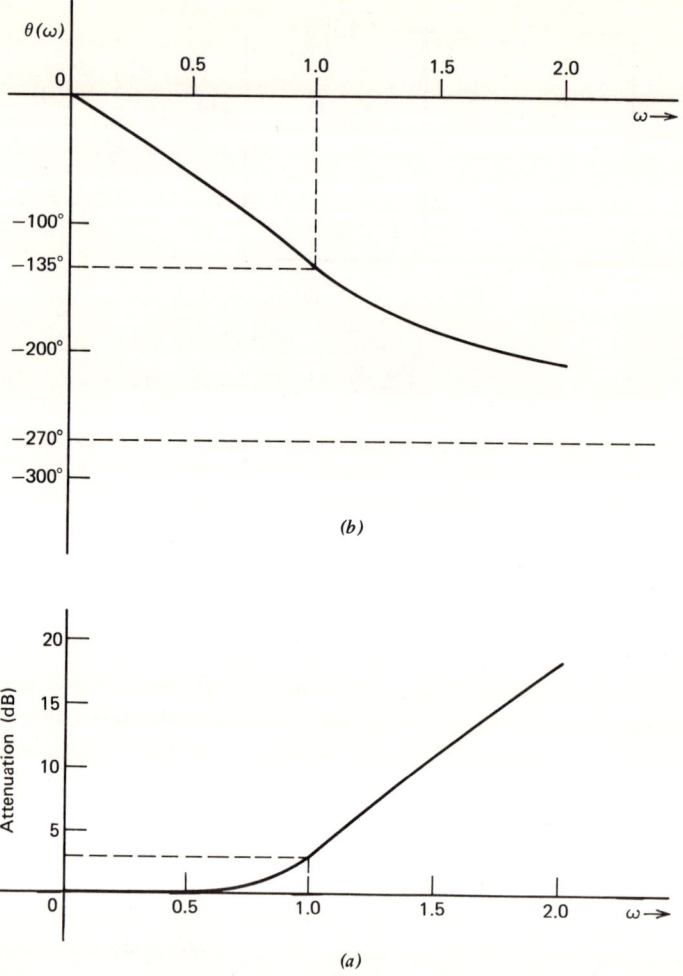

Figure 2-14. (a) Attenuation and (b) phase of network in Example 2-26.

Example 2-27. Determine the transfer function corresponding to the magnitude response of Example 2-26 (neglecting the gain constant),

$$|H(j\omega)|^2 = \frac{1}{1+\omega^6} \qquad (2.5\text{-}43)$$

1. Replace ω^2 by $-s^2$. Then, from (2.5-31),

$$H(s)H(-s) = \frac{1}{1-s^6} \qquad (2.5\text{-}44)$$

2. The six denominator roots are ± 1, $-\frac{1}{2} \pm j\sqrt{3}/2$, $+\frac{1}{2} \pm j\sqrt{3}/2$.
3. Associate the left-half plane roots with $H(s)$ to give

$$H(s) = \frac{1}{(s+1)(s+\frac{1}{2}+j\sqrt{3}/2)(s+\frac{1}{2}-j\sqrt{3}/2)} = \frac{1}{s^3+2s^2+2s+1} \qquad (2.5\text{-}45)$$

which agrees with (2.5-38).

The Transfer Function

Example 2-26 shows the computation of the magnitude and phase responses when $H(s)$ is expressed as a rational function. Often, however, the poles and zeros of $H(s)$ are available, and rather than first forming the numerator and denominator polynomials, and then computing $|H(j\omega)|$ and $\theta(\omega)$, we would like to express these responses in terms of the real and imaginary parts of these complex-valued poles and zeros. This formulation is suitable for machine computation, for in this form all calculations are performed with real numbers.

To arrive at these expressions we consider the transfer function composed of m finite zeros, $z_i = \gamma_i + j\delta_i$, and n poles $p_i = \alpha_i + j\beta_i$, $(m \leq n)$,

$$H(s) = H_0 \frac{(s - z_1)(s - z_2) \cdots (s - z_m)}{(s - p_1)(s - p_2) \cdots (s - p_n)} \tag{2.5-46}$$

In terms of the poles and zeros, $|H(j\omega)|$ is the absolute value of each factor evaluated at $s = j\omega$,

$$|H(j\omega)| = H_0 \frac{|j\omega - z_1||j\omega - z_2| \cdots |j\omega - z_m|}{|j\omega - p_1||j\omega - p_2| \cdots |j\omega - p_n|} \tag{2.5-47}$$

Consider the factor $|j\omega - p_i|$;

$$|j\omega - p_i| = |j\omega - \alpha_i - j\beta_i| = |-\alpha_i + j(\omega - \beta_i)| = \sqrt{\alpha_i^2 + (\omega - \beta_i)^2} \tag{2.5-48}$$

Since a similar expression exists for each numerator factor, $|H(j\omega)|$ can be written in terms of the real and imaginary parts of z_i and p_i as

$$|H(j\omega)| = H_0 \sqrt{\frac{\prod_{j=1}^{m} [\gamma_j^2 + (\omega - \delta_j)^2]}{\prod_{i=1}^{n} [\alpha_i^2 + (\omega - \beta_i)^2]}} \tag{2.5-49}$$

Now we determine the phase angle associated with the numerator factor $(j\omega - z_i)$. Then

$$j\omega - z_i = j\omega - \gamma_i - j\delta_i = -\gamma_i + j(\omega - \delta_i) \tag{2.5-50}$$

and the tangent of the phase angle is the imaginary part divided by the real part or

$$\theta_i = \tan^{-1} \frac{\delta_i - \omega}{\gamma_i} \tag{2.5-51}$$

Thus the phase response associated with $H(s)$ is

$$\theta(\omega) = \sum_{j=1}^{m} \tan^{-1} \frac{\delta_j - \omega}{\gamma_j} - \sum_{i=1}^{n} \tan^{-1} \frac{\beta_i - \omega}{\alpha_i} \tag{2.5-52}$$

The reason for adding the numerator factor phase angles and subtracting the denominator factor phase angles is explained in Section 2.5.4.

Example 2-28. We now determine $|H(j\omega)|$ and $\theta(\omega)$ from the third-order transfer function in (2.5-38) using the pole-zero formulations just discussed. There are no finite zeros and $H_0 = \frac{1}{2}$. The poles are -1 and $-\frac{1}{2} \pm j\sqrt{3}/2$. Therefore we make the identification

$$\begin{aligned} \alpha_1 &= -1, & \beta_1 &= 0 \\ \alpha_2 &= -\tfrac{1}{2}, & \beta_2 &= \sqrt{3}/2 \\ \alpha_3 &= -\tfrac{1}{2}, & \beta_3 &= -\sqrt{3}/2 \end{aligned} \tag{2.5-53}$$

Then, from (2.5-49),

$$|H(j\omega)| = \frac{\frac{1}{2}}{\sqrt{(1+\omega^2)[\frac{1}{4}+(\omega-\sqrt{3}/2)^2][\frac{1}{4}+(\omega+\sqrt{3}/2)^2]}} \quad (2.5\text{-}54)$$

which is reducible to (2.5-40). At $\omega = 1$, the magnitude response is computed as

$$|H(j1)| = \frac{\frac{1}{2}}{\sqrt{2[\frac{1}{4}+(1-\sqrt{3}/2)^2][\frac{1}{4}+(1+\sqrt{3}/2)^2]}} = \frac{1}{2\sqrt{2}} \quad (2.5\text{-}55)$$

The phase function, from (2.5-52), is

$$\theta(\omega) = -\tan^{-1}\omega + \tan^{-1}\frac{\sqrt{3}/2-\omega}{1/2} + \tan^{-1}\frac{-\sqrt{3}/2-\omega}{1/2} \quad (2.5\text{-}56)$$

Use of the trigonometric identity for the tangent of the sum of two angles yields the result in (2.5-42). Evaluation of (2.5-56) at $\omega = 1$ gives

$$\theta(1) = -45° - 15° - 75° = -135° \quad (2.5\text{-}57)$$

the value in Fig. 2-14b.

2.5.4 s-Plane Geometry

The magnitude and phase can be evaluated at values of s other than $j\omega$; moreover they can be determined from the s-plane geometry. Consider a general description for $H(s)$ as

$$H(s) = H_0 \frac{(s-s_2)(s-s_4)^q \cdots (s-s_m)}{(s-s_1)(s-s_3) \cdots (s-s_n)} \quad (2.5\text{-}58)$$

where the qth-order zero at $s = s_4$ is included for generality. Each factor of $H(s)$ may be considered a phasor $Ve^{j\phi}$, where V is the phasor length. The phasor angle is ϕ, which must lie between $-90°$ and $+90°$, where the reference line is the direction of the positive real axis. The angle is positive if measured in the counterclockwise sense and negative if measured in the clockwise sense. Then

$$H(s) = H_0 \frac{V_2 e^{j\phi_2} V_4^q e^{jq\phi_4} \cdots V_m e^{j\phi_m}}{V_1 e^{j\phi_1} V_3 e^{j\phi_3} \cdots V_n e^{j\phi_n}} \quad (2.5\text{-}59)$$

or

$$H(s) = H_0 \frac{V_2 V_4^q \cdots V_m}{V_1 V_3 \cdots V_n} \exp\{j[\phi_2 + q\phi_4 + \cdots + \phi_m - (\phi_1 + \phi_3 + \cdots + \phi_n)]\} \quad (2.5\text{-}60)$$

The magnitude of $H(s)$ is

$$|H(s)| = H_0 \frac{V_2 V_4^q \cdots V_m}{V_1 V_3 \cdots V_n} \quad (2.5\text{-}61)$$

and the phase angle is

$$\theta = \arg H(s) = \phi_2 + q\phi_4 + \cdots + \phi_m - (\phi_1 + \phi_3 + \cdots + \phi_n) \quad (2.5\text{-}62)$$

The magnitude of the function is the product of all phasor lengths resulting from the zeros, divided by the product of all phasor lengths resulting from the poles, where the phasors terminate at the point s at which an evaluation of the function is desired. The phase angle of the function is the sum of all phasor angles associated with the zeros minus the sum of all phasor angles associated with the

The Transfer Function

poles. We have now justified the subtraction operation in (2.5-52). Both magnitude and phase can therefore be measured in the s-plane with ruler and protractor. Also, general response shapes can often be determined by inspection.

Example 2-29. We now geometrically determine the magnitude and phase at $j\omega = j1$ for the third-order Butterworth transfer function given by (2.5-38). The magnitude and phase of the phasor for each pole are graphically computed in Fig. 2-15.

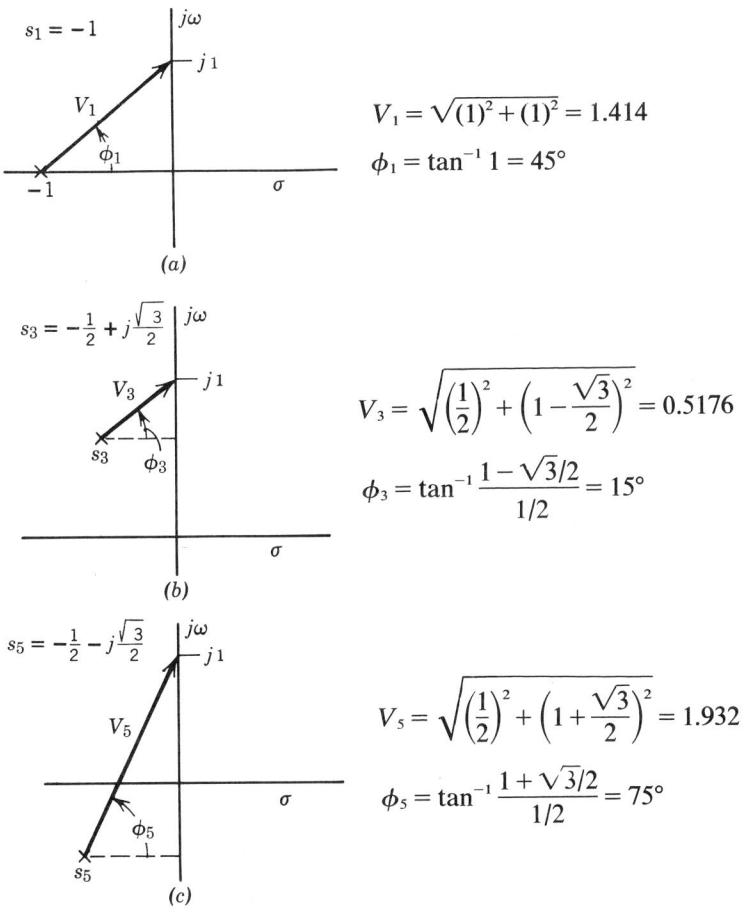

Figure 2-15. Graphical determination of magnitude and phase of third-order Butterworth response at $\omega = 1$.

Therefore with $H_0 = 1/2$, from (2.5-61),

$$|H(j1)| = \frac{1/2}{1.414 \times 0.5176 \times 1.932} = \frac{1/2}{\sqrt{2}} \tag{2.5-63}$$

and from (2.5-62)

$$\theta(1) = -(45° + 15° + 75°) = -135° \tag{2.5-64}$$

This agrees with the answers obtained analytically in Examples 2-26 and 2-28. The response at other frequencies is obtained in a similar manner.

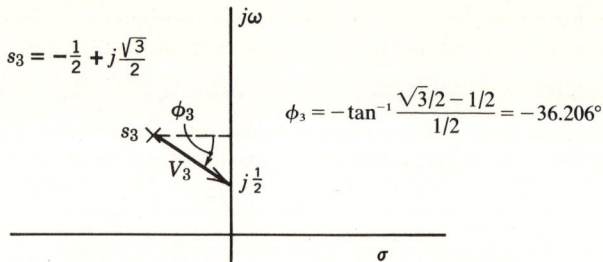

Figure 2-16. Example of a negative phasor angle.

To illustrate the computation of a negative phasor angle, we compute ϕ_3 at the frequency $j\omega = j\frac{1}{2}$. This geometry is represented in Fig. 2-16. The angle is negative because ϕ_3 is measured clockwise, whereas the positive direction is counterclockwise.

2.6 GROUP DELAY AND PHASE DELAY

The group delay function and the phase delay function are two steady-state responses that deserve special mention. Because they are the same in some cases and very different in others, misunderstanding often occurs with respect to their definition and applicability. Here we attempt to remove this confusion and additionally to consolidate the information about these two functions.

2.6.1 Definitions

If a signal is suddenly applied to a system input, the effect is usually not immediately observable at the output but occurs at a later time. Two useful functions for describing this delay are the group delay and the phase delay functions, which arise in the theoretic determination of signal transit time through linear time-invariant systems. However they also give meaningful values for signal delays through real-life networks. Furthermore, in practice the prediction of signal delays from the known system characteristics is more desirable than the time-consuming and costly process of analyzing the output signal to find these values. These two descriptions of delay enable us to predict the system delay.

In a general sense, the group delay describes the delay of a packet of frequencies, whereas the phase delay is the delay of a single sinusoid. The group delay $D(\omega)$† (often called envelope delay, as we learn later) is defined as

$$D(\omega) = -\frac{d\theta(\omega)}{d\omega} \qquad (2.6\text{-}1)$$

and the phase delay $\tau_p(\omega)$ is defined as

$$\tau_p(\omega) = -\frac{\theta(\omega)}{\omega} \qquad (2.6\text{-}2)$$

---

†$D(\omega)$ represents the general group delay function. Later we include subscripts to represent specific group delays [e.g., $D_l(\omega)$ to indicate the denormalized low-pass delay].

Group Delay and Phase Delay

where $\theta(\omega)$ is the phase function in radians given by (2.5-28). Thus $D(\omega)$ is the negative of the slope of the phase curve at the particular frequency ω, and $\tau_p(\omega)$ is the negative of the slope of the straight line from the origin to the point $[\omega, \theta(\omega)]$. These two delays have the same value if the phase function is linear with frequency and passes through the origin. Figure 2-17 illustrates computation of these delays at $\omega = \omega_1$. Here $D(\omega_1) = -\Delta\theta/\Delta\omega$ and $\tau_p(\omega_1) = \theta_1/\omega_1$. For this case the phase is nonlinear for $\omega > \omega_0$.

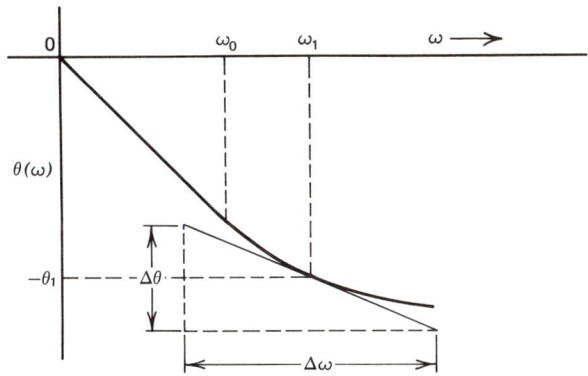

Figure 2-17. Illustration of phase delay and group delay.

The value of phase delay from (2.6-2) may be positive or negative depending on the sign of the phase angle θ. When θ is negative, τ_p is positive, and this is interpreted as the output sinusoid lagging the input sinusoid, a physically acceptable situation. However, a positive value of θ yields a negative value for τ_p, which is interpreted as the output sinusoid leading the input sinusoid. It is emphasized that the latter condition does not imply that a signal appears at the system output before a signal is applied to the system input. How then can the output lead the input but initially appear after the input is applied? The following example clarifies this apparent paradox.

Example 2-30. Consider the *RC* network in Fig. 2-18a with transfer function

$$H(s) = \frac{s}{s+1} \quad (2.6\text{-}3)$$

magnitude response (Fig. 2-18b)

$$|H(j\omega)| = \frac{\omega}{\sqrt{1+\omega^2}} \quad (2.6\text{-}4)$$

and phase response (Fig. 2-18c)

$$\theta = \tan^{-1}\frac{1}{\omega} \quad (2.6\text{-}5)$$

If at $t = 0$ we apply a sine wave of frequency $\omega = 0.6$, we expect the steady-state magnitude, from (2.6-4), to be 0.5145, and the phase angle, from (2.6-5), to be $+59.036°$. From (2.6-2) this corresponds to a negative phase delay of

$$\tau_p(0.6) = -\frac{\pi/180 \times 59.036}{0.6} = -1.7173 \text{ sec} \quad (2.6\text{-}6)$$

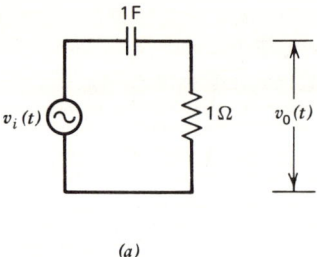

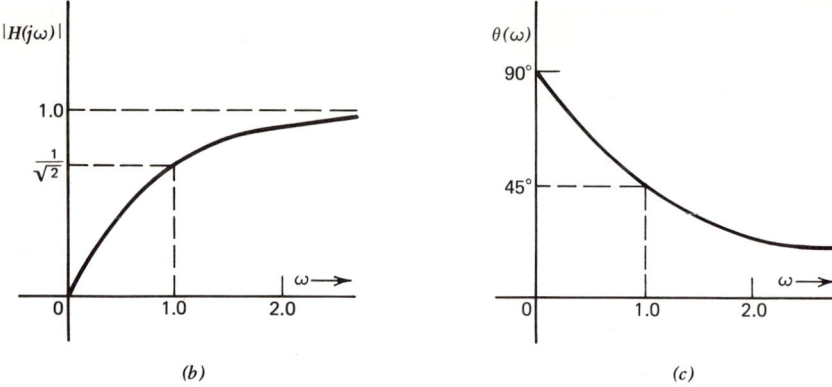

Figure 2-18. (*a*) *RC* network and its (*b*) magnitude response and (*c*) phase response.

Let us now evaluate and plot the output time response to determine the physical significance of this negative delay.

The Laplace transform of the output time function $u(t)$ is

$$U(s) = H(s)F(s) = \frac{s}{s+1} \cdot \frac{0.6}{s^2 + 0.36} = \frac{(15/34)s + 27/170}{s^2 + 0.36} - \frac{15/34}{s+1} \quad (2.6\text{-}7)$$

where $F(s)$ is the transform of $\sin 0.6t$. From Table 2-1,

$$u(t) = \frac{1}{34}[15 \cos 0.6t + 9 \sin 0.6t - 15e^{-t}]u_{-1}(t) \quad (2.6\text{-}8)$$

shown with the input function $f(t)$ in Fig. 2-19. By $t = 6$, $u(t)$ has reached steady state and we see $u(t)$ leading $f(t)$ by the predicted 1.7173 sec. This does not mean, for example, that the third peak of $u(t)$ ($t = 22$ sec) results from the third peak of $f(t)$ ($t = 23.5$ sec). Rather, the peak of $u(t)$ (the effect) results from the sharp rise in $f(t)$ at $t = 21$ sec (the cause). This is quite reasonable because this network approximates a differentiator; thus rapid changes of the input cause increased outputs. The occurrence of negative phase delay, therefore, arises from normal network behavior and does not violate the principle of causality.

Equation 2.6-2 indicates that the phase delay is zero if $\theta = 0$ and the output function is then in phase with the input function. This is verified by considering the network in Fig. 2-20, whose series resonance occurs at $\omega_0 = 0.6$ and whose phase angle there is $0°$. The response of this network to the sine wave $f(t)$ in Fig.

Group Delay and Phase Delay

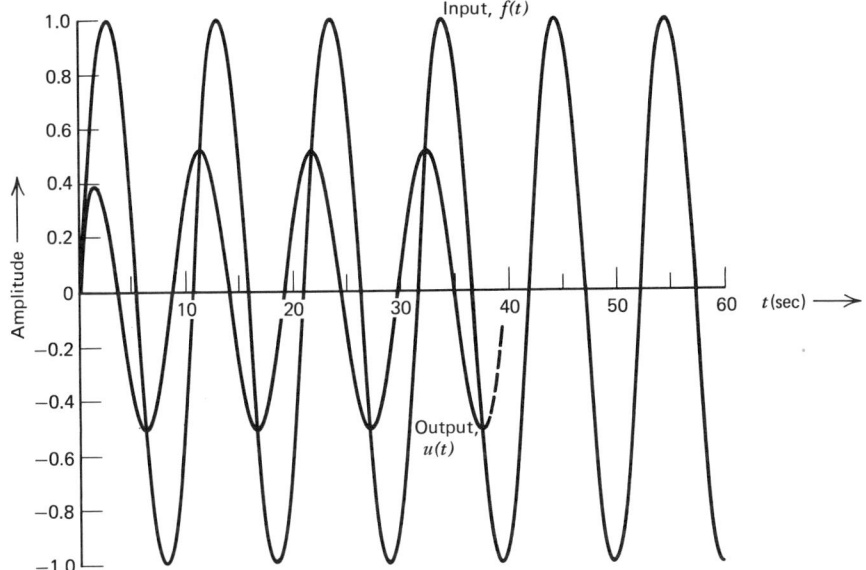

Figure 2-19. Excitation and response of network in Fig. 2-18a.

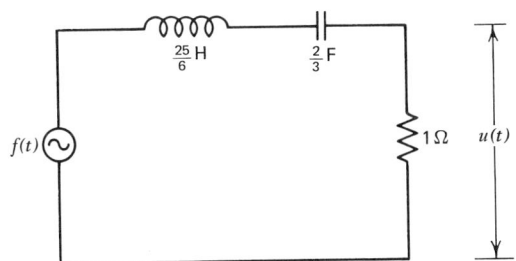

Figure 2-20. Network for which the phase angle is 0° at $\omega = 0.6$.

2-19 is plotted in Fig. 2-21. Now $u(t)$ takes longer to reach steady state because of the narrower bandwidth, but comparison with $f(t)$ in Fig. 2-19 indicates that the two are eventually in phase as predicted.

When the system transfer function is a rational function with m finite zeros $z_i = \gamma_i + j\delta_i$, and n finite poles $p_i = \alpha_i + j\beta_i$, the group delay, from (2.5-52) and (2.6-1), is

$$D(\omega) = \sum_{j=1}^{m} \frac{\gamma_j}{\gamma_j^2 + (\delta_j - \omega)^2} - \sum_{i=1}^{n} \frac{\alpha_i}{\alpha_i^2 + (\beta_i - \omega)^2} \tag{2.6-9}$$

which can also be expressed in terms of the poles and zeros as

$$D(\omega) = \sum_{j=1}^{m} \frac{z_j}{\omega^2 + z_j^2} - \sum_{i=1}^{n} \frac{p_i}{\omega^2 + p_i^2} \tag{2.6-10}$$

The phase delay, from (2.5-52) and (2.6-2), is

$$\tau_p(\omega) = -\frac{1}{\omega} \left[\sum_{j=1}^{m} \tan^{-1} \frac{\delta_j - \omega}{\gamma_j} - \sum_{i=1}^{n} \tan^{-1} \frac{\beta_i - \omega}{\alpha_i} \right] \tag{2.6-11}$$

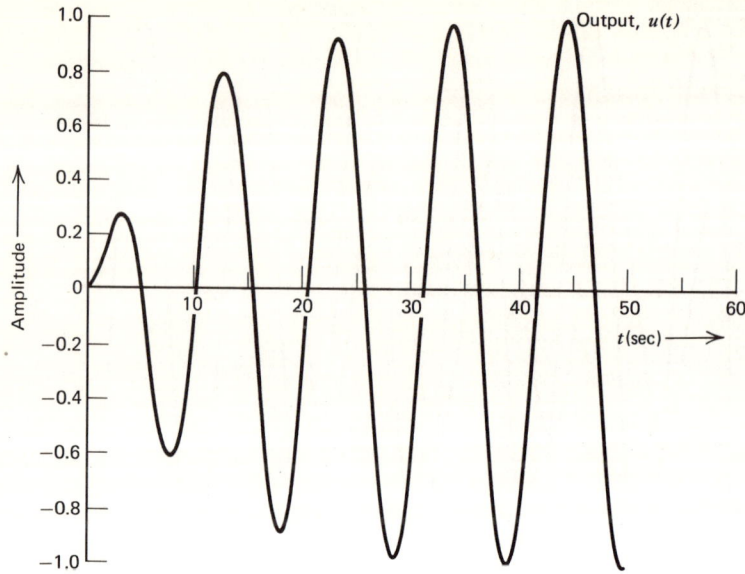

Figure 2-21. Response of network in Fig. 2-20 to sine wave $f(t)$ in Fig. 2-19.

Equations 2.6-9 and 2.6-11 are convenient for computation because only real numbers enter the calculations.

Example 2-31. We now determine the group delay and phase delay functions for the network in Fig. 2-13. From (2.5-42) and (2.6-1)

$$D(\omega) = \frac{2\omega^4 + \omega^2 + 2}{\omega^6 + 1} \qquad (2.6\text{-}12)$$

and from (2.5-42) and (2.6-2),

$$\tau_p(\omega) = \frac{1}{\omega} \tan^{-1} \frac{\omega(\omega^2 - 2)}{2\omega^2 - 1} \qquad (2.6\text{-}13)$$

Because the phase is reasonably linear from $\omega = 0$ to $\omega = 0.5$ (see Fig. 2-14b), the group delay and phase delay are approximately the same over this range. This is not to suggest that this is generallly true, indeed, we find in Chapter 3, for example, that Chebyshev filters, at a given frequency, can exhibit values of group delay much larger than the values of phase delay.

2.6.2 System Delay and Signal Distortion

Consider an ideal system as one with constant magnitude response and linear phase shift, $\theta(\omega) = -\omega T$. Figure 2-22 shows $u(t)$, the output of this system, for the input $f(t)$. The determination of the ideal-system delay presents no problem, for the output is simply a replica of the input delayed by T seconds,

$$u(t) = f(t - T) \qquad (2.6\text{-}14)$$

In fact (2.6-14) is true for any input $f(t)$. For this ideal system, group delay and phase delay are both equal to T seconds.

Group Delay and Phase Delay

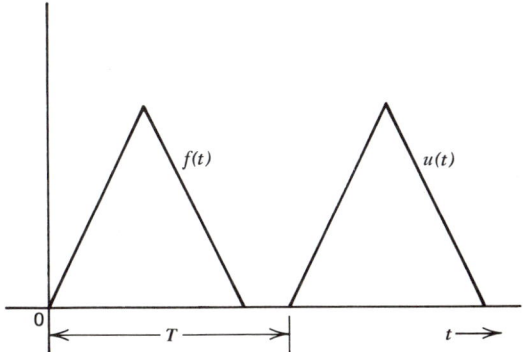

Figure 2-22. Input and output of ideal system.

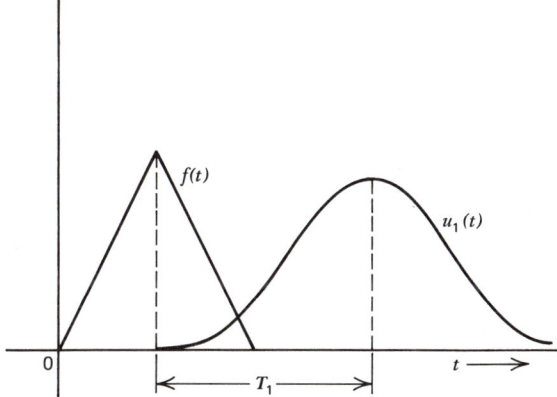

Figure 2-23. Input and output of nonideal system.

Suppose, however, that the magnitude response remains constant but the phase response is nonlinear. The output $u_1(t)$ may then resemble the curve in Fig. 2-23. Now $u_1(t)$ is not a delayed replica of $f(t)$ and the value of delay is not clearly defined. One might measure the time difference of the peak outputs and call T_1 the delay of the system, but this is arbitrary. For this system T_1 is a reasonable estimate of delay, but one can imagine phase nonlinearities for which the output signal bears little resemblance to the input signal. Then the concept of delay becomes meaningless. For this reason, group delay is useful for establishing an unambiguous value of delay only when the phase is linear with frequency over the frequency band of interest.† This constraint usually restricts the input signal to be narrow band and, of course, does not apply to the delay of a single-frequency waveform (given by (2.6-2)). However, a single sinusoid is rarely encountered in practice. In practical situations the magnitude response is not constant; but so long as its variation over the signal bandwidth is relatively small,

---

†However, in practice, meaningful values of delay are obtained when the phase approximates a linear phase response, and we encounter these designs in Chapter 3.

useful values of delay can be obtained. The effects of attenuation and delay interact and in general are difficult to express separately. Accordingly, throughout the remainder of this section the magnitude response is assumed constant, at least over the signal bandwidth.

We define a signal $f(t)$ to be distorted by the system if the output signal $u(t)$ is not a delayed version of the input signal. This degradation, known as delay distortion, arises when some portions of the signal are delayed more than others. Thus no distortion occurs when $u(t)$ is given by (2.6-14). A signal should not be declared useless because it is distorted in passing through a filter, however. In many practical cases the important attribute of the signal is not distorted, whereas the composite signal is distorted according to the preceding definition. Such is the case for amplitude-modulated signals and angle-modulated signals, where the information is contained in the amplitude function and the angle function, respectively. These signals are discussed in Section 2.6.4.

2.6.3 Phase-Intercept Distortion

If the phase function of a system is linear with frequency but does not equal zero radians or a multiple of 2π rads at zero frequency, the input signal is distorted in passing through the system. This distortion, known as phase-intercept distortion, is usually of academic interest only. However, it can affect modulated signals whose carrier frequency carries information (such as synchronization) in addition to being modulated.

We illustrate this distortion by considering the input signal $f(t)$ expressed as the sum of sinusoids

$$f(t) = \sum_{k=1}^{N} A_k \sin \omega_k t \qquad (2.6\text{-}15)$$

exciting the filter with unit magnitude response and the linear phase

$$\theta(\omega) = -T\omega - \theta_0 \qquad (2.6\text{-}16)$$

where $-\theta_0$ is the phase at $\omega = 0$. The output signal $u(t)$ is

$$u(t) = \sum_{k=1}^{N} A_k \sin[\omega_k(t-T) - \theta_0] \qquad (2.6\text{-}17)$$

but the identity $\sin(A-B) = \sin A \cos B - \cos A \sin B$ allows $u(t)$ to be partitioned as

$$u(t) = \cos\theta_0 \sum_{k=1}^{N} A_k \sin \omega_k(t-T) - \sin\theta_0 \sum_{k=1}^{N} A_k \cos \omega_k(t-T) \qquad (2.6\text{-}18)$$

Since the first summation is $f(t-T)$, $u(t)$ is expressible as

$$u(t) = \cos\theta_0 f(t-T) - \sin\theta_0 \sum_{k=1}^{N} A_k \cos \omega_k(t-T) \qquad (2.6\text{-}19)$$

The first term represents the input delayed by T seconds and scaled by $\cos\theta_0$. The second term represents the distortion term and arises because θ_0 is not a multiple of 2π. If $\theta_0 = 0, 2\pi, 4\pi, \ldots$ then $u(t)$ is the input delayed by T seconds, and no distortion exists. However there is distortion for other values of θ_0, and the

Group Delay and Phase Delay

following general condition is valid for signals composed of more than one sinusoid.

A system is distortionless (assuming constant magnitude response) if its phase response is constant or linear with frequency and its phase shift at $\omega = 0$ is a multiple of 2π rads. Then the output signal is a delayed replica of the input signal.

2.6.4 Modulated-Signal Delay

If the preceding definition for a distortionless system were the final word, every linear phase bandpass filter would require phase equalization to achieve the proper angle at $\omega = 0$, since it is very unlikely that the extension of the bandpass phase would intersect the ordinate at a multiple of 2π rads. However the term "distortion" takes on different meaning for various applications.

We now show that an amplitude-modulated (AM) signal and a frequency-modulated (FM) signal are each distorted in passing through a linear phase system with phase-intercept distortion. However the modulation, which contains the information, is not distorted, it is just delayed by the system group delay.

The system is assumed to have a constant magnitude response and a linear phase response given by

$$\theta(\omega) = -T\omega - \theta_0 \qquad (2.6\text{-}20)$$

where θ_0 is not necessarily a multiple of 2π rads. From (2.6-1) the system group delay is the constant value T,

$$D(\omega) = T \qquad (2.6\text{-}21)$$

and the phase delay, from (2.6-2), is

$$T_p(\omega) = T + \frac{\theta_0}{\omega} \qquad (2.6\text{-}22)$$

Amplitude-Modulated Signal. The AM input signal

$$f(t) = E(t) \cos \omega_c t = (1 + m \cos \omega_m t) \cos \omega_c t \qquad (2.6\text{-}23)$$

where

$E(t) = 1 + m \cos \omega_m t$ = envelope function
ω_c = angular carrier frequency
ω_m = angular modulating frequency
m = modulation index; $0 \leq m \leq 1$

can also be expressed as

$$f(t) = \cos \omega_c t + m \cos \omega_c t \cos \omega_m t \qquad (2.6\text{-}24)$$

or

$$f(t) = \cos \omega_c t + \frac{m}{2} \cos(\omega_c - \omega_m)t + \frac{m}{2} \cos(\omega_c + \omega_m)t \qquad (2.6\text{-}25)$$

Equation 2.6-25 shows that the AM waveform is composed of the carrier frequency and two sidebands, one located at $\omega_c - \omega_m$ and the other at $\omega_c + \omega_m$.

The system response to $f(t)$ is the superposition of these sinusoids shifted by the appropriate phase in (2.6-20). Then

$$u(t) = \cos[\omega_c(t-T) - \theta_0] + \frac{m}{2}\cos[(\omega_c - \omega_m)(t-T) - \theta_0]$$

$$+ \frac{m}{2}\cos[(\omega_c + \omega_m)(t-T) - \theta_0]$$

$$= \cos[\omega_c(t-T) - \theta_0] + m\cos[\omega_c(t-T) - \theta_0]\cos\omega_m(t-T) \quad (2.6\text{-}26)$$

The salient features of this response are exposed by expressing $u(t)$ as the AM signal

$$u(t) = [1 + m\cos\omega_m(t-T)]\cos\omega_c\left[t - \left(T + \frac{\theta_0}{\omega_c}\right)\right] \quad (2.6\text{-}27)$$

The carrier frequency is delayed by $\tau_p(\omega_c)$, the carrier phase delay from (2.6-22), and the envelope function is delayed by the constant group delay T in (2.6-21), which is why the group delay is often called the envelope delay. The input signal $f(t)$ is distorted in passing through the system because $\tau_p(\omega_c)$ differs from T. However the information content of the wave, contained in the envelope function, is not distorted by the system but is delayed by T.

Therefore no envelope distortion occurs if the system group delay is constant over the modulated signal bandwidth. In practice the system group delay is made to approximate a constant to reduce envelope distortion to an acceptable level. In Chapter 3 we discuss filters whose group delay approximates a constant.

The same output signal is obtained even if the phase shift over the entire band is not linear. Since only three frequencies comprise the AM waveform, only the phase shifts at $\omega_c - \omega_m$, ω_c, and $\omega_c + \omega_m$ enter into the computation of the output signal. Therefore it is only necessary that these three phase values lie on a straight line. The phase shift at other frequencies can vary widely without affecting the output signal.

Frequency-Modulated Signal. We now consider the FM input signal

$$f(t) = \cos\phi(t) = \cos(\omega_c t + \mu\sin\omega_m t) \quad (2.6\text{-}28)$$

where

ω_c = carrier frequency in rads/sec
$\mu = \Delta\omega/\omega_m$ = modulation index
$\Delta\omega$ = maximum frequency deviation from ω_c
ω_m = modulating frequency
$\phi(t) = \omega_c t + \mu\sin\omega_m t$ = instantaneous phase angle
$\omega(t) = \dfrac{d\phi(t)}{dt} = \omega_c + \Delta\omega\cos\omega_m t$ = instantaneous radian frequency

The Fourier series representation of $f(t)$ is

$$f(t) = \sum_{k=-\infty}^{\infty} J_k(\mu)\cos(\omega_c + k\omega_m)t \quad (2.6\text{-}29)$$

where $J_k(\mu)$ is the Bessel function of the first kind of kth order.[†] Thus the spectrum of $f(t)$ is the carrier frequency and an infinite number of sidebands, each

---

[†] As a historical note, Carson in his 1922 paper (reprinted in *Proc. IEEE*, June 1963) obtained a similar expression showing that the FM waveform occupies a wider bandwidth than the corresponding AM waveform. He concluded that the FM system was inferior to the AM system both in bandwidth and in signal distortion. Fifteen years later Armstrong showed the practical usefulness of FM.

The Hilbert Transform

separated from the carrier by integral multiples of the modulating frequency ω_m.

The system response to $f(t)$ is the superposition of these sinusoids shifted by the appropriate phase in (2.6-20). Then

$$u(t) = \sum_{k=-\infty}^{\infty} J_k(\mu) \cos\left[(\omega_c + k\omega_m)(t-T) - \theta_0\right] \qquad (2.6\text{-}30)$$

Again the output signal is distorted because θ_0 is not necessarily a multiple of 2π. However $u(t)$ can be rewritten as the FM signal

$$u(t) = \cos\left[\omega_c(t-T) - \theta_0 + \mu \sin \omega_m(t-T)\right] \qquad (2.6\text{-}31)$$

whose instantaneous phase is

$$\phi_1(t) = \omega_c(t-T) - \theta_0 + \mu \sin \omega_m(t-T) \qquad (2.6\text{-}32)$$

and whose instantaneous frequency is

$$\omega_1(t) = \frac{d\phi_1(t)}{dt} = \omega_c + \Delta\omega \cos \omega_m(t-T) \qquad (2.6\text{-}33)$$

As before, the signal information—in this case the instantaneous frequency—is not distorted but is delayed by T seconds, the system group delay. The value of θ_0 does not distort the transmission of $\omega(t)$. The important characteristic is the constant group delay over the signal bandwidth.

Example 2-32. A filter with unity magnitude response and the phase characteristic in (2.6-20) is excited by the AM signal

$$f(t) = (1 + m \cos 0.05t) \cos t \qquad (2.6\text{-}34)$$

We determine T and θ_0 in (2.6-20) to ensure that the carrier is delayed by 6 sec and the envelope is delayed by 4 sec. Since the envelope is delayed by the group delay, we know from (2.6-21) that $T = 4$ sec. The carrier ($\omega_c = 1$) is delayed by the phase delay; thus from (2.6-22)

$$T + \frac{\theta_0}{\omega_c} = 4 + \theta_0 = 6 \qquad (2.6\text{-}35)$$

and $\theta_0 = 2$ rads.

In this section we have attempted to consolidate the information about group delay and phase delay and to explain the important aspects of these functions. With the advent of satellite systems and long-range communication and navigation systems, the ability to understand and predict system group delay has become most important. The predictability of system delay is discussed in Chapter 4.

2.7 THE HILBERT TRANSFORM

The Hilbert transforms relate the real and imaginary parts of $F(j\omega)$, the Fourier transform of the causal time function $f(t)$, provided $F(s)$ is analytic within and on the boundary of the right-half of the s-plane and $F(\infty)$ is finite. In network theory the following functions are related by Hilbert transforms: (*a*) attenuation and phase of minimum-phase networks, (*b*) resistance and reactance of a driving point impedance function that is minimum reactance and minimum resistance.

We have previously related the real and the imaginary parts of $F(j\omega)$ by (2.1-22) and (2.1-23), but there each part is expressed by an unattractive double integral. Alternatively one can utilize the analytic properties of $F(s)$ in the right-half plane to obtain the following single-integral expressions known as Hilbert transforms [3, 5].

$$X(\omega) = -\frac{1}{\pi} \int_{-\infty}^{\infty} \frac{R(\lambda)}{\omega - \lambda} d\lambda \tag{2.7-1}$$

$$R(\omega) = R(\infty) + \frac{1}{\pi} \int_{-\infty}^{\infty} \frac{X(\lambda)}{\omega - \lambda} d\lambda \tag{2.7-2}$$

The constant $R(\infty)$ occurs because of possible impulses in $f(t)$ at $t = 0$, while the asymmetry between (2.7-1) and (2.7-2) is due to the assumption that $f(t)$ is real.

In Chapter 3 we use the Hilbert transforms to determine the unspecified attentuation or phase-lag angle of a minimum-phase network when ideal functions of each are specified.

Example 2-33. Evaluation of the transform integrals is now illustrated. Consider $f(t)$ in Example 2-1 given by

$$f(t) = u_{-1}(t) e^{-\alpha t} \tag{2.7-3}$$

We have already shown there that

$$R(\omega) = \frac{\alpha}{\alpha^2 + \omega^2}, \qquad X(\omega) = -\frac{\omega}{\alpha^2 + \omega^2} \tag{2.7-4}$$

Let us determine $X(\omega)$ from (2.7-1) as

$$X(\omega) = -\frac{1}{\pi} \int_{-\infty}^{\infty} \frac{\alpha}{\alpha^2 + \lambda^2} \cdot \frac{1}{\omega - \lambda} d\lambda \tag{2.7-5}$$

which is rewritten as

$$X(\omega) = -\frac{\alpha}{\pi(\alpha^2 + \omega^2)} \left[\int_{-\infty}^{\infty} \frac{\lambda \, d\lambda}{\lambda^2 + \alpha^2} + \omega \int_{-\infty}^{\infty} \frac{d\lambda}{\lambda^2 + \alpha^2} - \int_{-\infty}^{\infty} \frac{d\lambda}{\lambda - \omega} \right] \tag{2.7-6}$$

The first and third integrals are zero because the integrands are odd functions. The value of the second integral is π/α [2], and $X(\omega)$ in (2.7-4) is obtained. Similarly, from (2.7-2),

$$R(\omega) = -\frac{1}{\pi(\alpha^2 + \omega^2)} \left[\omega \int_{-\infty}^{\infty} \frac{\lambda \, d\lambda}{\lambda^2 + \alpha^2} - \alpha^2 \int_{-\infty}^{\infty} \frac{d\lambda}{\lambda^2 + \alpha^2} - \omega \int_{-\infty}^{\infty} \frac{d\lambda}{\lambda - \omega} \right] \tag{2.7-7}$$

Since $f(t)$ does not contain an impulse at $t = 0$, $R(\infty) = 0$. The integrals are the same as those in (2.7-6); hence $R(\omega)$ in (2.7-4) is obtained.

REFERENCES

1. Aseltine, J. A., *Transform Method in Linear System Analysis*, McGraw-Hill, New York, 1958.
2. Dwight, H. B., *Tables of Integrals and Other Mathematical Data*, Macmillan, New York, 1961.
3. Humpherys, D. S., *The Analysis, Design, and Synthesis of Electrical Filters*, Prentice-Hall, Englewood Cliffs, N.J., 1970.

PROBLEMS

2-1. Find the Fourier transform of
   (a) $f(t) = u_{-1}(t) - u_{-1}(t-1)$
   (b) $f(t) = e^{-a|t|}$

Compute $A(\omega)$ and plot it for each function.

2-2. Refer to Problem 1-8(c).
   (a) Determine the Fourier transform of the response to the rectangular pulse input.
   (b) Compute $A(\omega)$.
   (c) Compute the total energy dissipated in the resistor R in both the time and frequency domains. Remember that the derived expressions for energy assume a 1-ohm resistor.

2-3. The transfer function of a linear system has the following poles and constant multiplier (15): $-2.3222, -1.839 \pm j1.7544$.
   (a) Express the transfer function as a ratio of polynomials.
   (b) What is the differential equation describing the system?
   (c) Express the group delay as a ratio of polynomials in ω^2.
   (d) What is the inverse transform of $H(s)$ and its physical interpretation?
   (e) Find a set of linearly independent solutions of the homogeneous portion of the differential equation in part (b).

2-4. For the filter below

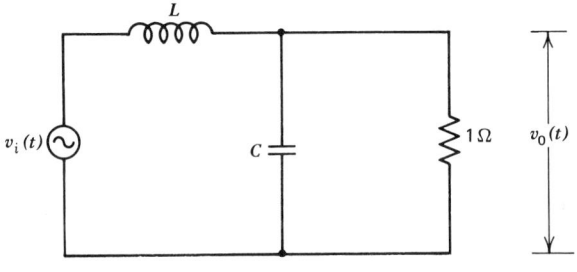

Figure P.2-4.

   (a) What values of L and C will ensure that the response to $\delta(t)$ is
$$h(t) = \left[\sqrt{2}\, e^{-(\sqrt{2}/2)t} \sin \frac{\sqrt{2}}{2} t \right] u_{-1}(t)$$
   (b) For these values of L and C, what is the steady-state response of the filter?
   (c) At what radian frequency is this response attenuated 3 dB from the value at $\omega = 0$?

2-5. For the *RC* filter below,

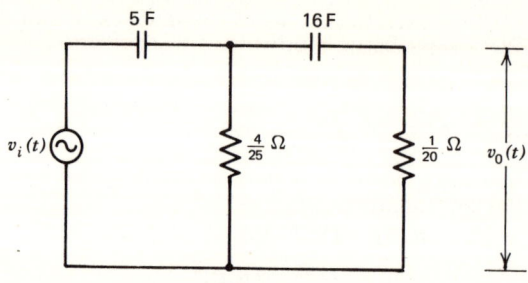

Figure P.2-5.

(a) What is the transfer function $H(s)$?
(b) What are the poles and zeros of $H(s)$?
(c) What is the impulse response?
(d) What is the response when $v_i(t) = u_{-1}(t)$?

2-6. The impulse response of a linear time-invariant system is $h(t) = \delta(t)$. What is the response of this system for the following input?

$$f(t) = \begin{cases} 0 & t < 0 \\ Ae^{-t} & t \geq 0 \end{cases}$$

2-7. Between $\omega = 0.5$ and $\omega = 1.5$, the linear group delay of a device is

$$D(\omega) = T_0 - \frac{T_0 \omega}{2}$$

What is the phase function of the device if the phase shift at $\omega = 1$ is θ_0? This characteristic is used to generate the linear FM chirp signal.

2-8. A filter with phase characteristic

$$\theta(\omega) = -(T\omega + \theta_0)$$

is excited by the modulated signal

$$v_i(t) = [1 + m \cos 0.02\, t] \cos t$$

The magnitude response is assumed constant over the signal bandwidth. What values of T and θ_0 will ensure that the carrier is delayed by 8 sec and the envelope is delayed by 2 sec?

2-9. The impulse response of a filter is

$$h(t) = \delta(t) - ae^{-at}u_{-1}(t)$$

(a) What is the filter transfer function?
(b) What is the filter group delay function?
(c) What is the filter step response?
(d) What is the filter response for the input $f(t) = e^{-at}u_{-1}(t)$?
(e) Sketch the filter attenuation response.

2-10. A filter's impulse response is

$$h(t) = tu_{-1}(t) - 2(t-1)u_{-1}(t-1) + (t-2)u_{-1}(t-2)$$

What is the response of this filter to the step function input, $u_{-1}(t)$?

Chapter Three

Linear System Responses

The filtering system response plays a major role in determining the overall system response. For this reason familiarity with the fundamental aspects of filter responses is not only a must for the filter designer, but it is advantageous for the system designer and those who specify and use filters. Appreciation of the factors affecting the various responses results in more realistic system and filter requirements. Consequently design time, cost, and problems are reduced.

Accordingly we here consider (a) restrictions in specifying arbitrary time- and frequency-domain responses, (b) the expected tradeoffs as ideal responses are approximated in practice, (c) three popular mathematical approximations and their application to filter responses, (d) the characteristics of responses that have already been obtained and tabulated, and (e) rise time and noise bandwidth, two important quantities in system design.

Before beginning these main topics, we introduce five basic filter types, each characterized by its attenuation response as a function of the radian frequency ω. These response characteristics are independent of the frequency range and apply equally well to light, X-rays, or the usual frequencies encountered in electronic engineering. The typical attenuation response in decibels is shown in terms of a passband, in which the attenuation is zero; a transition band; and a stop band that has a minimum attenuation there of A_{min} decibels.

1. Low pass (LP). The LP response (Fig. 3-1), has a passband between $\omega = 0$ and $\omega = \omega_c$, a transition band between ω_c and ω_s, and a stopband from ω_s to infinity. The attenuation A_{min} is first attained at ω_s.

2. High pass (HP). The HP response (Fig. 3-2) has a stopband from $\omega = 0$ to $\omega = \omega_s$, a transition band between ω_s and ω_c, and a passband extending from ω_c to infinity. Relative to the passband, the attenuation A_{min} is first attained at ω_s.

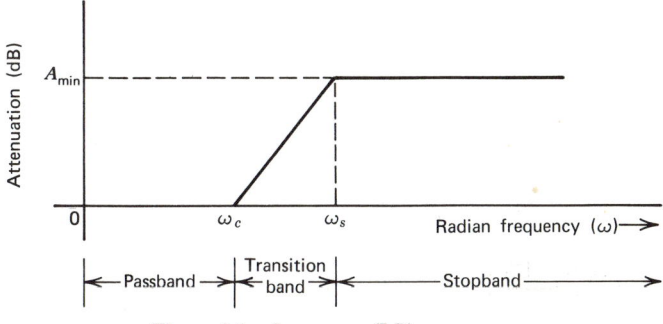

Figure 3-1. Low pass (LP) response.

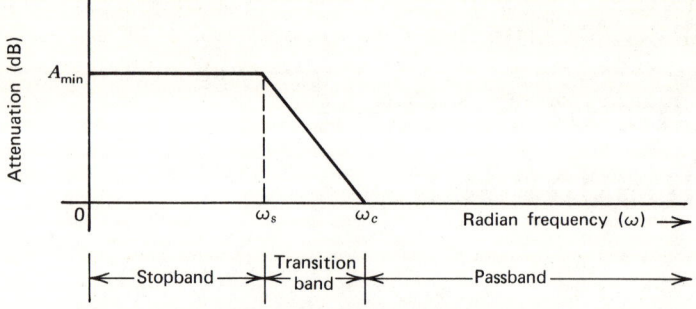

Figure 3-2. High pass (HP) response.

3. Bandpass (BP). The BP response (Fig. 3-3) has a passband extending from ω_1 to ω_2. The center frequency is $\omega_0 = \sqrt{\omega_1 \omega_2}$, the geometric mean of the two cutoff frequencies. There are two stopbands; the lower one goes from $\omega = 0$ to $\omega = \omega_{1s}$ and the upper one extends from ω_{2s} to infinity. Likewise there are two transition bands. The minimum attenuations may differ in each stopband and are indicated as A_{min1} and A_{min2}.

4. Bandstop (BS). The BS response (Fig. 3-4) has one stopband, extending from ω_{1s} to ω_{2s}, and two passbands. The first extends from $\omega = 0$ to ω_1 and the second from ω_2 to infinity. The center frequency is $\omega_0 = \sqrt{\omega_1 \omega_2}$.

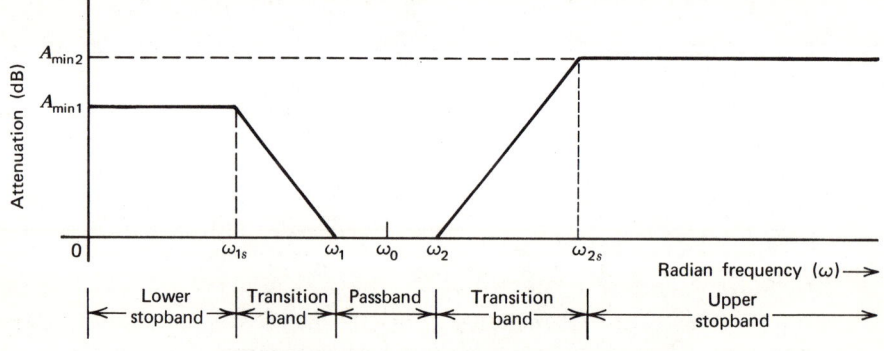

Figure 3-3. Bandpass (BP) response.

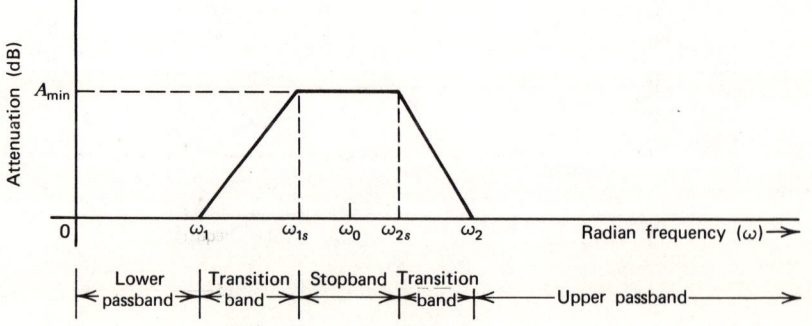

Figure 3-4. Bandstop (BS) response.

Ideal Low-Pass Responses

5. All pass (AP). The AP magnitude response is unity, corresponding to zero attenuation, for all frequencies. Thus the passband is infinite and there is no stopband. However the phase response is very useful for phase equalization, and this topic is discussed in Chapter 5.

3.1 IDEAL LOW-PASS RESPONSES

Physically unrealizable responses have an important place in the theory of filtering. The postulate of one such system response immediately predetermines another response of the same system, and these response pairs contain valuable information for both the theoretician and the practicing engineer. Each response pair can be considered the limit of many realizable responses, hence it tells us what can be expected in practice. Furthermore, the origin of various response characteristics is exposed by examination of these ideal responses.

We initially determine the impulse responses and step responses corresponding to two fundamental magnitude responses, namely, the rectangular response and the Gaussian response. This is followed by a discussion of the Paley–Wiener condition, which tells us when a given frequency function can be the Fourier spectrum of a causal function. We conclude by relating the attenuation and phase of minimum-phase networks, for these are extensively used in practice.

3.1.1 Rectangular Magnitude with Linear Phase

Often the rectangular magnitude with linear phase response is called the ideal frequency-domain response because the passband attenuation is zero (unity gain), there is no transition band, and the stopband attenuation is infinite (zero gain). That is, in this LP response (Fig. 3-5a),

$$|H(j\omega)| = \begin{cases} 1 & -\omega_c \leq \omega \leq \omega_c \\ 0 & \text{elsewhere} \end{cases} \quad (3.1\text{-}1)$$

The phase response is the linear function (Fig. 3-5b)

$$\theta(\omega) = -\frac{n\pi}{2}\frac{\omega}{\omega_c} \quad (3.1\text{-}2)$$

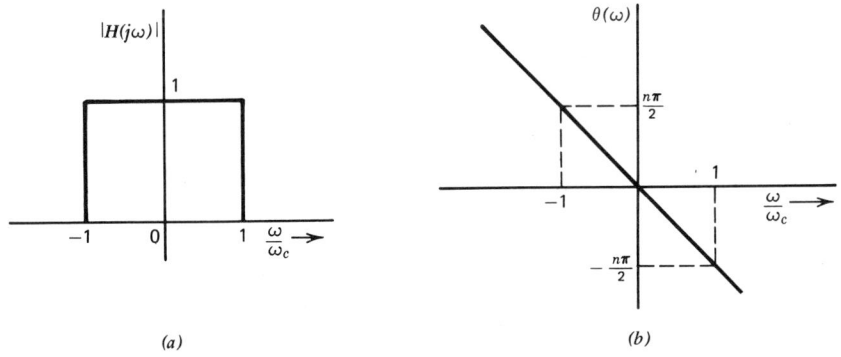

Figure 3-5. (a) Rectangular magnitude response, (b) linear phase response.

and consequently the group delay and phase delay are the same constant

$$D_l(\omega) = \tau_p(\omega) = \frac{n\pi}{2\omega_c} \qquad (3.1\text{-}3)$$

In practice one attempts to approximate this function by a rational function, and n can be considered the order of the approximating polynomial because each term in s gives a maximum phase shift of $\pi/2$ rads.

The impulse response and the step response corresponding to this frequency response are now determined.

3.1.2 Impulse Response

The impulse response $h(t)$ is obtained from the Fourier inversion integral in (2.1-6) as

$$h(t) = \frac{1}{2\pi} \int_{-\omega_c}^{\omega_c} 1 \cdot e^{j\omega(t - n\pi/2\omega_c)} \, d\omega = \frac{1}{\pi} \int_0^{\omega_c} \cos\left[\omega\left(t - \frac{n\pi}{2\omega_c}\right)\right] d\omega = \frac{\omega_c}{\pi} \frac{\sin(\omega_c t - n\pi/2)}{\omega_c t - n\pi/2} \qquad (3.1\text{-}4)$$

which appears in Fig. 3-6.†

Most important, the impulse response is anticipatory, hence physically unrealizable. The response begins at $t = -\infty$, but the impulse function is not applied until $t = 0$. We shall find that this anticipatory phenomenon results because the attenuation response is infinite over a frequency interval, in this case the entire stopband. As n increases, the response for $t < 0$ diminishes in amplitude, but for the anticipatory response to disappear, n must be infinite. This, however, results in an infinite system delay.

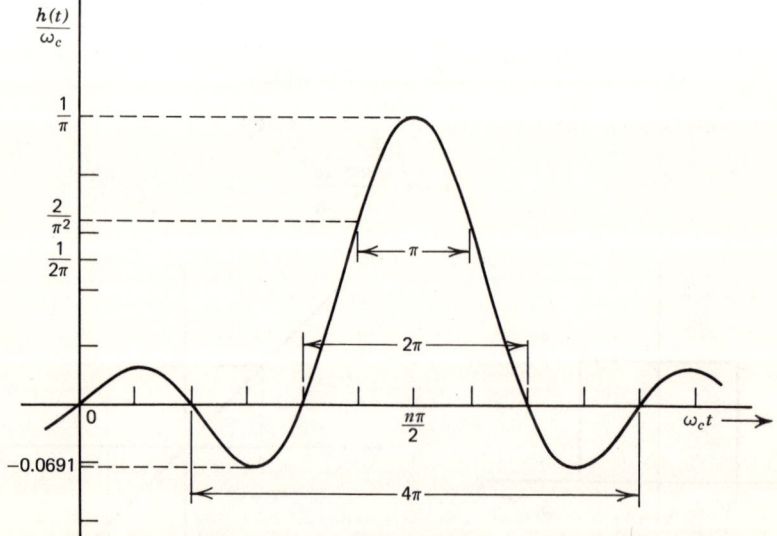

Figure 3-6. Impulse response corresponding to the rectangular magnitude response.

†The $\sin x / x$ function is often called the cardinal function.

Ideal Low-Pass Responses

The pulse width of $h(t)$ is taken to be π/ω_c, which is one-half the width measured between zero crossings on each side of the main lobe. We find in Section 7.4 that this is the usual value assigned in pulse compression. The pulse width is the reciprocal of the peak response and inversely proportional to the bandwidth ω_c. Hence the pulse width–amplitude product is the unit constant and is independent of the bandwidth. An increase in bandwidth causes the pulse to narrow and the peak to increase. As ω_c approaches infinity, $h(t)$ approaches the impulse function.

The symmetry of $h(t)$ results from the phase linearity (see Section 2.1.3), and the impulse-response peak occurs at $t = n\pi/2\omega_c$, which is the delay given in (3.1-3). We later show in Section 8.4.3 that the peak impulse response time is a reasonable measure for the average delay of the system to an impulse function, even when the phase response is not linear.

The first overshoot of $h(t)$ is -0.0691, and its absolute value relative to the peak response is attenuated 13.26 dB.

3.1.3 Step Response

The step response $g(t)$, from (1.8-12), is the integral of the impulse response. However, the impulse response is causal there, thus is zero for $t < 0$. Here the impulse response in noncausal, having been on for all time. The lower integral limit is then $-\infty$ rather than zero yielding

$$g(t) = \int_{-\infty}^{t} h(x)\, dx = \frac{\omega_c}{\pi} \int_{-\infty}^{t} \frac{\sin(\omega_c x - n\pi/2)}{\omega_c x - n\pi/2}\, dx$$

$$= \frac{1}{2} + \frac{1}{\pi} \int_{0}^{\omega_c t - n\pi/2} \frac{\sin y}{y}\, dy = \frac{1}{2} + \frac{1}{\pi} Si\left(\omega_c t - \frac{n\pi}{2}\right) \qquad (3.1\text{-}5)$$

which is plotted in Fig. 3-7. The sine integral is $Si(x)$ given by

$$Si(x) = \int_{0}^{x} \frac{\sin y}{y}\, dy \qquad (3.1\text{-}6)$$

It cannot be evaluated in closed form, but tabulations are available [1].

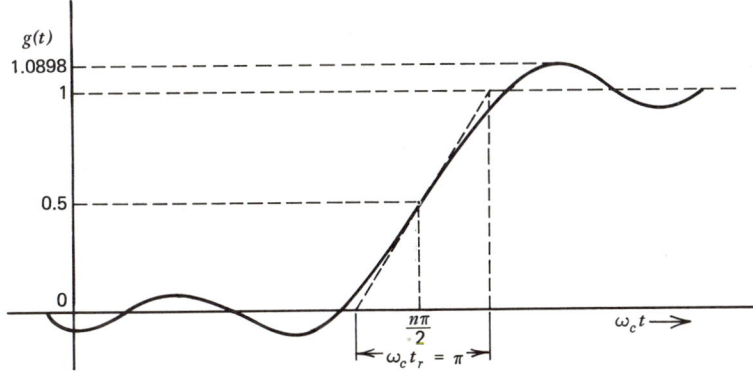

Figure 3-7. Step response corresponding to the rectangular magnitude response.

The step response, like the impulse response, is anticipatory, beginning at $t = -\infty$ for a step function applied at $t = 0$. Again an increase in n reduces the step response magnitude for $t < 0$, but n must be infinite for the response to disappear there.

Except for the constant $\frac{1}{2}$ in (3.1-5), $g(t)$ has odd symmetry about the time $t = n\pi/2\omega_c$. This is the same time at which the impulse response peaks and the step response is 0.5. The $g(t)$ overshoots its steady-state value by 9% and then oscillates with a pseudoperiod of $2\pi/\omega_c$.

An important characteristic of the step response is the rise time t_r, usually defined as the time required for $g(t)$ to go from 10 to 90% of its final value. Another definition, the one used here, is the time required to go from zero to unity with a slope equal to the maximum slope of the step response. This maximum slope is the peak amplitude of the impulse response, because, from (1.8-12), the impulse response is the derivative of the step response. Then, from Fig. 3-7,

$$t_r = \frac{\pi}{\omega_c} \tag{3.1-7}$$

and we find that the rise time–bandwidth product is a constant. Furthermore the rise time is equal to the pulse width of the impulse response, it is also equal to the reciprocal of the peak impulse response. If the bandwidth is expressed in hertz ($f_c = \omega_c/2\pi$) and the rise time in seconds, (3.1-7) becomes

$$t_r f_c = \tfrac{1}{2} \tag{3.1-8}$$

This is an accepted rule of thumb to determine the required LP system bandwidth to realize a specified pulse rise time.

3.1.4 Gaussian Magnitude with Linear Phase

We now consider an LP magnitude response that does not change abruptly at the cutoff frequency, and the associated linear phase characteristic in (3.1-2). One such magnitude response (Fig. 3-8) is the Gaussian error function

$$|H(j\omega)| = \exp\left[-\frac{\ln 2}{2}\left(\frac{\omega}{\omega_c}\right)^2\right] \tag{3.1-9}$$

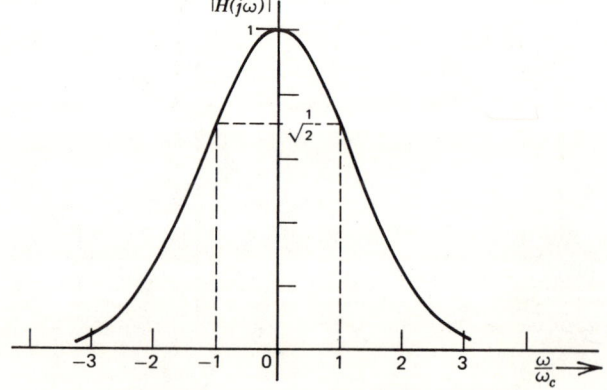

Figure 3-8. Gaussian magnitude response.

Ideal Low-Pass Responses

normalized so that $|H(j\omega_c)| = 1/\sqrt{2}$. The phase response appears in Fig. 3-5b, and the group delay is the constant $n\pi/2\omega_c$. This bandwidth normalization permits a meaningful comparison between the time-domain characteristics in Sections 3.1.2 and 3.1.3 and the time-domain characteristics that are now determined.

3.1.5 Impulse Response

The impulse response is obtained from (2.1-6), using the even and odd properties of the integrand, as

$$h(t) = \frac{1}{2\pi} \int_{-\infty}^{\infty} \exp\left[-\frac{\ln 2}{2}\left(\frac{\omega}{\omega_c}\right)^2\right] \exp\left[j\omega\left(t - \frac{n\pi}{2\omega_c}\right)\right] d\omega$$

$$= \frac{1}{\pi} \int_{0}^{\infty} \exp\left[-\frac{\ln 2}{2}\left(\frac{\omega}{\omega_c}\right)^2\right] \cos\left[\omega\left(t - \frac{n\pi}{2\omega_c}\right)\right] d\omega \qquad (3.1\text{-}10)$$

With $x = \omega(t - n\pi/2\omega_c)$,

$$h(t) = \frac{1}{\pi(t - n\pi/2\omega_c)} \int_{0}^{\infty} e^{-k^2 x^2} \cos x \, dx \qquad (3.1\text{-}11)$$

where

$$k^2 = \frac{\ln 2}{2\omega_c^2(t - n\pi/2\omega_c)^2}$$

The definite integral is $(\sqrt{\pi} e^{-(1/4k^2)})/2k$, yielding

$$h(t) = \frac{\omega_c}{\sqrt{2\pi \ln 2}} \exp\left[-\frac{(\omega_c t - n\pi/2)^2}{2 \ln 2}\right] \qquad (3.1\text{-}12)$$

which is shown in Fig. 3.9.

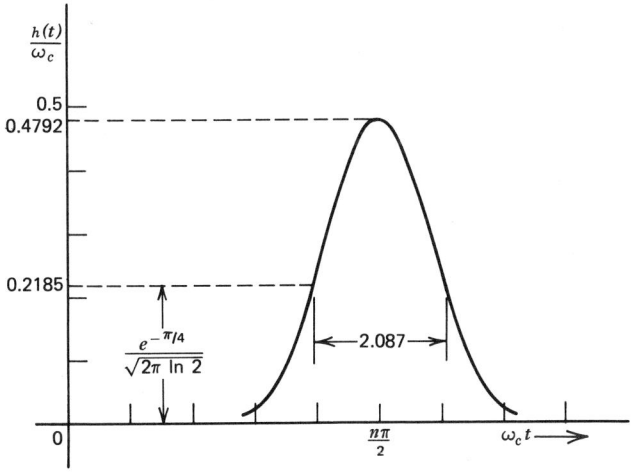

Figure 3-9. Impulse response corresponding to the Gaussian magnitude response.

The most significant property of $h(t)$ is that it is also a Gaussian function, hence there are no overshoots. This shape results because the Fourier transform of a Gaussian function is another Gaussian function. The $h(t)$ is physically unrealizable because the response begins at $t = -\infty$ for an impulse function applied at $t = 0$. Here this phenomenon is due to the behavior of the magnitude response at large values of ω, as we demonstrate in Section 3.1.8.

Consistent with the analysis in Section 3.1.2, we choose the pulse width of $h(t)$ to be $\sqrt{2\pi \ln 2}/\omega_c = 2.087/\omega_c$, the reciprocal of the peak response. The pulse width is again inversely proportional to the bandwidth, resulting in a pulse width–amplitude product of unity. The peak amplitude is $\omega_c/\sqrt{2\pi \ln 2} = 0.4792\omega_c$.

As expected, the phase linearity produces a symmetric impulse response about the group delay value $t = n\pi/2\omega_c$.

3.1.6 Step Response

The step response $g(t)$ is obtained by integrating the noncausal impulse response to give

$$g(t) = \int_{-\infty}^{t} h(x)\, dx = \frac{\omega_c}{\sqrt{2\pi \ln 2}} \int_{-\infty}^{t} \exp\left[-\frac{(\omega_c x - n\pi/2)^2}{2 \ln 2}\right] dx$$

$$= \frac{1}{2} + \frac{1}{\sqrt{\pi}} \int_{0}^{(\omega_c t - n\pi/2)/(\sqrt{2 \ln 2})} e^{-x^2}\, dx \quad (3.1\text{-}13)$$

This function is not obtainable in closed form but is expressible in terms of the tabulated error-function (erf) [1]

$$\operatorname{erf} y = \frac{2}{\sqrt{\pi}} \int_{0}^{y} e^{-x^2}\, dx \quad (3.1\text{-}14)$$

as

$$g(t) = \frac{1}{2}\left\{1 + \operatorname{erf}\left[\frac{\omega_c t - n\pi/2}{\sqrt{2 \ln 2}}\right]\right\} \quad (3.1\text{-}15)$$

which is plotted in Fig. 3-10. From this graph we see that $g(t)$ does not ring, that is, there are no response overshoots. The step response is physically unrealizable, beginning at $t = -\infty$. Except for the constant of $\frac{1}{2}$ in (3.1-15), $g(t)$ possesses odd symmetry about the delay time of $n\pi/2\omega_c$. This is the same time at which the impulse response peaks and the step response is 0.5.

Using the previous definition of rise time, we determine t_r as

$$t_r = \frac{\sqrt{2\pi \ln 2}}{\omega_c} \quad (3.1\text{-}16)$$

which is equal to the pulse width of $h(t)$. As expected, the rise time–bandwidth product is a constant. If the bandwidth is expressed in hertz ($f_c = \omega_c/2\pi$), and the rise time in seconds, (3.1-16) becomes

$$t_r f_c = \sqrt{\frac{\ln 2}{2\pi}} = 0.332 \quad (3.1\text{-}17)$$

Ideal Low-Pass Responses

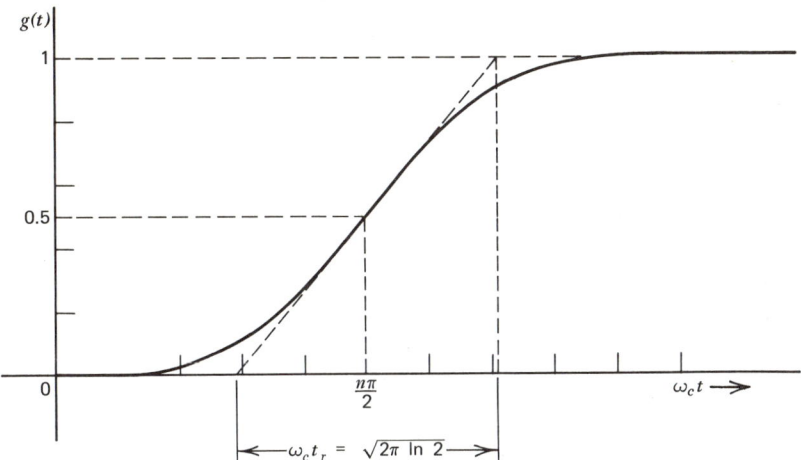

Figure 3-10. Step response corresponding to the Gaussian magnitude response.

Thus for a specified bandwidth, the rise time is approximately two-thirds of the rise time associated with the rectangular magnitude response.

3.1.7 Summary of Results

The rectangular magnitude response represents the ideal frequency domain response, for the transition band is zero and the stopband has infinite attenuation. The Gaussian magnitude response, on the other hand, represents the ideal time-domain response, in that no overshoots occur in the impulse response and the step response. Many of the responses attained in practice are approximations to these ideal ones, hence the information presented here is useful for predicting the general characteristics in the unspecified domain, given characteristics in the specified domain. The important results of their analyses are now summarized.

1. The impulse and step response associated with the rectangular response each contain overshoots, even though linear phase has been assumed. These overshoots must therefore be a consequence of the magnitude response, and indeed they result from the rapid change in attenuation at the cutoff frequency. This fact cannot be emphasized enough. Furthermore, the absence of overshoots in the time-domain responses associated with the Gaussian magnitude response is due to the gradual attenuation change at the cutoff frequency. Humpherys [10] shows that if the attenuation rate of the Gaussian response at the cutoff is increased, impulse response overshoots occur. He also shows that reducing the attenuation rate at the cutoff frequency of the rectangular response considerably reduces the impulse response overshoots.

2. Rise time and pulse width are both inversely proportional to the bandwidth, a consequence of the Fourier transform relationship between time and frequency. Thus narrow time functions have wide spectra, and vice versa. Papoulis [20] gives some insight into the relationship between the spreads of a function $h(t)$ and its spectrum $|H(j\omega)|$ by examining the product of the rms duration of each. This product is minimized when $|H(j\omega)|$ is the Gaussian function.

3. The stated definition of rise time yields a value equal to the impulse response pulse width. This relationship is useful for estimating realizable system characteristics. Also, the impulse response pulse width and peak amplitude are reciprocals.

4. If the phase response is linear with frequency, corresponding to a delay t_0, the impulse response is symmetric about $t = t_0$. We derived this property in Section 2.1.3. Phase distortion (nonlinear phase) produces a nonsymmetric impulse response and furthermore increases the impulse and step response overshoots [10, 20, 28].

3.1.8 Paley–Wiener Condition

We have just examined two magnitude responses, and associated with each is a noncausal impulse response. This raises the following question. When is a given frequency function the Fourier spectrum of a causal function?

Paley and Wiener answered this question in 1934 by proving that a necessary and sufficient condition for the square-integrable function $A(\omega)$ to be the Fourier spectrum of a causal function is that

$$I = \int_{-\infty}^{\infty} \frac{|\ln A(\omega)|}{1+\omega^2} d\omega < \infty \tag{3.1-18}$$

It does not follow that $F(\omega)$ has a causal inverse if $A(\omega)$ satisfies (3.1-18). Rather it implies that a suitable phase can be associated with $A(\omega)$ so that the resulting function has a causal inverse.

This integral is now evaluated for the $A(\omega)$'s in Sections 3.1.1 and 3.1.4. Since each magnitude response is square integrable, $A(\omega) = |H(j\omega)|$.

Example 3-1. For the magnitude response

$$|H(j\omega)| = \begin{cases} 1 & -\omega_c \leq \omega \leq \omega_c \\ \delta & \text{elsewhere} \end{cases} \tag{3.1-19}$$

the integral I from (3.1-18) is

$$I = \int_{-\infty}^{-\omega_c} \frac{|\ln \delta|}{1+\omega^2} d\omega + \int_{-\omega_c}^{\omega_c} \frac{\ln 1}{1+\omega^2} d\omega + \int_{\omega_c}^{\infty} \frac{|\ln \delta|}{1+\omega^2} d\omega = 2|\ln \delta| \left(\frac{\pi}{2} - \tan^{-1} \omega_c\right) \tag{3.1-20}$$

As $\delta \to 0$, $|H(j\omega)|$ approaches the rectangular magnitude response in Fig. 3-5a. It is clear from (3.1-20) that I does not converge as $\delta \to 0$. Thus for $\delta = 0$ there is no phase function that allows a causal inverse to be obtained. Furthermore (3.1-20) shows that any square-integrable $A(\omega)$ that is zero over a finite frequency interval, which includes all band-limited spectra, cannot be the spectrum of a causal function.

Example 3-2. For the Gaussian function in (3.1-9),

$$I = \lim_{B \to \infty} \int_{-B}^{B} \frac{|\ln e^{-c\omega^2}|}{1+\omega^2} d\omega = \lim_{B \to \infty} c \int_{-B}^{B} \frac{\omega^2}{1+\omega^2} d\omega = 2c \left[\lim_{B \to \infty} B - \frac{\pi}{2}\right] \tag{3.1-21}$$

Ideal Low-Pass Responses

where $c = \ln 2/(2\omega_c^2)$. Again I does not converge as $B \to \infty$, indicating that no phase function exists that allows the inverse transform to be causal. The Gaussian function is not bandlimited; nor is it zero over any frequency interval. The integral does not converge because of the high-frequency behavior of $e^{-c\omega^2}$. Consider the spectrum $|H(j\omega)| = e^{-c_1\omega^n}$. Then

$$I = c_1 \lim_{B \to \infty} \int_{-B}^{B} \frac{\omega^n}{1+\omega^2} d\omega \qquad (3.1\text{-}22)$$

When $n < 2$ the integrand approaches zero as $\omega \to \infty$. For this reason, the integral converges for $n < 2$ and diverges for $n \geq 2$. For the Gaussian response, $n = 2$.

3.1.9 Minimum-Phase Functions

A rational function of s is defined to be a minimum-phase function if its poles lie only in the left-half plane and its zeros lie either in the left-half plane or on the $j\omega$ axis. No right-half plane zeros are permitted.

The minimum-phase property is explained by considering the four pole-zero plots in Fig. 3-11, where in each diagram the pole set is the same. The magnitude response is also the same for each case, since the distance from each zero to a point on the $j\omega$ axis is the same whether the zero is located in the left- or right-half plane. The phase-lag angle, however, is different. For conciseness here, we refer to the pole-zero diagram as being a function, realizing, of course, that it only represents that function.

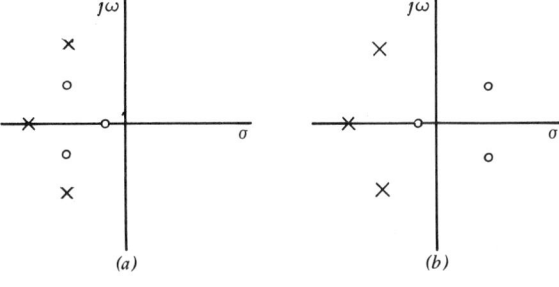

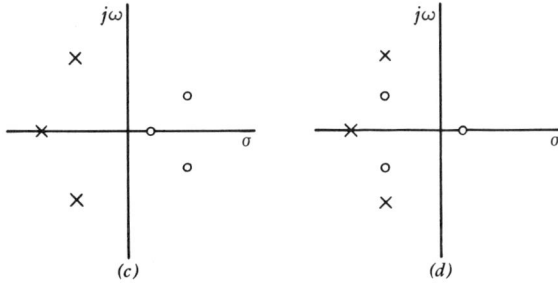

Figure 3-11. Four pole-zero diagrams yielding the same magnitude response.

Each function in Figs. 3-11b, c, and d can be considered as the function in Fig. 3-11a multiplied by the appropriate all-pass (AP) function. The AP function has constant magnitude response and thus does not affect the overall magnitude response. This property is achieved by accompanying each left-half-plane pole by its mirror-image zero in the right-half plane. In Fig. 3-12 the three AP functions that multiply the function in Fig. 3-11a yield the functions in Figs. 3-11b, c, and d, respectively. The AP function poles cancel the appropriate zeros in Fig. 3-11a.

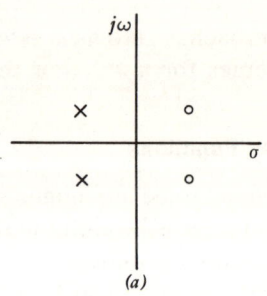

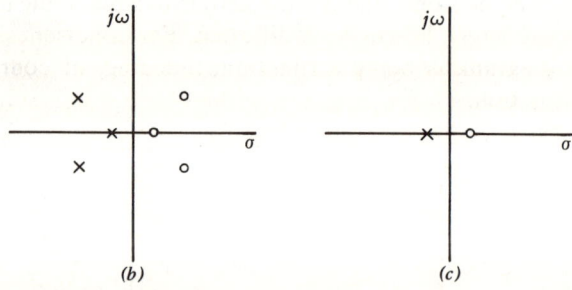

Figure 3-12. Pole-zero diagrams of the three AP functions that multiply the function in Fig. 3-11a to yield the remaining functions in Figs. 3-11b, c, and d, respectively.

The AP function has a phase-lag angle that is always positive; thus the phase-lag angle of each combination is always greater than the angle of the function in Fig. 3-11a. For this reason the function with only left-half-plane zeros (or on the $j\omega$ axis) is termed "minimum phase" and those with zeros in the right-half plane are termed "nonminimum phase." The function in Fig. 3-11a is minimum phase, but the other three functions in Fig. 3-11 are nonminimum phase.

Example 3-3. We now show that the phase-lag angle for the first-order AP transfer function given by

$$H(s) = \frac{\alpha - s}{\alpha + s} \tag{3.1-23}$$

is always positive. With $s = j\omega$,

$$H(j\omega) = \frac{\alpha - j\omega}{\alpha + j\omega} \tag{3.1-24}$$

Ideal Low-Pass Responses

and the phase-lag angle $\beta(\omega)$ is

$$\beta(\omega) = -\theta(\omega) = -\tan^{-1}\left(-\frac{\omega}{\alpha}\right) + \tan^{-1}\left(\frac{\omega}{\alpha}\right) = 2\tan^{-1}\frac{\omega}{\alpha} \qquad (3.1\text{-}25)$$

shown in Fig. 3-13. Since $\beta(\omega)$ is always positive and is two times the angle due to the pole only, the inclusion of this AP function always increases the phase-lag angle. This conclusion is also true for higher-order AP functions.

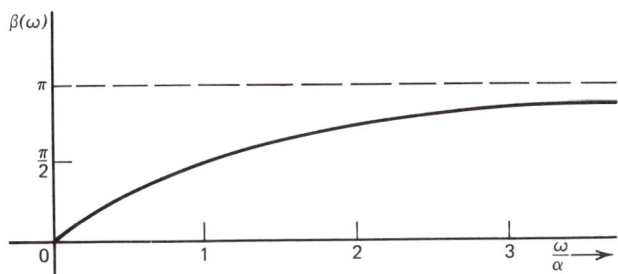

Figure 3-13. The phase-lag angle of the first-order AP function.

We have demonstrated qualitatively that the phase-lag angle for a given magnitude response is unique only when no zeros occur in the right-half plane. It is therefore reasonable to expect a mathematical relationship between $\alpha(\omega)$ and $\beta(\omega)$ for this case. The Hilbert transform discussed in Section 2.7 provides this relationship. We explore this subject in Sections 3.1.11 and 3.1.12.

3.1.10 Realizable Networks

What is so important about minimum-phase functions that we make special mention of them and then determine some attenuation-phase relationships for them? To better answer this question, we consider the four network realizations in Fig. 3-14, where each impedance is realizable.

The lattice (Fig. 3-14a) is a general network capable of realizing minimum-phase and nonminimum-phase transfer functions. It is easily converted to the bridge network in Fig. 3-14b. The bridged-Tee network in Fig. 3-14c is also a general network but offers the advantage of a common input-output connection. Because these three networks can realize nonminimum-phase transfer functions, they are called nonminimum-phase networks. A nonminimum-phase network must have more than one path from input to output, or it must include mutual inductors or a combination of both.

Practical designs often begin with the symmetric lattice; that is $Z_a = Z_d$ and $Z_b = Z_c$. The bridged-Tee in Fig. 3-15a and the differential bridge in Fig. 3-15b are two equivalents to the symmetric lattice. Each includes a unity-coupled, center-tapped transformer and two reactance arms, whereas the lattice contains four reactance arms. Consequently, these two equivalents are more economical than the symmetric lattice.

The unbalanced ladder network (it can also be realized as a balanced network) of Fig. 3-14d has only one input-output path, hence must be a minimum-phase

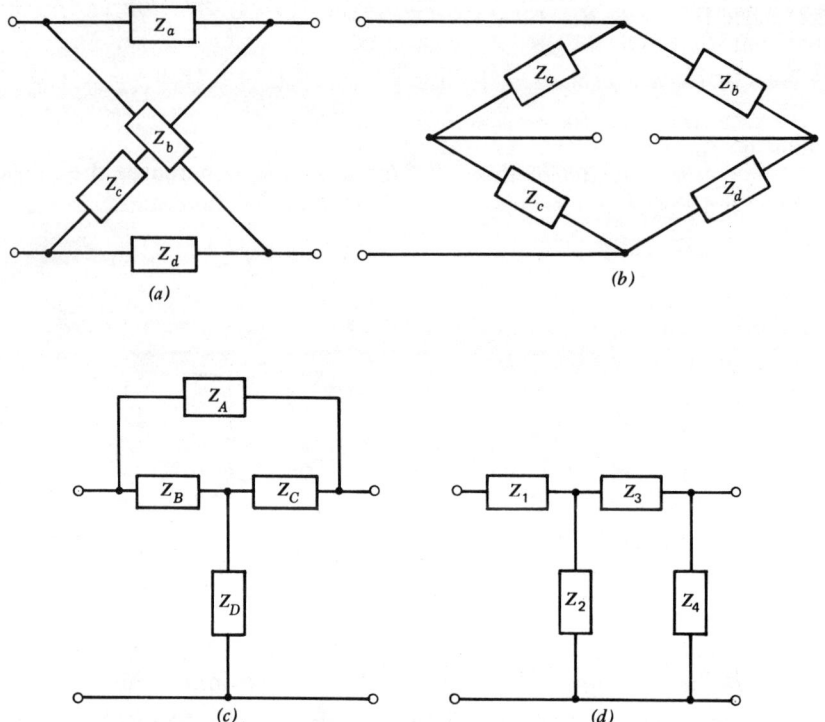

Figure 3-14. Four network realizations. (a) Lattice, (b) bridge, (c) bridged-Tee, (d) ladder.

network. Furthermore, finite transfer-function zeros are introduced by resonances (at complex frequencies) in the series and shunt arms and therefore must lie in the left-half plane or on the $j\omega$ axis. For example, if Z_2 is a series LC combination, a transfer-function zero occurs at the series resonance $\omega = (LC)^{-1/2}$. Ladder networks are more desirable than the previous three because element-value sensitivity is less (resonances in the other three result from impedance balancing), and network complexity is reduced. For these reasons ladder networks are more commonly used in practice, and it is worthwhile to know the expected tradeoffs between attenuation and phase for them. The following two

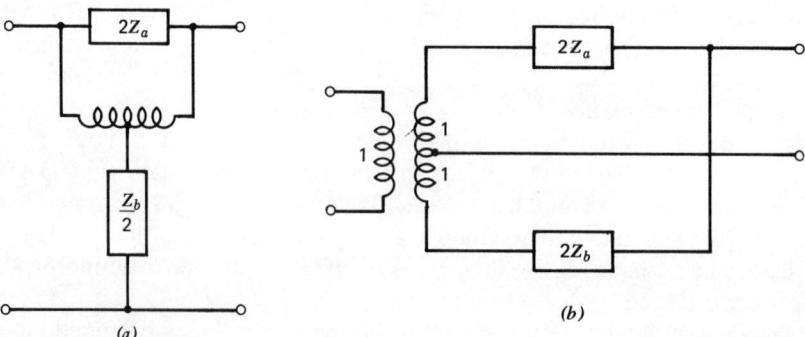

Figure 3-15. Two equivalents to the symmetric lattice. (a) Bridged-Tee, (b) differential bridge.

Ideal Low-Pass Responses

sections consider this relationship for two conditions—the rectangular magnitude response and the linear phase response.

3.1.11 Minimum-Phase Network with Rectangular Magnitude Response

We now determine the phase function for a minimum-phase network whose attenuation (Fig. 3-16a) is

$$\alpha(\omega) = \begin{cases} 0 & -\omega_c \leq \omega \leq \omega_c \\ \alpha_0 \text{ nepers} & \text{elsewhere} \end{cases} \quad (3.1\text{-}26)$$

From (2.5-33) the attenuation function $\alpha(\omega)$ is the real part of $-\ln H(j\omega)$ and the phase-lag angle $\beta(\omega)$ is the imaginary portion. We therefore associate these two functions with the Hilbert transforms of (2.7-2) and (2.7-1), respectively. Then

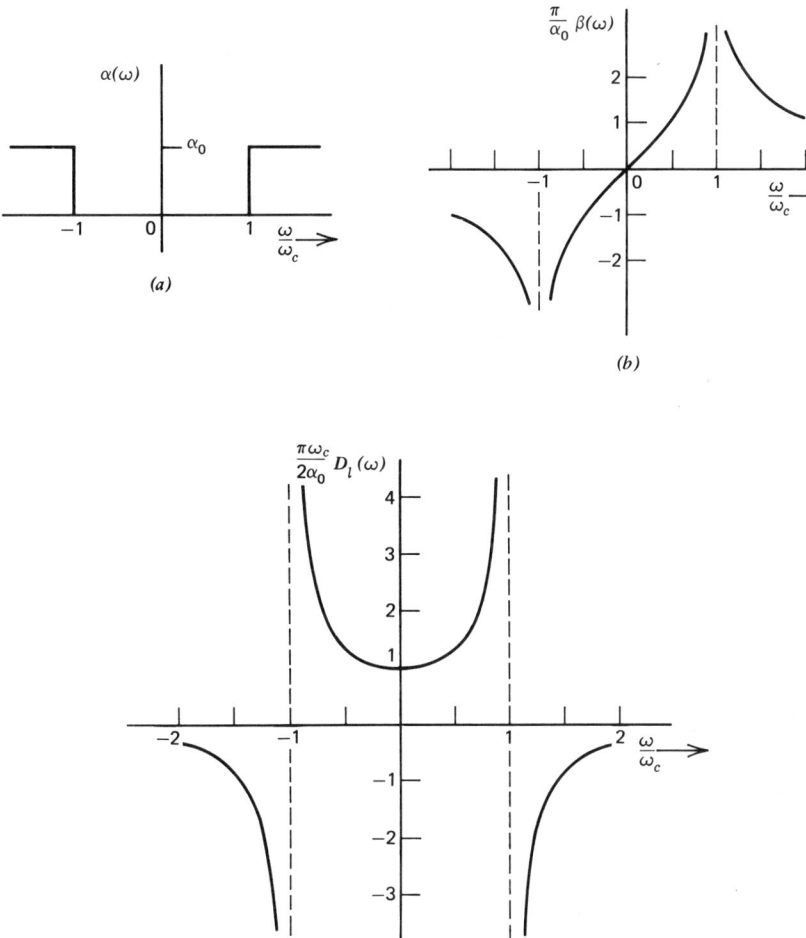

Figure 3-16. (a) Attenuation, (b) phase-lag angle, and (c) group delay for minimum-phase network in Section 3.1.11.

$\beta(\omega)$ is given by

$$\beta(\omega) = -\frac{1}{\pi} \int_{-\infty}^{\infty} \frac{\alpha(\lambda)}{\omega - \lambda} d\lambda = -\frac{\alpha_0}{\pi} \lim_{B \to \infty} \left[\int_{-B}^{-\omega_c} \frac{d\lambda}{\omega - \lambda} + \int_{\omega_c}^{B} \frac{d\lambda}{\omega - \lambda} \right] = \frac{\alpha_0}{\pi} \ln \left| \frac{\omega + \omega_c}{\omega - \omega_c} \right| \quad (3.1\text{-}27)$$

shown in Fig. 3-16b. The group delay function (Fig. 3-16c) is then

$$D_l(\omega) = \frac{d\beta}{d\omega} = \frac{2\alpha_0 \omega_c}{\pi(\omega_c^2 - \omega^2)} \quad (3.1\text{-}28)$$

The abrupt change of attenuation at the cutoff frequency causes an extremely nonconstant passband delay, which approaches infinity at the attenuation discontinuity. In practical designs the transition band is finite and the attenuation is smooth, rather than abrupt. Then the delay does not go to infinity but increases to some large finite value and then decreases. At $\omega = 0$ the delay is $2\alpha_0/\pi\omega_c$, which is inversely proportional to bandwidth. As the attenuation approaches the rectangular response of (3.1-1), the delay becomes infinite, and the system becomes anticipatory. The discontinuous phase in Fig. 3-16b and the negative stopband delay in Fig. 3-16c do not occur in the realizable approximations to $\alpha(\omega)$ introduced in Section 3.3.

The important result here is that a minimum-phase network cannot simultaneously realize a rapid-changing cutoff attenuation and a constant passband group delay. The two are not compatible!

3.1.12 Minimum-Phase Network with Linear Phase Response

Here we consider a phase function $\beta(\omega)$ that linearly increases from $-n\pi$ at $-\omega_c$ to $n\pi$ at ω_c and is constant elsewhere. The group delay corresponding to $\beta(\omega)$ is $n\pi/\omega_c$ in the passband and zero in the stopband. Both characteristics are presented in Fig. 3-17. The attenuation corresponding to this phase function is found from the Hilbert transform of (2.7-2) as

$$\alpha(\omega) = n \left[\ln |x^2 - 1| - x \ln \left| \frac{x-1}{x+1} \right| \right] \quad (3.1\text{-}29)$$

where $x = \omega/\omega_c$ [10].

The attenutation (Fig. 3-17c), has a rounded passband shape and an infinite slope at the cutoff frequencies. It is shown in Ref. 10 that $\alpha(\omega)$ is approximately Gaussian for $|\omega| \ll \omega_c$. Realizable approximations to $\beta(\omega)$ are smoothed at the cutoff frequencies, and the attenuation slope there is then finite.

Thus for a minimum-phase network, a constant passband delay goes hand in hand with a rounded passband attenuation characteristic rather than one that changes abruptly at the cutoff frequency.

3.2 MATHEMATICAL APPROXIMATIONS

The ideal responses discussed in Section 3.1 are unrealizable, hence cannot be achieved in practice. They are presented to illustrate the expected tradeoffs as these responses are approximated by realizable functions. Each ideal function has

Mathematical Approximations

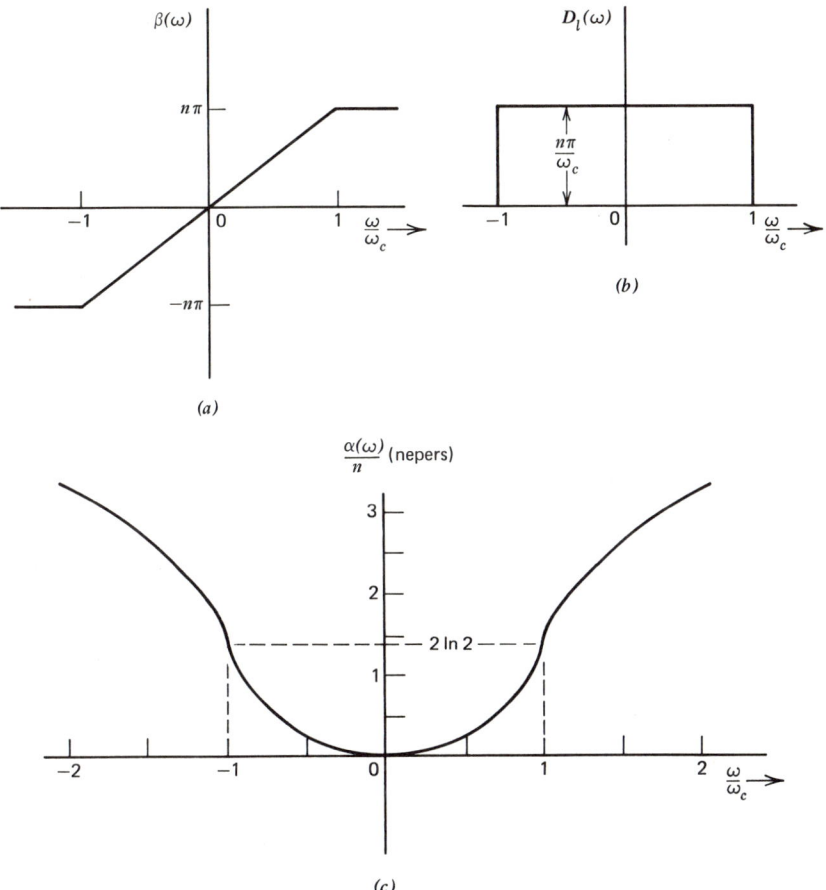

Figure 3-17. (a) Phase-lag angle, (b) group delay, and (c) attenuation for minimum-phase network in Section 3.1.12.

specific properties that we attempt to preserve by the approximation. These desired qualities usually dictate the "best" approximation technique.

Approximation theory is a specialty, the subject of numerous theoretical and practical studies. Because the derived equations often do not allow explicit solutions to be obtained, this task is performed on a digital computer by suitable search algorithms. Many of the sophisticated filter responses could not have been realized without the digital computer. It offers so many computational benefits that it is now a necessary ingredient of most approximation procedures.

Here we briefly examine three useful and popular approximations—the Taylor, the Chebyshev, and the least squares. The complexity of the resulting equations verifies that the digital computer is indispensable for carrying out the necessary computations, except for the low-order approximations.

The function to be approximated is represented by $g(x)$, the approximating function is represented by $f(x)$, and the interval of interest is $\Delta x = x_2 - x_1$. The error function $e(x)$ is the difference of these functions,

$$e(x) = f(x) - g(x) \qquad (3.2\text{-}1)$$

The usual procedure is to form an error criterion in terms of $e(x)$ and then adjust the parameters of $f(x)$ to satisfy this criterion. When this has been done, the approximation is said to be optimized.

The approximation of $g(x)$ by $f(x)$ does not necessarily ensure that the derivative of $g(x)$ is equally well approximated. Ripples in $f(x)$ can cause a marked difference between the derivative of $f(x)$ and the derivative of $g(x)$.

The independent variable x can represent time (t), radian frequency (ω), length (l), temperature (T), or any convenient quantity. Hence the approximation procedures are useful for a wide variety of scientific problems. The important variables for describing filtering systems are time and frequency.

3.2.1 Taylor Approximation

Assume that $g(x)$ and $f(x)$ are both well behaved in the interval Δx and furthermore can be expanded in a Taylor series there. Then if $x_1 \leq x_0 \leq x_2$,

$$f(x) = a_0 + a_1(x - x_0) + a_2(x - x_0)^2 + \cdots$$
$$g(x) = b_0 + b_1(x - x_0) + b_2(x - x_0)^2 + \cdots \qquad (3.2\text{-}2)$$

and the error function $e(x)$ from (3.2-1) is

$$e(x) = a_0 - b_0 + (a_1 - b_1)(x - x_0) + (a_2 - b_2)(x - x_0)^2 + \cdots \qquad (3.2\text{-}3)$$

The kth-order Taylor approximation is obtained by setting $a_0 = b_0$, $a_1 = b_1, \ldots, a_k = b_k$, yielding the resultant error

$$e(x) = (a_{k+1} - b_{k+1})(x - x_0)^{k+1} + (a_{k+2} - b_{k+2})(x - x_0)^{k+2} + \cdots \qquad (3.2\text{-}4)$$

What meaning can be attached to this adjustment of the a's? To answer this question, we refer to the definition of the a's and b's in (3.2-2), which according to Taylor's theorem are given by the rule

$$a_n = \frac{1}{n!} \frac{d^n f(x)}{dx^n}\bigg|_{x=x_0}, \qquad b_n = \frac{1}{n!} \frac{d^n g(x)}{dx^n}\bigg|_{x=x_0} \qquad (3.2\text{-}5)$$

The coefficients are identified with the derivatives of the function evaluated at $x = x_0$. By equating to zero the coefficients of successive powers of $x - x_0$ starting with the lowest term, we cause successive derivatives of $e(x)$ to vanish at $x = x_0$. Thus a kth-order Taylor approximation exists if the first k derivatives of $e(x)$ are zero at $x = x_0$.

All attention is given to the point x_0, and from (3.2-4) the resultant error there is zero. No consideration is given to other points in the interval. Consequently the approximation is very good in the vicinity of x_0 at the expense of a poorer approximation at points near the end of the interval.

Example 3-4. Consider the network in Example 2-22, with transfer function

$$H(s) = \frac{1}{LCs^2 + (L/R)s + 1} \qquad (3.2\text{-}6)$$

With $R = 1\,\Omega$, we determine the values of L and C so that $|H(j\omega)|^2$ approximates the constant unity at $\omega = 0$ in the Taylor sense. Furthermore $|H(j\omega)|$ is to be 3 dB down from unity at $\omega = 1$; thus the trivial solution $L = C = 0$ is eliminated.

Mathematical Approximations

Likewise, $|H(j\omega)|$ will approximate unity in the Taylor sense. When the first derivative of $|H(j\omega)|^2$ vanishes at $\omega = 0$,

$$\frac{d}{d\omega}|H(j\omega)|^2 \bigg|_{\omega=0} = 0$$

Then by the chain rule of differentiation,

$$2|H(j\omega)| \frac{d}{d\omega}|H(j\omega)| \bigg|_{\omega=0} = 0$$

but since $|H(j0)| \neq 0$,

$$\frac{d}{d\omega}|H(j\omega)| \bigg|_{\omega=0}$$

must be zero. Higher-order derivatives of $|H(j\omega)|$ for higher-order systems are similarly shown to be zero at $\omega = 0$ when the approximation is complete.

The magnitude function squared is

$$|H(j\omega)|^2 = \frac{1}{1 + L(L - 2C)\omega^2 + L^2 C^2 \omega^4} \tag{3.2-7}$$

The 3-dB constraint requires that

$$|H(j1)|^2 = \frac{1}{2}|H(j0)|^2 = 1/2 \tag{3.2-8}$$

yielding the relationship

$$L(L - 2C + LC^2) = 1 \tag{3.2-9}$$

The Taylor series expansion for $|H(j\omega)|^2$ about $\omega = 0$, obtained by long division, is

$$|H(j\omega)|^2 = 1 - L(L - 2C)\omega^2 + L^2[(L - 2C)^2 - C^2]\omega^4 + \cdots \tag{3.2-10}$$

and subtracting unity yields the Taylor series for the error function

$$e(\omega) = -L(L - 2C)\omega^2 + L^2[(L - 2C)^2 - C^2]\omega^4 + \cdots \tag{3.2-11}$$

The coefficient of ω^2 is zero when

$$L = 2C \tag{3.2-12}$$

which is simultaneously solved with (3.2-9) to yield

$$C = \frac{\sqrt{2}}{2}, \quad L = \sqrt{2} \tag{3.2-13}$$

Then (3.2-7) reduces to

$$|H(j\omega)| = \frac{1}{\sqrt{1 + \omega^4}} \tag{3.2-14}$$

which is known as the Butterworth response. Figure 3-18 shows the magnitude response for several values of L, with unity 3-dB radian bandwidth for each case. The improvement in the vicinity of $\omega = 0$ with $L = \sqrt{2}$ is apparent. The value $L = 1.3617$ results from a Taylor approximation to a constant group delay (Example 3-5), while $L = 1.3067$ corresponds to an equiripple approximation to a constant magnitude response.

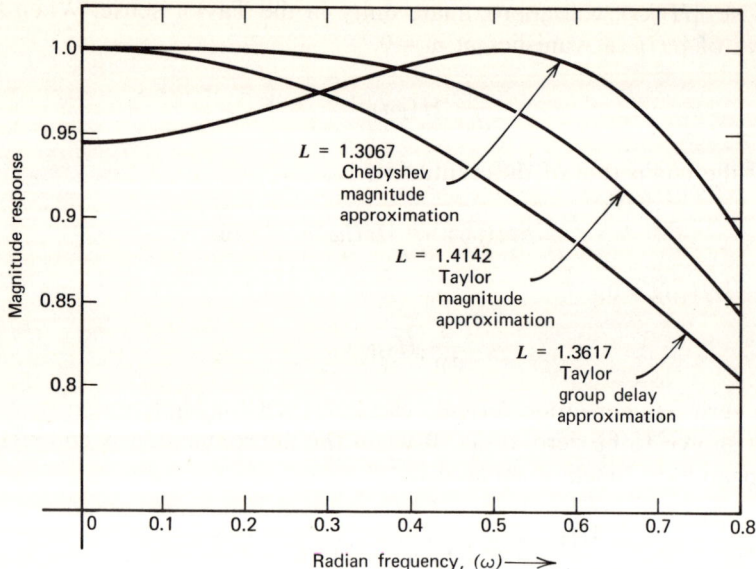

Figure 3-18. Magnitude response from (3.2-7) for three values of $L(\omega_{3dB} = 1)$.

Example 3-4 introduced the rational function as the approximating function. This case is of immediate interest to us because the magnitude function squared and group delay function of lumped networks are rational functions in ω^2.

We now discuss the Taylor approximation as applied to the rational function

$$\Phi(x) = K \frac{1 + c_1 x + c_2 x^2 + \cdots + c_m x^m}{1 + d_1 x + d_2 x^2 + \cdots + d_n x^n} \qquad m \leq n \qquad (3.2\text{-}15)$$

where the leading term of numerator and denominator is normalized to unity. Simple long division allows $\Phi(x)$ to be written as the Taylor series expansion about $x = 0$ as

$$\Phi(x) = K\left\{1 + (c_1 - d_1)x + [(c_2 - d_2) - d_1(c_1 - d_1)]x^2 + \cdots\right\} \qquad (3.2\text{-}16)$$

If $\Phi(x)$ is to approximate a constant, the bracketed term is required to approximate the unit constant; thus from (3.2-1), $g(x) = 1$, $f(x) = (1/K)\Phi(x)$, and from (3.2-3)

$$e(x) = \frac{1}{K}\Phi(x) - 1 = (c_1 - d_1)x + [(c_2 - d_2) - d_1(c_1 - d_1)]x^2 + \cdots \qquad (3.2\text{-}17)$$

The Taylor approximation then requires that the coefficients of successive powers of x vanish, yielding

$$c_k = d_k \qquad k = 1, 2, \ldots \qquad (3.2\text{-}18)$$

For this special case it is not necessary to form the Taylor expansions, but simply to express $\Phi(x)$ in the form of (3.2-15) and equate coefficients of like powers of x.

If (3.2-18) is satisfied for all k from $k = 1$ to $k = n - 1$, we say that we have a maximally flat approximation and $f(x)$ is a maximally flat function. Other adjustments of the coefficients may produce functions with a high degree of

Mathematical Approximations

flatness, but the word "maximal" is reserved for the complete correction. The degree of maximal flatness of a particular function depends on its complexity. For a given value of n, all adjustment potentialities are used when the conditions for maximal flatness are satisfied. The adjustment to maximal flatness reduces $e(x)$ to a function for which its first $(n-1)$ consecutive derivatives are zero at $x=0$. If one thinks of a constant as being represented by a function, all of whose derivatives are everywhere zero, one imagines the way in which a maximally flat function approximates a constant value in the vicinity of $x=0$.

Because $|H(j\omega)|^2$ and $D(\omega)$ are functions of ω^2, their odd-ordered derivatives are already zero at $\omega=0$. Thus matching coefficients of even powers of ω is equivalent to adjusting successive even-ordered derivatives to zero at $\omega=0$. The coefficient of ω^2 in Example 3-4 is therefore set equal to zero since no ω^2 term appears in the numerator of $|H(j\omega)|^2$ in (3.2-7). The result is then $L=2C$, in agreement with (3.2-12).

Example 3-5. Again consider the network in Example 2-22 with $R=1\Omega$, but we now determine L and C so that the group delay function is maximally flat at $\omega=0$, and $|H(j\omega)|$ is 3 dB down from unity at $\omega=1$. The group delay function, from (2.6-1), where

$$\theta(\omega) = \tan^{-1}\frac{\omega L}{LC\omega^2 - 1} \qquad (3.2\text{-}19)$$

is

$$D(\omega) = \frac{L + L^2 C\omega^2}{1 + (L^2 - 2LC)\omega^2 + L^2 C^2 \omega^4} \qquad (3.2\text{-}20)$$

The bandwidth relationship is given by (3.2-9), and the second equation is obtained by first rearranging (3.2-20) into the form of (3.2-15) and then equating the coefficients of ω^2, giving

$$L = 3C \qquad (3.2\text{-}21)$$

Thus if $L=2C$, the maximally flat magnitude response is obtained, while if $L=3C$ the maximally flat group delay function is obtained. Simultaneously solve (3.2-9) and (3.2-21) to give

$$C = \sqrt{\frac{\sqrt{5}-1}{6}} = 0.45388$$
$$L = 3C = 1.36165 \qquad (3.2\text{-}22)$$

The magnitude response for this case is shown in Fig. 3-18. As expected from our discussion of ideal responses for minimum-phase networks, the magnitude response becomes more rounded at the cutoff frequency as the delay approaches a constant value.

Note from (3.2-3) that the error function can be expanded about any point in the convergence interval, and for rational functions $e(x)$ can therefore be maximally flat at that point by adjusting the required number of coefficients to be zero. However, for this case, obtaining this condition by matching coefficients of like powers of x fails, for then the function is made maximally flat at $x=0$. This can easily be remedied by first replacing x by $y+x_0$, where $y=x-x_0$, and then matching coefficients of like powers of y. The derivatives of $\Phi(x)$ are then zero at $x=x_0$.

Example 3-6. The function

$$\Phi(x) = \frac{1 + Kx^2}{1 + 2Kx^4} \qquad (3.2\text{-}23)$$

is flattened at $x = 1$ by first replacing x by $y + 1$. Then, after normalizing the constant term in numerator and denominator to unity,

$$\Phi(y) = \frac{1+K}{1+2K} \cdot \frac{1 + \frac{2K}{1+K} y + \frac{K}{1+K} y^2}{1 + \frac{8K}{1+2K} y + \frac{12K}{1+2K} y^2 + \frac{8K}{1+2K} y^3 + \frac{2K}{1+2K} y^4} \qquad (3.2\text{-}24)$$

Because K is the only adjustable parameter, only the first derivative can be made to vanish, and $\Phi(y)$ cannot be made maximally flat at $y = 0$. However the flatness is improved by equating the coefficients of y and then solving to give $K = -3/2$. The trivial solution $K = 0$ is neglected.

3.2.2 Chebyshev Approximation

The Taylor approximation at $x = x_0$ assumes that behavior at this point is more important than behavior at any other point in the interval Δx. By choosing a different error criterion we can obtain much better characteristics throughout the entire interval. For example, suppose all points in the interval are equally important, and we wish to minimize $|e_{\max}(x)|$, the maximum deviation between the desired function and the approximating function at any point in the interval. This type of approximation is called an equiripple or Chebyshev approximation.

The minimization of $|e_{\max}(x)|$ in Δx is achieved when all peak errors are equal, for then no advantage is gained by increasing one error and decreasing another. The absolute error $|e(x)|$ will then be smaller at all other points in the interval. The error function before optimization and the error function when all peak errors have been made equal are given in Figs. 3-19a and b, respectively. The approximation interval is $x = 0$ to $x = 1$. Except in the very simplest cases, it is unlikely that the optimum parameters can be determined analytically; hence this task is usually performed on a computer by a search algorithm.

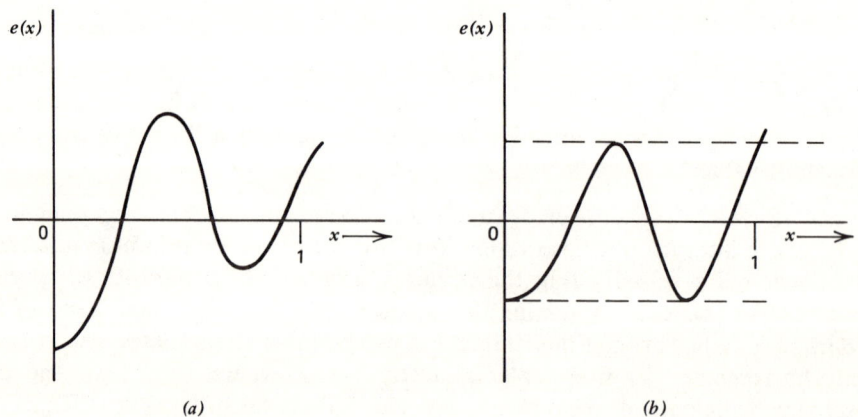

Figure 3-19. (a) Error function before optimization, (b) equiripple approximation after optimization.

Mathematical Approximations

Example 3-7. For the first-order system with transfer function

$$H(s) = \frac{a_1}{s + a_1} \tag{3.2-25}$$

we determine a_1 and ω_c so that in the range $\omega = 0$ to $\omega = \omega_c$ the group delay function approximates unity with an equiripple error δ as in Fig. 3-20a. The solution to this problem can be obtained analytically. The group delay function, from (2.6-10), is

$$D(\omega) = \frac{a_1}{\omega^2 + a_1^2} \tag{3.2-26}$$

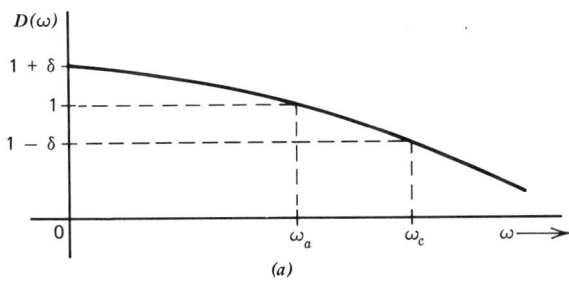

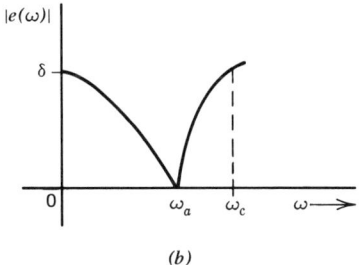

Figure 3-20. (a) Equiripple group delay and (b) error function for first-order system of Example 3-7.

At $\omega = 0$

$$D(0) = 1 + \delta = \frac{1}{a_1} \tag{3.2-27}$$

yielding

$$a_1 = \frac{1}{1 + \delta} \tag{3.2-28}$$

while at $\omega = \omega_c$

$$D(\omega_c) = 1 - \delta = \frac{a_1}{\omega_c^2 + a_1^2} \tag{3.2-29}$$

Substitute a_1 from (3.2-28) into (3.2-29) and solve for ω_c as

$$\omega_c = \frac{1}{1 + \delta} \sqrt{\frac{2\delta}{1 - \delta}} \tag{3.2-30}$$

At ω_a, $D(\omega_a) = 1$, the error is zero, and from (3.2-26) and (3.2-28),

$$\omega_a = \frac{\sqrt{\delta}}{1+\delta} = \sqrt{\frac{1-\delta}{2}}\, \omega_c \qquad (3.2\text{-}31)$$

For a given value of δ, the approximation interval ω_c is fixed and decreases as the tolerable error δ decreases. The absolute value of the error function is

$$|e(\omega)| = |1 - D(\omega)| = \frac{|(1+\delta)^2 \omega^2 - \delta|}{(1+\delta)^2 \omega^2 + 1} \qquad (3.2\text{-}32)$$

whose general shape appears in Fig. 3-20b.

The Chebyshev polynomials are a class of functions that possess the equiripple property, and $C_n(x)$ is the nth-order Chebyshev polynomial defined in terms of the real variable x by the equation

$$C_n(x) = \begin{cases} \cos(n \cos^{-1} x) & |x| \leq 1 \\ \cosh(n \cosh^{-1} x) & |x| \geq 1 \end{cases} \qquad (3.2\text{-}33)$$

Each polynomial oscillates between -1 and $+1$ in the interval $-1 \leq x \leq +1$ and increases in magnitude outside this interval. The $C_n(x)$ is indeed a polynomial, although this is not apparent from (3.2-33). Figure 3-21 shows these polynomials for $n = 0$ thru $n = 4$.

The Chebyshev polynomials can be found from the recursive formula

$$C_{n+1}(x) = 2xC_n(x) - C_{n-1}(x) \qquad (3.2\text{-}34)$$

For example,

$$C_4(x) = 2xC_3(x) - C_2(x) = 8x^4 - 8x^2 + 1 \qquad (3.2\text{-}35)$$

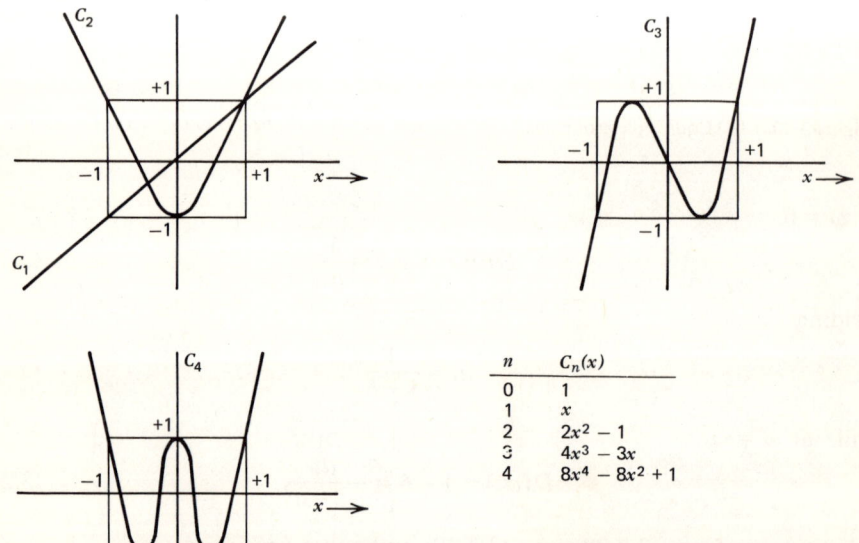

Figure 3-21. Chebyshev polynomials, $n = 0$ through $n = 4$.

Mathematical Approximations

In Section 3.4.2 we use these polynomials to formulate the magnitude response for the so-called Chebyshev filters.

Humpherys [10] gives a procedure for approximating any polynomial of specified degree by a rational function in an equiripple manner over a given interval. This immediately allowed him to obtain LP delay functions that approximate a constant delay in a Chebyshev sense.

3.2.3 Least-Squares Approximation

The Taylor approximation has the disadvantage of emphasizing only one point in the interval, whereas the equiripple approximation considers all points in the interval equally. The integrated least-squares approximation is a happy medium between these two extremes, for it includes a weighting function to consider some parts of the interval more heavily than others. This approximation technique was not very popular in the past because the computations are extremely messy, and analytic solutions are obtained only for the very simple cases. Now, however, high-speed computers perform the necessary computations, thus making this technique competitive with the previous two. We also discuss the special case of orthogonal approximating functions, which make it possible to obtain analytic solutions.

The error number is the integral of the weighted squared error $e(x)$ over $\Delta x = x_2 - x_1$, that is,

$$\varepsilon = \int_{x_1}^{x_2} w(x) e^2(x) \, dx = \int_{x_1}^{x_2} w(x)[g(x) - f(x)]^2 \, dx \tag{3.2-36}$$

where $w(x)$ is a weighting function that allows the accuracy to be concentrated into the region of greatest interest. Since it can be absorbed into $g(x)$ and $f(x)$, however, we need only consider the case $w(x) = 1$ without loss of generality. Then

$$\varepsilon = \int_{x_1}^{x_2} [G(x) - F(x)]^2 \, dx \tag{3.2-37}$$

where $G(x) = \sqrt{w(x)} g(x)$ and $F(x) = \sqrt{w(x)} f(x)$.

The approximating function $f(x)$ includes n constants a_i (which may be complex numbers) that are determined to minimize ε. Constraints on the coefficients may be included by additional equations involving the a_i's. A necessary condition for minimum ε (in the absence of constraints) is

$$\frac{\partial \varepsilon}{\partial a_i} = 0 \qquad i = 1, 2, \ldots, n \tag{3.2-38}$$

Applying this condition to (3.2-37) gives

$$\frac{\partial \varepsilon}{\partial a_i} = -2 \int_{x_1}^{x_2} [G(x) - F(x)] \frac{\partial F(x)}{\partial a_i} \, dx = 0 \tag{3.2-39}$$

or

$$\int_{x_1}^{x_2} G(x) \frac{\partial F(x)}{\partial a_i} \, dx = \int_{x_1}^{x_2} F(x) \frac{\partial F(x)}{\partial a_i} \, dx \qquad i = 1, 2, \ldots, n \tag{3.2-40}$$

This set of n nonlinear equations generally cannot be solved analytically; hence a computer is used to iteratively solve them. Since a computer must ultimately be used, the error can be computed from (3.2-37) and then, by a search algorithm, the a_i's are varied until ε is minimized. This approach is used in Section 4.6 to obtain a wide-band approximation to constant delay.

It is never certain that a global minimum is achieved, for the computer may "lock on" to one of the many local minima. However, if various starting values of the a_i's lead to the same minimum value of ε, this value can be assumed to be the global minimum.

Example 3-8. This example illustrates the least-squares technique and shows that even a first-order system may not allow an explicit solution for the adjustable parameter. For the system in Example 3-7, we now find the value of a_1 so that the group delay function is a least-squares approximation to unity from $\omega = 0$ to $\omega = \omega_c$. With $w(x) = 1$, (3.2-37) becomes

$$\varepsilon = \int_0^{\omega_c} \left[1 - \frac{a_1}{\omega^2 + a_1^2}\right]^2 d\omega \qquad (3.2\text{-}41)$$

and from (3.2-40)

$$\int_0^{\omega_c} \frac{\omega^2 - a_1^2}{(\omega^2 + a_1^2)^2} d\omega = a_1 \int_0^{\omega_c} \frac{\omega^2 - a_1^2}{(\omega^2 + a_1^2)^3} d\omega \qquad (3.2\text{-}42)$$

After integration

$$(4a_1 - 1)(\omega_c^2 + a_1^2) - 2a_1^2 = \frac{(\omega_c^2 + a_1^2)^2}{a_1 \omega_c} \tan^{-1} \frac{\omega_c}{a_1} \qquad (3.2\text{-}43)$$

and even for this first-order system an explicit expression for a_1 is not possible.

The three approximation techniques are now compared for $\omega_c = 1$. Equation 3.2-43 is solved numerically to yield $a_1 = 0.9$, while for an equiripple delay, from (3.2-28) and (3.2-30), $a_1 = 0.6478$. The corresponding delays are shown in Fig. 3-22, along with the maximally flat delay, obtained with $a_1 = 1$. It is clear that for a first-order system the equiripple approximation is not as good as the other two. Because the maximum possible delay at $\omega = 1$ is 0.5 (when $a_1 = 1$), the deviation at $\omega = 0$, which must be the same as that at $\omega = 1$, is slightly larger than 0.5 (0.543).

The approximation of a periodic function by a Fourier series is so commonly used that we sometimes forget that it is a least-squares approximation. Explicit expressions for the sine and cosine coefficients are obtained because these two functions are a member of the class of orthogonal functions. The functions $\phi_i(x)$ and $\phi_j(x)$ are said to be orthogonal with respect to the weight factor $w(x)$ over the interval $\Delta x = x_2 - x_1$ if

$$\int_{x_1}^{x_2} w(x) \phi_i(x) \phi_j^*(x) \, dx = \begin{cases} 0 & \text{if } i \neq j \\ \text{constant} & \text{if } i = j \end{cases} \qquad (3.2\text{-}44)$$

where ϕ_j^* is the complex conjugate of ϕ_j. Furthermore, if the constant is unity, the

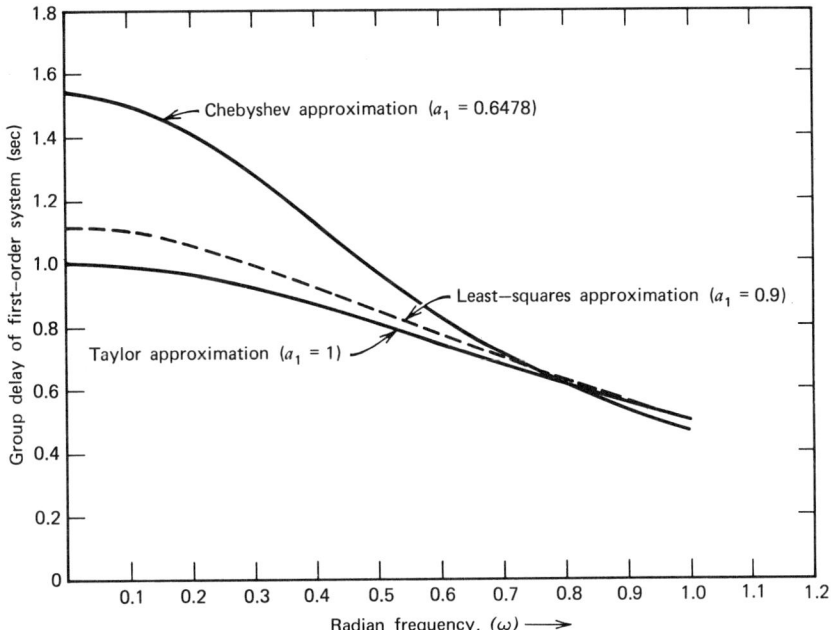

Figure 3-22. Approximations to unity by first-order system delay.

functions are then orthonormal and (3.2-44) becomes

$$\int_{x_1}^{x_2} w(x)\phi_i(x)\phi_j^*(x)\,dx = \delta_{ij} \qquad (3.2\text{-}45)$$

where δ_{ij} is the Kronecker delta

$$\delta_{ij} = \begin{cases} 0 & i \neq j \\ 1 & i = j \end{cases} \qquad (3.2\text{-}46)$$

Suppose we let $f(x)$ be a sum of n orthonormal functions $\phi_i(x)$, which may be complex. Then the error number, allowing for complex c_i's, is

$$\varepsilon = \int_{x_1}^{x_2} w(x)\left[g(x) - \sum_{i=1}^{n} c_i\phi_i(x)\right]\left[g(x) - \sum_{j=1}^{n} c_j^*\phi_j^*(x)\right] dx \qquad (3.2\text{-}47)$$

The values of the c_j's that minimize ε are those which satisfy

$$\frac{\partial \varepsilon}{\partial c_j^*} = \int_{x_1}^{x_2} w(x)\left[g(x) - \sum_{i=1}^{n} c_i\phi_i(x)\right]\left[-\phi_j^*(x)\right] dx = 0 \qquad (3.2\text{-}48)$$

or

$$\sum_{i=1}^{n} c_i \int_{x_1}^{x_2} w(x)\phi_i(x)\phi_j^*(x)\,dx = \int_{x_1}^{x_2} w(x)g(x)\phi_j^*(x)\,dx \qquad (3.2\text{-}49)$$

Use of (3.2-45) yields c_j as

$$c_j = \int_{x_1}^{x_2} w(x)g(x)\phi_j^*(x)\, dx \qquad (3.2\text{-}50)$$

which is independent of n.

The use of orthonormal functions results in explicit solutions for the c_i's, which remain the same value even if additional orthonormal functions are added to $f(x)$. This value of c_i reduces (3.2-47) to

$$\varepsilon = \int_{x_1}^{x_2} w(x)g^2(x)\, dx - \sum_{i=1}^{n} |c_i|^2 \qquad (3.2\text{-}51)$$

Since the first term is independent of the c_i's, the addition of another orthonormal function always reduces the error number and thereby improves the approximation.

In addition to the sine and cosine functions, other orthogonal sets of functions with respect to various weight factors include the Bessel, Legendre, Hermitian, and Laguerre polynomials. A set of linearly independent functions is not necessarily an orthogonal set. However, this set can be systematically orthogonalized by the Gram-Schmidt procedure [7].

Finally, the least-squares technique can be extended to the least-pth approximation [26], where p is any even number. Odd numbers for p allow positive and negative errors to cancel, hence are not suitable. Values of p between 4 and 10 result in approximations that are essentially equiripple [4]. Caution should be observed when p is large (>20), for then the extreme range of values can cause loss of significant figures.

3.3 REALIZABLE LOW-PASS RESPONSE CHARACTERIZATION

A convenient method of obtaining useful realizable system responses is to approximate the ideal responses in Section 3.1 using the techniques in Section 3.2. In the following sections we describe a group of LP responses obtained in this manner, along with others obtained by special techniques. Most of these responses have been synthesized as lumped-constant filters (see references in Ref. 18), and Zverev [30] tabulates their element values and gives normalized curves of the attenuation, group delay, impulse, and step responses. We often refer to these curves in our discussions.

These derived LP responses are additionally useful, for in Chapter 4 we use suitable frequency transformations to obtain LP, HP, BP, and BS filters. In this manner the approximation problem need not be repeated for each filter type. Furthermore LP denormalization information for these filter types is given there, allowing the LP responses in Ref. 30 to then characterize the HP, BP, and BS filters. Also, we find in Chapter 9 that many digital filters are designed to exhibit the same responses, thus the LP information has yet another benefit.

3.3.1 The Normalized Low-Pass Response

We define the normalized LP magnitude response to have a 3-dB radian frequency of unity ($\omega_c = 1$), and associated responses are indicated by the subscript "L." Then $D_L(\omega)$ is the normalized group delay, $h_L(t)$ is the normalized impulse response, and $g_L(t)$ is the normalized step response. The output power is referenced to the maximum output power, which usually occurs at $\omega = 0$. Two exceptions are the even-order Chebyshev and least-squares responses. The frequency axis for the normalized LP curves in Ref. 30 is labeled with a capital omega (Ω) rather than the commonly used lower case omega (ω). This allows us to express the frequency transformations in Chapter 4 as functions of ω given by $\Omega(\omega)$. The normalized time scale for the impulse and step responses is indicated by t_N.

3.3.2 All-Pole Networks

The nth-order LP transfer function

$$H(s) = \frac{K}{a_0 s^n + a_1 s^{n-1} + \cdots + a_n} \tag{3.3-1}$$

is termed all-pole because there are no finite zeros. The resulting filter, known as an all-pole filter, is realized by the network in Fig. 3-23a and the dual network in Fig. 3-23b. Arbitrary values of R_s are allowed except for the even-order Chebyshev and least-squares responses. In Section 3.7.4 we consider the Cauer filter, which has finite transfer-function zeros.

Except for special cases, such as the equally terminated Butterworth and Chebyshev filters, more than one set of element values realizes the same transfer

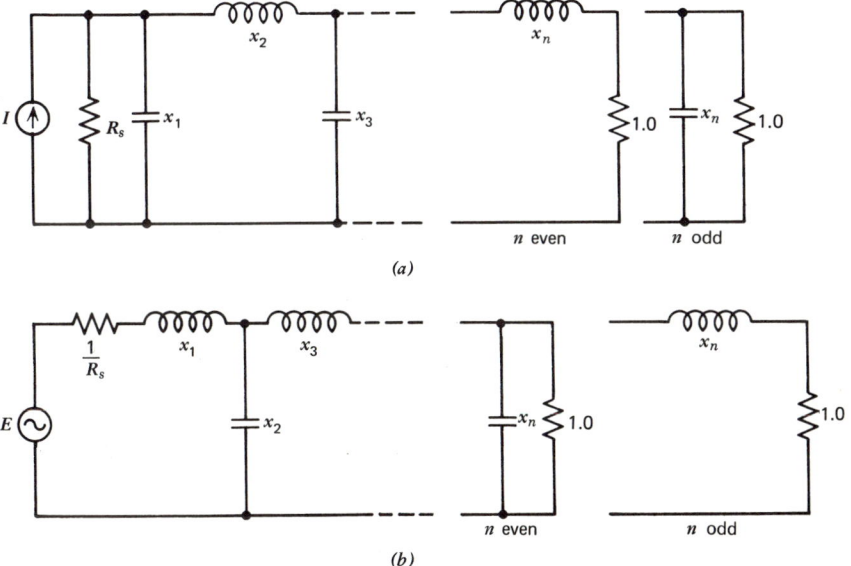

Figure 3-23. (a) Current source and (b) dual network with voltage source for all-pole network.

function. For $n > 2$ this can occur because the location of the reflection coefficient zeros is not unique. The all-pole element values listed in Ref. 30 and in most references result from placing all reflection zeros in one half of the s-plane or on the imaginary axis. This arrangement maximizes the gain-bandwidth product for a specified value of shunt capacitance. Another set of element values is obtained by alternating the reflection zeros in the left- and right-half planes [29]. This set can be advantageous when the maximum gain-bandwidth product is not necessary, for it may yield a smaller element-value spread. The choice of element values depends on the application. Regardless of these reflection-zero locations, the transfer function remains minimum phase and realizable by the ladder networks in Fig. 3-23.

3.3.3 Group Delay Function

The LP group delay is represented as $D_l(\omega)$, whereas normalized LP group delay is represented as $D_L(\omega)$. A useful quantity for characterizing the BP delay and insertion loss at band center is $D_l(0)$, the delay at $\omega = 0$. This delay is expressible in terms of the transfer-function coefficients, the transfer-function poles and zeros, or the network element values. Each representation is now examined.

Consider the LP transfer function

$$H(s) = H_0 \frac{s^m + b_1 s^{m-1} + \cdots + b_m}{s^n + a_1 s^{n-1} + \cdots + a_n} \quad (m < n) \qquad (3.3\text{-}2)$$

The phase shift is

$$\theta(\omega) = \tan^{-1} \frac{N_1(\omega)}{N_2(\omega)} - \tan^{-1} \frac{M_1(\omega)}{M_2(\omega)} \qquad (3.3\text{-}3)$$

where

$$N_1(\omega) = b_{m-1}\omega - b_{m-3}\omega^3 + \cdots$$
$$N_2(\omega) = b_m - b_{m-2}\omega^2 + \cdots$$
$$M_1(\omega) = a_{n-1}\omega - a_{n-3}\omega^3 + \cdots$$
$$M_2(\omega) = a_n - a_{n-2}\omega^2 + \cdots$$

and the group delay, from (2.6-1), is

$$D_l(\omega) = \frac{M_2 M_1' - M_1 M_2'}{M_1^2 + M_2^2} - \frac{N_2 N_1' - N_1 N_2'}{N_1^2 + N_2^2} \qquad (3.3\text{-}4)$$

where the prime indicates differentiation with respect to ω. At $\omega = 0$, $N_1 = M_1 = 0$, $N_2 = b_m$, $M_2 = a_n$, $N_2' = M_2' = 0$, $N_1' = b_{m-1}$, and $M_1' = a_{n-1}$. Then (3.3-4) becomes

$$D_l(0) = \frac{a_{n-1}}{a_n} - \frac{b_{m-1}}{b_m} \qquad (3.3\text{-}5)$$

For all-pole filters, $b_{m-1} = 0$ and

$$D_l(0) = \frac{a_{n-1}}{a_n} \qquad (3.3\text{-}6)$$

Thus the two coefficients a_n and a_{n-1} determine $D_l(0)$.

In terms of the transfer function poles p_k and zeros z_k, $D_l(0)$, from (2.6-10), is

$$D_l(0) = \sum_{k=1}^{m} \frac{1}{z_k} - \sum_{k=1}^{n} \frac{1}{p_k} \qquad (3.3\text{-}7)$$

and for the all-pole case

$$D_l(0) = -\sum_{k=1}^{n} \frac{1}{p_k} \qquad (3.3\text{-}8)$$

Equating (3.3-6) and (3.3-8) yields a result that can be used to verify the accuracy of numerically computed roots of a polynomial.

A useful expression for $D_l(0)$ in terms of the all-pole element values is [16]

$$D_l(0) = \frac{L_T + R_s R_L C_T}{R_s + R_L} \qquad (3.3\text{-}9)$$

where R_s is the source resistance, R_L is the load resistance, L_T is the sum of all series inductances, and C_T is the sum of all shunt capacitances.

The LP responses now examined are classified as (*a*) those approximating the rectangular magnitude response, (*b*) those approximating a constant group delay, (*c*) those approximating the ideal time-domain response, and (*d*) those chosen for special considerations.

3.4 RECTANGULAR MAGNITUDE RESPONSE APPROXIMATIONS

We choose the magnitude function associated with $H(s)$ in (3.3-1) to have the form

$$|H(j\omega)| = \frac{1}{\sqrt{1 + \epsilon^2 \psi_n^2(\omega)}} \qquad (3.4\text{-}1)$$

where ϵ is a positive, real number not greater than unity, and $\psi_n(\omega)$ is an nth-order polynomial containing only even or only odd powers of ω. If the value of $\psi_n(\omega)$ is small in the passband and large in the stopband, $|H(j\omega)|$ is a good approximation to the rectangular magnitude response in Fig. 3-5a. Four classes of filter responses obtained in this manner are now discussed.

3.4.1 Maximally Flat (Butterworth)

Setting $\epsilon = 1$ and $\psi_n(\omega) = \omega^n$ in (3.4-1) yields the Butterworth response

$$|H(j\omega)| = \frac{1}{\sqrt{1 + \omega^{2n}}} \qquad (3.4\text{-}2)$$

Reference to Section 3.2.1 shows that $|H(j\omega)|^2$ satisfies the maximally flat criterion at $\omega = 0$, since the required coefficients of powers of ω are zero. As we demonstrated in Example 3-4, a Taylor approximation by $|H(j\omega)|^2$ is also a Taylor approximation by $|H(j\omega)|$. Since $|H(j\omega)|$ is unity at $\omega = 0$ and $1/\sqrt{2}$ at $\omega = 1$ independent of the approximation order n, it is normalized.

The transfer function, obtained by the technique in Example 2-27, is listed in (3.4-3) for $n = 2, 3,$ and 4.

$$n = 2 \quad H(s) = \frac{1}{s^2 + \sqrt{2}s + 1}$$

$$n = 3 \quad H(s) = \frac{1}{s^3 + 2s^2 + 2s + 1} \quad (3.4\text{-}3)$$

$$n = 4 \quad H(s) = \frac{1}{s^4 + 2.613s^3 + 3.414s^2 + 2.613s + 1}$$

The denominator polynomials—the Butterworth polynomials—are tabulated in Refs. 10 and 30. The zeros of these polynomials $\sigma_i + j\omega_i$, which are the poles of $H(s)$, all lie on the unit circle in the left half of the s-plane and are given by

$$\begin{aligned} \sigma_i &= -\sin(2i-1)\frac{\pi}{2n} \\ \omega_i &= \cos(2i-1)\frac{\pi}{2n} \end{aligned} \quad i = 1, 2, \ldots, n \quad (3.4\text{-}4)$$

The normalized group delay function is obtained in closed form as [17]

$$D_L(\omega) = \frac{1}{1+\omega^{2n}} \sum_{l=1}^{n} \frac{\omega^{2(l-1)}}{\sin(2l-1)(\pi/2n)} \quad (3.4\text{-}5)$$

and the delay at $\omega = 0$ is

$$D_L(0) = \frac{1}{\sin(\pi/2n)} \quad (3.4\text{-}6)$$

The attenuation, group delay, impulse response, and step response for the Butterworth family are shown in Figs. 3-24 through 3-27, for $n = 2$ to $n = 10$. These figures are reprinted from Ref. 30, and similar sets exist there for most of the LP responses that we discuss. We use the curves for the Butterworth response in many examples here with the understanding that the same procedure applies to the other LP response curves.

The attenuation approaches the rectangular response of Fig. 3-5a as n gets very large. The abrupt attenuation change at $\omega = 1$, however, is accompanied by pronounced overshoots in the impulse and step responses, and a nonconstant group delay function that reaches a maximum in the vicinity of the cutoff frequency. These effects reflect the theoretical results obtained in Section 3.1.

The normalized filter element values corresponding to Fig. 3-23 with $R_s = 1$ ohm are

$$x_k = 2\sin(2k-1)\frac{\pi}{2n} \quad k = 1, 2, \ldots, n \quad (3.4\text{-}7)$$

and equal twice the negative value of each pole's real part, that is, from (3.4-4), $x_k = -2\sigma_k$.

As mentioned earlier, other sets of element values exist for $R_s \neq 1$, and we now illustrate the second-order element-value dependence on R_s. First equate coefficients of like powers of s in the second-order transfer function

$$H(s) = \frac{K}{s^2 + a_1 s + a_2} \quad (3.4\text{-}8)$$

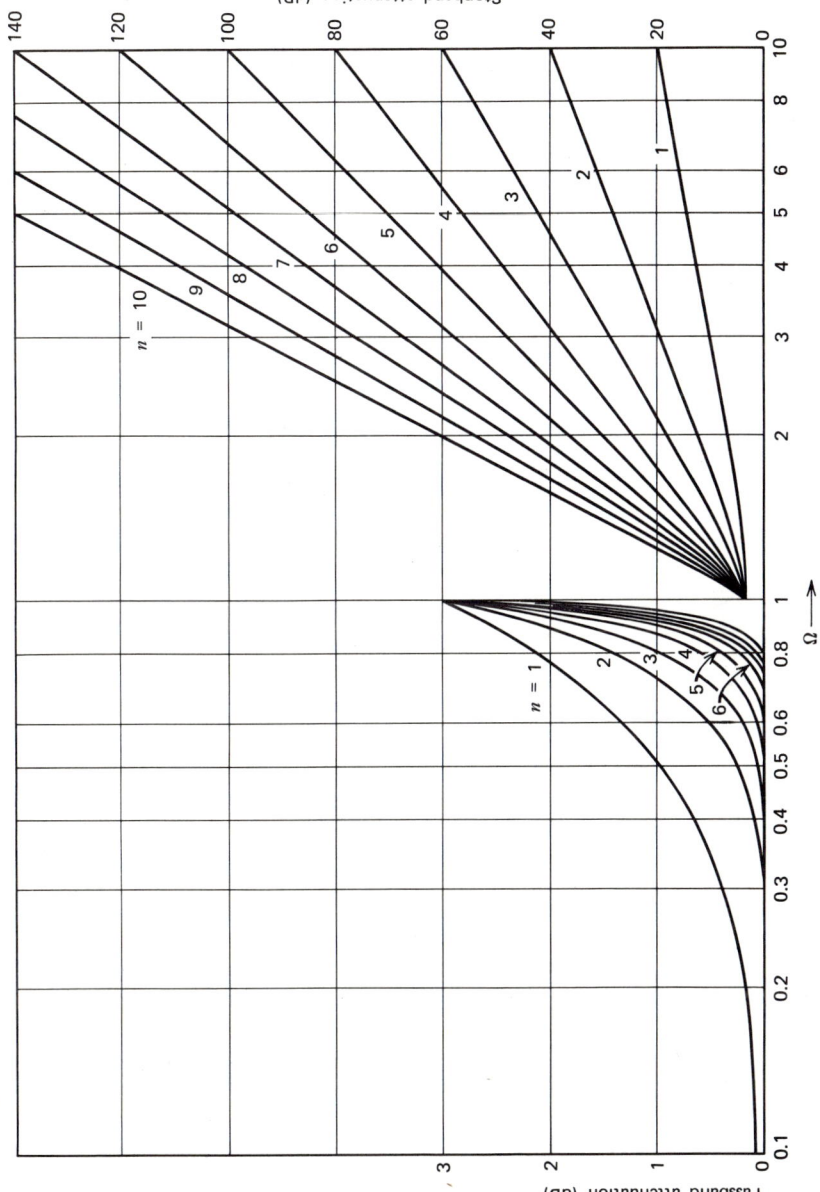

Figure 3-24. Normalized Butterworth LP attenuation function.

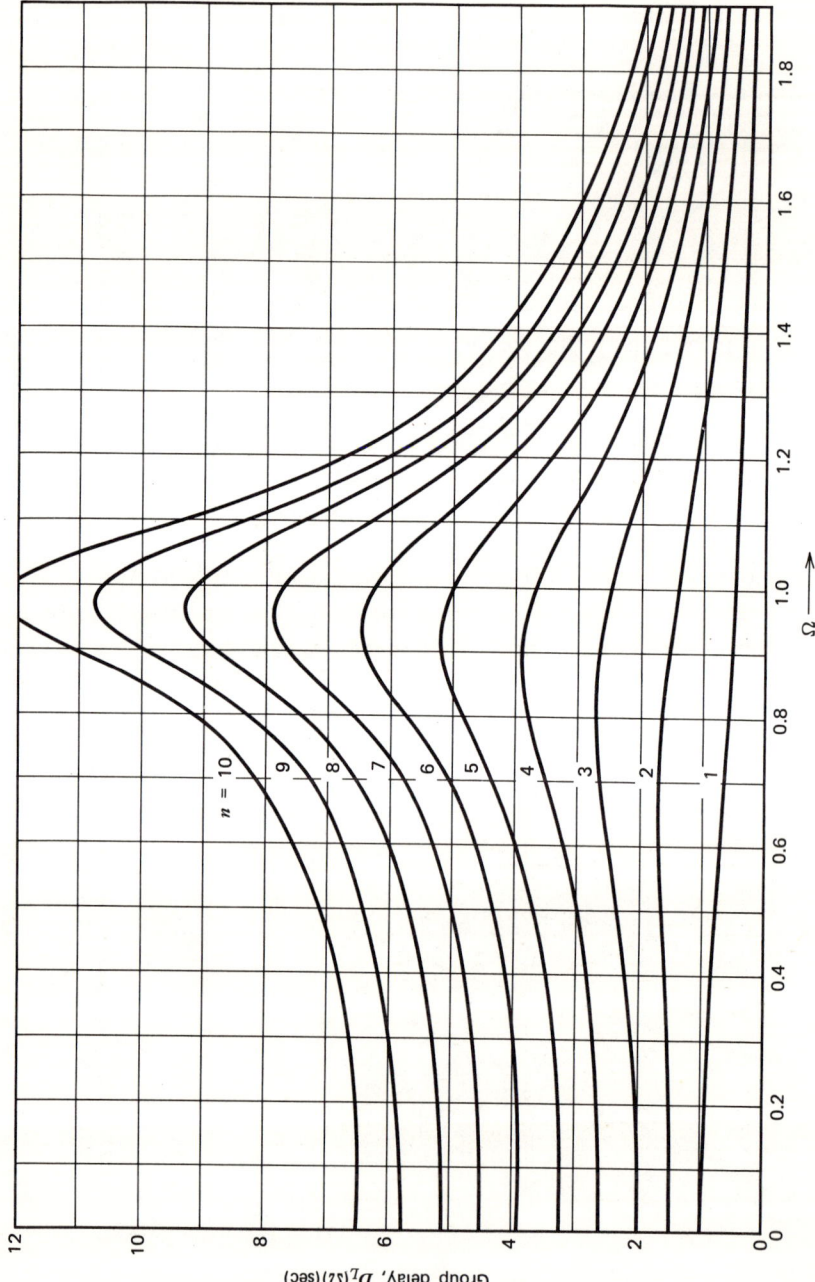

Figure 3-25. Normalized Butterworth LP group delay function.

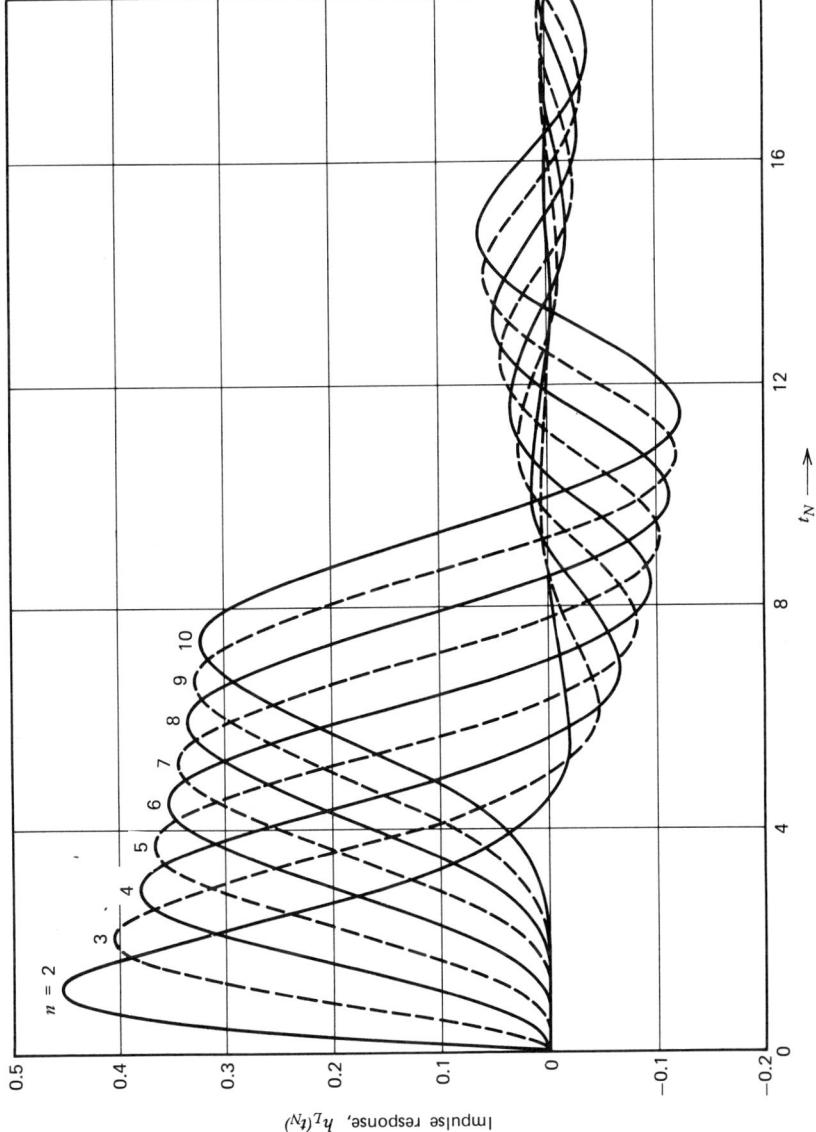

Figure 3-26. Normalized Butterworth LP impulse response.

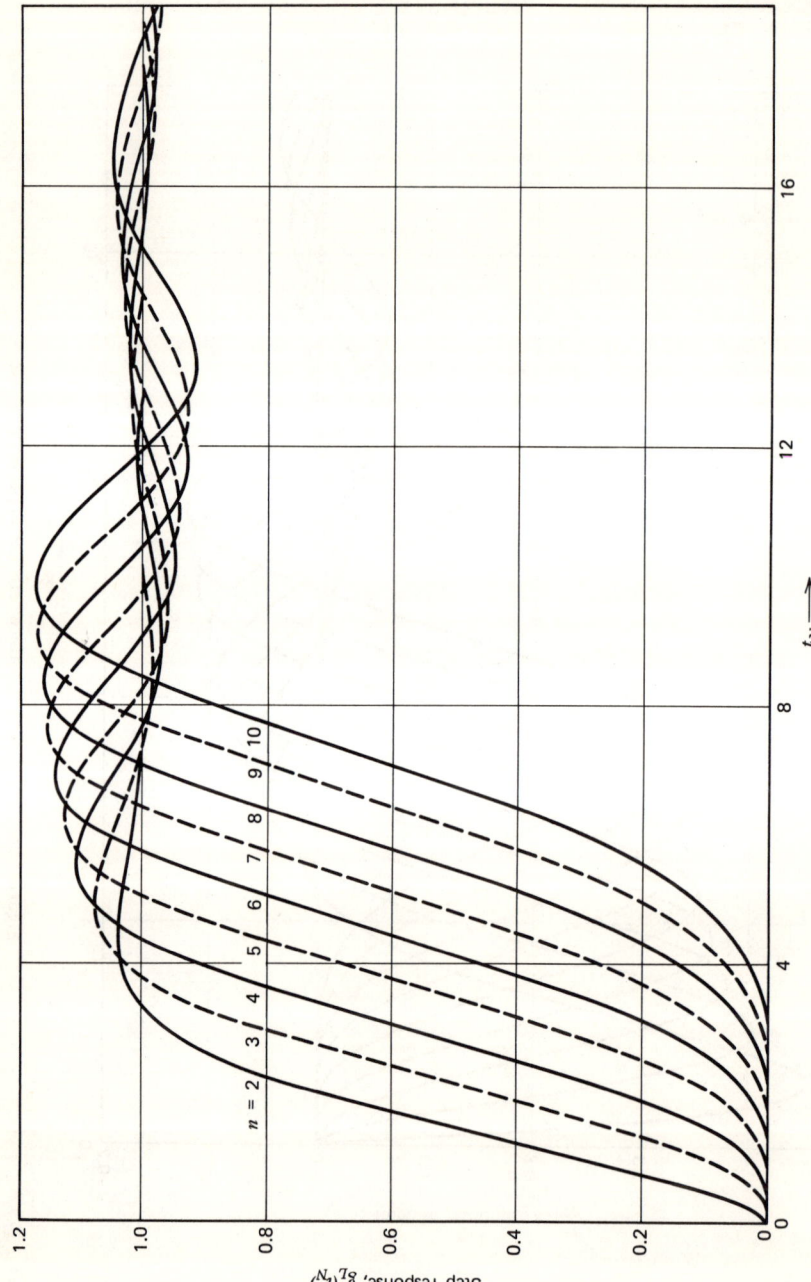

Figure 3-27. Normalized Butterworth LP step response.

Rectangular Magnitude Response Approximations

and the voltage transfer function of the second-order network in Fig. 3-23a, shown in Fig. 3-28. Then solve for the capacitance value x_1 and the inductance value x_2 as

$$x_1 = \frac{a_1(1+R_s)}{2a_2 R_s}\left[1 \pm \sqrt{1 - \frac{4a_2}{a_1^2(1+R_s)}}\right]$$
$$x_2 = \frac{1+R_s}{a_2 x_1 R_s}$$
(3.4-9)

For the Butterworth case, $a_2 = 1$, $a_1 = \sqrt{2}$ from (3.4-3), and x_1 reduces to

$$x_1 = \frac{1+R_s}{\sqrt{2}R_s}\left[1 \pm \sqrt{\frac{R_s - 1}{R_s + 1}}\right]$$
(3.4-10)

which is plotted in Fig. 3-28. For $R_s = 1$, there is one value for x_1 but for $R_s \neq 1$,

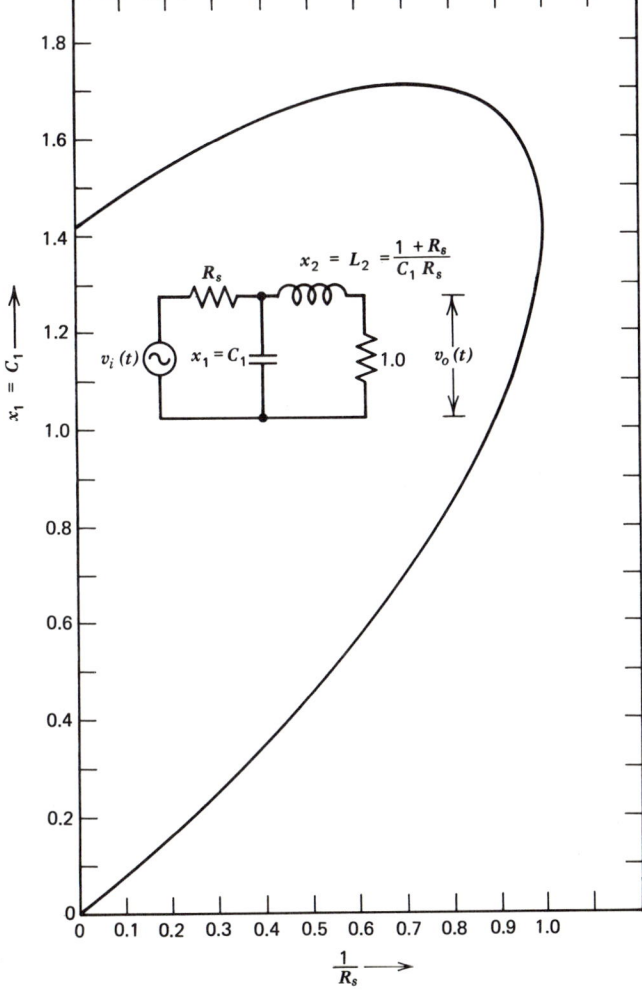

Figure 3-28. Element values for the $n = 2$ Butterworth response as a function of R_s.

there are two possible values for x_1. However, there is only one arrangement of the second order Butterworth reflection zeros [29].

In summary, the Butterworth filter has a tolerably good magnitude response except near the cutoff frequency. The group delay is reasonably constant from $\omega = 0$ to $\omega = 0.5$ but is nonconstant thereafter. The time-domain response overshoots are not compatible with good pulse transmission, and this filter is rarely used for such applications. It is used mainly when attenuation requirements are modest, where constant group delay and low time-domain overshoots are not needed, such as in the transmission of speech. The attractiveness of the Butterworth filter lies in its low sensitivity to element change of the various responses and its mathematical simplicity.

Example 3-9. We now determine the essential characteristics of the third-order Butterworth filter to illustrate use of the formulas and curves introduced here. From (3.4-2) the magnitude response is

$$|H(j\omega)| = \frac{1}{\sqrt{1+\omega^6}} \tag{3.4-11}$$

and the attenuation function, from (2.5-41) appears in Fig. 3-24 ($n = 3$). The frequency variable in Figs. 3-24 and 3-25 is Ω because these responses are normalized. The transfer-function determination has been demonstrated in Example 2-27. The poles of $H(s)$, from (3.4-4), are

$$\begin{aligned}
\sigma_1 &= -\sin\frac{\pi}{6} = -\frac{1}{2} & \omega_1 &= \cos\frac{\pi}{6} = \frac{\sqrt{3}}{2} \\
\sigma_2 &= -\sin\frac{\pi}{2} = -1 & \omega_2 &= \cos\frac{\pi}{2} = 0 \\
\sigma_3 &= -\sin\frac{5\pi}{6} = -\frac{1}{2} & \omega_3 &= \cos\frac{5\pi}{6} = -\frac{\sqrt{3}}{2}
\end{aligned} \tag{3.4-12}$$

which agree with the poles found in Example 2-27. They are shown in Fig. 3-29 and lie on the unit circle. The group delay function in Fig. 3-25 ($n = 3$) is obtained from (3.4-5) as

$$D_L(\omega) = \frac{2\omega^4 + \omega^2 + 2}{1 + \omega^6} \tag{3.4-13}$$

The zero frequency delay is 2, identical with the value obtained from (3.4-6). The impulse response, determined by the method in Example 2-24, is

$$h_L(t) = \left[e^{-t} - \frac{2}{\sqrt{3}} e^{-t/2} \cos\left(\frac{\sqrt{3}}{2} t + \frac{\pi}{6}\right) \right] u_{-1}(t) \tag{3.4-14}$$

and appears in Fig. 3-26 ($n = 3$). The time variable in Figs. 3-26 and 3-27 is t_N because these are normalized responses. The step response, obtained by finding the inverse transform of $H(s)/s$ or integrating the impulse response from 0 to t, is

$$g_L(t) = \left[1 - e^{-t} - \frac{2}{\sqrt{3}} e^{-t/2} \sin\frac{\sqrt{3}}{2} t \right] u_{-1}(t) \tag{3.4-15}$$

and appears in Fig. 3-27 ($n = 3$). The element values of the networks in Fig. 3-23

Rectangular Magnitude Response Approximations

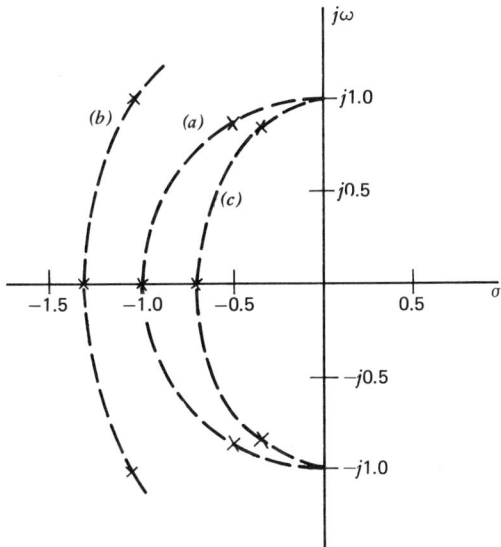

Figure 3-29. Poles of $n = 3$ (*a*) Butterworth, (*b*) Bessel, and (*c*) 0.1-dB ripple Chebyshev transfer functions.

with $R_s = 1$, from (3.4-7), are

$$x_1 = 2 \sin \frac{\pi}{6} = 1$$

$$x_2 = 2 \sin \frac{\pi}{2} = 2 \qquad (3.4\text{-}16)$$

$$x_3 = 2 \sin \frac{5\pi}{6} = 1$$

This completes the design of the third-order Butterworth filter and the calculation of the important responses that characterize it.

3.4.2 Equiripple (Chebyshev)

The equiripple passband approximation to unity is obtained by replacing $\psi_n(\omega)$ in (3.4-1) by the Chebyshev polynomials in (3.2-33) as

$$|H(j\omega)| = \frac{1}{\sqrt{1 + \epsilon^2 C_n^2(\omega)}} \qquad (3.4\text{-}17)$$

The value of ϵ determines the ripple height. Now all passband frequencies are weighted evenly, and the maximum passband deviation between unity and $|H(j\omega)|$ is minimized. The order n indicates the number of half-cycles (maximum to minimum) in the passband. Typical magnitude and attenuation responses are shown in Fig. 3-30 for $n = 5$, where the numbers on the response indicate each of the five half-cycles.

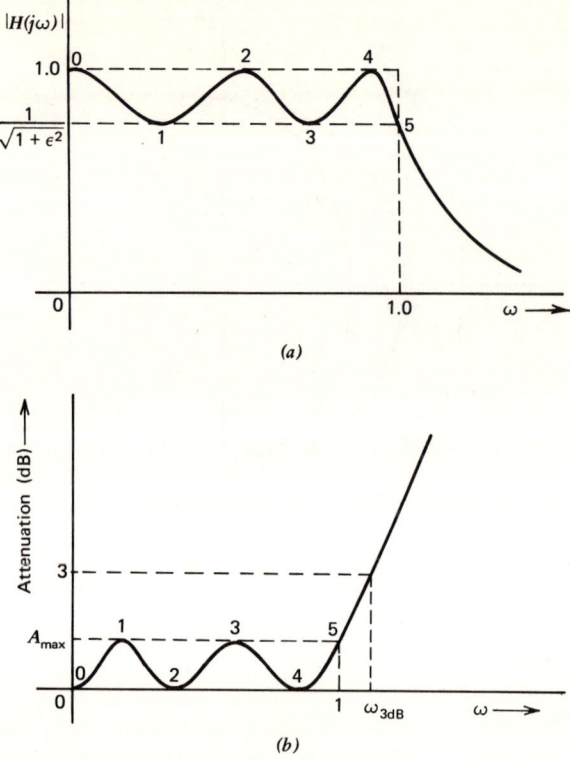

Figure 3-30. (a) Magnitude and (b) attenuation responses for $n = 5$ LP Chebyshev approximation.

The magnitude response at $\omega = 0$ is

$$|H(j0)| = \frac{1}{\sqrt{1 + \epsilon^2 C_n^2(0)}} \tag{3.4-18}$$

For n odd $C_n(0) = 0$ and $|H(j0)| = 1$. However for n even, $C_n(0) = \pm 1$ and

$$|H(j0)| = \frac{1}{\sqrt{1 + \epsilon^2}} \tag{3.4-19}$$

the maximum passband deviation. Because the magnitude response is not unity at $\omega = 0$ for n even, the Chebyshev response for these cases cannot be realized by equally terminated filters. The ripple attenuation from (3.4-19) is

$$A_{\max} = 10 \log (1 + \epsilon^2) \text{ decibels} \tag{3.4-20}$$

Thus ϵ, in terms of ripple attenuation, is

$$\epsilon = \sqrt{10^{0.1 A_{\max}} - 1} \tag{3.4-21}$$

The transfer-function poles $\sigma_k + j\omega_k$ are expressible as

$$\begin{aligned} \sigma_k &= -\sinh \xi \sin (2k - 1)\frac{\pi}{2n} \\ \omega_k &= \cosh \xi \cos (2k - 1)\frac{\pi}{2n} \end{aligned} \quad k = 1, 2, \ldots, n \tag{3.4-22}$$

Rectangular Magnitude Response Approximations

where

$$\xi = \frac{1}{n}\sinh^{-1}\frac{1}{\epsilon} \quad (3.4\text{-}23)$$

Divide σ_k by $\sinh\xi$, ω_k by $\cosh\xi$, square these expressions, and then add to give

$$\frac{\sigma_k^2}{\sinh^2\xi} + \frac{\omega_k^2}{\cosh^2\xi} = 1 \quad (3.4\text{-}24)$$

This is the equation of the ellipse that contains the transfer-function poles. The major axis of this ellipse lies along the $j\omega$ axis and the minor axis lies along the σ axis. This characteristic is illustrated in Example 3-10.

The frequency $\omega = 1$ corresponds to the edge of the ripple band rather than the customary 3-dB frequency. Since filter specifications are usually specified in terms of ω_{3dB}, however, we now obtain its value. Because the maximum response is unity,

$$|H(j\omega_{3dB})|^2 = \frac{1}{2} \quad (3.4\text{-}25)$$

requiring from (3.4-17) that

$$\epsilon^2 C_n^2(\omega_{3dB}) = 1 \quad (3.4\text{-}26)$$

Substitute $C_n(\omega)$ from (3.2-33), remembering that $\omega_{3dB} > 1$, into (3.4-26) and solve for ω_{3dB}, as

$$\omega_{3dB} = \cosh\left[\frac{1}{n}\cosh^{-1}\frac{1}{\epsilon}\right] \quad (3.4\text{-}27)$$

which is tabulated in Table 3-1 for selected values of passband ripple attenuation.

The normalized poles are obtained by dividing the values in (3.4-22) by ω_{3dB}, yielding

$$\begin{aligned}\sigma_k' &= -\frac{\sinh\left[1/n\,\sinh^{-1}1/\epsilon\right]}{\omega_{3dB}}\sin(2k-1)\frac{\pi}{2n} \\ \omega_k' &= \frac{\cosh\left[1/n\,\sinh^{-1}1/\epsilon\right]}{\omega_{3dB}}\cos(2k-1)\frac{\pi}{2n}\end{aligned} \quad (3.4\text{-}28)$$

As ϵ approaches zero (no passband ripples), the Chebyshev pole positions approach the Butterworth poles in (3.4-4); hence the Chebyshev response approaches the Butterworth response as $\epsilon \to 0$. This fact is not apparent from (3.4-17) because there $\omega = 1$ corresponds to the ripple bandwidth.

Closed-form expressions for the element values g_k corresponding to unity-ripple bandwidth are [3]

$$g_n = \frac{2a_n}{\eta}, \quad g_{n-k} = \frac{4a_{k+1}a_k}{b_k g_{n-k+1}} \quad k = 1, 2, \ldots, n-1 \quad (3.4\text{-}29)$$

where

$$\begin{aligned}\beta_1 &= \ln\left[\coth\frac{A_{max}}{40\log e}\right] \\ \eta &= \sinh\left(\frac{\beta_1}{2n}\right) \\ b_k &= \eta^2 + \sin^2\frac{k\pi}{n} \quad k = 1, 2, \ldots, n-1 \\ a_k &= \sin(2k-1)\frac{\pi}{2n} \quad k = 1, 2, \ldots, n\end{aligned} \quad (3.4\text{-}30)$$

Table 3-1 **3-dB Frequency for Chebyshev Responses**

ω_{3dB} (ratio of 3-dB bandwidth to ripple bandwidth)

A_{max} (dB)	ϵ	$n=2$	$n=3$	$n=4$	$n=5$	$n=6$	$n=7$	$n=8$	$n=9$	$n=10$
0.00001	0.001517	18.1667	5.52781	3.0957	2.22243	1.80663	1.5745	1.43103	1.33585	1.26934
0.0001	0.004799	10.2322	3.80181	2.36983	1.82053	1.54949	1.39491	1.29804	1.23318	1.18755
0.001	0.015175	5.78348	2.6427	1.84167	1.51559	1.34957	1.25313	1.19199	1.15071	1.12151
0.005	0.03394	3.90278	2.07401	1.56569	1.35109	1.23976	1.17437	1.13263	1.10432	1.08423
0.01	0.048013	3.30361	1.87718	1.4669	1.29122	1.19941	1.14527	1.11061	1.08706	1.07033
0.05	0.107608	2.26859	1.5121	1.2784	1.17537	1.12074	1.08824	1.06733	1.05308	1.04292
0.1	0.15262	1.94322	1.38899	1.2131	1.13472	1.09293	1.068	1.05193	1.04095	1.03313
0.2	0.217091	1.67427	1.28346	1.15635	1.09915	1.06852	1.05019	1.03835	1.03026	1.02449
0.25	0.243421	1.59814	1.25289	1.13977	1.08872	1.06134	1.04495	1.03435	1.02711	1.02194
0.3	0.267431	1.53936	1.22906	1.1268	1.08055	1.05571	1.04083	1.03121	1.02464	1.01994
0.4	0.310609	1.45249	1.19348	1.10736	1.06828	1.04725	1.03464	1.02649	1.02091	1.01693
0.5	0.349311	1.38974	1.16749	1.0931	1.05926	1.04103	1.03009	1.02301	1.01817	1.01471
0.6	0.384907	1.34127	1.14724	1.08196	1.0522	1.03616	1.02652	1.02028	1.01602	1.01297
0.7	0.418208	1.30214	1.13078	1.07288	1.04644	1.03218	1.02361	1.01806	1.01426	1.01154
0.8	0.449738	1.26955	1.11699	1.06526	1.0416	1.02883	1.02116	1.01618	1.01278	1.01035
0.9	0.479863	1.24176	1.10517	1.05872	1.03745	1.02596	1.01905	1.01457	1.01151	1.00932

1	0.508847	1.21763	1.09487	1.053	1.03381	1.02344	1.01721	1.01316	1.0104	1.00842
1.1	0.536889	1.19637	1.08576	1.04794	1.0306	1.02121	1.01557	1.01191	1.00941	1.00762
1.2	0.564142	1.17741	1.07761	1.04341	1.02771	1.01922	1.01411	1.01079	1.00853	1.0069
1.3	0.590731	1.16035	1.07025	1.03931	1.0251	1.01741	1.01278	1.00978	1.00773	1.00626
1.4	0.616753	1.14486	1.06355	1.03558	1.02272	1.01576	1.01157	1.00886	1.007	1.00566
1.5	0.642291	1.13069	1.0574	1.03216	1.02054	1.01425	1.01046	1.00801	1.00632	1.00512
1.6	0.667413	1.11766	1.05174	1.02899	1.01852	1.01285	1.00944	1.00722	1.00571	1.00462
1.7	0.692177	1.1056	1.04649	1.02606	1.01665	1.01156	1.00849	1.00649	1.00513	1.00415
1.8	0.716632	1.0944	1.0416	1.02333	1.01491	1.01035	1.0076	1.00582	1.00459	1.00372
1.9	0.740822	1.08394	1.03702	1.02077	1.01328	1.00921	1.00677	1.00518	1.00409	1.00331
2	0.764783	1.07414	1.03273	1.01837	1.01174	1.00815	1.00599	1.00458	1.00362	1.00293
2.1	0.788549	1.06493	1.02869	1.0161	1.0103	1.00715	1.00525	1.00402	1.00317	1.00257
2.2	0.81215	1.05624	1.02487	1.01396	1.00893	1.0062	1.00455	1.00348	1.00275	1.00223
2.3	0.83561	1.04803	1.02125	1.01194	1.00763	1.0053	1.00389	1.00298	1.00235	1.00191
2.4	0.858953	1.04024	1.01782	1.01001	1.0064	1.00445	1.00327	1.0025	1.00197	1.0016
2.5	0.882201	1.03284	1.01455	1.00818	1.00523	1.00363	1.00267	1.00204	1.00161	1.00131
2.6	0.905373	1.0258	1.01144	1.00643	1.00411	1.00286	1.0021	1.00161	1.00127	1.00103
2.7	0.928486	1.01907	1.00846	1.00476	1.00304	1.00211	1.00155	1.00119	1.00094	1.00076
2.8	0.951557	1.01265	1.00561	1.00316	1.00202	1.0014	1.00103	1.00079	1.00062	1.0005
2.9	0.9746	1.00649	1.00288	1.00162	1.00104	1.00072	1.00053	1.00041	1.00032	1.00026
3	0.997628	1.00059	1.00026	1.00015	1.0001	1.00007	1.00005	1.00004	1.00003	1.00002

Note from (3.4-7) that $2a_k$ is the Butterworth element value. The value of R_s in Fig. 3-23 is

$$R_s = \begin{cases} 1 & n \text{ odd} \\ \tanh^2\left(\dfrac{\beta_1}{4}\right) & n \text{ even} \end{cases} \quad (3.4\text{-}31)$$

The element value x_i in Fig. 3-23 corresponding to 3-dB attenuation at $\omega = 1$ is obtained by multiplying g_i by ω_{3dB} from (3.4-27). Element values and the major responses are given in Ref. 30 for 0.01, 0.1, and 0.5 dB passband ripple.

Consideration of the magnitude response throughout the entire passband has resulted in better characteristics near the cutoff frequency. The response in the stopband decreases very rapidly and this response is an excellent approximation to the rectangular response, especially for the higher orders. The Chebyshev response has the lowest complexity for yielding a prescribed maximum deviation in the passband and the fastest possible rate of attenuation near the cutoff frequency. In this sense it is optimum.

The abrupt attenuation change at the cutoff frequency causes pronounced ringing in the transient responses and nonconstant group delay, especially near the cutoff frequency. For this reason the Chebyshev filter is not recommended for pulse applications.

Example 3-10. The important parameters of the third-order Chebyshev filter with 0.1-dB passband ripple are now obtained for $\omega_{3dB} = 1$. From (3.4-21)

$$\epsilon = \sqrt{10^{0.01} - 1} = 0.15262 \quad (3.4\text{-}32)$$

and from (3.4-27), or Table 3-1,

$$\omega_{3dB} = \cosh\left(\frac{1}{3}\cosh^{-1}\frac{1}{\epsilon}\right) = 1.389 \quad (3.4\text{-}33)$$

The normalized transfer-function poles, from (3.4-28), are

$$\begin{aligned} \sigma_1' &= -0.34896 & \omega_1' &= 0.86836 \\ \sigma_2' &= -0.69792 & \omega_2' &= 0 \\ \sigma_3' &= -0.34896 & \omega_3' &= -0.86836 \end{aligned} \quad (3.4\text{-}34)$$

and they are shown in Fig. 3-29. The Butterworth poles lie on the unit circle, and the Chebyshev poles lie on an ellipse, as predicted by (3.4-24). Then from (3.4-30)

$$\beta_1 = \ln\left[\coth\left(\frac{0.1}{40 \log e}\right)\right] = 5.1574$$

$$\eta = 0.969406$$

$$b_1 = b_2 = \eta^2 + \sin^2\frac{\pi}{3} = 1.6897 \quad (3.4\text{-}35)$$

$$a_1 = a_3 = \sin\frac{\pi}{6} = \frac{1}{2}$$

$$a_2 = \sin\frac{\pi}{2} = 1$$

Rectangular Magnitude Response Approximations

The element values are obtained by multiplying the g's in (3.4-29) by ω_{3dB} as

$$x_3 = \frac{2a_3}{\eta} \omega_{3dB} = 1.4328$$

$$x_2 = \frac{4a_1 a_2}{b_1 g_3} \omega_{3dB} = 1.5937 \qquad (3.4\text{-}36)$$

$$x_1 = \frac{4a_2 a_3}{b_2 g_2} \omega_{3dB} = 1.4328$$

3.4.3 Legendre

Papoulis [19] and Fukada [8] independently derived the magnitude response having the greatest possible cutoff rate subject to the constraint that the passband response must decrease monotonically, that is, no passband ripples. The polynomials $\psi_n^2(\omega)$ that accomplish this are designated $L_n(\omega^2)$ and are related to the Legendre polynomials of the first kind $P_k(\omega)$. Hence the resulting filter is known as a Legendre or class L filter.

The magnitude response, from (3.4-1), is

$$|H(j\omega)| = \frac{1}{\sqrt{1 + \epsilon^2 L_n(\omega^2)}} \qquad (3.4\text{-}37)$$

with ϵ usually set to unity. The polynomials $L_n(\omega^2)$ and the poles of $H(s)$ are given in Ref. 10, and the four major responses and element values can be found in Ref, 30. This filter has achieved little popularity because the Chebyshev filter is superior in both the passband and stopband. However applications may arise in which passband ripple is intolerable, and stopband attenuation must be greater than that of the Butterworth filter. The Legendre filter satisfies the requirement.

3.4.4 Least-Squares

The least-squares magnitude response [10] is achieved by requiring $\psi_n(\omega)$ in (3.4-1) to approximate zero in the least-squares sense over the passband, subject to the constraint that $\psi_n(1) = 1$. If $\epsilon = 1$, the 3-dB radian frequency is then unity. The polynomial takes the form

$$\psi_n(\omega) = \begin{cases} a_0 + a_2 \omega^2 + \cdots + a_n \omega^n & n \text{ even} \\ a_1 \omega + a_3 \omega^3 + \cdots + a_n \omega^n & n \text{ odd} \end{cases} \qquad (3.4\text{-}38)$$

The a's are selected to minimize the integral [see (3.2-36) with $w(x) = 1$]

$$\varepsilon = \int_0^1 [\psi_n(\omega) - 0]^2 \, d\omega = \int_0^1 \psi_n^2(\omega) \, d\omega \qquad (3.4\text{-}39)$$

subject to $\psi_n(1) = 1$. The attenuation curves and poles of $H(s)$ for this approximation are given in Ref. 10.

The maximum passband ripple is approximately 0.1 dB but can be increased or decreased by choosing a value for ϵ other than unity. This filter, like the Legendre filter, has achieved little popularity because the Chebyshev filter for the same passband ripple gives slightly increased stopband attenuation. However, characteristic of the least-squares approximation, the passband ripple increases as the cutoff frequency is approached; hence its low-frequency behavior is superior to that of a Chebyshev response with the same passband ripple. This results in improved filtering for signals that concentrate most of their energies at low frequencies.

Example 3-11. The third-order response is now derived to illustrate the approximation procedure. From (3.4-38)

$$\psi_3(\omega) = a_1\omega + a_3\omega^3 \qquad (3.4\text{-}40)$$

The 3-dB constraint $\psi_3(1) = 1$ requires that $a_1 = 1 - a_3$, and from (3.4-39)

$$\varepsilon = \int_0^1 [(1-a_3)\omega + a_3\omega^3]^2 \, d\omega \qquad (3.4\text{-}41)$$

This integral is minimized when

$$\frac{d\varepsilon}{da_3} = 2\int_0^1 [(1-a_3)\omega + a_3\omega^3][\omega^3 - \omega] \, d\omega = 0 \qquad (3.4\text{-}42)$$

yielding $a_3 = 7/4$ and $a_1 = -3/4$. The magnitude function squared is then

$$|H(j\omega)|^2 = \frac{1}{1+\psi_3^2(\omega)} = \frac{1}{1+(\tfrac{7}{4}\omega^3 - \tfrac{3}{4}\omega)^2} \qquad (3.4\text{-}43)$$

and the transfer function, obtained by the method in Example 2-27, is

$$H(s) = \frac{0.5714}{s^3 + 1.3212s^2 + 1.3014s + 0.5714} \qquad (3.4\text{-}44)$$

whose poles are -0.66062 and $-0.33031 \pm j0.86942$.

3.5 CONSTANT GROUP DELAY APPROXIMATIONS

The constant group delay, which corresponds to linear phase, can also be approximated using the maximally flat, equiripple, and least-squares criteria. The delay function for the nth-order all-pole filter, obtained from (2.6-10), is

$$D_l(\omega) = -\sum_{i=1}^{n} \frac{p_i}{\omega^2 + p_i^2} \qquad (3.5\text{-}1)$$

Examples 3-7 and 3-8 have already considered these approximations for first-order systems, and we now turn to the general cases. The area under the delay curve—that is, the integral of $D_l(\omega)$—is $n\pi/2$, a constant independent of the pole locations.

3.5.1 Maximally Flat (Bessel) Delay

The first English language account of work on the maximally flat delay was reported by Thomson [27]; an earlier publication had appeared in Japan [13]. This

Constant Group Delay Approximations

filter is often called a Thomson filter or a Bessel filter. A direct method of obtaining maximally flat delay at $\omega = 0$ is to express $D_l(\omega)$ as a rational function in ω^2 and then equate like powers of ω^2 in the numerator and denominator, as discussed in Section 3.2.1. Then the first $2n - 1$ derivatives of $D_l(\omega)$ will be zero at $\omega = 0$.

Example 3-12. This technique is now illustrated for the third-order transfer function

$$H(s) = \frac{a_3}{s^3 + a_1 s^2 + a_2 s + a_3} \tag{3.5-2}$$

The delay function, in the form of (3.2-15), is

$$D_l(\omega) = \frac{a_2}{a_3} \frac{1 + \left(\frac{a_1}{a_3} - \frac{3}{a_2}\right)\omega^2 + \frac{a_1}{a_2 a_3}\omega^4}{1 + \left(\frac{a_1^2}{a_3^2} - \frac{2a_1}{a_3}\right)\omega^2 + \left(\frac{a_1^2 - 2a_2}{a_3^2}\right)\omega^4 + \frac{1}{a_3^2}\omega^6} \tag{3.5-3}$$

For maximal flatness,

$$\frac{a_1}{a_3} - \frac{3}{a_2} = \frac{a_1^2}{a_3^2} - \frac{2a_1}{a_3}$$

$$\frac{a_1}{a_2 a_3} = \frac{a_1^2 - 2a_2}{a_3^2} \tag{3.5-4}$$

There are two equations and three unknowns; thus as is customary the zero frequency delay is set to unity, implying $a_2 = a_3$. Then (3.5-4) is solved yielding the two sets of solutions

$$\begin{aligned} a_2 = a_3 = 0 & \quad a_1 = 1 \\ a_2 = a_3 = 15 & \quad a_1 = 6 \end{aligned} \tag{3.5-5}$$

The first set is of no practical value, and the second set yields the transfer function

$$H(s) = \frac{15}{s^3 + 6s^2 + 15s + 15} \tag{3.5-6}$$

and corresponding delay function

$$D_l(\omega) = \frac{225 + 45\omega^2 + 6\omega^4}{225 + 45\omega^2 + 6\omega^4 + \omega^6} \tag{3.5-7}$$

This method is straightforward, but the resulting nonlinear simultaneous equations for higher-order systems are difficult to solve analytically.

Storch [22] obtained the same results by cleverly approximating the ideal unit delay function

$$H(s) = e^{-s} \tag{3.5-8}$$

by a power series and terminating it after n terms. First he expressed $H(s)$ as

$$H(s) = \frac{1}{e^s} = \frac{1}{\sinh s + \cosh s} = \frac{1/\sinh s}{1 + \coth s} \tag{3.5-9}$$

and then he truncated the continued fraction expansion for coth s

$$\coth s = \frac{1}{s} + \cfrac{1}{\cfrac{3}{s} + \cfrac{1}{\cfrac{5}{s} + \cfrac{1}{\cfrac{7}{s} + \cdots}}} \qquad (3.5\text{-}10)$$

after n terms. The result is a rational function in s where the numerator is associated with $\cosh s$ and the denominator is associated with $\sinh s$. Then the sum of the numerator and denominator is the desired denominator polynomial for $H(s)$ in (3.5-9).

Example 3-13. Storch's method is now applied to the third-order system. From (3.5-10) the expansion, truncated after three terms, is

$$\coth s \approx \frac{1}{s} + \cfrac{1}{\cfrac{3}{s} + \cfrac{1}{\cfrac{5}{s}}} = \frac{6s^2 + 15}{s^3 + 15s} \qquad (3.5\text{-}11)$$

The sum of numerator and denominator is the same as the denominator polynomial in (3.5-6), verifying the technique. The numerator constant in (3.5-6) is selected for unity output at $\omega = 0$.

Storch's method yields maximally flat group delay at $\omega = 0$ no matter where the power series is truncated. Furthermore $B_n(s)$, the denominator polynomial of $H(s)$, is related to a class of Bessel polynomials, and its value can be determined from the recusion formula

$$B_n = (2n - 1)B_{n-1} + s^2 B_{n-2} \qquad (3.5\text{-}12)$$

For example, $B_0 = 1$ and $B_1 = s + 1$ are easily found by terminating (3.5-10) at $n = 0$ and $n = 1$, respectively. Then

$$B_2 = 3B_1 + s^2 B_0 = s^2 + 3s + 3 \qquad (3.5\text{-}13)$$

and

$$B_3 = 5B_2 + s^2 B_1 = s^3 + 6s^2 + 15s + 15 \qquad (3.5\text{-}14)$$

which agrees with (3.5-6).

There is another approach to obtain this response. In (3.5-1), $D_l(\omega)$ expressed as a power series in ω^2, is

$$D_l(\omega) = -\sum_{i=1}^{n} \left[\frac{1}{p_i} - \frac{\omega^2}{p_i^3} + \frac{\omega^4}{p_i^5} - \cdots \right] = -\sum_{i=1}^{n} \frac{1}{p_i} + \omega^2 \sum_{i=1}^{n} \frac{1}{p_i^3} - \cdots \qquad (3.5\text{-}15)$$

where coefficients of powers of ω^2 are proportional to the various derivatives of

Constant Group Delay Approximations

$D_t(\omega)$ at $\omega = 0$. Thus maximally flat delay of an nth-order LP system requires that

$$\sum_{i=1}^{n} p_i^{-3} = 0$$

$$\sum_{i=1}^{n} p_i^{-5} = 0 \qquad (3.5\text{-}16)$$

$$\vdots$$

$$\sum_{i=1}^{n} p_i^{-(2n-1)} = 0$$

This is the same set of equations that Huggins [9] obtained when he considered the semi-invariants associated with pole and zero distributions, defined as $\sum_{i=1}^{n} p_i^{-k}$. Equating the even semi-invariants (k even) to zero results in the maximally flat magnitude response while equating the odd semi-invariants (k odd) to zero yields the maximally flat delay response. It is noteworthy that Huggins' publication date is contemporary with Thomson's.

Storch's technique is computationally simpler than the other two, but Huggins' discussion is more general, and it is interesting that one of his special cases is indeed the maximally flat delay response.

The transfer functions derived here are normalized for unit delay at $\omega = 0$, but the normalization in Refs. 10 and 30 is for unity 3-dB bandwidth. Consistent with the constant delay analysis in Section 3.1.12, the price for this good delay characteristic is a slow rate of stopband attenuation near the cutoff frequency. As n approaches infinity, the magnitude response approaches the Gaussian magnitude response.

Comparing the time-domain responses of the Bessel and Butterworth filters shows lower overshoots for the former. The step response overshoot and rise time, defined in Section 3.1.3, are compared in Table 3-2. The rise time is further explored in Section 3.8.

Table 3-2 **Comparison of Butterworth and Bessel Step Response Characteristics**

Order	% Overshoot		Rise Time (sec)	
	Butterworth	Bessel	Butterworth	Bessel
2	4.3	0.43	2.193	1.947
3	8.1	0.75	2.472	2.151
4	10.8	0.84	2.621	2.220
5	12.8	0.77	2.728	2.236

The poles for the case of $n = 3$, ($\omega_c = 1$), -1.32268 and $-1.04741 \pm j0.99926$, are given in Fig. 3-29 for comparison with the Butterworth and Chebyshev transfer-function poles. The relative positions there are typical for any order n. The Chebyshev poles lie on an ellipse internal to the unit circle, whereas the Bessel poles lie on an ellipselike path outside the unit circle. The unit circle is occupied by the Butterworth poles.

3.5.2 Equiripple (Chebyshev) Delay

Similar to the Butterworth response, the Bessel response emphasizes only $\omega = 0$; thus it is not surprising that the Chebyshev approximation is used to realize equiripple delay error. Unfortunately there are no preexisting polynomials that give the desired result. Since determination of the optimum pole locations is computationally difficult for $n > 2$, an electronic computer is used to accomplish this task.

The results published in Refs. 10 and 30 are approximations to a linear phase with phase deviations from linearity of 0.05° and 0.5° for $n = 2$ to $n = 10$. The linear phase approximation was chosen because phase deviation has more physical meaning to an engineer than does group delay deviation. However the resulting delay, for all practical purposes, is also equiripple.

Because of the delay ripple, the stopband attenuation is greater than the corresponding Bessel filter attenuation, but so also are the time-domain overshoots. When the magnitude response steadily decreases with increasing frequency, as it does for the constant-delay type LP systems (see Section 3.1.12), the delay error at the higher frequencies (where the attenuation is greater) has less effect on the time response than the delay error at low frequencies. From this point of view equiripple delay is the worst offender because delay error occurs at the lower frequencies where the attenuation is small. However, there are many applications in which the maximum passband delay deviation is the principal requirement, and for these cases the equiripple delay responses are the optimum solution.

3.5.3 Least-Squares Delay

The least-squares approximation of a constant passband group delay T requires that the integral

$$\varepsilon = \int_0^1 [T - D_l(\omega)]^2 \, d\omega \tag{3.5-17}$$

be minimized subject to the constraint that the attenuation at $\omega = 1$ is 3 dB. The value of T for minimum error, after solving

$$\frac{d\varepsilon}{dT} = 0 \tag{3.5-18}$$

is

$$T = \int_0^1 D_l(\omega) \, d\omega \tag{3.5-19}$$

the average value of the filter's passband delay. For this value of T, ε reduces to

$$\varepsilon = \int_0^1 D_l^2(\omega) \, d\omega - \left[\int_0^1 D_l(\omega) \, d\omega\right]^2 \tag{3.5-20}$$

Computer minimization of ε for $n = 2, 3, 4$ yields the poles and values of T in Table 3-3. Also shown is the attenuation of the first impulse response overshoot

Ideal Time-Domain Approximations

Table 3-3 **Poles, Average Delay, and Impulse Response Overshoots for Least-Squares Delay Responses ($\omega_c = 1$)**

			First Overshoot of $h(t)$ (dB)		
n	Poles	T	Least Squares	0.05°	0.5°
2	$-0.889 \pm j0.694$	1.42	34.95	41.20	33.58
3	-1.043 $-0.853 \pm j1.074$	1.86	32.90	32.89	30.02
4	$-1.134 \pm j0.449$ $-0.851 \pm j1.344$	2.19	32.39	31.0	30.02

relative to the maximum value for the least-squares delay and the 0.05° and 0.5° equiripple phase responses. For $n > 4$, the least-squares delay rapidly approaches the maximally flat delay, for the latter is then essentially constant over the passband. Table 3-3 reveals that no significant time-domain improvements result from application of the least-squares criterion to the group delay. Characteristic of this approximation, the delay ripples increase as the cutoff is approached, and this feature may be beneficial if most of the signal's energy occurs at low frequencies, but like the least-squares magnitude approximation, this response has achieved little popularity.

3.6 IDEAL TIME-DOMAIN APPROXIMATIONS

The increased use of pulsed waveforms in electronics requires filters with low-overshoot transient responses. Figure 3-31a shows a typical series of RF pulse envelopes (carrier frequency is not shown) encountered in communication and radar systems. The essential information may be the phase or frequency of the RF signal, or its zero crossings. The response of the first pulse passing through an improperly designed filter is illustrated in Fig. 3-31b. Signal overshoots exist where the response should ideally be zero, and they may be appreciable at the time that the response of the second pulse appears at the filter output, indicated by the dotted curve. Consequently the two signals add, destroying the phase information of the carrier frequency. This undesirable condition is known as intersymbol interference.

In certain radar systems, a prescribed threshold (the dashed line in Fig. 3-31b) is specified, and any signal value greater than this threshold is considered a target return. Thus the first pulse overshoot indicates a target when none is present and creates a false alarm.

These are but two examples illustrating the desirability of filters with low-overshoot transient responses that rapidly decay with increasing time. The Gaussian time-domain response (no overshoots) is physically unrealizable as we showed in Section 3.1.5, but three realizable approximations to low-overshoot responses are now presented.

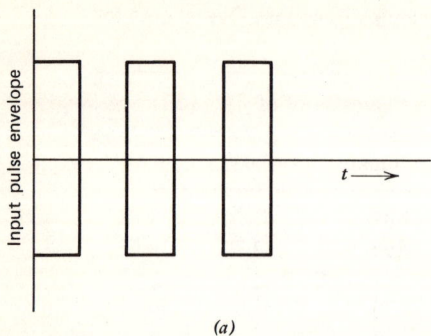

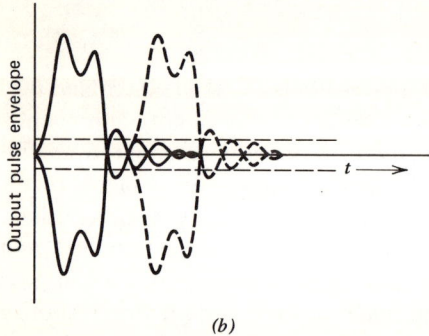

Figure 3-31. Pulse train illustrating intersymbol interference and false alarms.

3.6.1 Gaussian Response

An nth-order Taylor approximation to the Gaussian magnitude function

$$|H(j\omega)| = e^{-\omega^2/2} \tag{3.6-1}$$

is obtained by first squaring each side, replacing the exponential by its series expansion as

$$|H(j\omega)|^2 = \frac{1}{e^{\omega^2}} = \frac{1}{1 + \omega^2 + \frac{\omega^4}{2!} + \cdots + \frac{\omega^{2n}}{n!}} \tag{3.6-2}$$

and then truncating the denominator polynomial after n terms [5]. The transfer function is obtained in the usual way and the resulting poles are then normalized for unity 3-dB radian bandwidth.

Example 3-14. The second-order approximation to the Gaussian response is now illustrated. From (3.6-2)

$$|H(j\omega)|^2 = \frac{1}{1 + \omega^2 + \frac{\omega^4}{2}} \tag{3.6-3}$$

and

$$H(s)H(-s) = \frac{2}{s^4 - 2s^2 + 2} \tag{3.6-4}$$

Ideal Time-Domain Approximations

The left-half plane roots of the denominator polynomial are

$$-\left(\frac{\sqrt{2}+1}{2}\right)^{1/2} \pm j\left(\frac{\sqrt{2}-1}{2}\right)^{1/2}$$

yielding the transfer function

$$H(s) = \frac{\sqrt{2}}{s^2 + 2.19736s + \sqrt{2}} \qquad (3.6\text{-}5)$$

From (3.6-3), the 3-dB radian frequency is $(\sqrt{3}-1)^{1/2}$ and the poles, normalized for $\omega_{3dB} = 1$, are then

$$\text{poles} = -\left(\frac{\sqrt{2}+1}{2(\sqrt{3}-1)}\right)^{1/2} \pm j\left(\frac{\sqrt{2}-1}{2(\sqrt{3}-1)}\right)^{1/2} = -1.28411 \pm j0.531896 \qquad (3.6\text{-}6)$$

For these poles and unit magnitude response at $\omega = 0$, the impulse response, from Table 2-1, entry 4, is

$$h(t) = 3.632[e^{-1.28411t} \sin 0.531896t]u_{-1}(t) \qquad (3.6\text{-}7)$$

and the first overshoot is 65.9 dB down from the maximum value.

The Gaussian and Bessel filter performances are often compared. The passband delay of the former is not as flat as the Bessel delay, and the Bessel stopband attenuation is slightly greater than the Gaussian attenuation. However the transient overshoots of the Gaussian filter are less. Figure 3-32 shows the attenuation of the first impulse response overshoot for each filter. These results

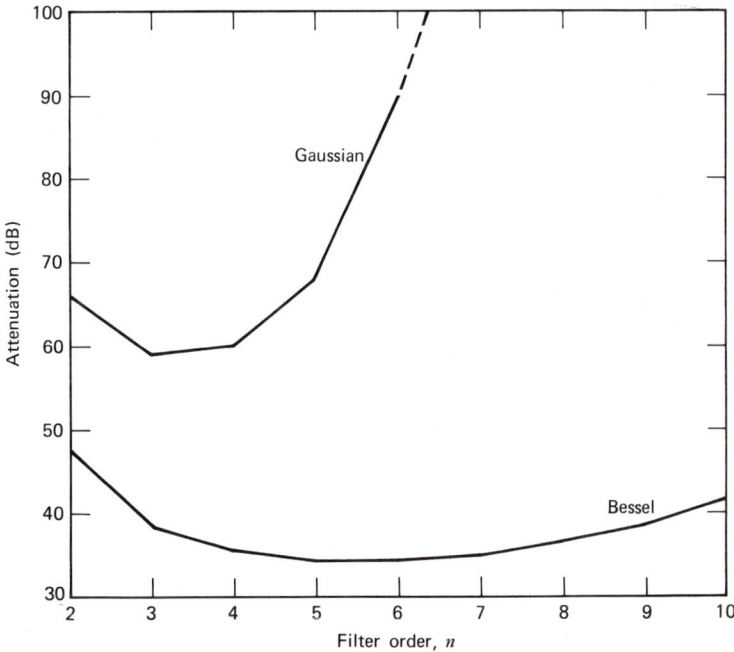

Figure 3-32. Attenuation of first impulse response overshoot for Bessel and Gaussian filters.

are expected, since each approximation considers different attributes of the response. The Bessel response realizes flat delay in the vicinity of $\omega = 0$, while the Gaussian approximation attempts to realize low time-domain overshoots. As the order n approaches infinity, both approximations approach the Gaussian response.

3.6.2 Equiripple Transient Response Overshoots

Jess and Schüssler [12] consider the minimization of a time-bandwidth product. The bandwidth ω_b is defined as

$$|H(j\omega)| \leq q_f |H(j0)| \quad \text{for} \quad \omega \geq \omega_b \tag{3.6-8}$$

as in Fig. 3-33a; q_f is a measure of the attenuation and may be specified as desired. The width of the impulse response in Fig. 3-33b is defined as

$$t_W = t_2 - t_1 \tag{3.6-9}$$

where

$$|h(t)| \leq q_t |h(t)|_{\max} \quad \text{for} \quad t \leq t_1 \text{ and } t \geq t_2 \tag{3.6-10}$$

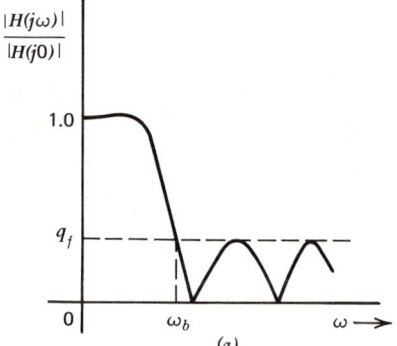

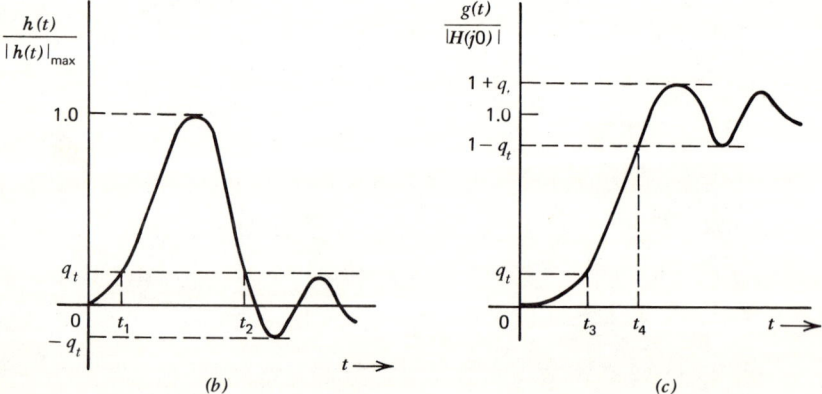

Figure 3-33. Time and frequency bandwidths used in describing filters with equiripple transient response overshoots.

Ideal Time-Domain Approximations

Similarly the rise time t_r of the step response $g(t)$, given in Fig. 3-33c, is defined as

$$t_r = t_4 - t_3 \tag{3.6-11}$$

where

$$g(t) \leq q_t |H(j0)| \quad \text{for} \quad t \leq t_3$$
$$|\Delta g(t)| = |g(t) - |H(j0)|| \leq q_t |H(j0)| \quad \text{for} \quad t \geq t_4 \tag{3.6-12}$$

Here q_t is a measure of the allowable step response overshoots and

$$H(j0) = \lim_{t \to \infty} g(t) \tag{3.6-13}$$

The time-bandwidth product is defined as

$$M_i = t_w \omega_b \tag{3.6-14}$$

for the impulse response, and

$$M_s = t_r \omega_b \tag{3.6-15}$$

for the step response. Jess and Schüssler found that each is minimized when the impulse response overshoots (for $t \geq t_2$), or $\Delta g(t)$ (for $t \geq t_4$), and $|H(j\omega)|$ (for $\omega \geq \omega_b$) simultaneously approximate zero in an equiripple manner. They did not prove that M is then a minimum, but the resulting responses are better than those of other pulse-forming systems when the same error criterion is used.

They considered transfer functions with finite zeros as well as the all-pole case, and samples of these are tabulated in Ref. 12 for $q_t = q_f = 0.01$. Other values of q_t and q_f are considered in Ref. 23 for a fifth-order transfer function with two finite zeros. Since transient overshoots 40 dB down are usually sufficient in practice, we present more information on filters with equiripple overshoots ($q_t = 0.01$).

For the all-pole case, Ref. 11 gives the poles for both the impulse and step response cases, while Ref. 10 gives the poles and attenuation for the former, normalized for $\omega_{3dB} = 1$. For both cases $q_f = 0.01$. The resulting stopband attenuation is slightly greater than the attenuation of other constant-delay type filters. Table 3-4 gives the element values for the equiripple impulse response filter for $n = 2$ to $n = 5$ and Fig. 3-34 presents the corresponding group delay responses.

An interesting feature of these delay curves is the stopband peak of the fifth-order curve. This seems to contradict the notion that low transient overshoots are associated with reasonably constant delay. Rather, these curves support the earlier statement here that if the delay ripples (in this case a large peak) occur where the attenuation is appreciable, their effect on the transient

Table 3-4 Ladder Network Element Values for Equiripple Impulse Response: Refer to Networks in Fig. 3-23 with $R_s = 1$ ohm

n	x_1	x_2	x_3	x_4	x_5
2	2.0964	0.6675			
3	2.3812	0.9322	0.4593		
4	2.1226	1.1377	0.7408	0.3947	
5	2.3511	1.0291	0.8899	0.6716	0.3518

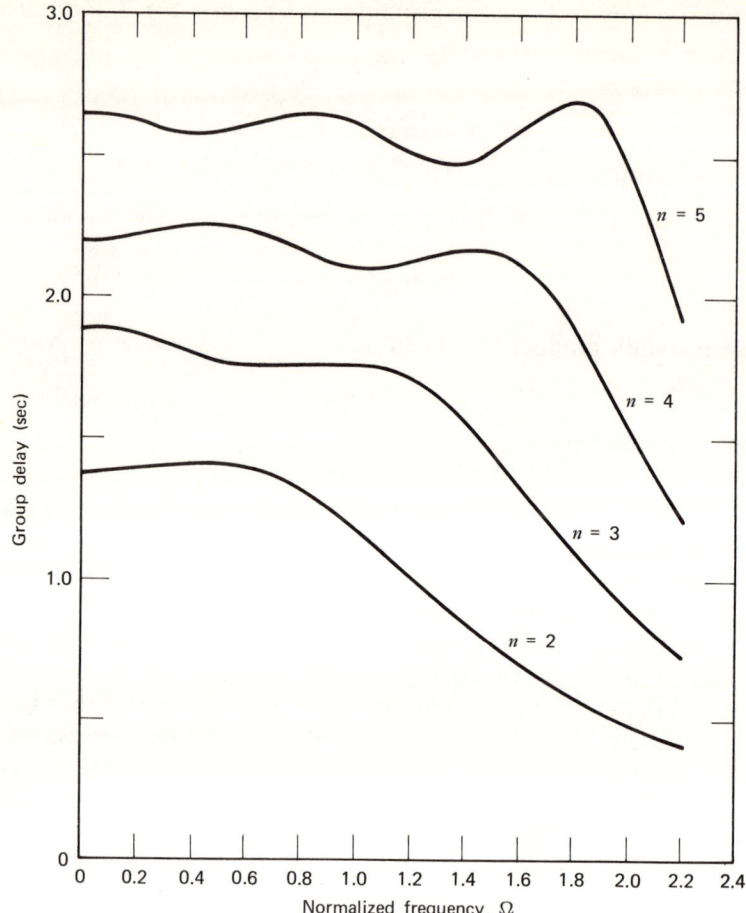

Figure 3-34. Group delay of filters with equiripple impulse response overshoots ($q_t = q_f = 0.01$).

response overshoots is minimized. This delay peak is even larger for the filter with equiripple step response overshoots, shown in Fig. 3-35 for $n = 4$, 5, and 6.

The Gaussian time-domain approximation was obtained by way of the frequency domain; the case just discussed is the first time-domain approximation encountered here. Remember that a prime consideration in resorting to the frequency domain is ease of obtaining specific attributes in the time waveform. Very often, however, this approach is not effective and the desired signal is achieved only by time-domain techniques. This subject is treated in Ref. 23, and we deal with it in Chapter 8.

3.6.3 Monotonic Step Response with Minimum Rise Time

In contrast to the equiripple case, the error criterion for monotonic step response requires that the step response have no overshoots or undershoots and that the rise time be minimum for a specified bandwidth. The filter that provides the fastest monotonic step response for a prescribed bandwidth is the prolate filter [24], so named because the filter transfer function is obtained from the autocorrelation function of an angular prolate spheroidal wave function.

Special Cases

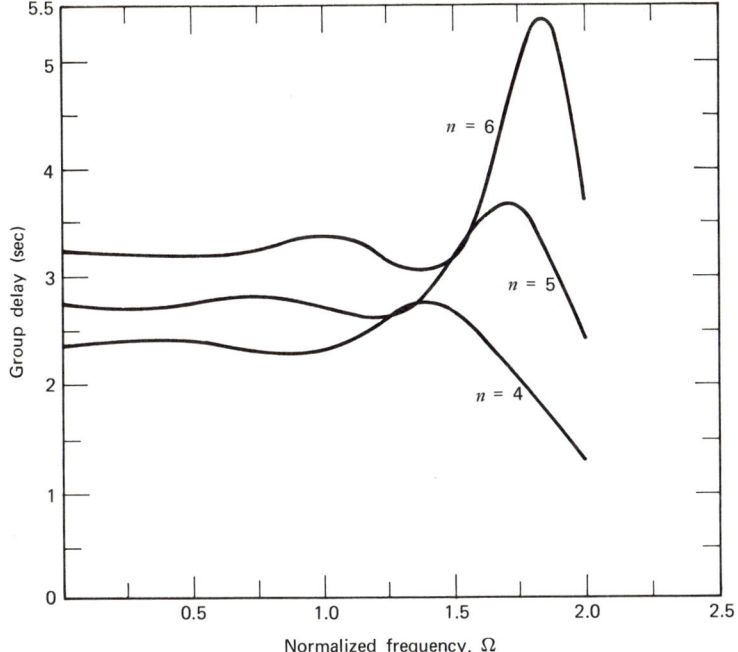

Figure 3-35. Group delay of filters with equiripple step response overshoots ($q_f = q_t = 0.01$).

Unfortunately the optimum magnitude function is band-limited and therefore must be approximated by a realizable function. The magnitude function with equiripple stopband behavior is suggested for this application. Forcing the step response to be monotonic results in rise times greater than those obtained for the equiripple overshoot step responses, hence the prolate filter is useful if no step response overshoots can be tolerated.

3.7 SPECIAL CASES

We now briefly examine filter responses derived from other considerations, including the transitional responses, which attempt to compromise between good delay and good attenuation; responses obtained by placing transfer-function poles on specific contours; the age-old synchronous response, and the magnitude response with equiripple passband and stopband behavior (Cauer filter).

3.7.1 Transitional Responses

Transitional responses are a compromise between constant passband delay and a rectangular magnitude response. From the ideal response analysis of Section 3.1.12 we know that constant passband delay requires a rounded passband attenuation function, whereas stopband behavior does not appreciably influence the delay. This fact is translated into two filter classes: the Butterworth–Thomson and the Gaussian–Chebyshev.

The first class [15] is obtained by mating the maximally flat magnitude and group delay approximations. The transfer-function poles incorporate a parameter m that is adjustable from 0 to 1. For $m = 0$, the Butterworth response is obtained and the Thomson response is found for $m = 1$. Intermediate values of m give responses whose characteristics lie somewhere between these two extremes. Transfer function poles for this class are given in Ref. 23 for $n = 2$ to $n = 10$, as well as 15 values of m for each value of n.

The second class [10] is a Gaussian approximation in a Chebyshev manner over a specified frequency interval after which the attenuation increases rapidly, characteristic of Chebyshev responses. Data for the Gaussian approximations to the 6 and to the 12-dB frequencies are given in Ref. 30.

The delay characteristics of each of these classes is not as good as the Bessel delay, nor are the transient overshoots as low as those of the Gaussian transient responses. However the particular application may not require the optimum properties of the Bessel or Gaussian responses, and the transitional filter, with its additional stopband attenuation, may be preferable.

3.7.2 Specific Pole Locations

To obtain approximations to a constant delay, investigators have empirically distributed the poles in the s-plane. In this manner the poles are obtained in closed form and the messy optimization procedure is avoided. Some pole distributions that have been proposed are linear contours, parabolic contours, elliptic contours, and catenary contours. The pros and cons of each are not discussed here, but the interested reader is referred to Ref. 23 and the original references listed there.

3.7.3 Synchronous Response

The transfer function for the synchronous response

$$H(s) = \frac{a^n}{(s+a)^n} \tag{3.7-1}$$

is the simplest of all. All poles lie at $s = -a$, where

$$a = \frac{1}{\sqrt{2^{1/n} - 1}} \tag{3.7-2}$$

for $\omega_{3dB} = 1$. The group delay, from (2.6-10), is

$$D(\omega) = \frac{na}{\omega^2 + a^2} \tag{3.7-3}$$

and the impulse response, from Table 2-1, entry 8, with $F(s) = 1/(s+a)$ is

$$h(t) = \frac{a^n}{(n-1)!} t^{n-1} e^{-at} \tag{3.7-4}$$

The impulse and step responses contain no overshoots, but this asset is offset by a magnitude response that offers poor selectivity [30]. The group delay is fairly constant over the lower portion of the passband, improving as n increases. For example, the passband delay is constant to within 7% for $n = 10$.

Special Cases

3.7.4 Cauer Filters

Except for some time-domain approximations, we have considered only all-pole filters, those having all transmission zeros at infinity. Many practical applications, however, do not need this high attenuation at frequencies far from the passband; instead, more attenuation next to the passband is called for. In this sense the Chebyshev filter, which is optimum in the passband, is not optimum in the stopband.

Equiripple passband and stopband attenuation is achieved by allowing finite transmission zeros in the transfer function. A typical response is shown in Fig. 3-36 for the case of $n = 5$. The filter order n again indicates the number of half-cycles in the passband.

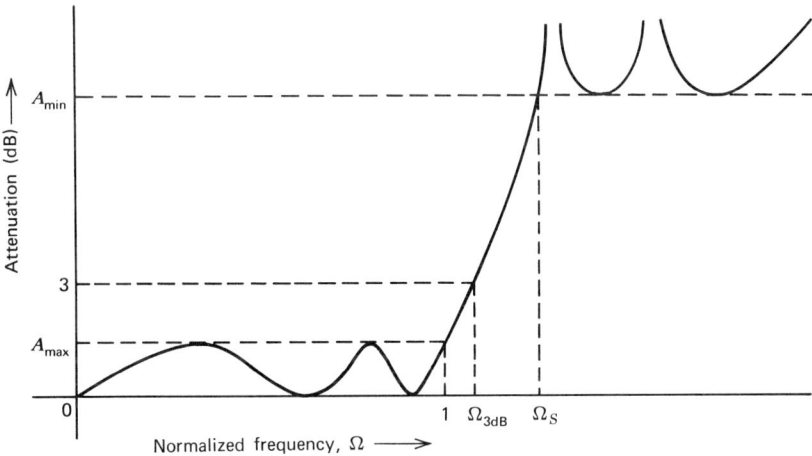

Figure 3-36. Equiripple passband and stopband for the fifth-order Cauer filter.

The filter realizing this optimum response is known as the Cauer filter in honor of the network theorist Prof. Wilhelm Cauer. This response is characterized by A_{max} the maximum passband ripple attenuation; A_{min}, the minimum stopband attenuation, identified by the modular angle θ; and Ω_S, the frequency where A_{min} is first attained.

The two standard networks for the Cauer filters are similar to the networks in Fig. 3-23. However the finite transmission zeros are provided by parallel resonant circuits in the series arms or by series resonant circuits in the shunt arms. Figure 3-37 shows the two networks for the fifth-order Cauer filter exhibiting the response in Fig. 3-36.

Element values, poles and zeros, and Ω_S for many combinations of A_{max} and A_{min} are tabulated in Refs. 21 and 30. The passband ripple there is represented by the reflection coefficient ρ, related to A_{max} by

$$A_{max} = -10 \log(1 - \rho^2) \text{ decibels} \quad (3.7\text{-}5)$$

and to the return loss A_e by

$$A_e = -20 \log \rho \text{ decibels} \quad (3.7\text{-}6)$$

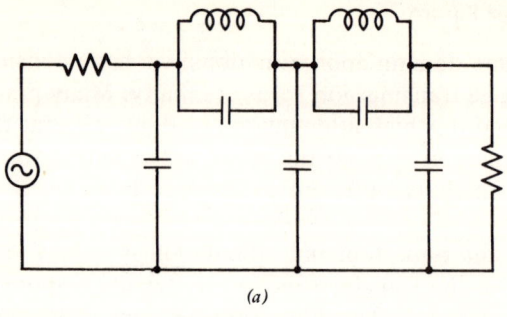

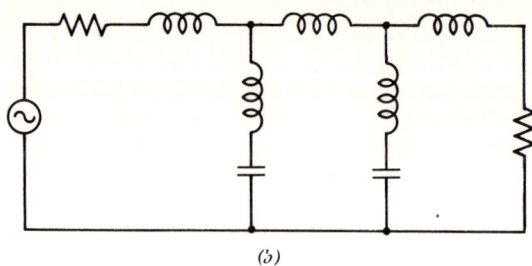

Figure 3-37. Standard networks for the fifth-order Cauer filter.

The voltage standing wave ratio (VSWR) and ρ are related by the well-known expression

$$\text{VSWR} = \frac{1+|\rho|}{1-|\rho|} \tag{3.7-7}$$

Similar to the Chebyshev filters, the normalized radian frequency $\Omega = 1$ corresponds to the edge of the ripple band; thus it is again worthwhile to know the half-power normalized frequency, Ω_{3dB}. Use of the frequency variable Ω is consistent with the notation in Ref. 30.

Example 3-15. The general procedure for finding Ω_{3dB} is now illustrated for the third-order response. From Ref. 30, the transfer-function zeros occur at $\pm j\Omega_2$, the poles occur at $-\sigma_0$ and $-\sigma_1 \pm j\Omega_1$; thus the transfer function is

$$H(S) = \frac{\sigma_0 \psi}{\Omega_2^2} \frac{S^2 + \Omega_2^2}{(S + \sigma_0)[(S + \sigma_1)^2 + \Omega_1^2]} \tag{3.7-8}$$

normalized to unity at $\Omega = 0$, where $\psi = \sigma_1^2 + \Omega_1^2$. When $S = j\Omega$, the magnitude response is

$$|H(j\Omega)| = \frac{\sigma_0 \psi}{\Omega_2^2} \frac{|\Omega_2^2 - \Omega^2|}{\sqrt{\sigma_0^2 + \Omega^2}\sqrt{(\psi - \Omega^2)^2 + 4\sigma_1^2 \Omega^2}} \tag{3.7-9}$$

At $\Omega = \Omega_{3dB}$, $|H(j\Omega_{3dB})| = 1/\sqrt{2}$. After some algebraic manipulations we obtain the following polynomial equation:

$$\Omega_{3dB}^6 + a\Omega_{3dB}^4 + b\Omega_{3dB}^2 - \sigma_0^2 \psi^2 = 0 \tag{3.7-10}$$

Rise Time and Noise Bandwidth

where

$$a = \sigma_0^2 + 4\sigma_1^2 - 2\psi\left(1 + \frac{\sigma_0^2\psi}{\Omega_2^4}\right) \qquad (3.7\text{-}11)$$

$$b = 2\sigma_0^2(2\sigma_1^2 - \psi) + \psi^2\left(1 + \frac{4\sigma_0^2}{\Omega_2^2}\right) \qquad (3.7\text{-}12)$$

For n even, equate the gain function $|H(j\Omega_{3\text{dB}})|$ to $1/\sqrt{2(1-\rho^2)}$, since then $1/\sqrt{1-\rho^2}$ is the maximum value of $|H(j\Omega)|$. Equation 3.7-10 is then solved on a digital computer for selected sets of poles and zeros. The results are given in Table 3-5. Tables 3-6 and 3-7 give $\Omega_{3\text{dB}}$ for $n = 4$ and $n = 5$, respectively. Intermediate values are obtained by interpolation. The angle 0° corresponds to the all-pole Chebyshev filter with all transmission zeros at infinity.

Table 3-5 Values of $\Omega_{3\text{dB}}$ for Third-Order ($n = 3$) Cauer Filters

Modular Angle, θ	\multicolumn{10}{c}{Reflection Coefficient, ρ}
	1%	2%	3%	4%	5%	8%	10%	15%	20%	25%	50%
0°	3.009	2.428	2.150	1.977	1.855	1.631	1.539	1.391	1.301	1.238	1.074
5°	2.964	2.406	2.136	1.966	1.847	1.626	1.534	1.388	1.299	1.237	1.074
10°	2.836	2.341	2.093	1.934	1.821	1.611	1.523	1.381	1.294	1.233	1.073
15°	2.647	2.241	2.025	1.883	1.781	1.586	1.503	1.369	1.285	1.227	1.071
20°	2.427	2.116	1.938	1.817	1.727	1.553	1.477	1.353	1.274	1.218	1.069
25°	2.202	1.978	1.839	1.739	1.664	1.512	1.445	1.332	1.259	1.207	1.066
30°		1.838	1.733	1.655	1.594	1.467	1.408	1.308	1.242	1.195	1.063
35°			1.627	1.568	1.520	1.417	1.368	1.281	1.223	1.180	1.059
40°				1.483	1.447	1.365	1.325	1.252	1.202	1.164	1.054
45°							1.281	1.221	1.179	1.146	1.050
50°							1.190	1.155	1.128	1.044
55°									1.131	1.109	1.039
60°									1.107	1.090	1.033

3.8 RISE TIME AND NOISE BANDWIDTH

The rise time and noise bandwidth are frequently important in the design and analysis of electronic systems. Their values depend on the system magnitude and phase characteristics, which are largely determined by the filter characteristics. Consequently the filter rise time and noise bandwidth are of considerable interest in practice. Unfortunately there are several definitions of each term and this has caused some confusion, which the following discussions should eliminate.

3.8.1 Rise Time

The rise time t_r has traditionally been defined as the time necessary for the step response to go from 10 to 90% of its final value; this is the definition most widely used in practice. Comparison of step responses by this rise time definition can be misleading because two very different responses can have the same rise time (Fig.

Table 3-6 Values of Ω_{3dB} for Fourth-Order ($n = 4$) Cauer Filters

Modular Angle, θ	Reflection Coefficient, ρ										
	1%	2%	3%	4%	5%	8%	10%	15%	20%	25%	50%
0°	2.013	1.739	1.603	1.517	1.456	1.341	1.292	1.214	1.166	1.132	1.041
5°	2.003	1.733	1.599	1.514	1.454	1.340	1.291	1.213	1.165	1.131	1.041
10°	1.980	1.720	1.590	1.507	1.447	1.335	1.288	1.211	1.163	1.130	1.041
15°	1.940	1.697	1.573	1.493	1.436	1.328	1.282	1.207	1.161	1.128	1.040
20°	1.884	1.664	1.550	1.475	1.421	1.318	1.274	1.202	1.157	1.125	1.039
25°	1.816	1.624	1.520	1.451	1.401	1.305	1.263	1.195	1.151	1.121	1.038
30°	1.737	1.575	1.484	1.423	1.378	1.289	1.250	1.186	1.145	1.116	1.037
35°	1.650	1.520	1.444	1.390	1.350	1.270	1.235	1.176	1.138	1.110	1.035
40°	1.560	1.461	1.398	1.354	1.319	1.249	1.218	1.164	1.129	1.104	1.033
45°		1.399	1.350	1.314	1.285	1.226	1.198	1.151	1.119	1.096	1.031
50°		1.336	1.300	1.272	1.249	1.200	1.177	1.136	1.108	1.088	1.029
55°			1.249	1.229	1.211	1.173	1.154	1.120	1.096	1.078	1.026
60°						1.144	1.130	1.102	1.083	1.068	1.023

Table 3-7 Values of Ω_{3dB} for Fifth-Order ($n = 5$) Cauer Filters

Modular Angle, θ	Reflection Coefficient, ρ										
	1%	2%	3%	4%	5%	8%	10%	15%	20%	25%	50%
0°	1.616	1.455	1.374	1.322	1.284	1.214	1.184	1.135	1.105	1.084	1.026
5°	1.611	1.452	1.372	1.320	1.283	1.213	1.183	1.135	1.104	1.083	1.026
10°	1.596	1.443	1.365	1.314	1.278	1.209	1.180	1.133	1.103	1.082	1.026
15°	1.573	1.428	1.353	1.305	1.270	1.204	1.176	1.131	1.101	1.080	1.025
20°	1.541	1.407	1.338	1.292	1.260	1.197	1.170	1.125	1.098	1.078	1.025
25°	1.501	1.382	1.318	1.277	1.246	1.187	1.162	1.120	1.093	1.075	1.024
30°	1.456	1.352	1.296	1.258	1.230	1.176	1.152	1.113	1.089	1.071	1.023
35°	1.407	1.319	1.270	1.236	1.212	1.163	1.143	1.106	1.083	1.066	1.021
40°	1.355	1.283	1.242	1.213	1.191	1.148	1.129	1.097	1.076	1.061	1.020
45°	1.302	1.246	1.212	1.188	1.170	1.133	1.116	1.088	1.069	1.056	1.018
50°	1.249	1.207	1.181	1.162	1.147	1.116	1.102	1.077	1.061	1.050	1.016
55°	1.199	1.169	1.149	1.135	1.123	1.098	1.087	1.067	1.053	1.043	1.014
60°	1.151	1.132	1.118	1.108	1.099	1.080	1.071	1.055	1.044	1.036	1.012

3-38a), whereas two very similar responses can have very different rise times (Fig. 3-38b). This fact seldom causes appreciable problems in practice, since the additional specification of overshoot or settling time usually determines the acceptable response.

Figure 3-39 plots the rise times of the Butterworth, Gaussian, and Bessel filters using the 10 to 90% definition with $\omega_{3dB} = 1$. The ordinate is labeled $\omega_c t_r$, where ω_c is the actual LP filter cutoff frequency. We discuss the denormalization process in Chapter 4. As an example, the rise time of a fourth-order LP Bessel filter with a cutoff frequency of 100 kHz is $2.2/(2\pi \times 10^5) = 3.5$ μsec.

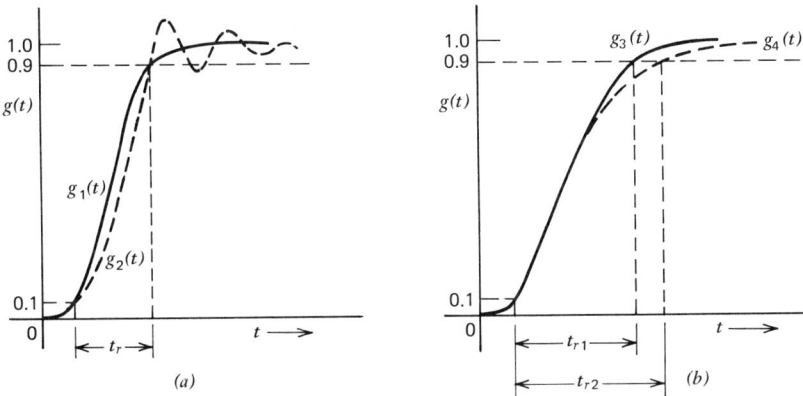

Figure 3-38. Comparison of 10 to 90% rise times for (*a*) two dissimilar responses and (*b*) two similar responses.

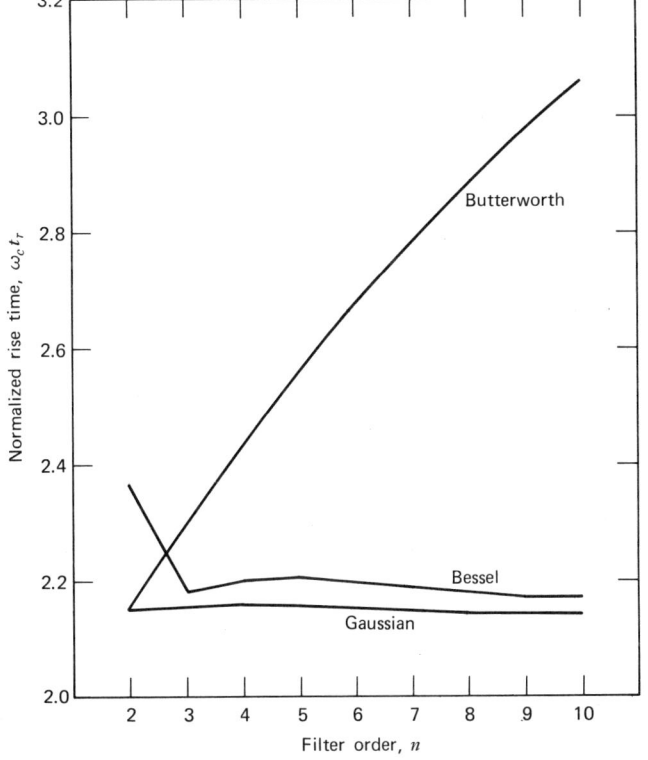

Figure 3-39. Filter rise times based on 10 to 90% definition.

The rise time of the constant-delay type networks (such as Bessel and Gaussian) is smaller than the rise time of the constant magnitude types (such as Butterworth and Chebyshev). The Chebyshev rise times are larger than the Butterworth's. For a given value of n, the rise time of the former increases as the passband ripple increases.

Another rise-time definition (used in Section 3.1) is the time required for the

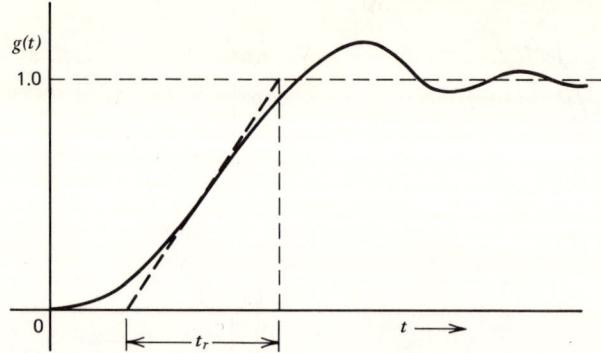

Figure 3-40. Rise time by maximum slope definition.

step response to go from zero to its final value (unity for most cases) with a slope equal to its maximum slope, as in Fig. 3-40. This maximum slope is $h_{\max}(t)$, the peak amplitude of the impulse response; thus from Fig. 3-40

$$t_r = \frac{1}{h_{\max}(t)} \tag{3.8-1}$$

This value of rise time is shown in Fig. 3-41 for various filters, and the curves are in the same relative positions as those of Fig. 3-39. Furthermore, for $n > 2$, the corresponding values by each definition differ at most by 8%.

Su [23] gives four other definitions of rise time and concludes that none of the existing formulations allows for a satisfactory comparison of the constant-delay type networks. He proposes that the standard step response $g_p(t)$ (Fig. 3-42) is the goal of all step responses, and their failure is due to the physical restrictions of the network. He then proposes that for each step response $g(t)$, t_1 and t_2 be varied until $g_p(t)$ approximates $g(t)$ in the least-squares sense; that is, the integral

$$\varepsilon = \int_0^\infty [g_p(t) - g(t)]^2 \, dt \tag{3.8-2}$$

is minimized. Then the rise time t_r and delay time t_d are defined as

$$t_r = t_2 - t_1$$
$$t_d = \frac{1}{2}(t_1 + t_2) \tag{3.8-3}$$

Su carried out this minimization for a variety of responses, and the rise times for the more pertinent ones are displayed in Fig. 3-43. The normalization is $\omega_{3dB} = 1$. The various filter types are in the same relative positions as before, but their values are larger. If they are multiplied by 0.8 (this rise time is measured from 0 to 1.0), they agree very closely with the rise times by the two previous definitions.

In summary, the three definitions of rise time reveal the same important fact. For a specified bandwidth and value of $n > 2$, the Gaussian response rise time is less than the Bessel rise time, and these rise times are less than the rise time of the Butterworth and Chebyshev responses. As the Chebyshev ripple attenuation increases, so does the rise time.

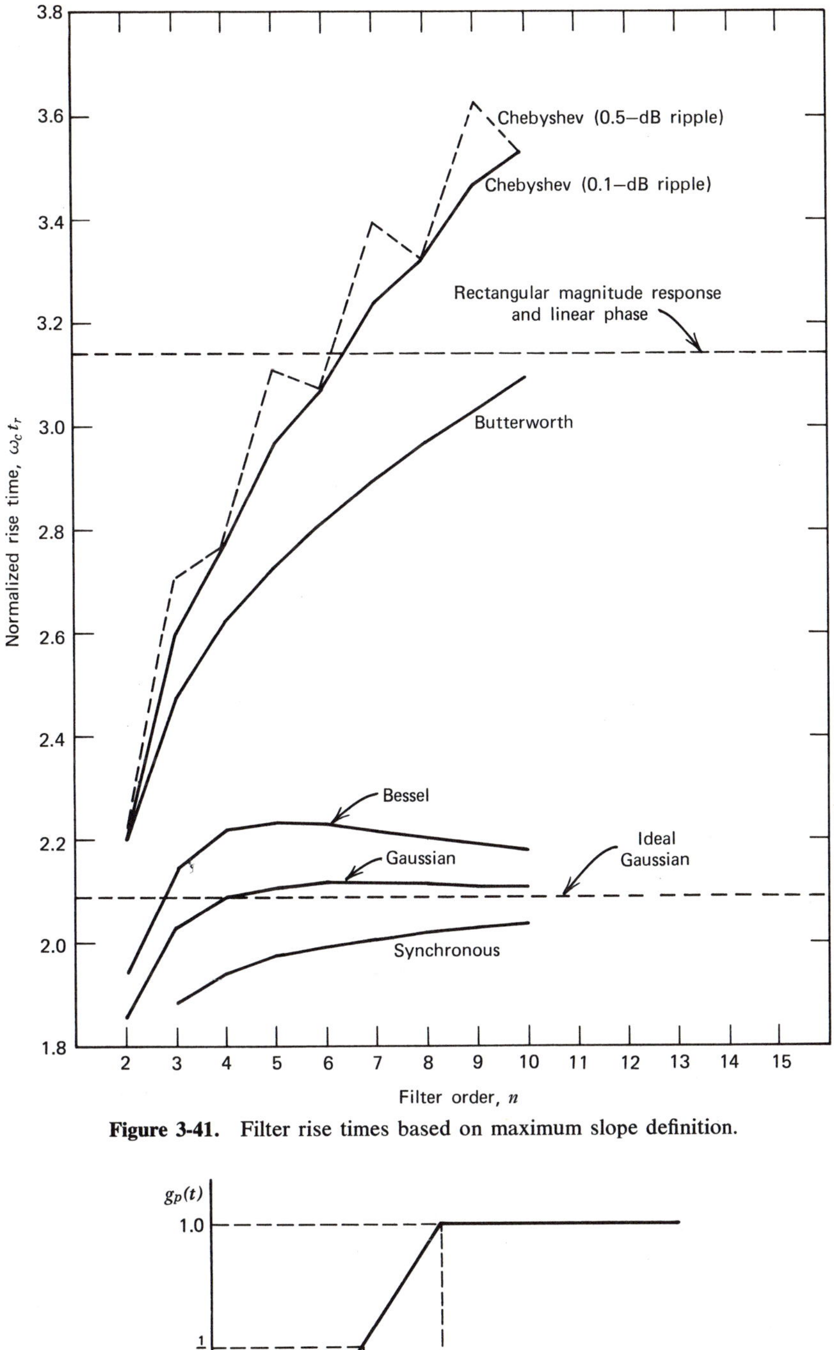

Figure 3-41. Filter rise times based on maximum slope definition.

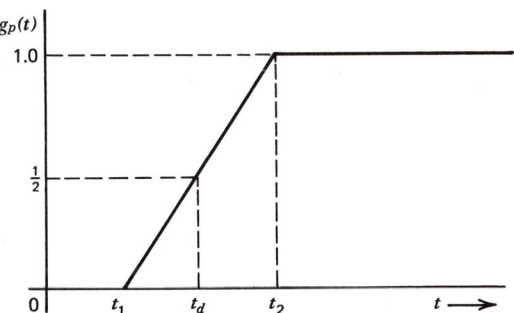

Figure 3-42. Standard step response, $t_r = t_2 - t_1$.

143

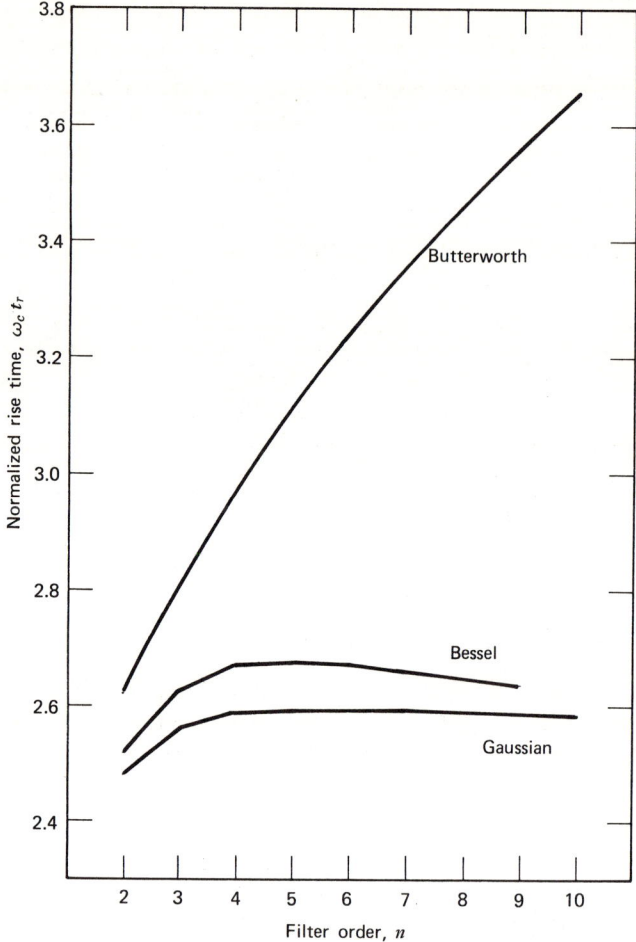

Figure 3-43. Rise times based on Su's definition (Fig. 3-42).

This discussion appears to contradict the conclusion of some that for a given order n, the Butterworth (or Chebyshev) filter rise time is less than the Gaussian (or Bessel) filter rise time. This is not a contradiction provided the reference frequency is specified, as we now explain.

Often a system must simultaneously be band-limited and have a small rise time. Since no realizable system is truly band-limited, a reference attentuation, such as 40 dB, can be postulated as the band edge. Thus it is not unreasonable in comparing filter rise times to normalize all filters to the 40-dB frequency instead of the 3-dB frequency.

Example 3-16. The rise time normalized to a 40-dB bandwidth of unity is now determined for the Butterworth family. From Section 3.1 and Figs. 3-39, 3-41, and 3-43 the rise time–bandwidth product is a constant; once the new cutoff frequency is obtained, therefore, the rise time is easily determined. Since the ratio corresponding to 40 dB is 0.01, from (3.4-2)

$$1 + \omega_{40\text{dB}}^{2n} = 10^4 \tag{3.8-4}$$

Rise Time and Noise Bandwidth

or

$$\omega_{40\text{dB}} \approx (100)^{1/n} = K_n \qquad (3.8\text{-}5)$$

Since $\omega_{3\text{dB}} = 1$, the ratio of the 40-dB frequency to the 3-dB frequency is also K_n. We now translate $\omega_{40\text{dB}}$ to unity, which means that the new 3-dB frequency ω_r is

$$\omega_r = \frac{1}{K_n} = (100)^{-1/n} \qquad (3.8\text{-}6)$$

Let Q_n be the rise time–bandwidth product, which is constant for a specified value of n. Then

$$\omega_r t_r = \frac{t_r}{K_n} = Q_n \qquad (3.8\text{-}7)$$

or

$$t_r = K_n Q_n = (100)^{1/n} Q_n \qquad (3.8\text{-}8)$$

Using the maximum slope definition, we obtain Q_n from Fig. 3-41 and then compute t_r for $n = 2$ to $n = 10$. Table 3-8 lists K_n and t_r, and Fig. 3-44 shows t_r as a function of the order n. The rise times for the other responses were recomputed in

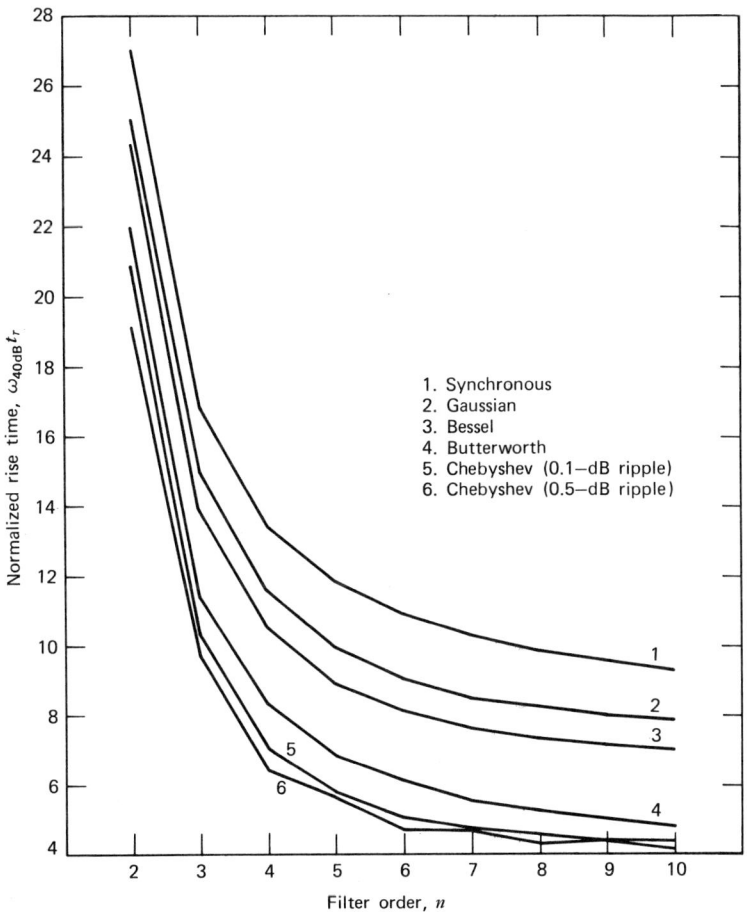

Figure 3-44. Rise times by maximum slope definition normalized to $\omega_{40\text{dB}} = 1$.

Table 3-8 **Butterworth Rise Times Normalized for** $\omega_{40dB} = 1$

n	K_n	Rise time, t_r
2	9.9997	21.932
3	4.6415	11.476
4	3.1622	8.287
5	2.5119	6.852
6	2.1544	6.067
7	1.9307	5.588
8	1.7783	5.272
9	1.6681	5.055
10	1.5848	4.901

a similar manner for $\omega_{40dB} = 1$, and these also appear in Fig. 3-44. Now the Butterworth and Chebyshev filter rise times are less than those of the constant-delay type filters. We conclude that the reference frequency must be specified before filter rise times can be compared in a meaningful manner.

3.8.2 Noise Bandwidth

The noise bandwidth of a filter ω_N is of theoretical and practical interest because it indicates the noise power at the filter output, hence often serves as a performance measure for comparing filters. Consider the LP filter with magnitude response $|H_l(j\omega)|$ (Fig. 3-45a) excited by white noise with spectral density N_0 watts per hertz. The total power at the filter output delivered to a one-ohm resistor, including the contribution from the negative frequencies, is

$$\text{output power} = \frac{N_0}{2\pi} \int_{-\infty}^{\infty} |H_l(j\omega)|^2 \, d\omega \tag{3.8-9}$$

while the output power of the rectangular magnitude filter in Fig. 3-45b for the same white noise input is

$$\text{output power} = \frac{N_0 K^2 \omega_N}{\pi} \tag{3.8-10}$$

The noise bandwidth ω_N is defined as the bandwidth for which the power output of the rectangular magnitude filter is the same as the power output of the specified LP filter. Equate the power from (3.8-9) and (3.8-10) and note that $|H_l(j\omega)|^2$ is an even function. Then

$$\omega_N = \frac{1}{K^2} \int_0^{\infty} |H_l(j\omega)|^2 \, d\omega \tag{3.8-11}$$

Here K is the value of $|H_l(j\omega)|$ at a convenient frequency, the two most common being $\omega = 0$ and the frequency where $|H_l(j\omega)|$ is maximum. Except for a few all-pole filters, the maximum value of $|H_l(j\omega)|$ occurs at $\omega = 0$. Accordingly we use

Rise Time and Noise Bandwidth

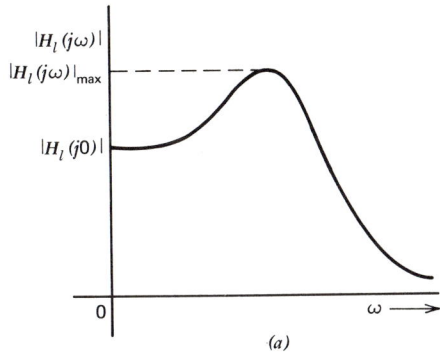

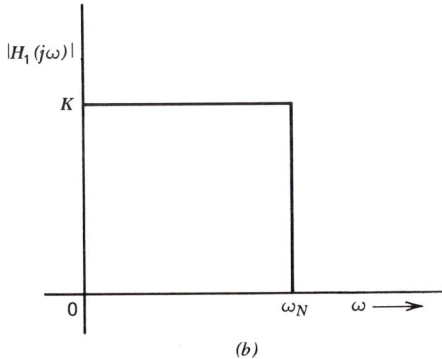

Figure 3-45. (a) Arbitrary LP response and (b) rectangular magnitude response.

the maximum value as the reference, and then

$$\omega_N = \frac{\int_0^\infty |H_l(j\omega)|^2 \, d\omega}{|H_l(j\omega)|^2_{\max}} \tag{3.8-12}$$

The tedium of evaluating this integral for each filter order is eliminated by use of the simple relationship [10, 14]

$$\int_0^\infty |H_l(j\omega)|^2 \, d\omega = \frac{\pi}{2C_1} \tag{3.8-13}$$

where C_1 is the input shunt capacitance for the LP filter singly terminated in one ohm. Refer to Fig. 3-23a with $R_s = \infty$ and $x_1 = C_1$. The values for C_1 are tabulated in Ref. 30, and ω_N is easily obtained. For all filters there, except the even-order Chebyshev filters, the 3-dB radian bandwidth and maximum gain are both normalized to unity. Thus

$$\omega_N = \frac{\pi}{2C_1} \tag{3.8-14}$$

and for the even-order Chebyshev filters

$$\omega_N = \frac{\pi \cdot 10^{-0.1 A_{\max}}}{2C_1} \tag{3.8-15}$$

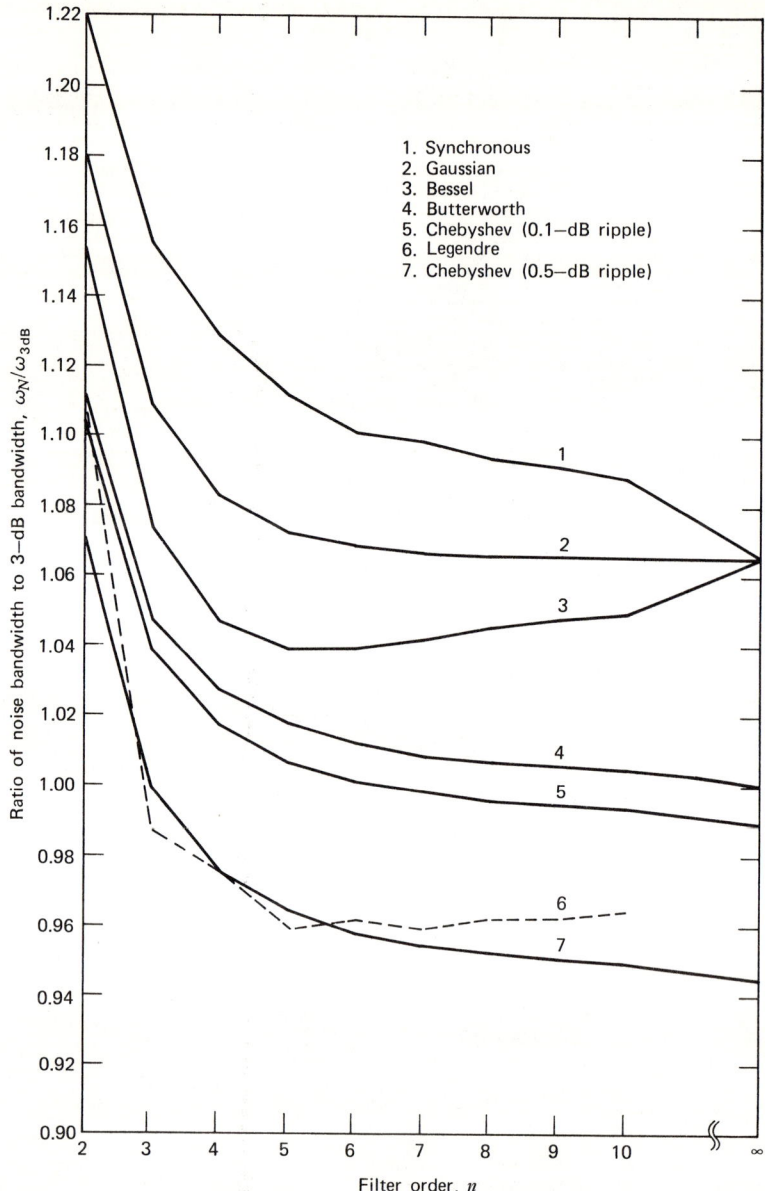

Figure 3-46. Filter noise bandwidths.

The noise bandwidth relative to the 3-dB bandwidth (since $\omega_{3dB} = 1$) of seven different filters is shown in Fig. 3-46. The Butterworth and Chebyshev filters, because of their rapid transition from passband to stopband, have a smaller noise bandwidth than do the constant-delay type filters. For a given order n, the Chebyshev filter noise bandwidth increases as the ripple decreases, the limit being the Butterworth noise bandwidth.

3.8.3 Relationship Between Noise Bandwidth and Impulse Response Energy

The noise bandwidth is also related to E_l, the LP filter impulse response energy dissipated in a one-ohm resistor. By Parseval's theorem (2.1-31)

$$\omega_N = \int_0^\infty |H_l(j\omega)|^2 \, d\omega = \pi \int_0^\infty h_l^2(t) \, dt = \pi E_l \tag{3.8-16}$$

hence E_l is obtained by dividing ω_N by π. The curves of Fig. 3-46 can be used to determine the normalized energy E_L by replacing the ordinate label by πE_L. For example, the impulse response energy of a fifth-order Bessel filter is $1.039/\pi = 0.3307$ Joules.

Example 3-17. The noise bandwidth is now expressed in closed form for three responses. Substitute the Butterworth function from (3.4-2) into (3.8-12) to yield [6, item 856.05]

$$\omega_N = \int_0^\infty \frac{d\omega}{1 + \omega^{2n}} = \frac{\pi}{2n \sin(\pi/2n)} = \frac{\pi D_L(0)}{2n} \tag{3.8-17}$$

where $D_L(0)$, from (3.4-6), is the normalized group delay at $\omega = 0$. Similarly the Chebyshev noise bandwidth is

$$\omega_N = \frac{\pi \cosh\left[\frac{1}{n} \ln\left(\frac{1 + \sqrt{1 + \epsilon^2}}{\epsilon}\right)\right]}{\omega_{3dB}\sqrt{1 + \epsilon^2} \, 2n \sin\left(\frac{\pi}{2n}\right)} \tag{3.8-18}$$

where ω_{3dB} is obtained from (3.4-27).

For the synchronous filter, we substitute $|H(j\omega)|^2$ from (3.7-1) into (3.8-12), yielding

$$\omega_N = a^{2n} \int_0^\infty \frac{d\omega}{(a^2 + \omega^2)^n} = \frac{\sqrt{\pi} \, \Gamma(n - \tfrac{1}{2})}{2a(n-1)!} \tag{3.8-19}$$

where $\Gamma(x)$ is the gamma function [1] and a is given by (3.7-2).

REFERENCES

1. Abramowitz, M., and I. Stegun, *Handbook of Mathematical Functions*. (No. 55) Washington, D.C., Government Printing Office, 1964.
2. Blinchikoff, H. J., "High-Pass Filter Step-Response Energy: A New Performance Measure," *IEEE Trans. Circuit Theory*, Vol. CT-20, pp. 593–596, September 1973.
3. Cohn, S. B., "Direct-Coupled Resonator Filters," *Proc. IRE*, Vol. 45, pp. 187–196, February 1957.
4. Director, S. W., "Survey of Circuit-Oriented Optimization Techniques," *IEEE Trans. Circuit Theory*, Vol. CT-18, pp. 3–10, January 1971.
5. Dishal, M., "Gaussian-Response Filter Design," *Electr. Commun.*, Vol. 36, pp. 3–26, March 1959.
6. Dwight, H. B., *Tables of Integrals and Other Mathematical Data*, Macmillan, New York, 1961.
7. Friedman, B., *Principles and Techniques of Applied Mathematics*, Wiley, New York, 1956.
8. Fukado, M. "Optimum Filters of Even Orders with Monotonic Response," *IRE Trans. Circuit Theory*, Vol. CT-6, pp. 277–281, September 1959.

9. Huggins, W. H., *Network Approximation in the Time Domain*, Report E 5048A, Air Force Cambridge Research Laboratories, Cambridge, Mass., October 1949.
10. Humpherys, D. S., *The Analysis, Design, and Synthesis of Electrical Filters*, Prentice-Hall, Englewood Cliffs, N.J., 1970.
11. Jess, J., and H. W. Schüssler, "A Class of Pulse-Forming Networks," *IEEE Trans. Circuit Theory*, Vol. CT-12, pp. 296–299, June 1965.
12. Jess, J., and H. W. Schüssler, "On the Design of Pulse-Forming Networks," *IEEE Trans. Circuit Theory*, Vol. CT-12, pp. 393–400, September 1965.
13. Kiyasu, Z., "On a Design Method of Delay Networks," *J. Inst. Electr. Commun. Eng., Japan*, Vol. 26, pp. 598–610, August 1943.
14. Ku, W. H., "Noise Margin and Gain-Bandwidth Limitations of Legendre Filters," *IEEE Trans. Circuit Theory*, Vol. CT-17, pp. 670–672, November 1970.
15. Peless, Y., and T. Murakomi, "Analysis and Synthesis of Transitional Butterworth-Thomson Filters and Bandpass Amplifiers," *RCA Rev.*, Vol. 18, pp. 60–94, March 1957.
16. O'Meara, T. R., "Band-Center Group Delay and Incidental Dissipation," *IRE Trans. Circuit Theory*, Vol. CT-9, p. 192, June 1962.
17. Orchard, H. J., "The Phase and Envelope Delay of Butterworth and Tchebycheff Filters," *IRE Trans. Circuit Theory*, Vol. CT-7, pp. 180–181, June 1960.
18. Orchard, H. J., and Temes G. C., "Filter Design Using Transformed Variables," *IEEE Trans. Circuit Theory*, Vol. CT-15, pp. 385–408, December 1968.
19. Papoulis, A., "Optimum Filters with Monotonic Response," *Proc. IRE*, Vol. 46, pp. 606–609, March 1958.
20. Papoulis, A., *The Fourier Integral and Its Applications*, McGraw-Hill, New York, 1962.
21. Saal, R., *Der Entwurf von Filtern mit Hilfe des Kataloges normierter Tiefpasse*, Telefunken GmbH, Backnang, West Germany, 1963.
22. Storch, L., "Synthesis of Constant-Time Delay Ladder Networks Using Bessel Polynomials," *Proc. IRE*, Vol. 42, pp. 1666–1675, November 1954.
23. Su, K. L., *Time-Domain Synthesis of Linear Networks*, Prentice-Hall, Englewood Cliffs, N.J., 1971.
24. Temes, G. C., "The Prolate Filter: An Ideal Lowpass Filter with Optimum Step-Response, "*J. Franklin Inst.*, Vol. 293, pp. 77–103, February 1972.
25. Temes, G. C., and D. A. Calahan, "Computer-Aided Network Optimization, the State of the Art," *Proc. IEEE*, Vol. 55, pp. 1832–1863, November 1967.
26. Temes, G. C., and D. Y. F. Zai, "Least-pth Approximation," *IEEE Trans. Circuit Theory*," Vol. CT-16, pp. 235–237, May 1969.
27. Thomson, W. E., "Delay Networks Having Maximally Flat Frequency Characteristics," *Proc. IEE*, Vol. 96, pp. 487–490, November 1949.
28. Weber, E., *Linear Transient Analysis*, Vol. II, Wiley, New York, 1956, Chap. 4.
29. Weinberg, L., "Tables of Networks Whose Reflection Coefficients Possess Alternating Zeros," *IRE Trans. Circuit Theory*, Vol. CT-4, pp. 313–320, December 1957.
30. Zverev, A. I., *Handbook of Filter Synthesis*, Wiley, New York, 1967.

PROBLEMS

3-1. For the filter below, determine the values of L and C so that

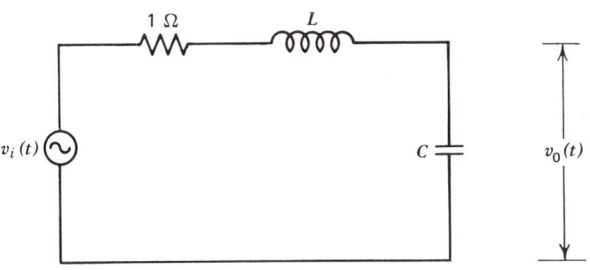

Figure P.3-1.

(a) The magnitude response is maximally flat at $\omega = 0$.
(b) The group delay response is maximally flat at $\omega = 0$.
For each case $|H(j\omega)|$ is to be 3 dB down at $\omega = 1$.

3-2. The transfer function of an all-pass filter (Fig. 3-12a) is

$$H(s) = \frac{(s-\alpha)^2 + \beta^2}{(s+\alpha)^2 + \beta^2}$$

What is the ratio α/β for maximally flat delay at $\omega = 0$?

3-3. Determine a_1 and a_2 so that $f(x) = a_1 x + a_2 x^2$ approximates the function $g(x) = \sqrt{x}$ in the least-squares sense over the interval $0 \leq x \leq 2$ ($w(x) = 1$).

3-4. For the Butterworth response, what is the order n necessary to obtain A_S decibels at ω_s, when the response is 3 dB down at $\omega = 1$?

3.5 Consider a Chebyshev filter ($n = 5$) normalized for unity ripple-bandwidth. The passband ripple is 0.35 dB.
  (a) What is the filter attenuation at $\omega = 2$?
  (b) What is the 3-dB frequency?

3-6. Consider the third-order Chebyshev filter (0.1-dB passband ripple). What is the attenuation at twice the 3-dB frequency?

3-7. Use the recursion formula to determine the fifth-order Bessel filter transfer function normalized for unit delay and unit gain at $\omega = 0$.

3-8. For the filter in Problem 3-1,
  (a) Determine the set of values for L and C to produce the synchronous response, and to ensure that $|H(j\omega)|$ is 3 dB down at $\omega = 1$ from its value at $\omega = 0$.
  (b) For these values, how much is the input signal $\sin(t/2)$ delayed by the filter?

3-9. Filtering with passive lossless networks is achieved by the degree of impedance mismatch between the filter input impedance and the source resistor. Aided by (3.7-5) and (3.7-7), show that a VSWR there of $S(\omega)$ causes an output attenuation of

$$A = -10 \log \left[\frac{4S}{(S+1)^2}\right] \text{ decibels}$$

Therefore, for the equally terminated LP filter

$$|H(j\omega)|^2 = \frac{4S}{(S+1)^2}$$

3-10. Show that the rise time of the second-order system characterized by

$$H(s) = \frac{\alpha^2 + \beta^2}{(s+\alpha)^2 + \beta^2}$$

using the definition in (3.8-1), is

$$t_r = \frac{\exp[(\alpha/\beta) \tan^{-1}(\beta/\alpha)]}{\sqrt{\alpha^2 + \beta^2}}$$

3-11. Show that the rise time of the synchronous filter characterized by (3.7-1), using the definition in (3.8-1), is

$$t_r = \frac{(n-1)! e^{n-1}}{a(n-1)^{n-1}}$$

3-12. What is the noise bandwidth of the Gaussian filter characterized by (3.1-9)?
3-13. Derive the expression in (3.7-2).

Chapter Four

Frequency Transformations

In this chapter we discuss the realization of low-pass (LP), high-pass (HP), bandpass (BP), and bandstop (BS) filters by applying suitable frequency transformations to the normalized LP filter. These transformations generally preserve the LP magnitude response (attenuation); other LP characteristics are often retained, however, especially in the case of the narrow-band BP filter. We also give the necessary equations (where possible) for denormalizing the LP responses, allowing them to characterize the appropriate filter type. When the transformation preserves the important attributes of the LP response, the tedious and time-consuming approximation step in the design sequence is eliminated. This is the big advantage of the frequency-transformation approach. Furthermore, if the transformation function is of the same form as a reactance function, the filter element values are easily determined.

Unfortunately certain LP characteristics are distorted by the transformation, and to avoid these distortions we either re-solve the approximation problem or use a different transformation. We present an example of each approach. The first results in the wide-band constant-delay filter and the second results in the low-transient HP filter. This chapter also includes narrow-band BP filter design, which is applicable not only to lumped-constant filters but also to transmission line and waveguide filters.

4.1 NORMALIZED PARAMETERS

We assume that the desired LP responses normalized for 3-dB attenuation at $\Omega = 1$ rad/sec (magnitude, phase, group delay, and transients), and the corresponding element values impedance normalized to one ohm, can be obtained from a catalog such as Ref. 12. The filter normalized in this manner is referred to as the prototype or normalized LP filter. To denormalize the impedance to R ohms, we must multiply all inductance values by R and divide all capacitance values by R.

We distinguish between frequency ω and normalized frequency Ω, where

$\omega = \text{Im } s = $ imaginary part of the complex frequency s

$\Omega = \text{Im } S = $ imaginary part of the complex normalized frequency S

(4.1-1)

For each filter type (LP, HP, BP, and BS) the transformation $\Omega(\omega)$ is defined, thus providing the proper denormalization of the frequency scale.

In general, a function with a capital letter subscript is a normalized quantity, whereas a function with a lowercase subscript is a denormalized quantity. For example, $g_H(t)$ is the step response of the normalized HP filter ($\omega_{3dB} = 1$) and $g_h(t)$ is the step response of the denormalized HP filter.

The transformed group delay is indicated here by $D_T(\omega)$. However when we consider each filter type, the subscript "T" is changed to a subscript identifying that particular filter type. The normalized LP delay $D_L(\omega)$, from (2.6-1), is

$$D_L(\omega) = -\frac{d\theta_L(\omega)}{d\omega} \qquad (4.1\text{-}2)$$

and $D_T(\omega)$, after replacing ω by $\Omega(\omega)$, is

$$D_T(\omega) = -\frac{d\theta_L[\Omega(\omega)]}{d\omega} \qquad (4.1\text{-}3)$$

where $\theta_L[\Omega(\omega)]$ is the normalized LP phase evaluated at $\Omega(\omega)$. By the chain rule of differentiation, $D_T(\omega)$ is

$$D_T(\omega) = -\frac{d\theta_L(\Omega)}{d\Omega} \cdot \frac{d\Omega}{d\omega} = D_L(\omega)|_{\omega=\Omega} \cdot \frac{d\Omega}{d\omega} \qquad (4.1\text{-}4)$$

revealing the two parts:

a) $D_L(\Omega)$, the transformed LP delay, such as displayed in Fig. 3-25.
b) $d\Omega/d\omega$, a function that may scale or distort the normalized LP delay.
   The ordinate of the normalized group delay curve is replaced by $D_T(\omega)/(d\Omega/d\omega)$, where $d\Omega/d\omega$ is evaluated for each transformation.

The time scale for the normalized impulse and step responses is denoted by t_N, the normalized time (e.g., Figs. 3-26 and 3-27). If these denormalized responses apply to the filter type under discussion, we include the denormalization of the time scale.

4.2 LOW-PASS FILTER

The transformation for frequency scaling the normalized LP filter to one with 3-dB cutoff frequency ω_c is

$$S = \frac{s}{\omega_c} \qquad (4.2\text{-}1)$$

from which

$$\Omega = \frac{\omega}{\omega_c} \qquad (4.2\text{-}2)$$

The frequency scale is thus expanded or contracted as desired, with behavior at ω_c the same as the LP prototype at $\Omega = 1$. Attenuation and phase are not affected by the transformation, but from (4.1-4) and (4.2-2) group delay is altered by

$$\frac{d\Omega}{d\omega} = \frac{1}{\omega_c} \qquad (4.2\text{-}3)$$

and the normalized LP group delay ordinate is replaced by $\omega_c D_l(\omega)$, where $D_l(\omega)$ is the denormalized LP delay.

Low-Pass Filter

4.2.1 Transient Responses

The denormalized pole s_i (or zero, if present) from (4.2-1), is

$$s_i = \omega_c S_i \qquad (4.2\text{-}4)$$

where S_i is the normalized pole. Denormalization of the transient responses is obtained by remembering that the normalized impulse and step responses can always be written in the form

$$\sum_{i=1}^{n} A_i e^{S_i t_N} \qquad (4.2\text{-}5)$$

for simple transfer-function poles. Substituting from (4.2-4) into (4.2-5) changes the exponent from $S_i t_N$ to

$$s_i \left(\frac{t_N}{\omega_c} \right) = s_i t \qquad (4.2\text{-}6)$$

where t is the denormalized time. The time scale is then replaced by

$$t_N = \omega_c t \qquad (4.2\text{-}7)$$

a result that is true even when multiple transfer-function poles occur.

The ordinate of the step response remains the same, but the impulse response ordinate is replaced by $(1/\omega_c) h_l(t)$, where $h_l(t)$ is the denormalized impulse response. This is explained by evaluating (2.4-25) at $s = 0$, giving $H(0)$ as the area under the impulse response. For a given LP filter response, therefore, the impulse response area is independent of the cutoff frequency and equals the magnitude response at $\omega = 0$. Since the normalized time scale is divided by ω_c, the impulse response scale must be multiplied by ω_c for constant area.

4.2.2 Element Values

The normalized inductive reactance, using S from (4.2-1), is

$$SL_N = s \frac{L_N}{\omega_c} = sL \qquad (4.2\text{-}8)$$

which yields the denormalized inductance value L as

$$L = \frac{L_N}{\omega_c} \qquad (4.2\text{-}9)$$

The normalized capacitive reactance is

$$\frac{1}{SC_N} = \frac{1}{s(C_N/\omega_c)} = \frac{1}{sC} \qquad (4.2\text{-}10)$$

yielding the denormalized capacitance value C as

$$C = \frac{C_N}{\omega_c} \qquad (4.2\text{-}11)$$

The denormalized element values are thus obtained by dividing the normalized values by the radian cutoff frequency.

4.2.3 Example of Low-pass Calculation

The third-order Butterworth prototype LP filter is schematized in Fig. 4-1, whereas the Butterworth LP filter with cutoff frequency $\omega_c = 2 \times 10^6$ rads/sec, and source and load resistors of 100 ohms, is shown in Fig. 4-2. The denormalized element values, obtained from the normalized values by frequency scaling

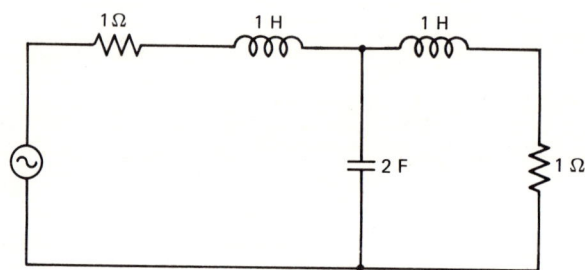

Figure 4-1. Third-order Butterworth prototype LP filter ($\omega_c = 1$).

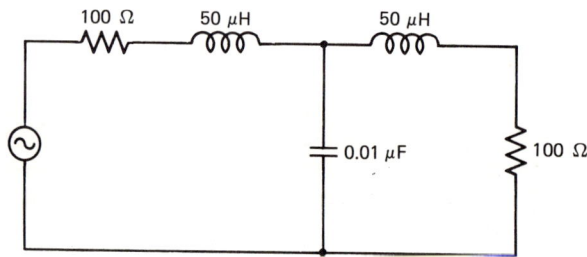

Figure 4-2. Third-order Butterworth LP filter with $\omega_c = 2 \times 10^6$.

according to (4.2-9) or (4.2-11), and impedance scaling by 100 ohms, are

$$L = \frac{RL_N}{\omega_c} = \frac{100 \times 1}{2 \times 10^6} = 50 \ \mu\text{H}$$

$$C = \frac{C_N}{\omega_c R} = \frac{2}{2 \times 10^6 \times 100} = 0.01 \ \mu\text{F}$$

(4.2-12)

The normalized pole positions, from (3.4-12), are -1 and $-\tfrac{1}{2} \pm j\sqrt{3}/2$; and the denormalized poles, from (4.2-4), are

$$s_1 = -1 \times 2 \times 10^6 = -2 \times 10^6$$

$$s_{2,3} = \left(-\frac{1}{2} \pm j\frac{\sqrt{3}}{2}\right) \times 2 \times 10^6 = (-1 \pm j\sqrt{3}) \times 10^6$$

(4.2-13)

Sample calculations of attenuation and passband group delay are listed in Table 4-1. The normalized values are obtained from Figs. 3-24 and 3-25 or from (3.4-11) and (3.4-13), with Ω given by (4.2-2) as

$$\Omega = \frac{\omega}{2 \times 10^6}$$

(4.2-14)

High-Pass Filter

Table 4-1 Attenuation and Group Delay of LP Filter, Fig. 4-2

ω (rads/sec)	Ω	Attenuation (dB)	$D_L(\Omega)$ (sec)	$D_l(\omega) = \dfrac{D_L(\Omega)}{\omega_c}$ (μsec)
0	0	0	2	1
1.0×10^6	0.5	0.07	2.34	1.17
1.4×10^6	0.7	0.5	2.66	1.33
2.0×10^6	1	3	2.5	1.25
4.0×10^6	2	18		
10.0×10^6	5	42		
20.0×10^6	10	60		

The impulse and step responses (Figs. 3-26 and 3-27, respectively) are denormalized in Table 4-2 for two values of time. The normalized time scale for both responses is given by (4.2-7), and the impulse response ordinate is replaced by $h_l(t)/\omega_c$. The step response ordinate remains the same.

Table 4-2 Denormalization of Impulse Response and Step Response of Filter in Fig. 4-2

t (μsec)	$t_N = \omega_c t$ (sec)	$\dfrac{h_l(t)}{\omega_c}$	$h_l(t)$	$g_l(t)$
1	2	0.404	0.808×10^6	0.45
2	4	0.12	0.24×10^6	1.03

4.3 HIGH-PASS FILTER

Two transformations are discussed here. The first is the conventional transformation, which preserves the LP magnitude characteristics, the second transformation preserves the quality (in terms of shape, overshoot, etc.) of the LP transient responses.

4.3.1 Conventional Transformation

The HP filter with 3-dB cutoff frequency ω_c is obtained from the LP prototype filter by the transformation

$$S = \frac{\omega_c}{s} \qquad (4.3\text{-}1)$$

and the normalized frequency, which is always negative, is

$$\Omega = -\frac{\omega_c}{\omega} \qquad (4.3\text{-}2)$$

Attenuation, phase, and group delay of the normalized LP filter are not plotted for negative values of Ω because no additional information is contained there.

Therefore we use $|\Omega|$ and note that the attenuation and phase of the HP filter at ω/ω_c is the same as that of the LP filter at ω_c/ω, except that the sign of the phase is reversed. Thus the LP attenuation and phase characteristics are preserved.

From (4.1-4) and (4.3-2), the LP delay is altered by

$$\frac{d\Omega}{d\omega} = \frac{\omega_c}{\omega^2} = \frac{\Omega^2}{\omega_c} \qquad (4.3\text{-}3)$$

and the group delay ordinate is replaced by $(\omega_c/\Omega^2)D_h(\omega)$, where $D_h(\omega)$ is the denormalized HP delay. The HP passband delay does not resemble the LP passband shape, which is distorted by the factor Ω^2/ω_c. This is not surprising because the HP bandwidth is infinite, and (2.6-10) shows that the group delay approaches zero at high frequencies. Thus it is impossible for the passband delay of the HP filter to retain, for example, the constant passband delay of the Bessel LP filter.

4.3.2 Transient Responses

The LP transient responses are not simply related to the HP transient responses, hence are not preserved. This is seen by first determining the HP denormalized poles s_i from (4.3-1) as

$$s_i = \frac{\omega_c}{S_i} \qquad (4.3\text{-}4)$$

and substituting from (4.3-4) into (4.2-5). The exponential is then $\exp(\omega_c t_N/s_i)$, and for a simple denormalization of the time scale, the exponential must be expressible as $\exp(s_i t)$, where the relationship between t and t_N is independent of the subscript "i." Since this is clearly not possible, each HP transient response must be evaluated by inverse Laplace transforming the appropriate frequency-domain function. Figures 4-3 through 4-6 show the HP step responses $g_H(t)$ obtained in this manner for different filter families, each with $\omega_c = 1$ rad/sec. All exhibit appreciable overshoots, and the low overshoots of the LP Bessel filter are not preserved. The normalized time is indicated by t_H. If the HP cutoff frequency is ω_c rather than unity, the time scale t_H is replaced by $\omega_c t$, as we did for the LP case in (4.2-7).

4.3.3 Relationship Between LP and HP Transient Responses

The LP and HP transient responses are related, although not by a simple time-scale denormalization. The normalized HP filter step response energy E_H dissipated in a one-ohm resistor is equal to the normalized LP filter impulse response energy E_L of (3.8-16) dissipated in a one-ohm resistor [5].

From Parseval's theorem (2.1-31)

$$E_L = \int_0^\infty h_L^2(t)\, dt = \frac{1}{\pi}\int_0^\infty |H_L(j\omega)|^2\, d\omega \qquad (4.3\text{-}5)$$

Likewise the energy E_H in the normalized HP filter step response $g_H(t)$ is

$$E_H = \int_0^\infty g_H^2(t)\, dt = \frac{1}{\pi}\int_0^\infty \frac{1}{\omega^2}|H_H(j\omega)|^2\, d\omega \qquad (4.3\text{-}6)$$

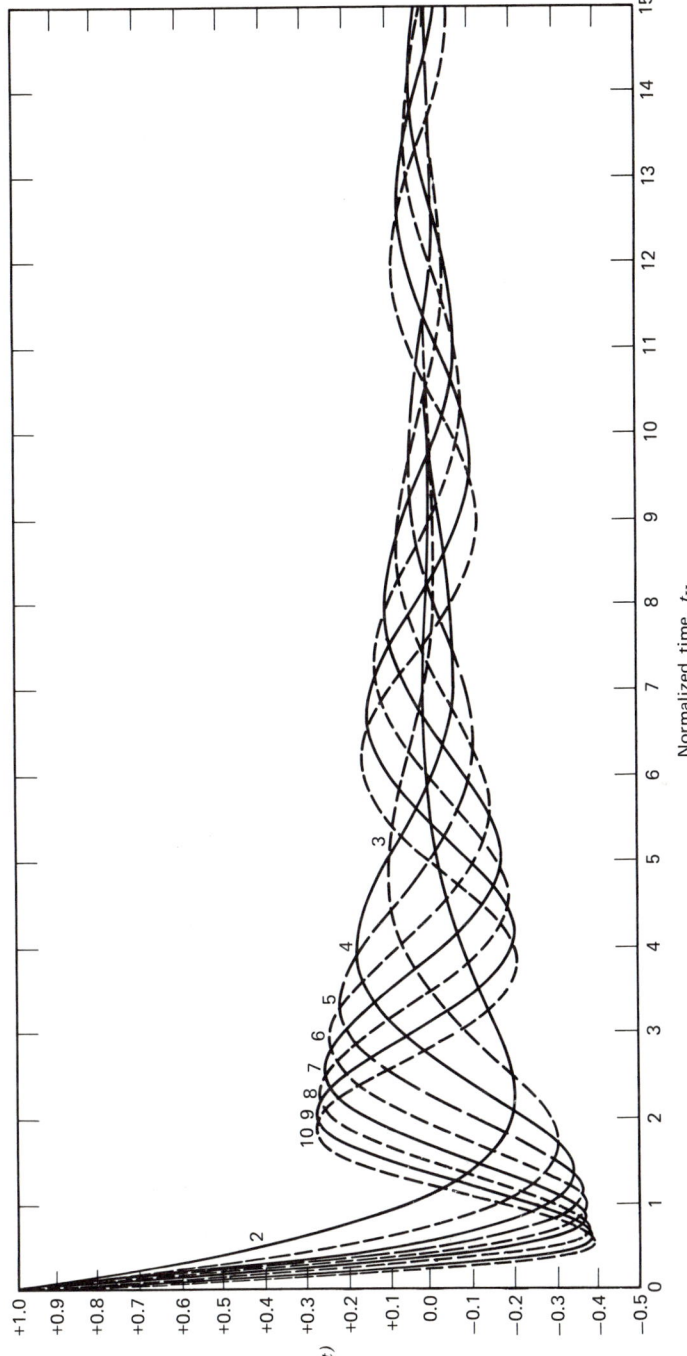

Figure 4-3. Step response of HP Butterworth filters obtained by conventional transformation.

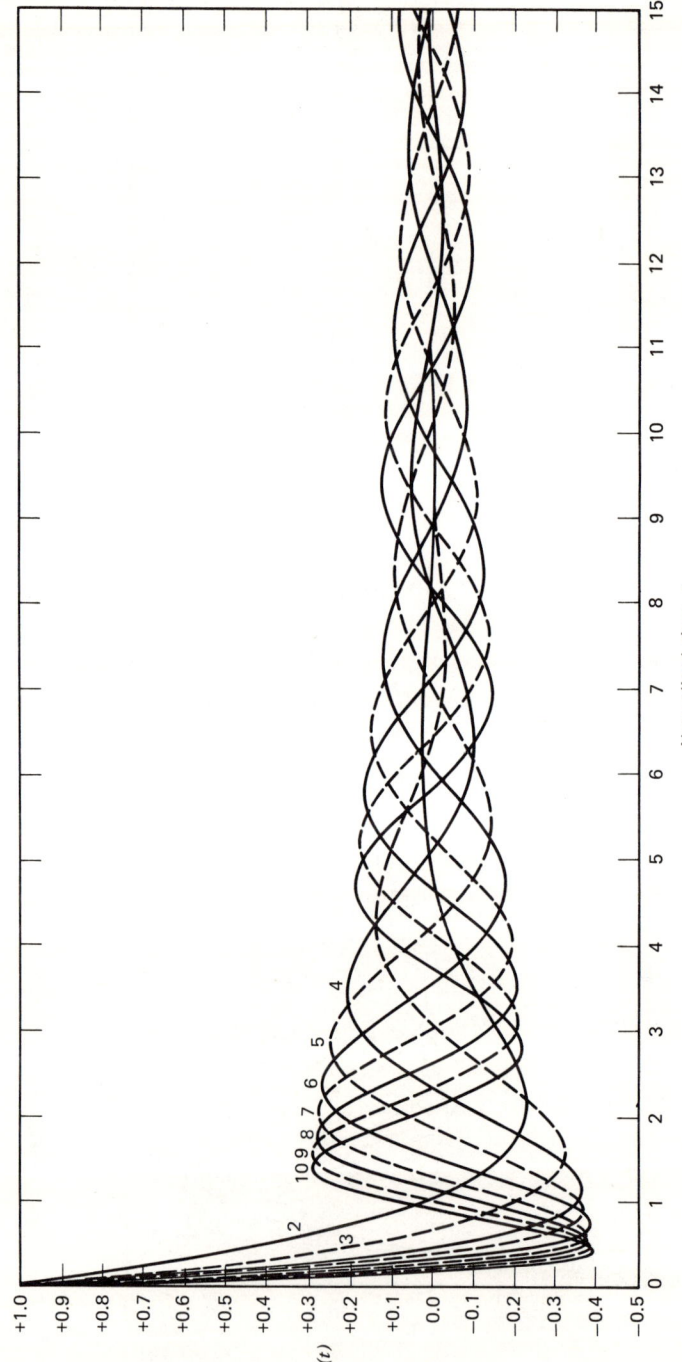

Figure 4-4. Step response of HP Chebyshev filters (0.1-dB ripple) obtained by conventional transformation.

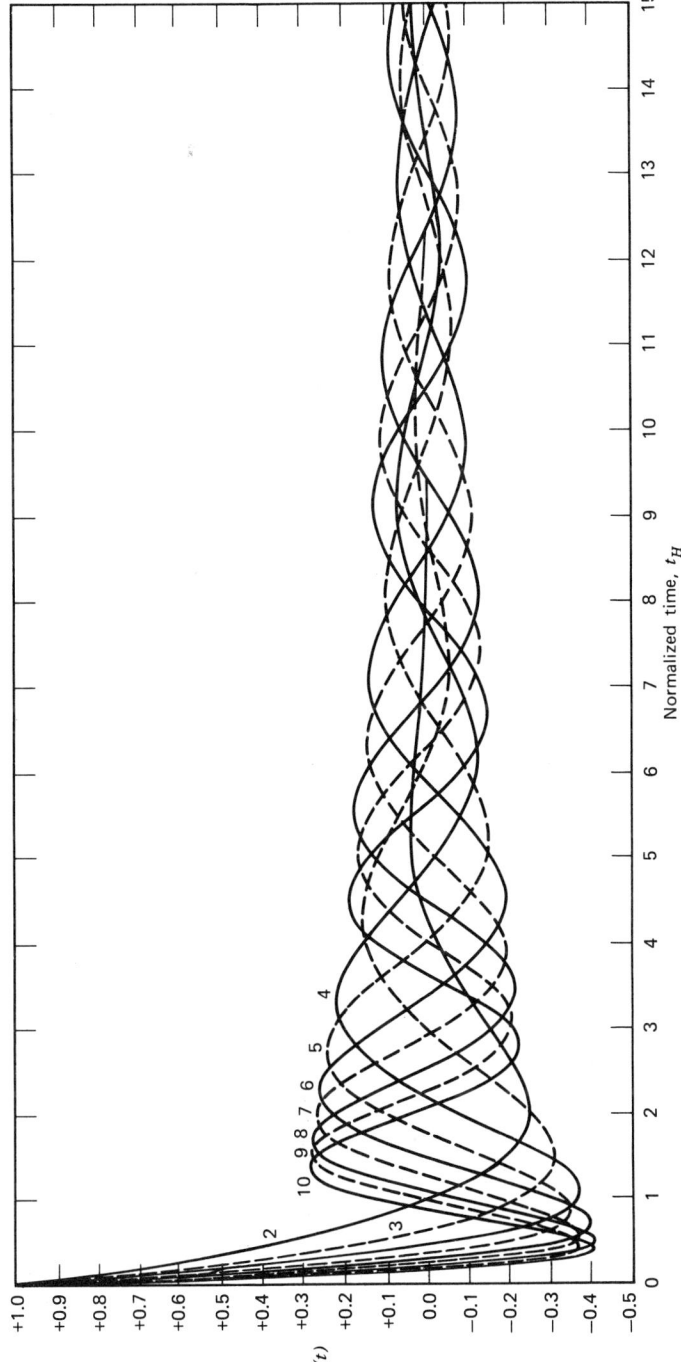

Figure 4-5. Step response of HP Chebyshev filters (0.5-dB ripple) obtained by conventional transformation.

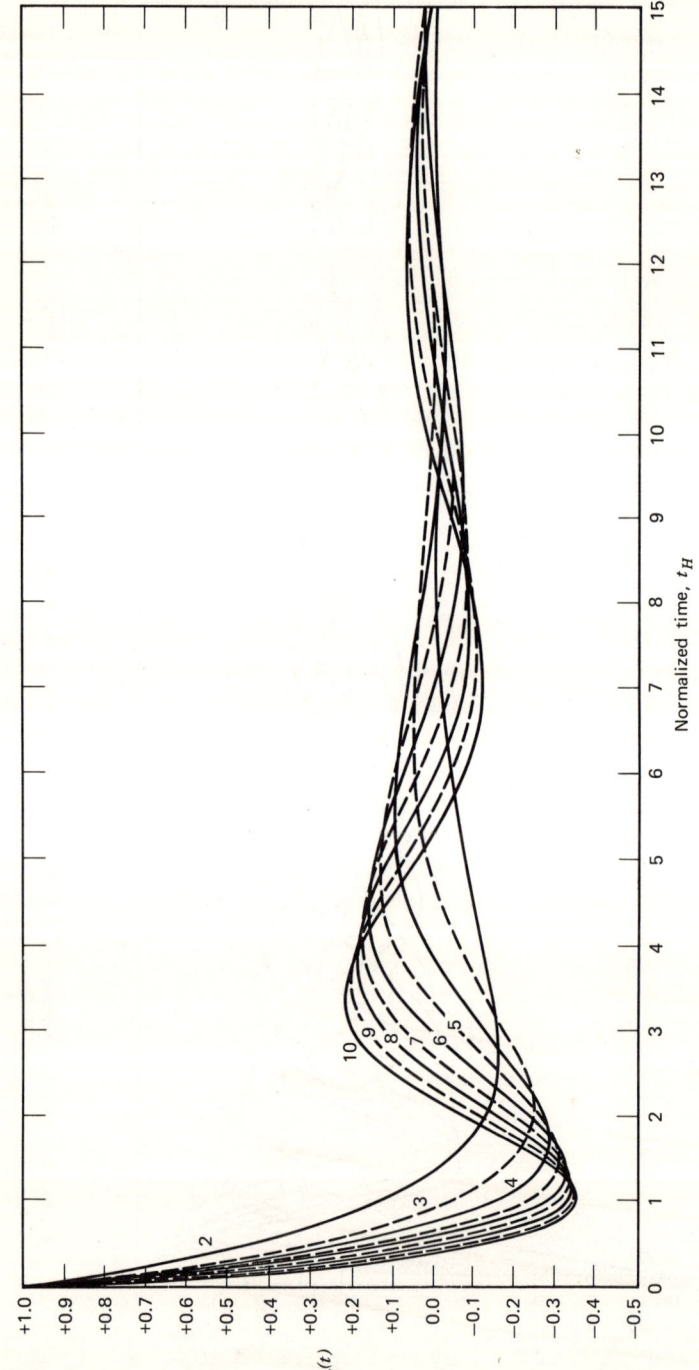

Figure 4-6. Step response of HP Bessel filters obtained by conventional transformation.

High-Pass Filter

where $H_H(s)$ is the normalized HP filter transfer function and $(1/s)H_H(s)$ is the Laplace transform of $g_H(t)$. Apply the frequency transformation (4.3-1) (with $\omega_c = 1$) to give

$$H_H(s) = H_L\left(\frac{1}{s}\right) \qquad (4.3\text{-}7)$$

and (4.3-6) becomes

$$E_H = \frac{1}{\pi}\int_0^\infty \frac{1}{\omega^2}\left|H_L\left(\frac{1}{j\omega}\right)\right|^2 d\omega \qquad (4.3\text{-}8)$$

Let $x = -1/\omega$, $dx = (1/\omega^2)\,d\omega$ and remember that $|H_L(jx)|^2$ is an even function. Then, referring to (4.3-5),

$$E_H = \frac{1}{\pi}\int_0^\infty |H_L(jx)|^2\,dx = E_L \qquad (4.3\text{-}9)$$

This energy (discussed in Section 3.8.3) has been shown to be related to the noise bandwidth by (3.8-16). Consequently E_H and E_L are obtainable from Fig. 3-46.

4.3.4 Element Values

The filter element values are found by applying (4.3-1) to the normalized inductive reactance and then to the normalized capacitive reactance. For the former,

$$SL_N = \frac{\omega_c}{s}L_N = \frac{1}{s(1/\omega_c L_N)} = \frac{1}{sC} \qquad (4.3\text{-}10)$$

revealing that each normalized inductor L_N is replaced by the denormalized capacitor

$$C = \frac{1}{\omega_c L_N} \qquad (4.3\text{-}11)$$

The normalized capacitive reactance is

$$\frac{1}{SC_N} = s\frac{1}{\omega_c C_N} = sL \qquad (4.3\text{-}12)$$

Thus each normalized capacitor C_N is replaced by the denormalized inductor

$$L = \frac{1}{\omega_c C_N} \qquad (4.3\text{-}13)$$

4.3.5 Example of High-Pass Calculation

An HP filter with radian cutoff frequency $\omega_c = 2 \times 10^6$, derived from the third-order Butterworth LP filter in Fig. 4-1, appears in Fig. 4-7. The element values, obtained from the normalized values by frequency scaling according to (4.3-11) or (4.3-13) and impedance scaling by 100 ohms, are

$$\begin{aligned}C &= \frac{1}{\omega_c L_N R} = \frac{1}{2 \times 10^6 \times 1 \times 100} = 5000\text{ pF} \\ L &= \frac{R}{\omega_c C_N} = \frac{100}{2 \times 10^6 \times 2} = 25\ \mu\text{H}\end{aligned} \qquad (4.3\text{-}14)$$

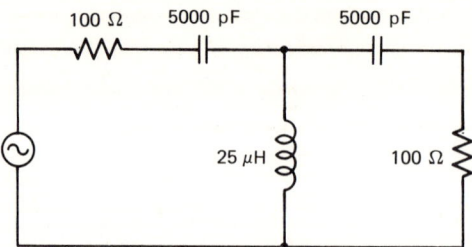

Figure 4-7. Third-order Butterworth HP filter with $\omega_c = 2 \times 10^6$.

The denormalized poles, from (4.3-4), are

$$s_1 = \frac{2 \times 10^6}{-1} = -2 \times 10^6$$

$$s_{2,3} = \frac{2 \times 10^6}{-1/2 \pm j(\sqrt{3}/2)} = (-1 \mp j\sqrt{3}) \times 10^6$$

(4.3-15)

The three zeros at infinity transform to three zeros at $s = 0$, yielding the HP transfer function

$$H_h(s) = \frac{s^3}{(s - s_1)(s - s_2)(s - s_3)}$$

(4.3-16)

The attenuation and passband group delay $D_h(\omega)$ are given in Table 4-3 for the normalized values in Table 4-1, and each normalized response is presented in Fig.

Table 4-3 Attenuation and Group Delay of HP Filter in Fig. 4-7

ω	$\|\Omega\|$	Attenuation (dB)	$D_L(\Omega)$ (sec)	$D_h(\omega) = \frac{\Omega^2}{\omega_c} D_L(\Omega)$ (μsec)
0.2×10^6	10.0	60		
0.4×10^6	5.0	42		
1.0×10^6	2.0	18		
2.0×10^6	1.0	3	2.5	1.25
2.86×10^6	0.7	0.5	2.66	0.652
4.0×10^6	0.5	0.07	2.34	0.293

4-8. The attenuation, as expected, preserves the LP characteristic, while the passband group delay bears little resemblance to the LP passband delay in Fig. 3-25 ($n = 3$).

However the group delay in Fig. 4-8 is identical to the third-order curve in Fig. 3-25. For cutoff frequencies other than $\omega_c = 1$, the Butterworth HP group delay is a scaled version of $D_L(\Omega)$. This unusual feature arises for the Butterworth family because the HP poles also lie on a circle in the s-plane. The poles in (4.3-15) lie on a circle of radius 2×10^6, the radian cutoff frequency. Except for a scale factor, the Butterworth group delay is invariant under the LP-to-HP frequency transformation.

High-Pass Filter

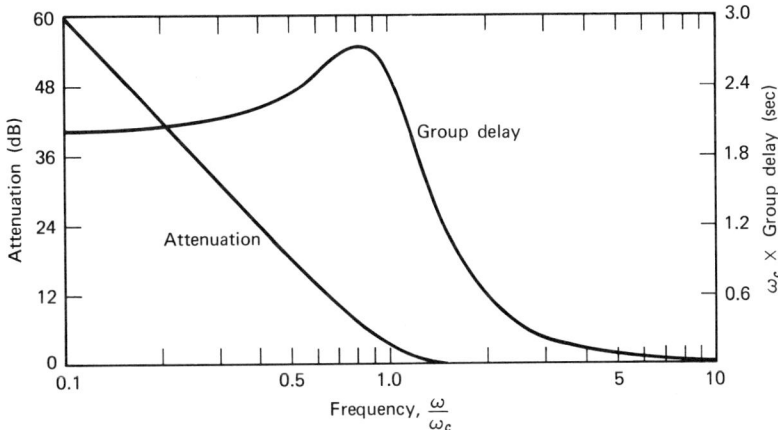

Figure 4-8. Normalized attenuation and group delay for the HP filter of Section 4.3.5 ($\omega_c = 2 \times 10^6$).

We determine the step response to illustrate a transient calculation. The unit-radian cutoff filter is considered and the actual response is then obtained by a time-scale denormalization. The HP poles computed from (4.3-4) are the same as the Butterworth LP poles; thus the HP transfer function is

$$H_H(s) = \frac{s^3}{(s+1)[(s+\tfrac{1}{2})^2 + \tfrac{3}{4}]} \qquad (4.3\text{-}17)$$

and the step response $g_H(t)$ is

$$g_H(t) = \mathscr{L}^{-1}\left[\frac{1}{s} H_H(s)\right] = \mathscr{L}^{-1}\left[\frac{s^2}{(s+1)[(s+\tfrac{1}{2})^2 + \tfrac{3}{4}]}\right] \qquad (4.3\text{-}18)$$

The partial function expansion of $(1/s)H_H(s)$ is

$$\frac{1}{s} H_H(s) = \frac{1}{s+1} - \frac{1}{(s+\tfrac{1}{2})^2 + \tfrac{3}{4}} \qquad (4.3\text{-}19)$$

From Table 2-1,

$$g_H(t) = \left[e^{-t} - \frac{2}{\sqrt{3}} e^{-t/2} \sin \frac{\sqrt{3}}{2} t\right] u_{-1}(t) \qquad (4.3\text{-}20)$$

as in Fig. 4-3 ($n = 3$). The time scale is denormalized by replacing t_H by

$$t_H = \omega_c t = 2 \times 10^6 t \qquad (4.3\text{-}21)$$

Thus the first undershoot reaches its minimum at $t_H = 1.7$ or $t = 0.85$ μsec.

4.3.6 Preservation of LP Transient Characteristics

Low-overshoot HP filters are of interest because they are the prototypes for the bandstop filters with low-overshoot transient responses. This characteristic of the LP constant-delay type filters is not preserved by the previous transformation, but it is preserved by the transformation now described [2].

Let the HP transfer function $H_h(s)$ be the frequency-domain complement of the LP transfer function $H_l(s)$; that is,

$$H_h(s) = 1 - H_l(s) \qquad (4.3\text{-}22)$$

Then the HP impulse response $h_h(t)$, which is the Laplace transform of $H_h(s)$, is

$$h_h(t) = \delta(t) - h_l(t) \qquad (4.3\text{-}23)$$

and $g_h(t)$, the integral of $h_h(t)$, is

$$g_h(t) = 1 - g_l(t) \qquad (4.3\text{-}24)$$

The principal aspect of this transformation is the preservation of the LP transient responses' quality (in terms of shape, overshoot, etc.). The impulse response of the HP filter, except for the impulse at $t = 0$, is the negative of the LP filter impulse response. Thus for $t > 0$, the absolute value of the impulse response is invariant under the transformation. The HP filter step response is the time-domain complement of the LP filter step response.

The frequency-domain characteristics are obtained by substituting the all-pole LP transfer function from (3.3-1), with $K = a_n$, into (4.3-22) to yield the HP transfer function

$$H_h(s) = \frac{a_0 s^n + a_1 s^{n-1} + \cdots + a_{n-1} s}{a_0 s^n + a_1 s^{n-1} + \cdots + a_n} \qquad (4.3\text{-}25)$$

Only one of the n zeros of $H_h(s)$ occurs at $s = 0$, whereas the previous transformation resulted in all n zeros occurring there. This new transformation distorts the magnitude response, but such distortion does not eliminate the practical importance of these filters because a passband ripple of a few decibels is acceptable in many cases. For small values of ω, $|H_h(j\omega)| \approx (a_{n-1}/a_n)\omega$, and it is clear that the maximum attenuation rate of this HP filter frequency response near $\omega = 0$ is 6 dB/octave, independent of the filter order n.

Networks realizing the transfer function of (4.3-25) are not lossless but contain resistors. Section 4.3.7 gives one such network.

4.3.7 Application of New High-Pass Transformation

The second-order LP Bessel transfer function derived from (3.5-13) is

$$H_l(s) = \frac{3}{s^2 + 3s + 3} \qquad (4.3\text{-}26)$$

Substituting $H_l(s)$ into (4.3-22) and normalizing for a unity 3-dB radian bandwidth yields the HP transfer function

$$H_H(s) = \frac{s(s + 3.9469)}{s^2 + 3.9469s + 5.1926} \qquad (4.3\text{-}27)$$

The magnitude response, shown in Fig. 4-9 along with the magnitude response obtained by the conventional transformation, exhibits a passband ripple of 1.5 dB and the aforementioned 6 dB/octave asymptotic behavior near $\omega = 0$.

The step response is the inverse Laplace transform of $(1/s)H_H(s)$,

$$g_H(t) = e^{-1.9734t}[\cos 1.1394t + 1.732 \sin 1.1394t]u_{-1}(t) \qquad (4.3\text{-}28)$$

Bandpass Filter

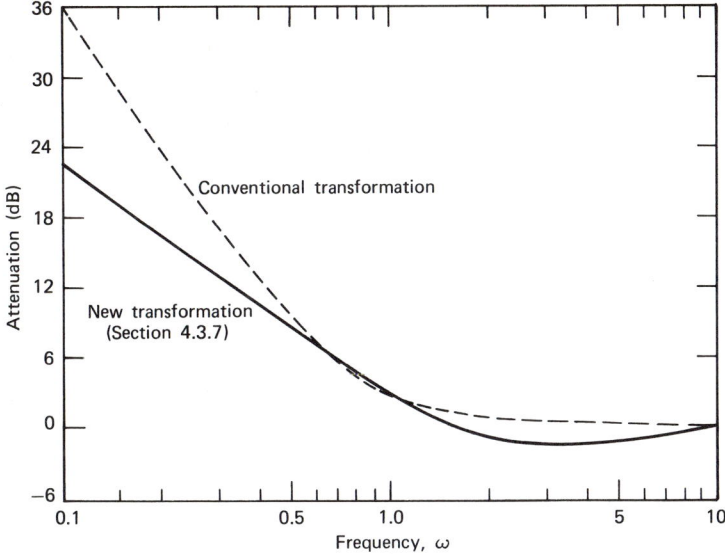

Figure 4-9. Magnitude response for $n = 2$ HP filters derived from LP Bessel prototype.

and is given in Fig. 4-10, with the second-order curve from Fig. 4-6. The low overshoots of the LP Bessel filter are preserved, in accordance with (4.3-24).

A network realization operating between 1-ohm source and load resistors is diagrammed in Fig. 4-11.

4.4 BANDPASS FILTER

The BP filter is the filter type most used in practice, consequently the majority of this chapter is devoted to it. First we consider the BP filter obtained from the LP prototype by the conventional transformation, which preserves the magnitude response on a geometric basis but distorts the LP group delay. In Section 4.5 we show how to systematically design narrow-band BP filters realized as coupled resonators, starting with the LP prototype. For this class the magnitude and group delay responses are preserved on an arithmetic basis. Finally, in Section 4.6, we realize a BP filter whose group delay is a least-squares approximation to a constant delay over the passband for large relative bandwidths. Since there is no simple transformation for obtaining this filter, an electronic computer is used to determine the optimum filter parameters.

4.4.1 Conventional Transformation

The BP filter with center frequency ω_0, lower 3-dB frequency ω_1, and upper 3-dB frequency ω_2 is obtained from the LP prototype filter by the transformation

$$S = \frac{1}{\gamma}\left(\frac{s}{\omega_0} + \frac{\omega_0}{s}\right) \tag{4.4-1}$$

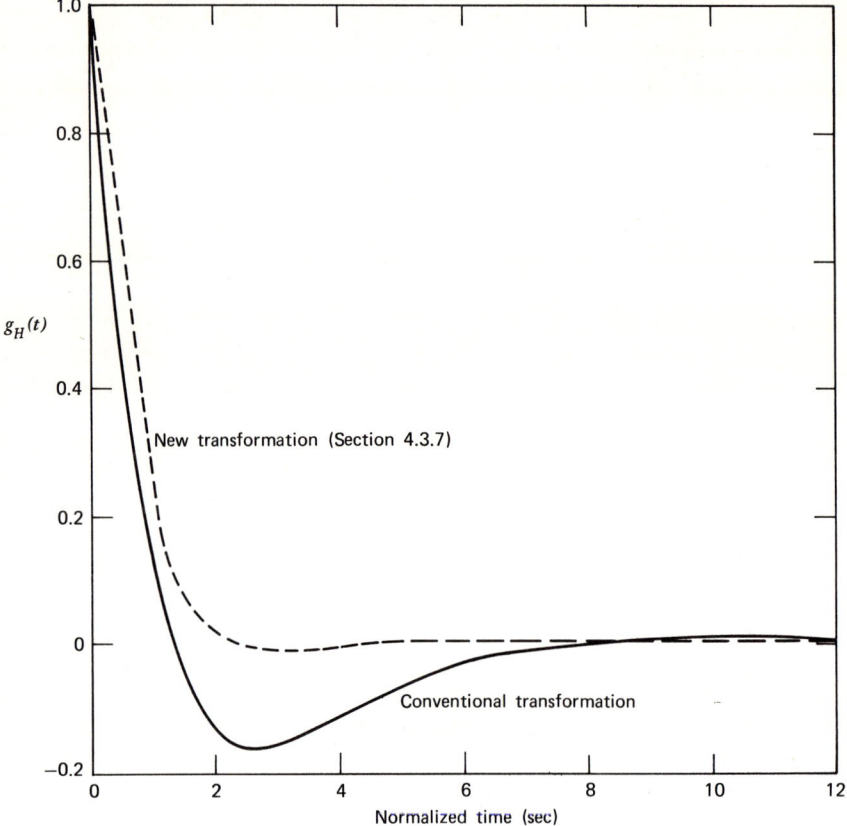

Figure 4-10. Step response for $n = 2$ HP filters derived from LP Bessel prototype.

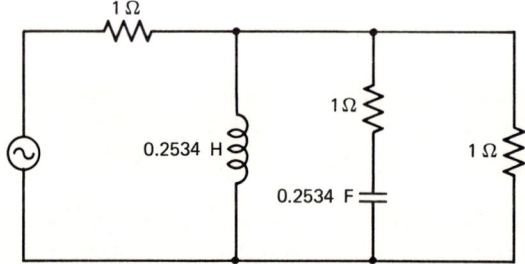

Figure 4-11. A network for the $n = 2$ HP Bessel filter in Section 4.3.7.

where

$$\gamma = \frac{\omega_2 - \omega_1}{\omega_0}, \text{ the relative bandwidth} \qquad (4.4\text{-}2)$$

The normalized frequency is then

$$\Omega = \frac{1}{\gamma}\left(\frac{\omega}{\omega_0} - \frac{\omega_0}{\omega}\right) \qquad (4.4\text{-}3)$$

Bandpass Filter

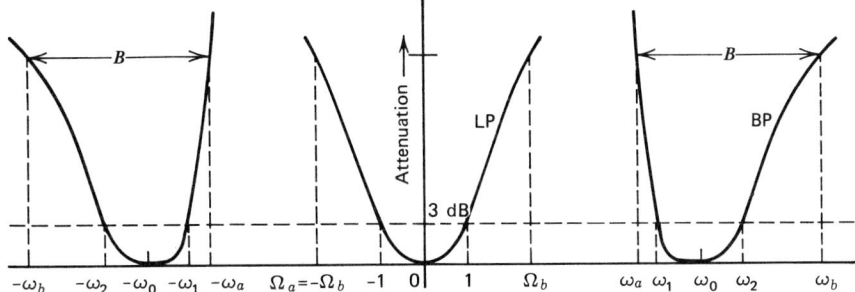

Figure 4-12. LP and BP attenuation curves obtained by the transformation in (4.4-1).

and as a consequence of this transformation, ω_0 is the geometric mean of ω_1 and ω_2,

$$\omega_0 = \sqrt{\omega_1 \omega_2} \tag{4.4-4}$$

Figure 4-12 illustrates a typical LP attenuation response and the BP responses at positive and negative frequencies resulting from the transformation. Use of (4.4-3) yields the values in Table 4-4 for Ω at the important BP frequencies. At positive frequencies Ω is negative for $\omega < \omega_0$, but as in the HP case, $|\Omega|$ is used to enter the LP normalized curves.

Table 4-4 Values of Ω at Important BP Frequencies

ω	Ω
$-\omega_2, \omega_1$	$\Omega_1 = -1$
ω_0	0
$-\omega_1, \omega_2$	$\Omega_2 = +1$

We now demonstrate that a frequency ratio in the LP case corresponds to a bandwidth ratio in the BP case, thus relating the shape factors of the LP and BP attenuation responses. Consider the frequencies ω_a and ω_b in Fig. 4-12 (corresponding to the LP frequencies Ω_a and Ω_b, respectively) related to ω_0 by

$$\omega_0 = \sqrt{\omega_a \omega_b} \tag{4.4-5}$$

and forming the bandwidth

$$B = \omega_b - \omega_a \tag{4.4-6}$$

At ω_b, from (4.4-3),

$$\Omega_b = \frac{1}{\gamma}\left(\frac{\omega_b}{\omega_0} - \frac{\omega_0}{\omega_b}\right) \tag{4.4-7}$$

and at ω_2

$$\Omega_2 = \frac{1}{\gamma}\left(\frac{\omega_2}{\omega_0} - \frac{\omega_0}{\omega_2}\right) \tag{4.4-8}$$

The ratio Ω_b/Ω_2, after use of (4.4-4) and (4.4-5), is

$$\frac{\Omega_b}{\Omega_2} = \frac{\omega_2(\omega_b^2 - \omega_0^2)}{\omega_b(\omega_2^2 - \omega_0^2)} = \frac{\omega_2(\omega_b^2 - \omega_a \omega_b)}{\omega_b(\omega_2^2 - \omega_1 \omega_2)} = \frac{\omega_b - \omega_a}{\omega_2 - \omega_1} \tag{4.4-9}$$

In a similar manner,

$$\frac{\Omega_a}{\Omega_1} = \frac{\Omega_b}{\Omega_2} = \frac{\omega_b - \omega_a}{\omega_2 - \omega_1} \qquad (4.4\text{-}10)$$

but from Table 4-4 $|\Omega_1| = |\Omega_2| = 1$, and (4.4-10) reduces to

$$|\Omega_a| = \Omega_b = \frac{\omega_b - \omega_a}{\omega_2 - \omega_1} \qquad (4.4\text{-}11)$$

Equation 4.4-11 relates the LP and BP shape factors. The attenuation at Ω_b is the same as the attenuation at the frequencies whose bandwidth is Ω_b times the BP 3-dB bandwidth. Thus the BP attenuation response is easily computed on a bandwidth basis, and the absolute frequencies ω_a and ω_b are obtained as follows. Substitute $\omega_a = \omega_0^2/\omega_b$ from (4.4-5) into (4.4-6) to give

$$\omega_b^2 - B\omega_b - \omega_0^2 = 0 \qquad (4.4\text{-}12)$$

The solution to (4.4-12) yields a positive and a negative root. The former is associated with the response at positive frequencies, the latter with the response at negative frequencies. The positive root

$$\omega_b = \omega_0 \left[\sqrt{1 + \left(\frac{B}{2\omega_0}\right)^2} + \frac{B}{2\omega_0} \right] \qquad (4.4\text{-}13)$$

is then substituted into (4.4-6) to yield

$$\omega_a = \omega_0 \left[\sqrt{1 + \left(\frac{B}{2\omega_0}\right)^2} - \frac{B}{2\omega_0} \right] \qquad (4.4\text{-}14)$$

A consequence of (4.4-5) is an attenuation response that is geometrically symmetric about ω_0 rather than arithmetically symmetric, (Fig. 4-12). Therefore the response below ω_0 is more compressed than the response above that frequency. For $\omega > \omega_0$ the phase angle is the same as that of the LP filter, but for $\omega < \omega_0$ the sign of the phase angle is changed from lagging to leading.

Example 4-1. A BP filter derived from the fifth-order Butterworth LP filter has a 3-dB bandwidth of 30 MHz and a center frequency of 60 MHz. What are the 3-dB frequencies and the frequencies at which the attenuation is 60.2 dB? We find the 3-dB frequencies f_2 and f_1 from (4.4-13) and (4.4-14), respectively, where here B is the 3-dB radian bandwidth. These values are

$$f_b = f_2 = 60 \times 10^6 \left[\sqrt{1 + \frac{1}{16}} + \frac{1}{4} \right] = 76.8466 \text{ MHz}$$

$$f_a = f_1 = 60 \times 10^6 \left[\sqrt{1 + \frac{1}{16}} - \frac{1}{4} \right] = 46.8466 \text{ MHz}$$
$$(4.4\text{-}15)$$

If the response were arithmetically symmetric, the cutoff frequencies would be 45 and 75 MHz. From Fig. 3-24, the attenuation is 60.2 dB at $\Omega_b = 4$; thus from (4.4-11)

$$(f_b - f_a) = 4(f_2 - f_1) = 120 \text{ MHz} \qquad (4.4\text{-}16)$$

Bandpass Filter

The actual frequencies f_a and f_b are again found from (4.4-13) and (4.4-14), where B is now the 60.2-dB bandwidth. Then

$$\frac{B}{2\omega_0} = \frac{2\pi \times 120 \times 10^6}{2 \times 2\pi \times 60 \times 10^6} = 1 \tag{4.4-17}$$

and

$$\begin{aligned} f_b &= 60 \times 10^6 (\sqrt{2} + 1) = 144.853 \text{ MHz} \\ f_a &= 60 \times 10^6 (\sqrt{2} - 1) = 24.853 \text{ MHz} \end{aligned} \tag{4.4-18}$$

4.4.2 Group Delay Behavior

The BP group delay $D_b(\omega)$, from (4.1-4) and (4.4-3), is

$$D_b(\omega) = D_L(\Omega) r(\omega) \tag{4.4-19}$$

where

$$r(\omega) = \frac{d\Omega}{d\omega} = \frac{1}{\omega_2 - \omega_1}\left(1 + \frac{\omega_0^2}{\omega^2}\right) \tag{4.4-20}$$

This transformation distorts the delay characteristics of the LP filter; the distortion increases as the deviation from center frequency increases. Asymmetry appears, with the delay at frequencies below ω_0 being emphasized and the delay at frequencies above ω_0 deemphasized. The delay of a wide-bandwidth filter is so distorted that it no longer possesses the attributes of the LP delay. At band center,

$$D_b(\omega_0) = \frac{2 D_L(0)}{\omega_2 - \omega_1} \tag{4.4-21}$$

which is the LP delay at $\omega = 0$ divided by one-half the radian 3-dB bandwidth. This value is independent of the center frequency.

The derivative of the delay curve at band center is found by differentiating both sides of (4.4-19) with respect to ω and evaluating the result at $\omega = \omega_0$ [1]. Then

$$\frac{dD_b(\omega)}{d\omega} = \frac{dD_L(\Omega)}{d\omega} r(\omega) + D_L(\Omega) \frac{dr(\omega)}{d\omega} \tag{4.4-22}$$

and aided by (4.1-4) and (4.4-19),

$$\frac{dD_b(\omega)}{d\omega} = r^2(\omega) \frac{dD_L(\Omega)}{d\Omega} + D_b(\omega) \frac{1}{r(\omega)} \frac{dr(\omega)}{d\omega} \tag{4.4-23}$$

We find from (2.6-10) with ω replaced by Ω that $dD_L(\Omega)/d\Omega$ is proportional to Ω; hence the first term in (4.4-23) is zero because $\omega = \omega_0$ corresponds to $\Omega = 0$. From (4.4-20)

$$\left.\frac{1}{r(\omega)} \frac{dr(\omega)}{d\omega}\right|_{\omega=\omega_0} = -\frac{1}{\omega_0} \tag{4.4-24}$$

and (4.4-23) reduces to the surprisingly simple formula

$$\left.\frac{dD_b(\omega)}{d\omega}\right|_{\omega=\omega_0} = -\frac{D_b(\omega_0)}{\omega_0} \tag{4.4-25}$$

The derivative of the delay at band center is the negative of the delay there divided by the radian center frequency. For frequencies at which the definition of group delay has physical interpretation, delay must be positive; hence the derivative at band center is always negative. Substitute from (4.4-21) into (4.4-25) to give

$$\left.\frac{dD_b(\omega)}{d\omega}\right|_{\omega=\omega_0} = -\frac{2D_L(0)}{\omega_0(\omega_2-\omega_1)} \quad (4.4\text{-}26)$$

The LP delay at $\omega = 0$ is found from (3.3-5), (3.3-7), or (3.3-9), allowing easy computation of the BP delay and its derivative at band center.

4.4.3 Pole-Zero Locations

The denormalized BP pole s_i (or zero) is related to the normalized LP pole S_i (or zero) by (4.4-1):

$$S_i = \frac{1}{\gamma}\left(\frac{s_i^2 + \omega_0^2}{\omega_0 s_i}\right) \quad (4.4\text{-}27)$$

Solving for s_i gives

$$s_i = \omega_0\left[\frac{\gamma}{2}S_i \pm j\sqrt{1-\left(\frac{\gamma S_i}{2}\right)^2}\right] \quad (4.4\text{-}28)$$

The real value $S_i = -\alpha$, provided $(\gamma\alpha/2)^2 < 1$, transforms to the complex conjugate pair

$$s_{1,2} = -\frac{\omega_2-\omega_1}{2}\alpha \pm j\omega_0\sqrt{1-\left(\frac{\gamma\alpha}{2}\right)^2} \quad (4.4\text{-}29)$$

as in Fig. 4-13a. Zeros on the $j\Omega$ axis, $S_i = \pm j\beta$, transform to the two conjugate pairs on the $j\omega$ axis,

$$s_{1,2} = \pm j\left[\frac{\omega_2-\omega_1}{2}\beta \pm \omega_0\sqrt{1+\left(\frac{\gamma\beta}{2}\right)^2}\right] \quad (4.4\text{-}30)$$

as in Fig. 4-13b.

A complex conjugate pair in the S-plane, $-\alpha \pm j\beta$, transforms to two pairs of complex conjugate s-plane locations, having the property that a straight line from the origin to each singularity makes the same angle with the $j\omega$ axis (Fig. 4-13c). The equations for the s-plane pairs depend on the sign of α, and are given in (4.4-31).

	$\alpha > 0$	$\alpha < 0$
Pair 1:	$\omega_0\left[-\frac{\gamma}{2}\alpha + b \pm j\left(\frac{\gamma}{2}\beta - a\right)\right]$	$\omega_0\left[-\frac{\gamma}{2}\alpha + b \pm j\left(\frac{\gamma}{2}\beta + a\right)\right]$
Pair 2:	$\omega_0\left[-\frac{\gamma}{2}\alpha - b \pm j\left(\frac{\gamma}{2}\beta + a\right)\right]$	$\omega_0\left[-\frac{\gamma}{2}\alpha - b \pm j\left(\frac{\gamma}{2}\beta - a\right)\right]$

(4.4-31)

where

$$A = 1 - \frac{\gamma^2}{4}(\alpha^2 - \beta^2) \qquad B = \frac{-\alpha\beta\gamma^2}{2}$$

$$a = \sqrt{\frac{\sqrt{A^2+B^2}+A}{2}} \qquad b = \sqrt{\frac{\sqrt{A^2+B^2}-A}{2}}$$

From (4.4-27) we see that each LP zero at infinity produces a BP zero at $s = 0$; thus an nth-order all-pole LP prototype transforms to a BP filter with n zeros at

Bandpass Filter

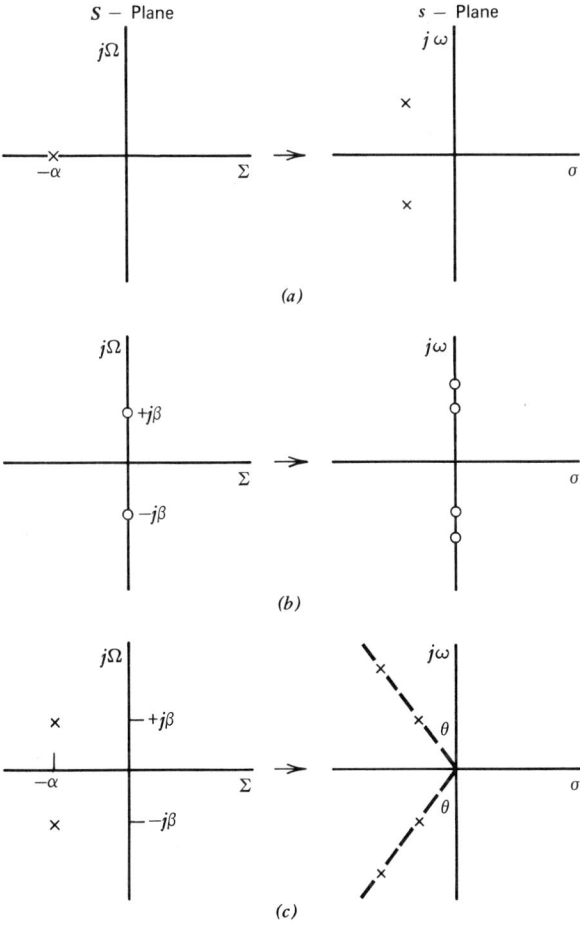

Figure 4-13. s-Plane singularities resulting from conventional transformation (4.4-1).

$s = 0$ and n zeros at infinity. To appreciate this, replace "s" in the LP transfer function by the function in (4.4-1).

There is no simple relationship between the LP and wide-bandwidth BP transient responses, although methods for relating the two are given in Refs. 8 and 9. However the transient responses of LP and narrow-band BP filters are simply related, and this topic is discussed in Section 4.5.7.

4.4.4 Element Values

The BP network parameters are obtained by replacing the LP frequency variable in the LP impedances by the BP frequency variable in (4.4-1). The impedance of the LP inductor SL_N transforms to

$$SL_N = \frac{\omega_0 L_N}{\omega_2 - \omega_1}\left(\frac{s}{\omega_0} + \frac{\omega_0}{s}\right) = s\frac{L_N}{\omega_2 - \omega_1} + \frac{1}{s(\gamma/L_N\omega_0)} \tag{4.4-32}$$

Each LP inductor L_N is replaced by a series LC combination resonating at ω_0, for which the inductor value is $L_N/(\omega_2 - \omega_1)$ and the capacitor value is $\gamma/L_N\omega_0$. The impedance of the LP capacitor $1/SC_N$ transforms to

$$\frac{1}{SC_N} = \frac{1}{\dfrac{C_N\omega_0}{\omega_2 - \omega_1}\left(\dfrac{s}{\omega_0} + \dfrac{\omega_0}{s}\right)} = \frac{1}{\dfrac{C_N}{\omega_2 - \omega_1}s + \dfrac{1}{s(\gamma/\omega_0 C_N)}} \quad (4.4\text{-}33)$$

Each LP capacitor C_N is replaced by a parallel LC combination resonating at ω_0, for which the inductor value is $\gamma/C_N\omega_0$ and the capacitor value is $C_N/(\omega_2 - \omega_1)$. Figure 4-14 gives the BP elements after the transformation.

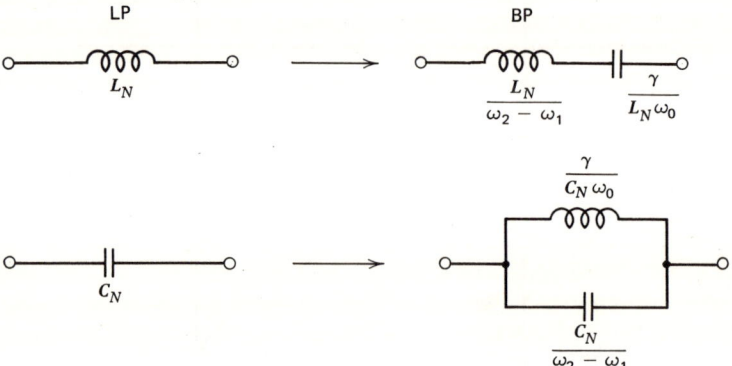

Figure 4-14. BP element values derived from LP prototype.

The ratio of the BP inductor values is $L_N C_N/\gamma^2$ and the ratio of the BP capacitor values is $\gamma^2/L_N C_N$. For narrow bandwidths, these ratios achieve extreme values, resulting in impractical element values. The element-value ratio for a relative bandwidth of 2% is approximately 2500. These situations call for the narrow-band schematics in Section 4.5.8.

4.4.5 Example of Bandpass Calculation

Using the third-order Butterworth LP filter in Fig. 4-1 as the prototype, we now design a BP filter with $f_1 = 40$ MHz and $f_2 = 90$ MHz. Source and load resistors are each 100 ohms. The center frequency, from (4.4-4), is

$$\omega_0 = 2\pi\sqrt{40 \times 10^6 \times 90 \times 10^6} = 120\pi \times 10^6 \text{ rads/sec} \quad (4.4\text{-}34)$$

and

$$\gamma = \frac{\omega_2 - \omega_1}{\omega_0} = \frac{5}{6} \quad (4.4\text{-}35)$$

The element values corresponding to the BP schematic in Fig. 4-15 are obtained

Bandpass Filter

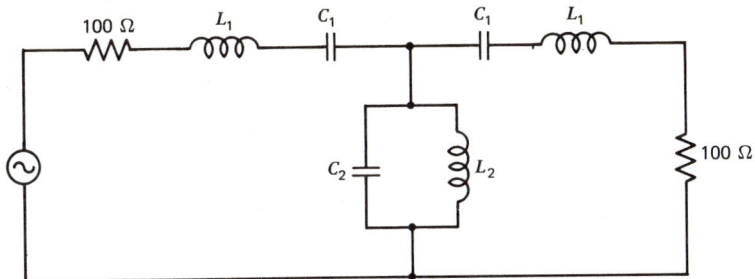

Figure 4-15. BP schematic for filter of Section 4.4.5.

from the equations in Fig. 4-14 as

$$L_1 = \frac{RL_N}{\omega_2 - \omega_1} = \frac{100 \times 1}{100\pi \times 10^6} = 0.3183 \ \mu\text{H}$$

$$C_1 = \frac{\gamma}{RL_N\omega_0} = \frac{5}{6 \times 100 \times 1 \times 120\pi \times 10^6} = 22.1 \ \text{pF}$$

$$C_2 = \frac{C_N}{R(\omega_2 - \omega_1)} = \frac{2}{100 \times 100\pi \times 10^6} = 63.7 \ \text{pF} \qquad (4.4\text{-}36)$$

$$L_2 = \frac{\gamma R}{C_N\omega_0} = \frac{5 \times 100}{6 \times 2 \times 120\pi \times 10^6} = 0.1105 \ \mu\text{H}$$

The normalized poles, from (3.4-12), are -1 and $-1/2 \pm j\sqrt{3}/2$. The denormalized poles are obtained from (4.4-29) with $\alpha = 1$ as

$$s_{1,2} = 120\pi \times 10^6 \left[-\frac{5}{12} \pm j\frac{\sqrt{119}}{12} \right] = (-50 \pm j109.087)\pi \times 10^6 \qquad (4.4\text{-}37)$$

and from (4.4-31) with $\alpha = 1/2$ and $\beta = \sqrt{3}/2$ as

$$\begin{aligned}\text{pole pair 1} &= (-16.367 \pm j82.096)\pi \times 10^6 \\ \text{pole pair 2} &= (-33.633 \pm j168.699)\pi \times 10^6\end{aligned} \qquad (4.4\text{-}38)$$

There are also three zeros at the origin because the LP prototype has three zeros at infinity. The pole-zero plot in Fig. 4-16 reveals that the straight line drawn from the origin to the poles of (4.4-38) does indeed make the same angle θ with the $j\omega$ axis. The poles in the upper half plane form a cluster about $j\omega_0$ while their complex conjugates form a cluster about $-j\omega_0$. This arrangement is typical for BP pole locations.

The BP attenuation, group delay, and phase delay are summarized in Table 4-5 for several frequencies to illustrate use of the normalized curves. From (4.4-3)

$$\Omega = 1.2 \left(\frac{f}{f_0} - \frac{f_0}{f} \right) . \qquad (4.4\text{-}39)$$

and the normalized LP attenuation and group delay are given in Figs. 3-24 and 3-25, respectively; the normalized phase response can be taken from (2.5-42). The BP group delay $D_b(\omega)$ is given by (4.4-19) and the BP phase delay $\tau_p(\omega)$ is found from (2.6-2). Note that 30 and 120 MHz satisfy (4.4-5); hence the attenuation at

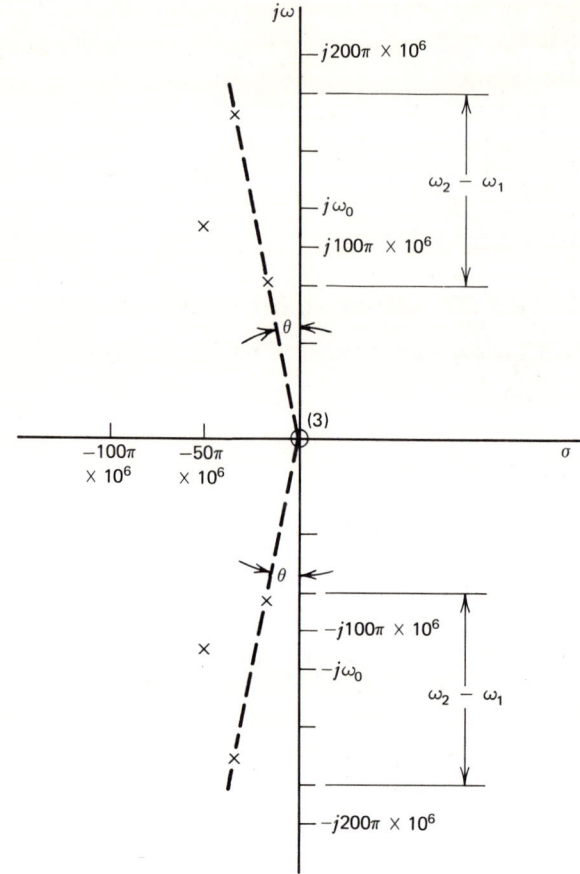

Figure 4-16. Pole-zero diagram for BP filter of Section 4.4.5.

Table 4-5 Attenuation, Group Delay, and Phase Delay of Filter in Fig. 4-15

f(MHz)	Ω	Attenuation (dB)	$D_L(\Omega)$	$1+\left(\dfrac{\omega_0}{\omega}\right)^2$	$D_b(\omega)$ (nsec)	BP Phase Angle	$\tau_p(\omega)$ (nsec)
30	−1.8	15.4	0.75	5	11.93	202.2°	−18.72
40	−1	3	2.5	3.25	25.86	135°	−9.38
60	0	0	2	2	12.73	0°	0
90	1	3	2.5	1.44	11.49	−135°	4.17
120	1.8	15.4	0.75	1.25	2.98	−202.2°	4.68

these frequencies is the same. Attenuation and group delay are represented in Fig. 4-17a, the former displaying the expected geometric symmetry about f_0, whereas the latter possesses no symmetry and is a distorted version of the LP delay. In the absence of distortion the BP delay is symmetric about f_0. This characteristic is discussed in Section 4.5.4.

As an example of the phase delay computation, consider the frequency 40 MHz. From (4.4-39), $|\Omega| = 1$, and the phase of the LP prototype there is −135° (see Fig.

Bandpass Filter

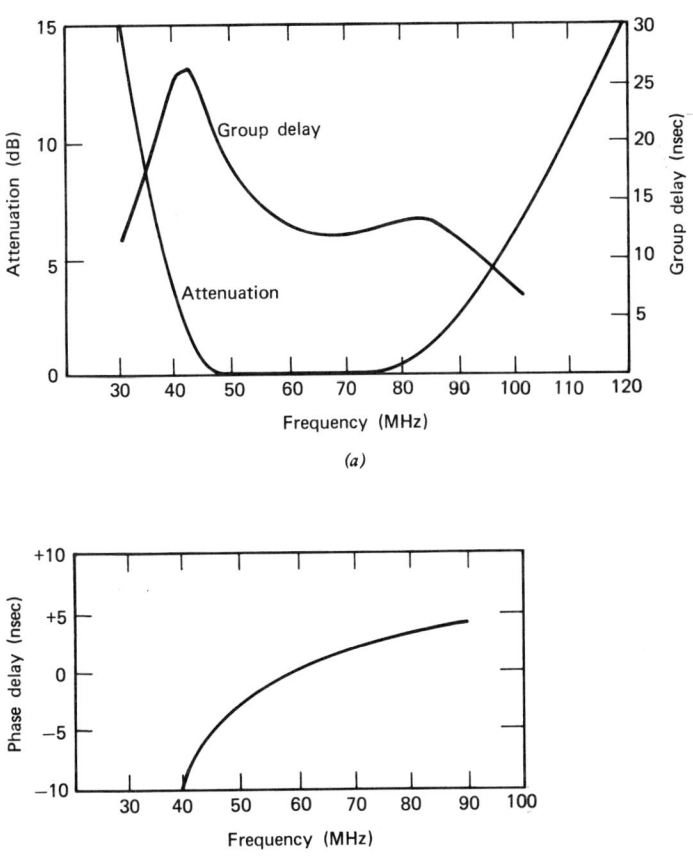

Figure 4-17. (a) Attenuation and group delay, (b) passband phase delay of filter in Section 4.4.5.

2-14b). Since 40 MHz is below the center frequency, the sign of the LP phase is reversed to give a BP phase there of $+135°$. Then, from (2.6-2), we convert $+135°$ to rads and divide by $2\pi \times 40 \times 10^6$ rads/sec, yielding a -9.375-nsec phase delay.

The passband phase delay (Fig. 4-17b) is negative for frequencies below f_0, zero at f_0, and positive for frequencies greater than f_0. We should not be disturbed by a negative delay, for we discussed its physical significance in Section 2.6.1. A positive phase angle produces a negative delay, which is interpreted as the output sinusoid leading the input sinusoid. The important point here is that the output value at a given instant of time is not the result of a future input value. Causality is not violated. For frequencies greater than f_0, the output sinusoid lags the input sinusoid, whereas a sinusoidal input at 60 MHz is in phase with the output sinusoid.

Figure 4-17 indicates that for a given passband frequency, the group delay is larger than the phase delay. This condition always holds for moderate-bandwidth BP filters, and it can be appreciated by examining a typical BP phase response and then applying the definition of each delay. The slope of the BP phase curve is steeper than the slope obtained by dividing the phase angle by the frequency.

4.5 NARROW-BAND BANDPASS FILTER

In Section 4.4.1 we presented the conventional transformation, which preserves the attenuation characteristics of the LP filter independent of the relative bandwidth. Furthermore it allows the BP element values to be easily determined from the LP element values. However several shortcomings accompany this transformation: (1) the BP group delay is a distorted version of the LP delay, (2) the attenuation is geometrically symmetric about ω_0 rather than arithmetically symmetric, (3) the BP s-plane locations are not a simple function of the LP S-plane locations, and (4) the BP transient responses are not simply related to the LP transient responses. We now impose the narrow-band condition ($\gamma < 0.05$) and reexamine the aforementioned quantities.

4.5.1 Basic Definitions

The center frequency ω_0 is now approximated by the arithmetic mean of ω_1 and ω_2, as

$$\omega_0 = \frac{\omega_1 + \omega_2}{2} \tag{4.5-1}$$

rather than the geometric mean of (4.4-4). The two means are related by

$$\text{arithmetic mean} = \frac{\omega_1 + \omega_2}{2\sqrt{\omega_1 \omega_2}} \times \text{geometric mean} \tag{4.5-2}$$

and approach each other as $\omega_1 \to \omega_2$. Even with $\omega_1 = 0.9$ and $\omega_2 = 1.1$ ($\gamma \approx 0.2$), the two means differ by only 0.5%.

An arbitrary frequency ω within the approximation range can be expressed as

$$\omega \approx \omega_0 + \frac{\Delta \omega}{2} \tag{4.5-3}$$

where

$$\Delta \omega = 2(\omega - \omega_0) \tag{4.5-4}$$

is positive for $\omega > \omega_0$ and negative for $\omega < \omega_0$. Substitute ω from (4.5-3) into (4.4-3) to give

$$\Omega = \frac{(\omega + \omega_0)(\omega - \omega_0)}{\omega(\omega_2 - \omega_1)} \approx \frac{(2\omega_0 + \Delta\omega/2)\Delta\omega/2}{(\omega_2 - \omega_1)(\omega_0 + \Delta\omega/2)} = \frac{\Delta\omega}{\omega_2 - \omega_1} \times \frac{1 + \Delta\omega/4\omega_0}{1 + \Delta\omega/2\omega_0} \approx \frac{\Delta\omega}{\omega_2 - \omega_1} \tag{4.5-5}$$

Again Ω is expressed as a bandwidth ratio [see (4.4-11)], but now the actual frequencies are known, and (4.4-13) and (4.4-14) are obviated. As before, we enter the LP curves with $|\Omega|$; that is, we consider $\Delta\omega$ to be the positive bandwidth of interest. We can now replace the abscissa label of the LP attenuation curves by $\Delta\omega/(\omega_2 - \omega_1)$. The attenuation is then arithmetically symmetric about ω_0.

4.5.2 Example of Attenuation Calculation

The attenuation of a BP filter obtained from the fifth-order Butterworth LP filter is determined using both the narrow-band approximation and the exact expressions. The BP cutoff frequencies are $\omega_1 = 0.975$ and $\omega_2 = 1.025$, thus $\omega_2 - \omega_1 =$

Narrow-Band Bandpass Filter

0.05. For better accuracy the attenuation is computed from the Butterworth function

$$A = 10 \log (1 + \Omega^{10}) \text{ decibels} \qquad (4.5\text{-}6)$$

rather than taken from Fig. 3-24, and the results appear in Table 4-6, where A_x is the attenuation using the exact expression for Ω, and A_a is the attenuation using the approximate expression in (4.5-5). The approximate technique results in a deviation of only 1.5 dB at the 49-dB level, thus verifying its accuracy for relatively narrow bandwidths. Figure 4-18 shows the two attenuation responses.

Table 4-6 Attenuation Comparison Using (4.4-3) and (4.5-5)

ω	$\|\Omega\|$ (4.4-3)	$\|\Delta\omega\|$ (4.5-4)	$\|\Omega\|$ (4.5-5)	A_x (dB)	A_a (dB)
0.975	1	0.05	1	3.01	3.01
1.025	1	0.05	1	3.01	3.01
0.9625	1.516	0.075	1.5	18.14	17.68
1.0375	1.485	0.075	1.5	17.25	17.68
0.95	2.040	0.1	2	30.96	30.11
1.05	1.964	0.1	2	29.32	30.11
0.925	3.108	0.15	3	49.25	47.71
1.075	2.907	0.15	3	46.34	47.71

4.5.3 Pole-Zero Locations

The narrow-band s-plane locations, obtained from (4.4-28) with the condition

$$\left|\frac{\gamma S_i}{2}\right| \ll 1$$

are

$$s_i = \frac{\omega_2 - \omega_1}{2} S_i \pm j\omega_0 \qquad (4.5\text{-}7)$$

and are now simply related to the LP S-plane locations. Equation 4.5-7 is known as the narrow-band transformation. One just scales the LP poles (or zeros) by one-half the radian bandwidth and slides them up to $+j\omega_0$ and down to $-j\omega_0$. The BP poles now have the same relative position about $+j\omega_0$ and $-j\omega_0$ as the LP poles have about $\omega = 0$. This transformation for the third-order Butterworth poles is presented in Fig. 4-19.

The response behavior in the vicinity of $j\omega_0$ is essentially independent of the pole cluster about $-j\omega_0$ and the resulting zeros at $\omega = 0$, because phasors (see Section 2.5.4) from these poles and zeros to frequencies near $+j\omega_0$ change very little and can be considered constant (which may be a complex number). Thus the response near $+j\omega_0$ is an excellent BP version of the LP response when γ is less than about 0.05.

4.5.4 Group Delay Behavior

The passband group delay for narrow bandwidths is obtained by substituting ω from (4.5-3) into (4.4-20), retaining the first two terms of the power series

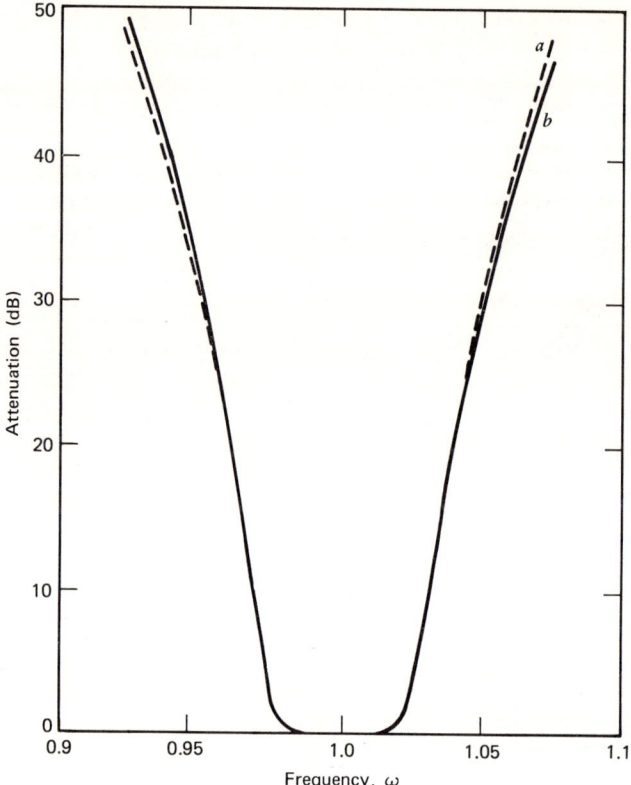

Figure 4-18. Comparison of attenuation for filter in Section 4.5.2 using (*a*) narrow-band approximation (4.5-5) and (*b*) exact expression (4.4-3).

expansion for $r(\omega)$, and substituting the result into (4.4-19) to give

$$D_b(\omega) = \frac{2}{\omega_2 - \omega_1}\left[1 - \frac{\Delta\omega}{2\omega_0}\right]D_L(\Omega) \qquad (4.5\text{-}8)$$

The distortion is now given by the decreasing linear term in brackets (Fig. 4-20). Here $\Delta\omega$ takes the sign obtained from (4.5-4).

For $\gamma = 0.1$ the maximum passband distortion is only 5%; thus the distortion term is neglected, yielding the simpler expression

$$D_b(\omega) = \frac{2}{\omega_2 - \omega_1} D_L\left(\frac{\Delta\omega}{\omega_2 - \omega_1}\right) \qquad (4.5\text{-}9)$$

The BP delay is now a scaled version of the LP delay and can be obtained easily from the LP normalized curves by replacing the abscissa label by $\Delta\omega/(\omega_2 - \omega_1)$ and the ordinate label by $[(\omega_2 - \omega_1)/2]D_b(\omega)$. The resulting BP delay is symmetric about ω_0 and the LP delay characteristics are essentially preserved. Therefore for narrow bandwidths the constant delay of the LP prototype is transformed to the BP delay with negligible distortion.

Although derived for the narrow-band condition, (4.5-7) has surprisingly good delay-preserving properties for wide bandwidths [3], and we can explicitly show this delay's dependence on relative bandwidth. Consider the BP pole clusters at

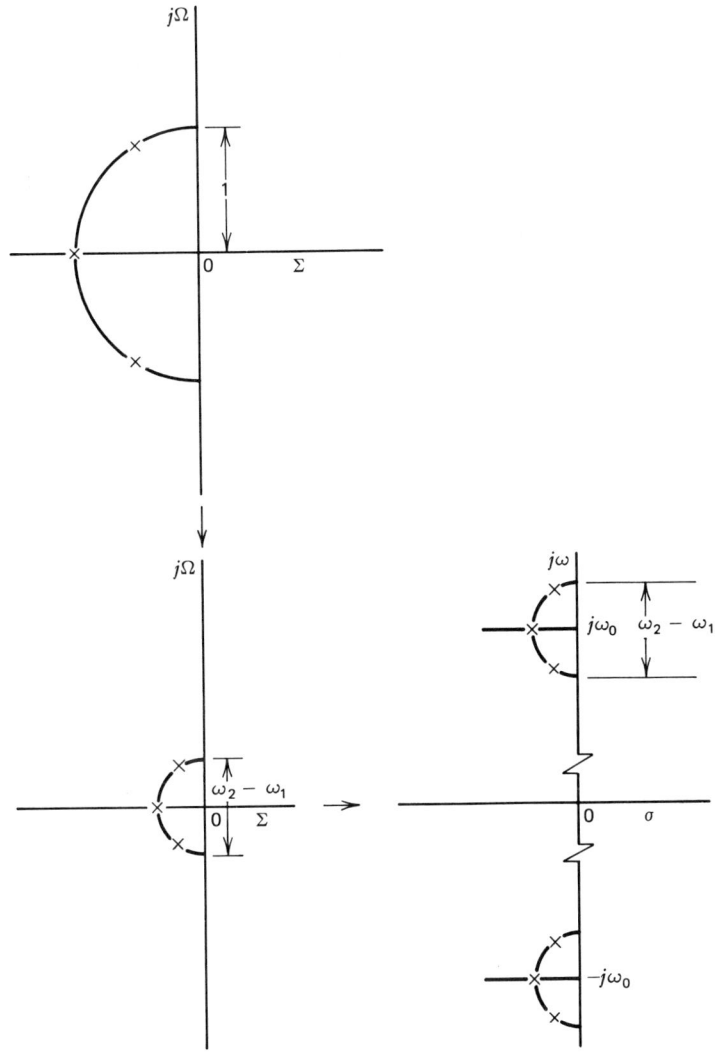

Figure 4-19. Narrow-band transformation of $n = 3$ Butterworth poles.

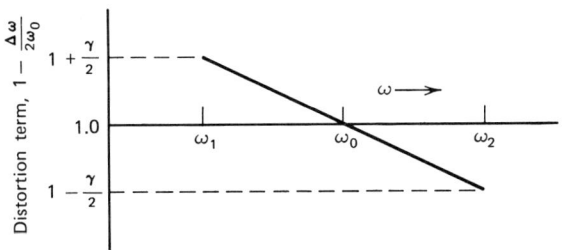

Figure 4-20. Linear distortion term in delay expression (4.5-8).

±$j\omega_0$ as each resulting from different narrow-band transformations. The one at +$j\omega_0$ results from

$$\Omega_+(\omega) = 2\frac{\omega - \omega_0}{\omega_2 - \omega_1} \quad (4.5\text{-}10)$$

and the one at $-j\omega_0$ results from

$$\Omega_-(\omega) = 2\frac{\omega + \omega_0}{\omega_2 - \omega_1} \quad (4.5\text{-}11)$$

Then the BP transfer function is

$$H_b(j\omega) = H_L(j\Omega_+(\omega))H_L(j\Omega_-(\omega)) \quad (4.5\text{-}12)$$

and the BP delay, from (4.4-19), is

$$D_b(\omega) = \frac{2}{\omega_2 - \omega_1}[D_L(\Omega_+) + (D_L(\Omega_-)] \quad (4.5\text{-}13)$$

The approximations in (4.5-3) and (4.5-5) are applied to (4.5-10) and (4.5-11) with the results

$$\Omega_+ = \Omega$$
$$\Omega_- = \Omega + \frac{4}{\gamma} \quad (4.5\text{-}14)$$

and (4.5-13) is then

$$D_b(\omega) = \frac{2}{\omega_2 - \omega_1}\left[D_L(\Omega) + D_L\left(\Omega + \frac{4}{\gamma}\right)\right] \quad (4.5\text{-}15)$$

where Ω takes on the negative sign for $\omega < \omega_0$. Remember, however, that $D_L(-\omega) = D_L(+\omega)$.

Equation 4.5-15 clearly shows why this transformation has such good delay-preserving properties. The first term $[2/(\omega_2 - \omega_1)]D_L(\Omega)$ is the transformed LP delay given by (4.5-9). The second term is the error delay due to the narrow-band approximation. It may be neglected for surprisingly large bandwidths because the term $4/\gamma$ appreciably increases the value of the argument and, from (2.6-10), $D_L(\omega) \to 0$ as $\omega \to \infty$.

Since the LP delay at high frequencies is seldom included on delay curves, we now present an easily computed approximation to this delay. For large values of ω, the LP delay in (2.6-10) approaches

$$D_l(\omega) \approx \frac{1}{\omega^2}\left[\sum_{j=1}^{m} z_j - \sum_{i=1}^{n} p_i\right] \quad (4.5\text{-}16)$$

For the LP transfer function in (3.3-2) it can be shown that

$$b_1 = -\sum_{j=1}^{m} z_j \quad (4.5\text{-}17)$$

and

$$a_1 = -\sum_{i=1}^{n} p_i \quad (4.5\text{-}18)$$

Then, from (4.5-16),

$$D_l(\omega) \approx \frac{a_1 - b_1}{\omega^2} \quad (4.5\text{-}19)$$

Narrow-Band Bandpass Filter

but for all-pole LP filters, $b_1 = 0$, and

$$D_l(\omega) \approx \frac{a_1}{\omega^2} \quad (4.5\text{-}20)$$

Most synthesis books give values for a_1 for the classical filter polynomials.

The error delay $D_L(\Omega + 4/\gamma)$ is easily computed from (4.5-20) and compared to $D_L(\Omega)$. Usually the term $4/\gamma$ is large enough to result in a small additional delay, which may be neglected.

4.5.5 Example of Group Delay Calculation

To determine the passband delay for the filter of Section 4.5.2, we use the approximate expressions and the exact expressions. The $D_L(\Omega)$ is obtained from (3.4-5) with $n = 5$, and Table 4-7 gives the results for some selected frequencies. The slight difference is due to the absence of the linear distortion term which, from Fig. 4-20, is a maximum of 2.5% at the band edges. For all practical purposes the LP delay is preserved and the use of (4.5-9) is justified.

Table 4-7 Group Delay of the Narrow-Band Filter of Section 4.5.2 Using Exact and Approximate Expressions

ω	$\|\Omega\|$ (4.4-3)	$\|\Delta\omega\|$ (4.5-4)	$\|\Omega\|$ (4.5-5)	$D_b(\omega)$ (4.4-19)	$D_b(\omega)$ (4.5-9)
0.975	1	0.05	1	203.985	198.885
0.9775	0.8976	0.045	0.9	210.507	205.929
0.99	0.3894	0.02	0.4	139.447	138.650
1	0	0	0	129.443	129.443
1.01	0.4104	0.02	0.4	137.812	138.650
1.0225	0.9023	0.045	0.9	201.510	205.929
1.025	1	0.05	1	194.035	198.885

4.5.6 Example of Wide-Band Group Delay Calculation

The narrow-band transformation is now used to realize the BP filter in Section 4.4.5. There $f_1 = 40$ MHz, $f_2 = 90$ MHz, and from (4.5-1), $f_0 = 65$ MHz, yielding $\gamma = 50/65 = 0.7692$. This is clearly a very-wide-band filter, and we now compute the BP delay from (4.5-15) to show that this transformation introduces negligible error. A value for Ω is computed from (4.5-5), observing the negative sign for $\omega < \omega_0$, and $a_1 = 2$ for use in (4.5-20). Table 4-8 lists the results at band center and the cutoff frequencies. The high-frequency delay approximation from (4.5-20) is excellent. Figure 4-21 plots the transformed LP delay and the delay from Fig. 4-17a obtained by the standard LP to BP transformation. Now the delay is essentially symmetric about 65 MHz; that is, the delays above and below f_0 are mirror images of each other and each is a scaled version of the LP delay. This example clearly demonstrates the delay-preserving properties of the narrow-band transformation.

Table 4-8 Computation of Group Delay for Filter in Section 4.4.5 Using Narrow-Band Transformation

f (MHz)	Ω	$\Omega + 4/\gamma$	$D_L(\Omega)$	$D_L(\Omega + 4/\gamma)$	$\dfrac{a_1}{(\Omega + 4/\gamma)^2}$	$D_b(\omega)$ (nsec)
40	-1	4.2	2.5	0.1169	0.1134	16.66
65	0	5.2	2	0.0754	0.0740	13.21
90	$+1$	6.2	2.5	0.0527	0.0520	16.25

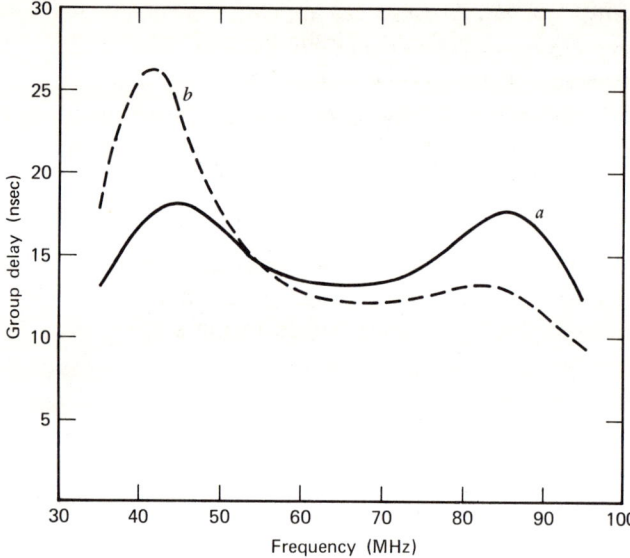

Figure 4-21. Comparison of group delay for filter in Section 4.4.5 using (*a*) narrow-band transformation and (*b*) standard transformation.

4.5.7 Transient Responses

For narrow-band filters with symmetric attenuation responses, the LP and BP transient responses are simply related.† Recall that the LP impulse and step responses take the form of (4.2-5). Substitute from (4.5-7) into (4.2-5) to yield the exponent

$$(s_i \pm j\omega_0)\frac{2t_N}{\omega_2 - \omega_1} = (s_i \pm j\omega_0)t \quad (4.5\text{-}21)$$

where t is the denormalized time. Thus the LP time scale is replaced by

$$t_N = \frac{\omega_2 - \omega_1}{2} t \quad (4.5\text{-}22)$$

---

†The precise conditions for which the relationships to be described are valid are given in Refs. 8 and 9. However the class of narrow-band filters satisfies these conditions to the extent that accurate results are obtained for cases of practical interest.

Narrow-Band Bandpass Filter

Comparison of (4.2-7) and (4.5-22) shows that for a given time interval, such as the pulse rise time, the BP bandwidth is twice the LP bandwidth. Furthermore, the BP impulse response $h_b(t)$ is the amplitude-modulated waveform

$$h_b(t) = (\omega_2 - \omega_1) h_L(t_N) \cos(\omega_0 t + \theta) \qquad (4.5\text{-}23)$$

whose envelope is $h_L(t_N)$, the LP impulse response ($\omega_{3dB} = 1$), and the waveform's carrier frequency is ω_0, the filter center frequency [8, 9]. The filter phase at band center is θ, and $(\omega_2 - \omega_1)$ is the scaling factor. The LP impulse response curves are denormalized by replacing the ordinate label by $[h_b(t)]/(\omega_2 - \omega_1)$. This proportionality constant, $(\omega_2 - \omega_1)$, reflects the fact that as the bandwidth increases, more of the impulse energy passes through the filter, resulting in a higher peak output signal. As $(\omega_2 - \omega_1)$ approaches infinity, the impulse response is the impulse, whose peak is likewise infinite (assuming the filter phase is linear).

For a general amplitude-modulated input signal

$$f(t) = f_i(t) \cos \omega_0 t \qquad (4.5\text{-}24)$$

the BP output is the amplitude-modulated signal [8, 9]

$$u_b(t) \approx u_i(t) \cos(\omega_0 t + \theta) \qquad (4.5\text{-}25)$$

where the envelope response $u_i(t)$ is the response of the LP filter with bandwidth $(\omega_2 - \omega_1)/2$ to the input $f_i(t)$. Consequently the normalized LP step response $g_L(t_N)$ is the envelope of the BP response to the step-modulated carrier

$$f(t) = u_{-1}(t) \cos \omega_0 t \qquad (4.5\text{-}26)$$

where the time scale of $g_L(t_N)$ is replaced by $[(\omega_2 - \omega_1)/2]t$.

Example 4-2. The impulse and step responses are now determined for the Butterworth BP filter in Section 4.5.2 using the narrow-band approximation ($\omega_2 - \omega_1 = 0.05$, $\omega_0 = 1$). From (4.5-23), assuming $\theta = 0°$, the BP impulse response

$$h_b(t) = 0.05 h_L(t_N) \cos t \qquad (4.5\text{-}27)$$

appears in Fig. 4-22a. The dashed line, indicating the envelope function $h_L(t_N)$, is the fifth-order LP impulse response in Fig. 3-26. The carrier frequency is shown (not to scale), the time scale is given by (4.5-22). The denormalized impulse response peak is $0.05 \times 0.366 = 0.0183$ V and it occurs at

$$t = \frac{3.7}{0.025} = 148 \text{ sec} \qquad (4.5\text{-}28)$$

From (4.5-25), the step-modulated carrier response is

$$g_b(t) = g_L(t_N) \cos t \qquad (4.5\text{-}29)$$

shown in Fig. 4-22b, where $g_L(t_N)$, indicated by the dashed line, is the fifth-order LP step response in Fig. 3-27, with the time scale replaced by $0.025t$.

In summary, the scaled LP transient responses are the envelopes of the BP transient responses, and the LP time scale is replaced by $[(\omega_2 - \omega_1)/2]t$. It is fortunate that these simple results apply to narrow-band symmetric filters, for these are widely used in practice. When the filter is not symmetric or the input carrier frequency is not the filter center frequency, the BP response is less easily

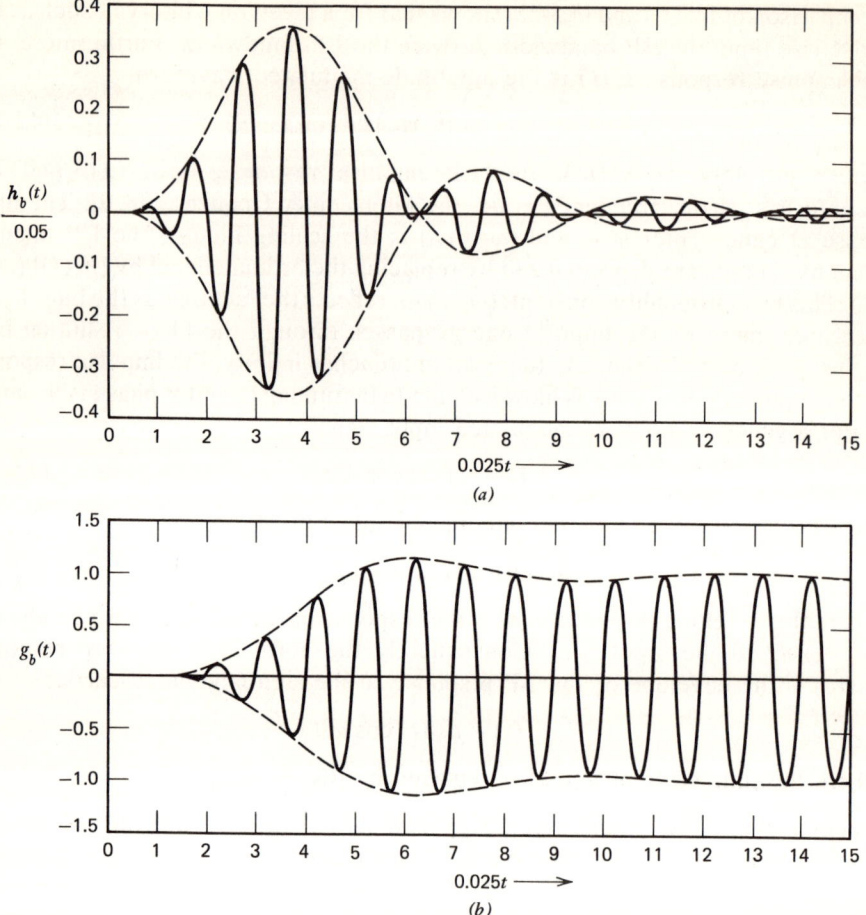

Figure 4-22. (a) Impulse response and (b) step-modulated carrier response of BP filter in Example 4-2.

expressed in a meaningful form, but methods are given in Refs. 8 and 9 for formulating this response.

4.5.8 Realizations With Nodal and Mesh Networks

In Section 4.4-4 we showed that the BP element values obtained by the conventional transformation attain impractical values when the bandwidth is relatively narrow. For this case, which occurs most frequently in practice, we use different networks that provide flexibility in selecting practical element values.

The narrow-band filters discussed here were introduced by Dishal [6] and are achieved by coupling together resonators of similar types. The resonator may be a resonant or antiresonant circuit and the coupling mechanism may be capacitive or inductive. The previously discussed wide-band filters incorporate resonators as coupling devices rather than single reactances. Those who have designed magnetically-coupled, tuned circuits exhibiting the familiar overcoupled and undercoupled responses will recognize the similarity between those filters and the filters discussed here.

Narrow-Band Bandpass Filter

For $\gamma < 0.05$, the transfer-function poles for these narrow-band networks are good approximations to those obtained by the narrow-band transformation; the approximation improves as γ decreases. For $0.1 > \gamma > 0.05$, the BP responses reasonably preserve those LP responses approximating the rectangular magnitude response (Butterworth, Chebyshev, etc.); however the BP attenuation and delay symmetry is distorted when the LP response is the constant-delay type (equiripple delay, Bessel, etc.). The basic assumption underlying the coupled-resonator theory is that the coupling reactance is constant with frequency, but it is not. For narrow bandwidths it is a good assumption, but as the relative bandwidth increases the approximation becomes poorer.

Coupled-resonator filters are also the equivalent lumped network for the helical, cavity, stripline, and wave guide filters. Furthermore, the design parameters for these distributed element filters are the same as for the lumped constant filters, namely, the coupling coefficient between resonators and the loaded Q of the end resonators. For the high-frequency filters these quantities are converted to input and output coupling loops, iris dimensions, shield heights, and so on. The narrow-band filter design theory thus applies from audio frequencies to microwave frequencies, although the hardware is vastly different for the various frequency ranges. Furthermore this design allows the application of a simple resonator tuning method [7], greatly easing the alignment procedure, particularly at the higher frequencies.

The general design prescription presented here permits the systematic realization of high-order networks with predictable responses. The BP design parameters are obtained from the LP prototype in Fig. 3-23, namely, the normalized coupling coefficients k_{ij} and the normalized quality factors q_i.

The quality factor is defined as

$$q = \frac{\omega L}{R} \qquad (4.5\text{-}30)$$

for an inductor L in series with a resistor R, and

$$q = R\omega C \qquad (4.5\text{-}31)$$

for a capacitor C in parallel with a resistor R. For the lossless networks in Fig. 3-23, the quality factor of each element evaluated at $\omega = \omega_{3dB} = 1$ is

$$q_1 = x_1 R_s; \qquad q_2 = q_3 = \cdots q_{n-1} = \infty; \qquad q_n = x_n \qquad (4.5\text{-}32)$$

The coupling coefficient $k_{ij}(j = i + 1)$ is defined as

$$k_{ij} = \frac{1}{\omega_{3dB}\sqrt{x_i x_j}} \qquad (4.5\text{-}33)$$

and with $\omega_{3dB} = 1$,

$$k_{12} = \frac{1}{\sqrt{x_1 x_2}}, \qquad k_{23} = \frac{1}{\sqrt{x_2 x_3}}, \qquad \text{etc.} \qquad (4.5\text{-}34)$$

The denormalized values are

$$Q_i = \frac{q_i}{\gamma}; \qquad K_{ij} = \gamma k_{ij} \qquad \left(\gamma = \frac{\omega_2 - \omega_1}{\omega_0}\right) \qquad (4.5\text{-}35)$$

The nodal and mesh forms of the coupled-resonator filter exhibiting the three types of coupling appear in Fig. 4-23.

In the element-value determination for the nodal schematic, for capacitive coupling between the ith and jth nodes, the coupling capacitance $C_{ij}(j = i + 1)$ is

$$C_{ij} = K_{ij}\sqrt{C_i C_j} \qquad (4.5\text{-}36)$$

where K_{ij} is given by (4.5-35) and C_i is the total shunt (node) capacitance from the ith node to ground when the $(i-1)$th and $(i+1)$th node are each shorted to

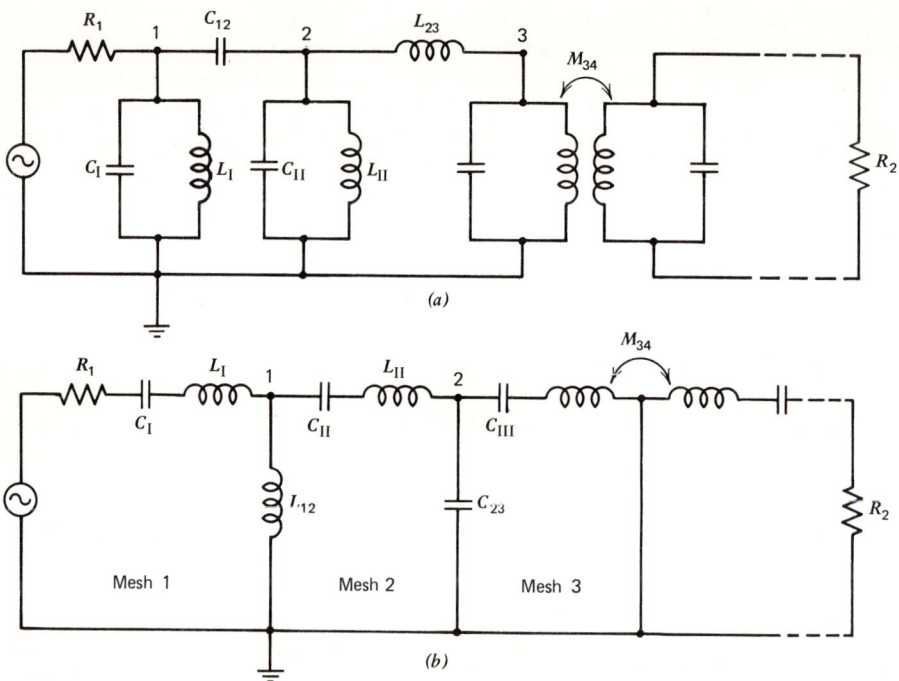

Figure 4-23. Coupled-resonator filter schematics. (a) Nodal, (b) mesh.

ground. For example, in Fig. 4-23a the total capacitance C_1 from node 1 to ground when a short is placed from node 2 to ground, is

$$C_1 = C_I + C_{12} \qquad (4.5\text{-}37)$$

Similarly C_2, the total capacitance from node 2 to ground when a short is placed from node 1 to ground and from node 3 to ground, is

$$C_2 = C_{II} + C_{12} \qquad (4.5\text{-}38)$$

For inductive coupling, the coupling inductance $L_{ij}(j = i + 1)$ is

$$L_{ij} = \frac{\sqrt{L_i L_j}}{K_{ij}} \qquad (4.5\text{-}39)$$

where L_i is the total shunt (node) inductance from the ith node to ground when

Narrow-Band Bandpass Filter

the $(i-1)$th and $(i+1)$th node are each shorted to ground. From Fig. 4-23a,

$$L_1 = L_I \quad \text{and} \quad L_2 = \frac{L_{II} L_{23}}{L_{II} + L_{23}} \qquad (4.5\text{-}40)$$

For magnetic coupling we first design a filter with inductive coupling and replace the network of inductors by its transformer equivalent, as illustrated in Fig. 4-24 for a BP filter derived from the third-order LP filter. The equivalent transformer for the π network is shown in Fig. 4-25.

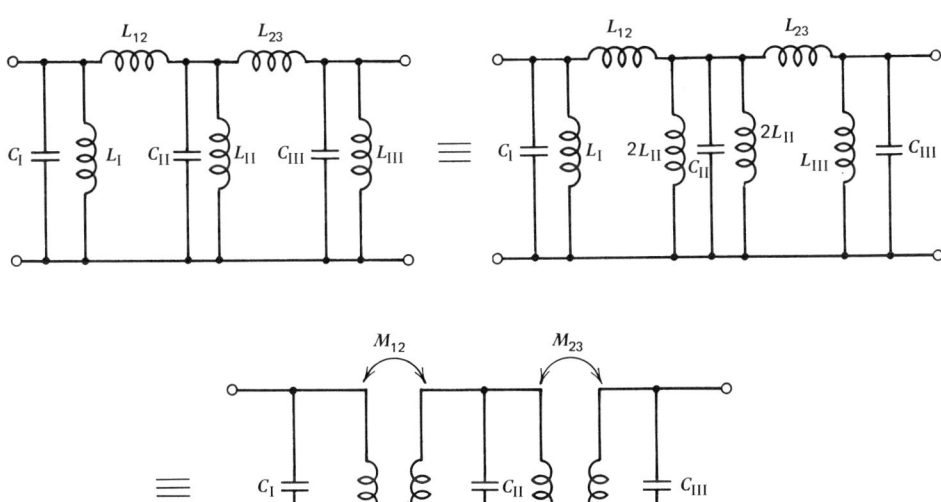

Figure 4-24. Sequence for obtaining nodal BP filter with mutual coupling.

With the $(i-1)$th node and the $(i+1)$th node each shorted to ground, the nodal inductor L_i and nodal capacitor C_i resonate at f_0, that is,

$$f_0 = \frac{1}{2\pi\sqrt{L_i C_i}} \qquad (4.5\text{-}41)$$

For doubly loaded networks, the Q's for each node are

$$Q_1 = \omega_0 C_1 R_1 = \frac{R_1}{\omega_0 L_1}$$

$$Q_2, \ldots, Q_{n-1} = \infty \qquad (4.5\text{-}42)$$

$$Q_n = \omega_0 C_n R_2 = \frac{R_2}{\omega_0 L_n}$$

A dual set of relationships exists for the mesh network in Fig. 4-23b. The value of the coupling capacitor C_{ij} $(j = i+1)$ is

$$C_{ij} = \frac{\sqrt{C_i C_j}}{K_{ij}} \qquad (4.5\text{-}43)$$

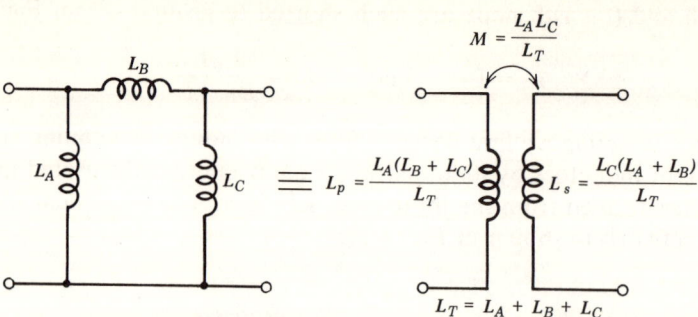

Figure 4-25. Transformer equivalent of the π network.

where C_i is the total series capacitance of the ith mesh when the $(i-1)$th mesh and $(i+1)$th mesh are open circuited. For example, in Fig. 4-23b the total series capacitance C_1 in mesh 1 when mesh 2 is open circuited (L_{II} disconnected from node 2), is

$$C_1 = C_I \qquad (4.5\text{-}44)$$

whereas C_2, the total series capacitance in mesh 2 when mesh 1 and mesh 3 are each open circuited (L_I disconnected from node 1 and C_{III} disconnected from node 2), is

$$C_2 = \frac{C_{II} C_{23}}{C_{II} + C_{23}} \qquad (4.5\text{-}45)$$

For inductive coupling, the coupling inductance L_{ij} ($j = i + 1$) is

$$L_{ij} = K_{ij} \sqrt{L_i L_j} \qquad (4.5\text{-}46)$$

where L_i is the total series inductance of the ith mesh when the $(i-1)$th mesh and $(i+1)$th mesh are open circuited. From Fig. 4-23b

$$L_1 = L_I + L_{12}, \qquad L_2 = L_{II} + L_{12} \qquad (4.5\text{-}47)$$

For magnetic coupling we again design a filter with inductive coupling and replace the Tee network of inductors by its transformer equivalent, as illustrated in Fig. 4-26 for a BP filter derived from the third-order LP filter. Figure 4-27 diagrams the equivalent transformer for the Tee network. Also, each mesh, with adjacent meshes open circuited, resonates at f_0 according to (4.5-41), where L_i and C_i are defined for the mesh circuit. The necessary Q relationships are

$$Q_1 = \frac{\omega_0 L_1}{R_1} = \frac{1}{\omega_0 C_1 R_1}$$

$$Q_2, \ldots, Q_{n-1} = \infty \qquad (4.5\text{-}48)$$

$$Q_n = \frac{\omega_0 L_n}{R_2} = \frac{1}{\omega_0 C_n R_2}$$

The flexibility of these networks allows the introduction of practical considerations, such as selection of equal inductors for optimum Q in the capacitively coupled case or selection of standard-valued coupling capacitors. Furthermore, the nodal schematic allows for stray capacitance from node to ground and inductor distributed capacitance to be absorbed by the resonator capacitor.

Narrow-Band Bandpass Filter

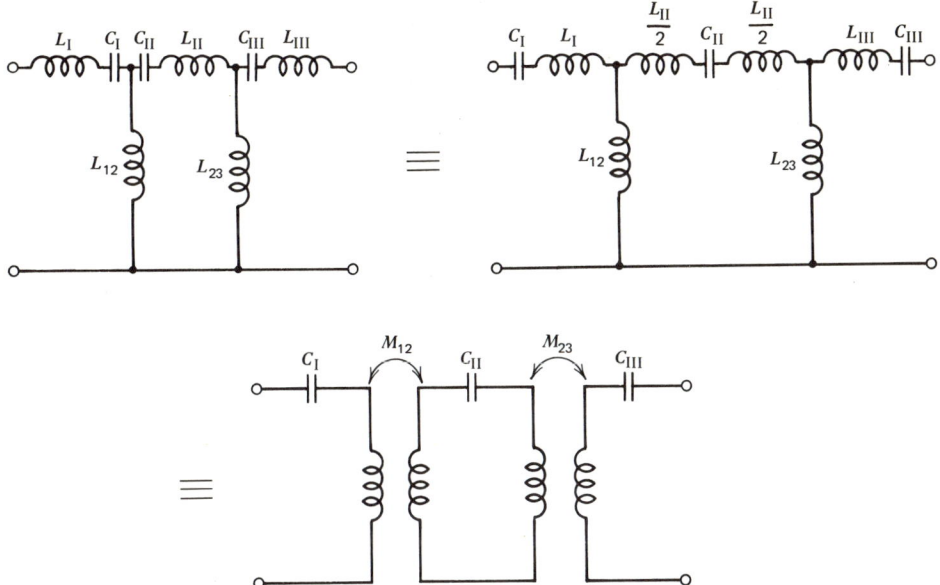

Figure 4-26. Sequence for obtaining mesh BP filter with mutual coupling.

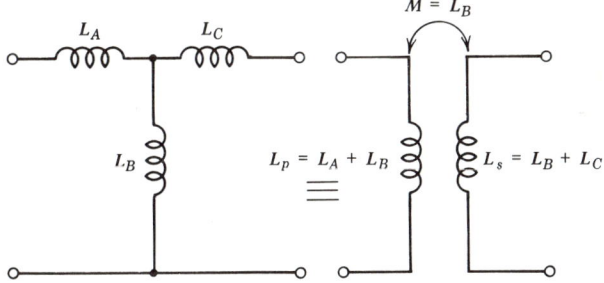

Figure 4-27. Transformer equivalent of the Tee network.

4.5.9 Example of Network Design

A capacitively coupled nodal and mesh BP network are now obtained from the third-order Butterworth LP filter. The center frequency is $\omega_0 = 10^7$, $\omega_2 - \omega_1 = 5 \times 10^5$, and source and load termination are each 500 ohms. The LP element values for the equally terminated condition, from (3.4-7), are $x_1 = 1$, $x_2 = 2$, and $x_3 = 1$. The normalized q's from (4.5-32) are $q_1 = 1$, $q_2 = \infty$, $q_3 = 1$; the normalized coupling coefficients from (4.5-34) are $k_{12} = k_{23} = \sqrt{2}/2$. The denormalized K's and Q's, from (4.5-35), are

$$K_{12} = K_{23} = \gamma k_{12} = \frac{1}{20} \times \frac{\sqrt{2}}{2} = 0.03536$$

$$Q_1 = Q_3 = \frac{q_1}{\gamma} = 20 \times 1 = 20$$

(4.5-49)

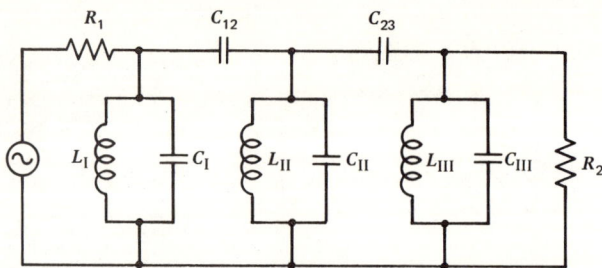

Figure 4-28. Nodal BP schematic derived from a third-order LP filter.

The nodal schematic in Fig. 4-28 is obtained first. To demonstrate the network's flexibility, we set each inductance to 10 μH, which is known to have a high Q value at these frequencies. For capacitive coupling, this is the nodal inductance and, from (4.5-41), the nodal capacitance is

$$C_1 = C_2 = C_3 = \frac{1}{\omega_0^2 L_1} = 1000 \text{ pF} \qquad (4.5\text{-}50)$$

The coupling capacitors, from (4.5-36), are

$$C_{12} = C_{23} = K_{12}\sqrt{C_1 C_2} = K_{23}\sqrt{C_2 C_3} = 35.36 \text{ pF} \qquad (4.5\text{-}51)$$

and C_I, from (4.5-37), is

$$C_\text{I} = C_1 - C_{12} = (1000 - 35.36) \times 10^{-12} = 964.64 \text{ pF} \qquad (4.5\text{-}52)$$

After shorting the first node and the third node to ground, we determine C_II as

$$C_\text{II} = C_2 - C_{12} - C_{23} = 929.28 \text{ pF} \qquad (4.5\text{-}53)$$

Shorting the second node determines C_III as

$$C_\text{III} = C_3 - C_{23} = 964.64 \text{ pF} \qquad (4.5\text{-}54)$$

The source and load resistors must satisfy the loaded Q conditions of (4.5-42); yielding

$$R_1 = R_2 = Q_1 \omega_0 L_1 = 20 \times 10^7 \times 10^{-5} = 2000 \text{ ohms} \qquad (4.5\text{-}55)$$

Because we initially selected a practical inductance value, the required terminating resistance values differ from the specified source and load value of 500 ohms. This situation is frequently encountered in practice and is easily solved by matching the 2000 and 500 ohms with a secondary winding or tap on L_I and L_III. If the 10-μH inductor requires N_1 turns, the proper tap N_T is

$$N_T = N_1 \sqrt{\frac{500}{2000}} = \frac{N_1}{2} \qquad (4.5\text{-}56)$$

Other impedance transformations exist, and the interested reader is referred to Refs. 8 and 12. The final network is shown in Fig. 4-29; Fig. 4-30 contains the computed response.

The mesh realization in Fig. 4-31 is now obtained, and for this case we initially select $C_{12} = C_{23} = 270$ pF, which is a standard commercial and military value. The

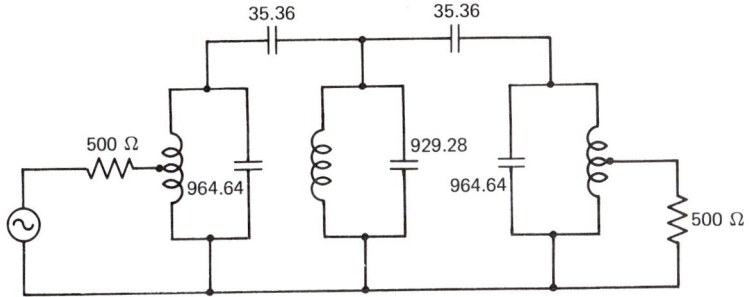

Figure 4-29. Nodal schematic for BP filter of Section 4.5.9. All inductance values are 10 μH and capacitances are in picofarads.

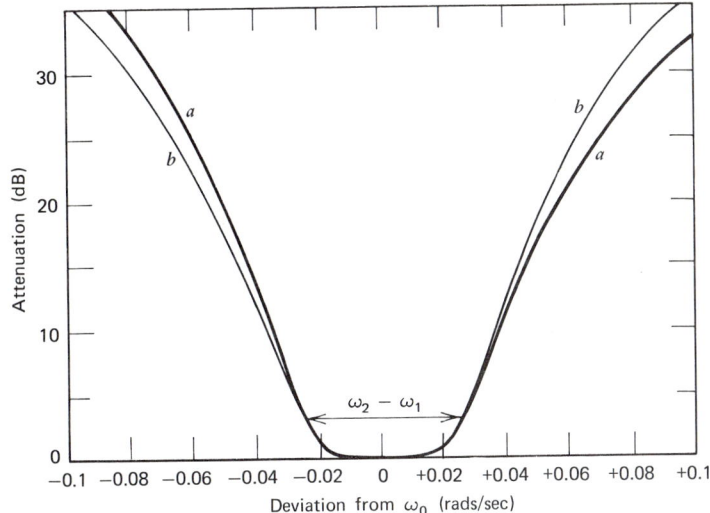

Figure 4-30. Computed responses of (a) nodal and (b) mesh realizations of Butterworth response in Section 4.5.9.

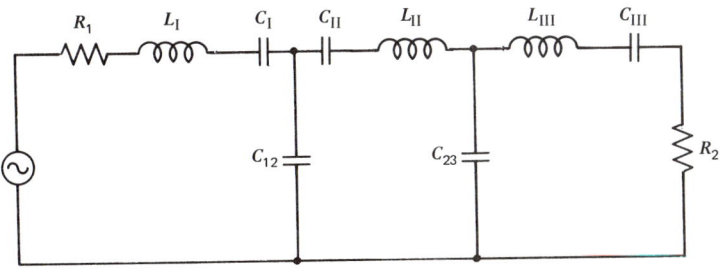

Figure 4-31. Mesh BP schematic derived from a third-order LP filter.

first and third mesh inductances, from (4.5-48), are

$$L_1 = L_3 = \frac{R_1 Q_1}{\omega_0} = \frac{500 \times 20}{10^7} = 1 \text{ mH} \tag{4.5-57}$$

which is also the value for L_I and L_III, since no other inductors are in either mesh. The corresponding mesh capacitances, from (4.5-41), are

$$C_1 = C_3 = \frac{1}{\omega_0^2 L_1} = 10 \text{ pF} \tag{4.5-58}$$

and from (4.5-43) C_2 is

$$C_2 = \frac{C_{12}^2 K_{12}^2}{C_1} = \frac{(270)^2}{10} \times \frac{1}{800} \times 10^{-12} = 9.1125 \text{ pF} \tag{4.5-59}$$

and

$$L_2 = \frac{1}{\omega_0^2 C_2} = 1.097 \text{ mH} \tag{4.5-60}$$

The mesh capacitance of meshes 1 and 3 is the series combination of C_1 and C_{12}; therefore C_I and C_III are

$$C_\text{I} = C_\text{III} = \frac{C_{12} C_1}{C_{12} - C_1} = 10.385 \text{ pF} \tag{4.5-61}$$

The mesh capacitance of mesh 2 is the series combination of C_II, C_{12}, and C_{23}; thus

$$C_\text{II} = \frac{1}{1/C_2 - 1/C_{12} - 1/C_{23}} = 9.772 \text{ pF} \tag{4.5-62}$$

For the computed response of this network, refer to Fig. 4-30.

4.5.10 Features of Nodal and Mesh Networks

Two features of the networks in Section 4.5.9 merit special mention. The first concerns the relative magnitude of element values. The mesh network inductances are approximately 100 times the optimum value of the nodal network, suggesting that the mesh network is more practical when the source and load resistances are relatively low, and the nodal network is more practical when these resistances are relatively high. And indeed this is the rule that is followed in practice.

The second feature is the asymmetry that begins to appear in the responses in Fig. 4-30. This property is due to the geometric symmetry associated with lumped-constant elements and the number of transmission zeros at $\omega = 0$. In Fig. 4-30 the asymmetry appears in the stopband, but for wider bandwidths the asymmetry is apparent in the passband, particularly in the group delay responses.

For a BP filter derived from an nth-order LP filter, the number of zeros at $\omega = 0$ and infinity total $2n$. The conventional LP to BP transformation yields n zeros at $\omega = 0$ and n zeros at infinity, whereas the number of transfer-function zeros at $\omega = 0$ for the networks described here depends on the coupling between resonators. For the mesh network with capacitive coupling, there are n zeros at $\omega = \infty$ due to the series inductors and $(n - 1)$ zeros at $\omega = \infty$ due to the shunt capacitors. This leaves only one zero at $\omega = 0$ accounting for the increased

Narrow-Band Bandpass Filter

attenuation on the high side of the passband. As the bandwidth increases, this asymmetry is more apparent and overrides the opposite effect due to geometric symmetry. For the nodal network with capacitive coupling, there is only one zero at $\omega = \infty$ and $2n - 1$ zeros at $\omega = 0$. The combined effect of this and the geometric symmetry causes the attenuation to be greater on the low side of the passband than on the upper side.

Since the number of zeros at $\omega = 0$ dictates which side of the passband has steeper attenuation, it is certainly worthwhile to find the proper number of zeros there for a good approximation to a symmetrical response. We know it will be less than n because the conventional transformation results in increased attenuation on the low side. We now quantitatively determine the number of zeros at $\omega = 0$ to ensure that the distortion factor of the BP delay is as flat as possible at $\omega = \omega_0$, and this aids in producing symmetry. For minimum-phase networks, a symmetrical delay response leads to a symmetrical attenuation response.

Consider the frequency transformation

$$S = \frac{k_0(s^2 + \omega_0^2)}{s^{1-x}} \tag{4.5-63}$$

with imaginary component

$$\Omega = \frac{k_1(\omega_0^2 - \omega^2)}{\omega^{1-x}} \tag{4.5-64}$$

where k_1 may be a complex number depending on the value of x. This is not a legitimate transformation because of the possible fractional power of s; however, it yields a result that is meaningful in realistic designs. The BP transfer function obtained by transforming an nth-order LP filter using (4.5-63) contains $n(1-x)$ zeros at $\omega = 0$. The BP delay distortion component in (4.1-4) is obtained from (4.5-64) as

$$\frac{d\Omega}{d\omega} = -k_1 \frac{(1+x)\omega^2 + (1-x)\omega_0^2}{\omega^{2-x}} \tag{4.5-65}$$

To reduce delay distortion in the vicinity of ω_0, we wish to approximate a constant there. To accomplish this we now find the value of x that causes the first derivative of $d\Omega/d\omega$ to vanish at ω_0. Then

$$\frac{d}{d\omega}\left(\frac{d\Omega}{d\omega}\right)\bigg|_{\omega=\omega_0} = -k_1 \frac{x(1+x)\omega^2 - (1-x)(2-x)\omega_0^2}{\omega^{3-x}}\bigg|_{\omega=\omega_0} = 0 \tag{4.5-66}$$

for which

$$x = \tfrac{1}{2} \tag{4.5-67}$$

For symmetry the number of zeros at $\omega = 0$ should be $n/2$, a value not always realizable with the mesh or nodal schematics even if different types of coupling are used in the same network. However symmetry improves as the number of zeros at $\omega = 0$ approaches $n/2$. For n odd, $n/2$ is not an integer; therefore the closest integer is selected. A few networks that do realize $n/2$ zeros for n even are covered in Section 4.6.

Because the networks discussed in this section are inherently narrow-band devices, their use for wider bandwidths (5–20% relative bandwidths) incorporating both inductive and capacitive coupling may yield unpredictable response irregularities. Thus it is suggested that the response first be computed (there are many network analysis programs) before the filter is constructed.

4.6 WIDE-BAND CONSTANT-DELAY BANDPASS FILTER

In Section 4.4.2 we demonstrated that for wide-bandwidth filters the standard LP to BP transformation introduces delay distortion so that the BP delay no longer possesses the attributes of the LP delay. This is unfortunate because wide-bandwidth filters are essential components of electronic systems such as high-speed data transmission systems. A constant delay is often necessary to ensure distortionless pulse transmission. Also intermodulation distortion in FM systems is further increased if the group delay is asymmetric about the carrier frequency in addition to being nonconstant. Thus a class of wide-band BP filters whose passband delay approximates a constant delay and is arithmetically symmetrical about the center frequency is of great value.

In Section 4.5.4 we demonstrated that the narrow-band transformation essentially preserves the LP delay for astonishingly large bandwidths. However the BP delay, although predictable, is not optimized; the element values are not tabulated in the literature for the constant-delay type filters, and the 3-dB bandwidth is known only approximately. These shortcomings are obviated by the filter class now described [4].

We first preset the bandwidth of each filter for arithmetic symmetry about ω_0 and optimally determine the transfer-function poles by a digital computer, causing the filter's group delay to be a least-squares approximation to a constant delay over the 3-dB bandwidth. The value of this mean delay is optimized and the approximation error is quantitatively known. Distortions inherent in LP-to-BP transformations are eliminated because the BP rather than the LP parameters are optimized. The Darlington synthesis method [11] is then used to find ladder networks exhibiting this response, and their element values, attenuation responses, and delay responses are presented for two filter orders.

4.6.1 Transfer Function and Least-Squares Approximation

The transfer function of the BP filters considered here is

$$H(s) = \frac{H_0 s^{n/2}}{\prod_{i=1}^{n} [(s + \alpha_i)^2 + \beta_i^2]} \tag{4.6-1}$$

where $n/2$ zeros at $s = 0$ are chosen for symmetrical attenuation and delay responses, as explained in Section 4.5.10. Zeros on the imaginary axis contribute nothing to the group delay because they contribute piecewise constant angles there; hence the BP delay, from (2.6-9), is

$$D_b(\omega) = 2 \sum_{i=1}^{n} \frac{\alpha_i(\omega^2 + \alpha_i^2 + \beta_i^2)}{\omega^4 + 2(\alpha_i^2 - \beta_i^2)\omega^2 + (\alpha_i^2 + \beta_i^2)^2} \tag{4.6-2}$$

To approximate the constant delay T over the 3-dB bandwidth $\omega_2 - \omega_1$ by $D_b(\omega)$, we determine T and the $\{\alpha_i, \beta_i\}$† to ensure that the value of the integral ε, from (3.2-37),

$$\varepsilon = \int_{\omega_1}^{\omega_2} [T - D_b(\omega)]^2 \, d\omega \tag{4.6-3}$$

†The symbol { } indicates "set of." Thus $\{\alpha_i, \beta_i\}$ is the set of α_i's and β_i's.

Wide-Band Constant-Delay Bandpass Filter

is minimized. The least-squares criterion allows T_M, the optimum value of T, to be explicitly expressed in terms of a specified $\{\alpha_i, \beta_i\}$ and the error to be quantitatively given by the value of ε.

For a specified $\{\alpha_i, \beta_i\}$, a necessary condition for minimum ε is

$$\frac{d\varepsilon}{dT} = 0 \tag{4.6-4}$$

Then

$$2\int_{\omega_1}^{\omega_2} [T_M - D_b(\omega)]\, d\omega = 0 \tag{4.6-5}$$

and

$$T_M = \frac{1}{\omega_2 - \omega_1} \int_{\omega_1}^{\omega_2} D_b(\omega)\, d\omega \tag{4.6-6}$$

Here T_M is the average value of the filter's delay over the bandwidth expressed in terms of the $\{\alpha_i, \beta_i\}$. Substitute T_M into (4.6-3) to give

$$\varepsilon = \int_{\omega_1}^{\omega_2} D_b^2(\omega)\, d\omega - (\omega_2 - \omega_1) T_M^2 \tag{4.6-7}$$

Unfortunately, in this representation ε becomes the difference of large, almost-equal numbers as it approaches its minimum value. It is then very susceptible to round-off error; therefore (4.6-3) rather than (4.6-7) is used for computations. An analytic solution for the optimum $\{\alpha_i, \beta_i\}$ is not possible because of their nonlinear relationship; hence this task is performed on the computer by a search algorithm that additionally maintains the desired 3-dB bandwidth at each step of the minimization.

4.6.2 Tables of Optimum Parameters

Table 4-9 ($n = 2$) and Table 4-10 ($n = 4$) give the optimum parameters for various values of γ, the relative bandwidth. In the tables H_0 is the value for unity gain at the normalized band center (one rad/sec), and the mean delay T_M is denormalized by dividing it by ω_0, the actual radian center frequency. Figure 4-32 compares the optimum $\{\alpha_i, \beta_i\}$ for $n = 2$ to the BP poles obtained by applying the

Table 4-9 Optimum Parameters for Wide-Band Constant Delay ($n = 2$)

γ	α_1	β_1	α_2	β_2	T_M	H_0	ε_{min}
0.1	0.044511	0.965277	0.044508	1.034715	28.3775	0.01275	0.115044
0.2	0.089329	0.930344	0.089278	1.069526	14.1879	0.05131	0.056034
0.3	0.134836	0.894882	0.134567	1.104443	9.4569	0.11667	0.035701
0.4	0.181546	0.858420	0.180657	1.139378	7.0897	0.21060	0.025056
0.5	0.230179	0.820255	0.227872	1.174122	5.6670	0.33592	0.018346
0.6	0.281785	0.779306	0.276597	1.208308	4.7155	0.49689	0.013725
0.7	0.337958	0.733796	0.327288	1.241328	4.0322	0.69996	0.010502
0.8	0.401267	0.680511	0.380447	1.272188	3.5152	0.95506	0.008475
0.9	0.476159	0.612773	0.436512	1.299242	3.1075	1.27828	0.007726

Table 4-10 Optimum Parameters for Wide-Band Constant Delay ($n = 4$)

γ	α_1	β_1	α_2	β_2	α_3	β_3	α_4	β_4	T_M	H_0	$\varepsilon_{min} \times 10^5$
0.2	0.114619	1.044764	0.114697	0.954882	0.085827	1.134242	0.085978	0.865360	21.9757	0.006175	0.484363
0.3	0.172351	1.067484	0.172637	0.930875	0.129240	1.202886	0.129922	0.795038	14.6196	0.032093	0.304757
0.4	0.235782	1.088517	0.237162	0.905318	0.175754	1.270874	0.178543	0.722063	10.8891	0.109165	0.167842
0.5	0.300384	1.109270	0.303271	0.875795	0.223361	1.340451	0.230216	0.640595	8.6508	0.287063	0.120428
0.6	0.370460	1.125351	0.378336	0.837932	0.273742	1.409428	0.292384	0.544766	7.1355	0.662010	0.102398
0.7	0.443841	1.138139	0.457506	0.787494	0.326699	1.479135	0.368296	0.411812	6.0414	1.387289	0.110114

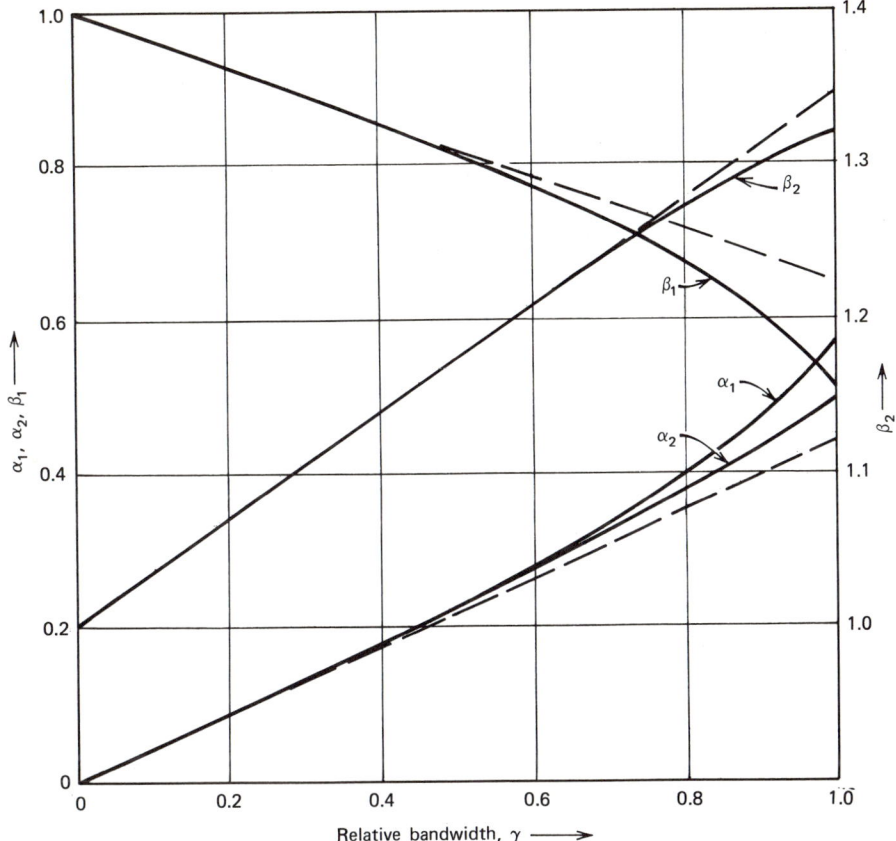

Figure 4-32. Pole positions versus γ ($n = 2$): solid curves, optimum parameters; dashed curves, narrow-band transformation.

narrow-band transformation (4.5-7) to the second-order least-squares delay LP poles in Table 3-3. The approximation is excellent for $\gamma < 0.5$, but for wider bandwidths the new data yield more-constant delay characteristics than are obtained by the narrow-band transformation. However, these curves support the preservation of the LP delay by the narrow-band transformation for reasonably large bandwidths.

4.6.3 Normalized Attenuation and Group Delay Curves

Figures 4-33 and 4-34 present the normalized attenuation curves, which exhibit the expected symmetry for narrow bandwidths, with increasing asymmetry as γ increases. The passband response is rounded, characteristic of constant-delay type filters. The relative attenuation for frequencies greater than ω_0 is relatively immune to changes in γ.

As expected, the normalized delay characteristics (Figs. 4-35 and 4-36) are also reasonably symmetric for $\gamma < 0.5$. The delays for $n = 4$ are extremely constant, and even for $\gamma = 0.7$ the maximum deviation from T_M is less than 0.12%, more

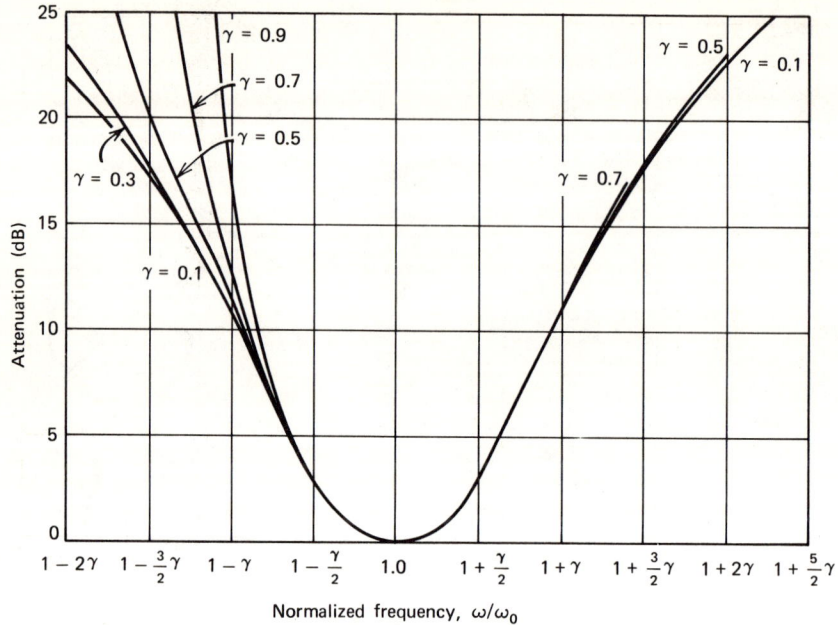

Figure 4-33. Normalized attenuation for wide-band constant delay ($n = 2$).

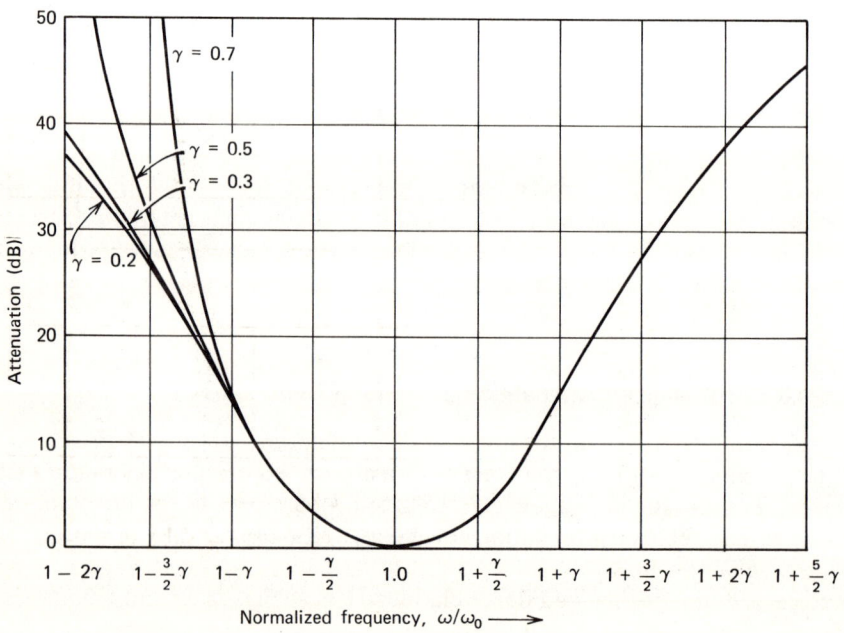

Figure 4-34. Normalized attenuation for wide-band constant delay ($n = 4$).

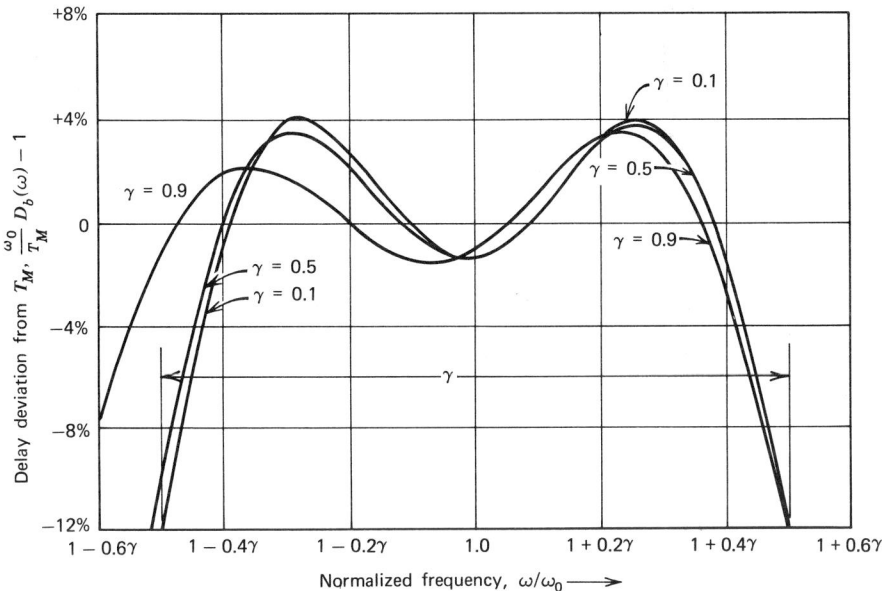

Figure 4-35. Normalized group delay for wide-band constant delay ($n = 2$).

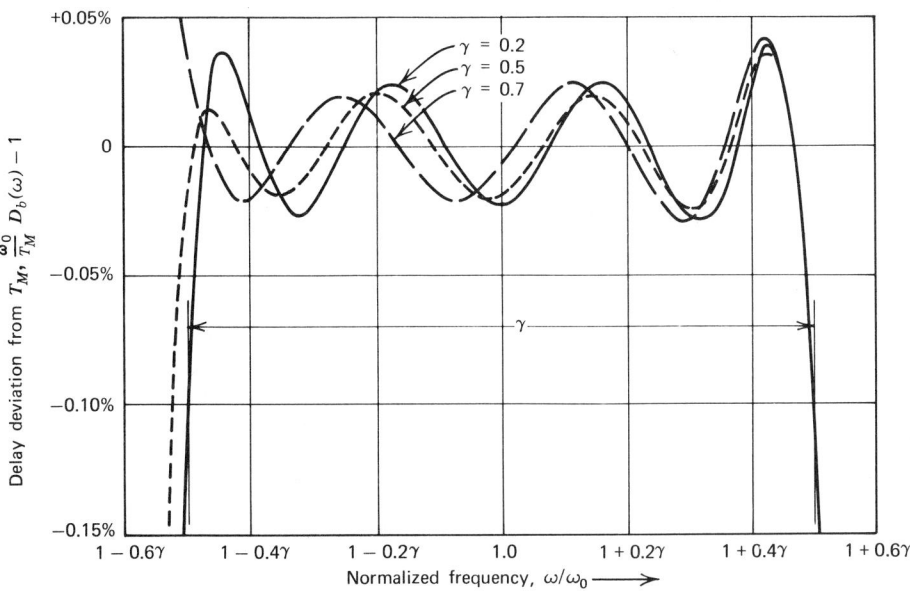

Figure 4-36. Normalized group delay for wide-band constant delay ($n = 4$).

than adequate for most applications. Over most of the band, however, this deviation is less than 0.05%. The ripple height increases as the frequency deviation from band center increases. This characteristic of the least-squares approximation is advantageous for data transmission applications, where the delay tolerance is less important near the band edge.

4.6.4 Element Values

Tables 4-11 and 4-12 give ladder networks and element values that realize the optimum responses. Inductance in henrys and capacitance in farads are normalized for a source and load resistance of 1 ohm and a center frequency of one rad/sec. The denormalized element values are

$$L_D = \frac{RL_k}{\omega_0} \quad \text{and} \quad C_D = \frac{C_k}{R\omega_0} \tag{4.6-8}$$

where L_k and C_k are the normalized values, R is the new impedance level, and ω_0 is the radian center frequency. The narrow-band schematics in Table 4-11 realize $n/2$ zeros at $\omega = 0$ for $n = 2$, but this network is not applicable for $n = 4$. The networks in Table 4-12 satisfy this requirement, but other equivalent networks can be generated.

Example 4-3. The denormalization of the network parameters and attenuation and delay responses is now illustrated by the design of the second-order filter with center frequency $f_0 = 100$ kHz and $\gamma = 0.5$. Source and load resistances are each 1000 ohms. The network in Table 4-11a is selected, and the denormalized element values from (4.6-8), where

$$\frac{R}{\omega_0} = 1.592 \times 10^{-3} \qquad \frac{1}{R\omega_0} = 1.592 \times 10^{-9} \tag{4.6-9}$$

are given in Fig. 4-37. This filter's attenuation and delay responses are the $\gamma = 0.5$ response in Figs. 4-33 and 4-35, respectively. For example, the normalized frequency $\omega/\omega_0 = 1 - 0.2\gamma$ corresponds to

$$\frac{\omega}{\omega_0} = \frac{f}{f_0} = 1 - (0.2)(0.5) = 0.9 \tag{4.6-10}$$

or

$$f = 0.9f_0 = 90 \text{ kHz} \tag{4.6-11}$$

The delay at 90 kHz is determined to illustrate denormalization. From Fig. 4-35

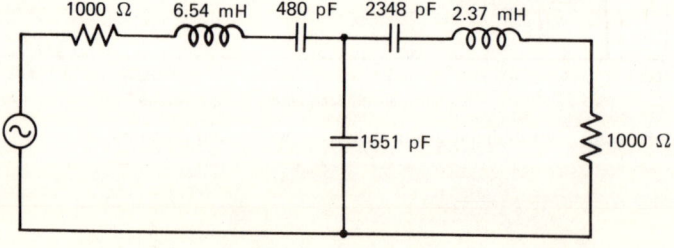

Figure 4-37. Denormalized BP filter in Example 4-3.

Table 4-11 BP Networks and Element Values for Wide-Band Constant Delay ($n = 2$)

Diagram (a)

γ	L_1	C_1	L_2	C_2	C_3
0.1	20.89	0.05022	7.683	0.1491	0.9776
0.2	10.17	0.1080	3.862	0.3398	0.9920
0.3	6.782	0.1693	2.555	0.5964	0.9893
0.4	5.113	0.2338	1.891	0.9537	0.9822
0.5	4.112	0.3017	1.486	1.475	0.9744
0.6	3.458	0.3704	1.208	2.327	0.9633
0.7	2.995	0.4393	1.003	3.969	0.9508
0.8	2.653	0.5064	0.8429	8.630	0.9369
0.9	2.375	0.5734	0.7121	96.78	0.9254

Diagram (b)

γ	C_1	L_1	C_2	L_2	L_3

Table 4-12 BP Networks and Element Values for Wide-Band Constant Delay ($n = 4$)

Diagram (a)

γ	L_1	C_1	L_2	C_2	L_3	C_3	L_4	C_4	C_5
0.3	7.721	0.1430	2.453	0.6317	0.3529	3.823	0.7714	0.9270	1.204
0.4	5.708	0.2042	2.413	0.7123	0.7489	2.191	0.8488	0.6760	1.058
0.5	4.535	0.2701	2.169	0.9196	1.302	1.567	0.8574	0.5280	0.9981
0.6	3.783	0.3384	1.888	1.301	2.097	1.249	0.8265	0.4227	0.9691
0.7	3.238	0.4106	1.617	2.036	3.262	1.069	0.7725	0.3468	0.9623

(a)

Diagram (b)

γ	C_1	L_1	C_2	L_2	C_3	L_3	C_4	L_4	L_5

(b)

Bandstop Filter

the value of delay deviation at $\omega/\omega_0 = 1 - 0.2\gamma$ is $+2.5\%$. Therefore

$$\frac{\omega_0}{T_M} D_b(\omega) - 1 = 0.025 \tag{4.6-12}$$

and with $T_M = 5.667$ from Table 4-9

$$D_b(\omega) = \frac{1.025 T_M}{\omega_0} = 9.245 \ \mu\text{sec} \tag{4.6-13}$$

whereas the mean delay is $T_M/\omega_0 = 9.019 \ \mu\text{sec}$.

Advantages may be obtained if zeros of the transfer function are not restricted to lie at $s = 0$. For example, the optimized delay obtained with zeros along the $j\omega$ axis may compare favorably with the delays presented here, but there is a beneficial increase in stopband attenuation because the peak frequencies are closer to the passband. Judicious placement of these zeros also results in improved symmetry in the responses. Complex zeros in either the left or right half of the s-plane may provide useful advantages, although the network realization will not display the elementary forms given here. Ladder networks incorporating mutual inductance or the networks in Fig. 3-15 are possible ways to realize these zeros with unbalanced networks.

4.7 BANDSTOP FILTER

We now consider the conventional transformation and the narrow-band transformation for obtaining a BS filter from the LP prototype.

4.7.1 Conventional Transformation

As the HP transformation is the inverse of the LP transformation, the BS transformation is the inverse of the BP transformation, that is,

$$S = \frac{\gamma}{s/\omega_0 + \omega_0/s} \tag{4.7-1}$$

and the normalized frequency is

$$\Omega = -\frac{\gamma}{\omega/\omega_0 - \omega_0/\omega} \tag{4.7-2}$$

The 3-dB frequencies are ω_1 and ω_2, and ω_0 is the geometric mean frequency of (4.4-4). Again $|\Omega|$ is used to enter the normalized curves. For $\omega < \omega_0$, the phase response is identical to that of the corresponding LP filter, but for $\omega > \omega_0$ the sign of the LP phase is reversed.

The BS group delay $D_s(\omega)$, from (4.1-4) and (4.7-2), is

$$D_s(\omega) = \frac{\Omega^2}{\omega_2 - \omega_1} \left(1 + \frac{\omega_0^2}{\omega^2}\right) D_L(\Omega) \tag{4.7-3}$$

and again the LP delay is considerably distorted. Also this transformation does not preserve the LP transient responses.

The BS poles (or zeros) s_i are found from (4.7-1), similar to the BP case, as

$$s_i = \omega_0 \left[\frac{\gamma}{2S_i} \pm j \sqrt{1 - \left(\frac{\gamma}{2S_i}\right)^2} \right] \qquad (4.7\text{-}4)$$

Each LP pole produces two BS poles, and LP complex-conjugate poles transform to poles that lie on a straight line drawn from the origin (Fig. 4-13c). Substitution from (4.7-1) into the nth-order all-pole LP transfer function results in n zeros at $+j\omega_0$ and n zeros at $-j\omega_0$.

The BS filter element values are summarized in Fig. 4-38. Each LP inductor is replaced by an antiresonant circuit, and each LP capacitor is replaced by a series resonant circuit. Each circuit resonates at ω_0.

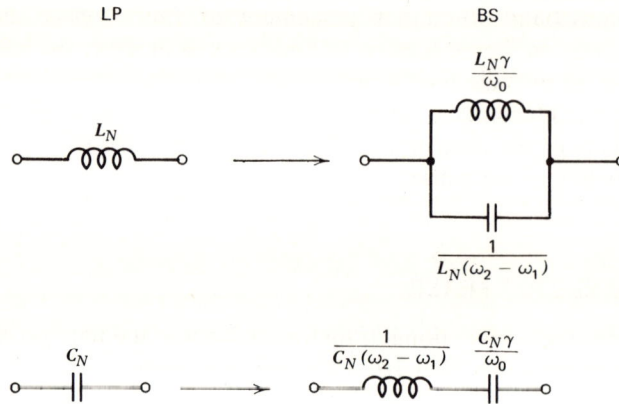

Figure 4-38. BS element values derived from LP prototype.

4.7.2 Example of Bandstop Calculation

A BS filter with 3-dB bandwidth $\omega_2 - \omega_1 = 10^5$ and a center frequency $\omega_0 = 10^6$ is derived from the third-order Butterworth LP filter. Source and load resistors are each 10,000 ohms. Application of the transformation in Fig. 4-38 to the LP prototype in Fig. 4-1 leads to the BS schematic in Fig. 4-39, whose element values are

$$\begin{aligned} L_A &= \frac{RL_N\gamma}{\omega_0} = \frac{10^4 \times 1 \times 0.1}{10^6} = 1 \text{ mH} \\ C_A &= \frac{1}{L_A\omega_0^2} = 1000 \text{ pF} \\ L_B &= \frac{R}{C_N(\omega_2 - \omega_1)} = \frac{10^4}{2 \times 10^5} = 50 \text{ mH} \\ C_B &= \frac{1}{L_B\omega_0^2} = 20 \text{ pF} \end{aligned} \qquad (4.7\text{-}5)$$

To use the normalized attenuation curve, we compute the normalized frequency at

Bandstop Filter

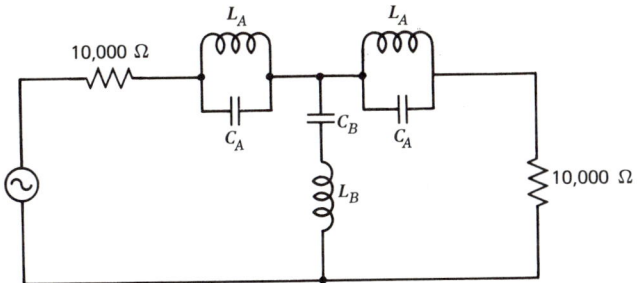

Figure 4-39. BS filter derived from $n = 3$ Butterworth LP prototype.

$\omega = 1.02 \times 10^6$ from (4.7-2), as

$$\Omega = -\frac{0.1}{(1.02 - 1/1.02)} = -2.52 \qquad (4.7\text{-}6)$$

Enter Fig. 3-24 with $|\Omega| = 2.52$ and from the third-order curve obtain 24 dB. Continuation of this process yields the attenuation response in Fig. 4-40.

4.7.3 Narrow-Band Transformation

The narrow-band condition ($\gamma < 0.05$) is now imposed by substituting ω from (4.5-3) into (4.7-2), yielding

$$\Omega \approx -\frac{\omega_2 - \omega_1}{\Delta\omega} \qquad (4.7\text{-}7)$$

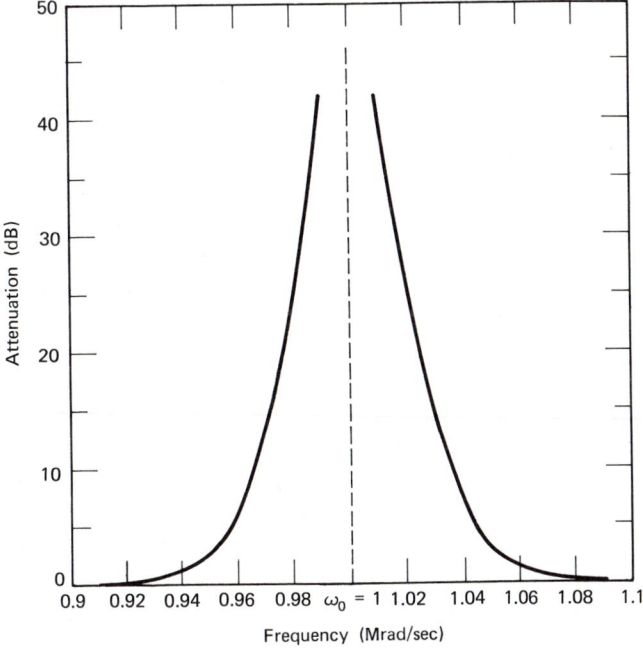

Figure 4-40. Attenuation response of BS filter in Section 4.7.2.

The narrow-band delay, from (4.7-3), is then

$$D_s(\omega) \approx \frac{2\Omega^2}{\omega_2 - \omega_1} D_L(\Omega) \tag{4.7-8}$$

but the LP delay characteristics are still not preserved. The s-plane locations from (4.7-4), with the condition $|\gamma/2S_i| \ll 1$, are approximately

$$s_i \approx \frac{\omega_2 - \omega_1}{2S_i} \pm j\omega_0 \tag{4.7-9}$$

The BS poles (or zeros) are obtained by scaling the inverse of the LP poles (or zeros) by one-half the radian bandwidth and sliding them to $+j\omega_0$ and $-j\omega_0$. Even this simplification, however, does not allow us to simply relate the LP transient responses to the BS transient responses. Evaluating the exponents of the impulse and step responses as we did in (4.5-21) results in the functional form that we experienced in relating the LP and HP transient responses. The normalized LP poles appear in the denominator of the expression, thus preventing a simple time-scale denormalization.

By now we should have noticed some similarities in the frequency transformations. The LP-to-HP transformation in (4.3-1) is the inverse of the LP denormalization in (4.2-1), whereas that for LP to BS in (4.7-1) is the inverse of the LP to BP in (4.4-1). This observation and the transient response discussion of the previous paragraph suggest that the HP and BS filters might be related in the same way that the LP and BP filters are. This idea is now investigated.

The conventional LP-to-HP transformation changes the LP inductor L to a capacitor of value $1/L$ [see (4.3-11) with $\omega_c = 1$]. The LP-to-BP transformation in Fig. 4-14 then changes this capacitor to a parallel circuit whose inductance is $L\gamma/\omega_0$ and whose capacitance is $1/[L(\omega_2 - \omega_1)]$. But these values are the same as those obtained from the conventional LP-to-BS transformation in Fig. 4-38. Similarly an LP capacitor transforms to an HP inductor, and the LP-to-BP transformation then changes it to the series circuit of Fig. 4-38. Thus the BS filter is obtainable by two transformations: (1) the LP-to-BS transformation applied to the LP prototype, and (2) the LP-to-BP transformation applied to the HP prototype. The latter is preferable because the HP transient response characteristics are preserved for the narrow-band case. The following discussion justifies this statement.

The response of a narrow-band BP filter excited by an amplitude-modulated signal expressed in terms of the equivalent LP filter is given by (4.5-25). There it is shown that if the input to the BP filter is $f_l(t) \cos \omega_0 t$, where ω_0 is the center frequency of the filter, the filter response is $u_l(t) \cos(\omega_0 t + \theta)$, where the envelope response $u_l(t)$ is the response of the equivalent LP filter to $f_l(t)$. Since we can likewise use the LP-to-BP transformation to transform the HP filter to a BS filter, these results are also applicable, where now the envelope response of the BS filter to $f_h(t) \cos \omega_0 t$ is $u_h(t)$, the response of the equivalent HP filter to $f_h(t)$. The time-scale denormalization is again given by (4.5-22).

The element values for the BS filter attain the same impractical values for narrow bandwidths as did the element values of the BP filter. Accordingly narrow-band schematics similar to the coupled resonator BP filters are available and they are also characterized by the k and q values [10, p. 217].

Example 4-4. The response of the filter in Section 4.7.2 to the step-modulated carrier of (4.5-26) is now determined. Since this is a relatively narrow-band filter ($\gamma = 0.1$), the response to this waveform is an amplitude-modulated signal whose envelope is the step response of the prototype HP filter and whose carrier frequency is ω_0. In the BS response (Fig. 4-41), the envelope, indicated by the dashed line, is the third-order Butterworth HP step response in Fig. 4-3. The carrier frequency shown is not to scale. If the sinusoid of frequency ω_0 had been on for all time, the filter output would be zero, but this signal is "turned on" at $t = 0$. Consequently the output signal does not instantaneously go to zero but varies as shown in Fig. 4-41.

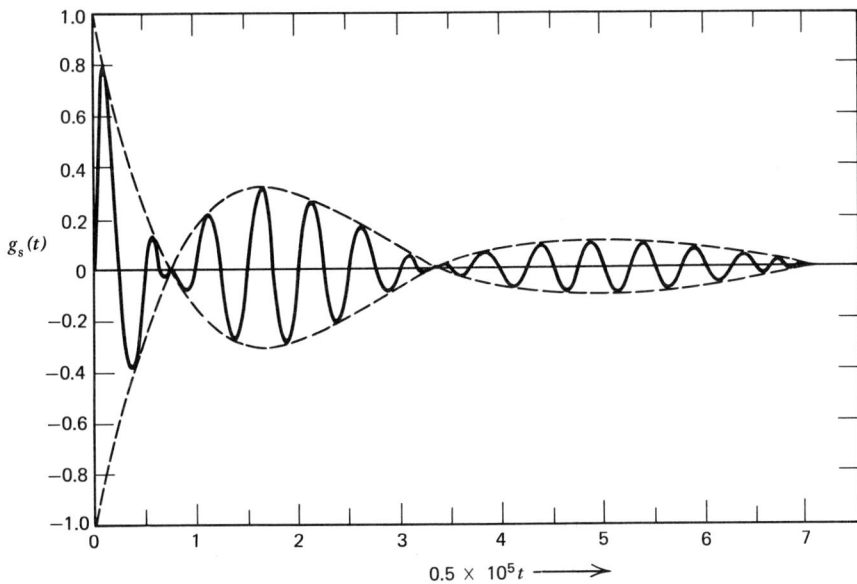

Figure 4-41. Response of BS filter in Example 4-4 to step-modulated carrier.

Narrow-band BS filters are often used to reject unwanted telemetry signals in communication systems and clutter signals in pulse-Doppler radar systems while passing signals returned from slow-moving targets. For the latter application the energy content in the response in Fig. 4-41 beyond a specified value of t is a consideration in selecting the optimum BS response. Example 4-4 illustrates a simple method of obtaining this information from the prototype HP filter step response.

REFERENCES

1. Blinchikoff, H. J., "Derivative of Group Delay at Band Center," *IEEE Trans. Circuit Theory*, Vol. CT-17, pp. 636–637, November 1970.
2. Blinchikoff, H. J., "Low Transient High-Pass Filters," *IEEE Trans. Circuit Theory*, Vol. CT-17, pp. 663–667, November 1970.

3. Blinchikoff, H. J., "A Note on Wide-Band Group Delay," *IEEE Trans. Circuit Theory*, Vol. CT-18, pp. 577–578, September 1971.
4. Blinchikoff, H. J., and M. Savetman, "Least-Squares Approximation to Wide-Band Constant Delay," *IEEE Trans. Circuit Theory*, Vol. CT-19, pp. 387–389, July 1972.
5. Blinchikoff, H. J., "High-Pass Filter Step-Response Energy: A New Performance Measure," *IEEE Trans. Circuit Theory*, Vol. CT-20, pp. 593–596, September 1973.
6. Dishal, M., "Design of Dissipative Band-Pass Filters Producing Desired Exact Amplitude-Frequency Characteristics," *Proc. IRE*, Vol. 37, pp. 1050–1069, September 1949.
7. Dishal, M., "Alignment and Adjustment of Synchronously Tuned Multiple Resonant-Circuit Filters," *Proc. IRE*, Vol. 39, pp. 1448–1455, November 1951.
8. Humpherys, D. S., *The Analysis, Design, and Synthesis of Electrical Filters*, Prentice-Hall, Englewood Cliffs, N.J., 1970.
9. Papoulis, A., *The Fourier Integral and Its Applications*, McGraw-Hill, New York, 1962.
10. *Reference Data for Radio Engineers*, 4th ed., ITT Corp., New York, 1959.
11. Van Valkenburg, M. E., *Introduction to Modern Network Synthesis*, Wiley, New York, 1960.
12. Zverev, A. I., *Handbook of Filter Synthesis*, Wiley, New York, 1967.

PROBLEMS

4-1. Use the schematic of Fig. 3-23a and design a fifth-order Butterworth LP filter with a 3-dB frequency of 100 kHz. The source and load resistors are each 1000 ohms.

4-2. Use the Butterworth responses in Fig. 3-25 and Fig. 3-26.
  (a) What order LP filter with a 3-dB cutoff frequency of 3000 Hz has a group delay of 159 μsec at 1500 Hz?
  (b) At what time does the impulse response of this filter reach its maximum value, and what is this maximum value?

4-3. A seventh-order Gaussian LP filter has a 3-dB cutoff frequency of 100 kHz. How long does it take its step response to go from 10 to 90% of its final value? Refer to Fig. 3-39.

4-4. An HP filter, derived from the fifth-order Butterworth LP filter, has 60-dB attenuation at 2500 Hz.
  (a) What is the HP filter 3-dB frequency?
  (b) What is its group delay at 10 kHz?

4-5. Design a BP filter derived from the third-order Butterworth LP filter of Fig. 3-23a, using the standard LP-to-BP transformation. The 3-dB frequencies are 40 and 90 MHz; the source and load resistor are each 100 ohms. Compute the group delay at 40, 50, 60, 70, 80, and 90 MHz. Note the asymmetry.

4-6. A BP filter with a 3-dB bandwidth of 2 MHz is derived from the third-order Butterworth LP filter. The center frequency is 60 MHz. Compute and plot the group delay at 59, 59.2, 59.5, 60, 60.5, 60.8, and 61 MHz. Note that this delay is essentially a scaled version of the LP delay and is accurately obtained from the narrow-band expression. What is the maximum value of the envelope of the impulse response, and at what time does it occur?

4-7. A BP filter is derived from the fifth-order Butterworth LP filter. The BP 60-dB bandwidth is 30 MHz.

Problems

(a) What is the 3-dB bandwidth?

(b) If the upper 60-dB frequency is 45 MHz, what is the center frequency f_o?

(c) What is the group delay at f_o?

4-8. Using (3.4-7), (4.5-32), and (4.5-34), compute the normalized Butterworth k and q values for $n = 2, \ldots, n = 5$. A similar table can be constructed for each response (Chebyshev, Bessel, etc.). See Refs. 8, 10, and 12.

4-9. Design a BP filter with the following schematic

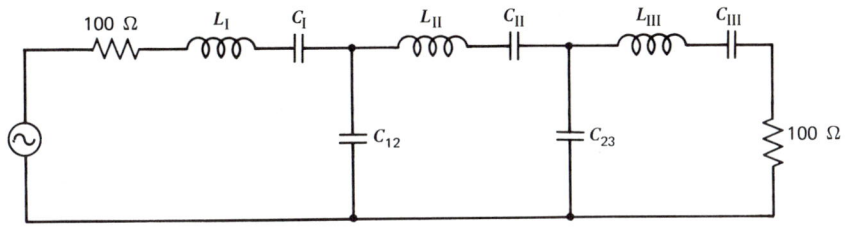

Figure P.4-9.

to have a Butterworth response, $\omega_0 = 10^6$, $\omega_2 - \omega_1 = 0.25 \times 10^5$. Assume that all mesh capacitances are the same. What is the group delay at ω_0?

4-10. Consider the narrow-band filter below, designed in accordance with Dishal's technique to operate between a source resistor R_1 and the load resistor R_2. What is the filter input impedance Z at the center frequency ω_0?

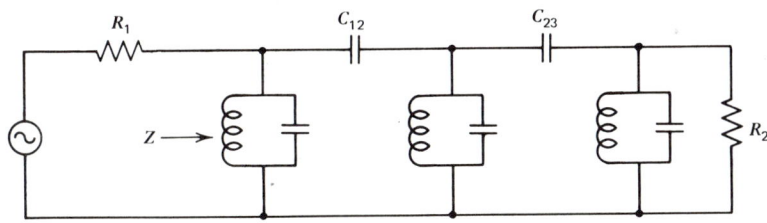

Figure P.4-10.

4-11. A BS filter is derived from the second-order Butterworth LP filter. Its 3-dB bandwidth is 2 MHz centered at 60 MHz. Source and load resistors are each 100 ohms.

(a) What are the schematic and element values? Use the LP schematic in Fig. 3-23a.

(b) What is the attenuation at 59.5 MHz?

(c) What is the attenuation at 60 MHz?

Chapter Five

All-Pass Functions

The magnitude response of an all-pass (AP) filter is unity for all frequencies, thus all frequencies are passed without attenuation. The associated phase response, however, is useful for approximating a specified phase characteristic, and the AP group delay function is useful for approximating a specified delay characteristic. Moreover, if we indicate that a parameter α can be adjusted for a specific phase response, it is understood that α can likewise be adjusted for a specific group delay response.

Approximating a linear phase by the AP phase corresponds to approximating a constant delay by the AP delay. Theoretically these approximations are not necessarily the same, but in practice the difference is often negligible. The phase (delay) properties of AP filters are so important for achieving a specified phase (delay) response of the electronic system that we devote an entire chapter to them, and we hope to consolidate much of the scattered information on this subject.

5.1 APPLICATIONS OF ALL-PASS FILTERS

5.1.1 Delay Equalization

The most common and oldest use of AP filters is to reduce delay distortion in a transmission channel. The group delay of a transmission channel, whose bandwidth is essentially dictated by a selective BP filter, should ideally be constant to ensure distortionless pulse propagation for radar, television, telegraph, and high-speed data communication systems. Unfortunately the passband delay associated with these filters (which are usually minimum phase) is nonconstant, and signal frequencies in the channel are delayed by different amounts of time. The resulting signal degradation is the familiar delay distortion.

Delay distortion can be reduced by inserting an AP filter, which introduces a controlled amount of group delay to approximate an overall constant passband delay. This process is known as group delay (or phase) equalization, and the AP filter is then called a delay (or phase) equalizer. Although there are several forms of delay distortion, group delay distortion is of most interest because the important attributes of a modulated signal are affected by nonconstant group delay (see Section 2.6.4), and virtually all useful signals in information-bearing systems are modulated signals. Furthermore, only the difference in delay values in the channel causes signal distortion.

Applications of All-Pass Filters

5.1.2 Synthesis of Linear Delay for Pulse Compression

In addition to delay equalization, AP filters are used to realize prescribed delays, such as the constant or the linear delay, both within a specified frequency interval. The pulse compression technique (one type of matched filtering) requires that the expansion and compression filter possess a linear delay. This characteristic permits a time expansion of a narrow pulse (the pulse energy remains constant), which can then be transmitted with a peak power that does not saturate the transmitter tube. However, increased system-resolving capability, both in range and velocity, is still obtained because the received pulse is recompressed into a narrow pulse. Sideband inversion allows the identical AP filter to be used in both the transmitter and receiver. Realization of this linear delay with lumped-constant AP filters was very popular in the 1950s and 1960s, but these devices are gradually being replaced by bulk wave metallic strip delay lines and surface wave delay lines, which approximate AP filters. The advantages of the pulse compression technique are also achieved with coded waveforms (such as the Barker code) processed digitally and acoustically (surface wave devices).

5.1.3 Digital System Delay Equalization

All-pass filters also find application in digital filter design. Presampling (guard) filters are often necessary to limit the digital filter's input frequency spectrum to one-half the Nyquist frequency. Consequently the filter group delay should be reasonably constant to minimize delay distortion. All-pass filters may be used for compensating the nonconstant delay introduced by the guard filter.

One popular technique for designing a digital filter is to obtain a suitable digital approximation to a known analog transfer function. If delay equalization is necessary for the analog filter, a suitable AP filter must be determined. The design of the AP filter is thus an intermediate step in the design of the digital filter.

5.1.4 Phase-Splitting Networks

A useful scheme in single-sideband communication systems is known as signal phase splitting. Two networks are so designed that with their inputs paralleled, the

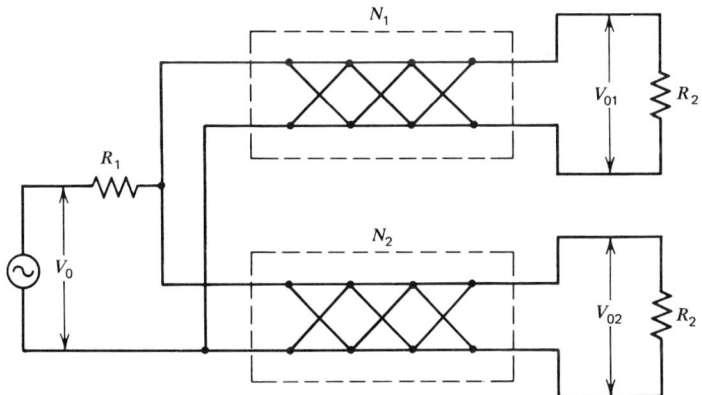

Figure 5-1. Phase-splitting network using AP filters.

individual output voltages are equal but their phases are in quadrature, within a specified frequency interval. These networks can be realized by AP filters and are indicated in Fig. 5-1 by N_1 and N_2.

5.1.5 Approximation Techniques

For all the aforementioned applications of AP filters, the filter parameters must be carefully selected. Again, the most popular approximations are the maximally flat, equiripple, and least-squares designs. The equiripple design provides a somewhat better response near the edge of the band than does the least-squares design, but at the expense of a considerably poorer response over the remainder of the band. This is a trade that is usually undesirable, since for data transmission applications, the delay tolerance is less important near the band edge. On the other hand, the maximally flat design is unnecessarily good over the lower part of the band and very poor at the band edge. The least-squares design is a happy medium between these extremes. Figure 5-2 represents these three delay approximations to a constant LP delay.

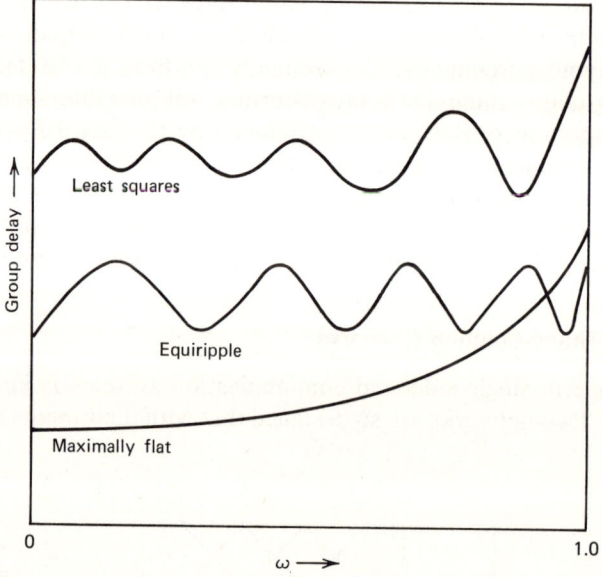

Figure 5-2. Three major delay approximations to a constant LP delay.

A good bibliography on delay approximations is given by Crane [1], who describes a computer algorithm for designing AP networks to approximate the desired delay in an equiripple manner. Humpherys [2] considers the general problem of approximating a polynomial by a rational function to give equiripple error, hence his technique is applicable to delay equalization. In Section 5.3 we discuss AP filter design using the least-squares criterion.

5.2 ALL-PASS TRANSFER FUNCTION

The AP transfer function is achieved by accompanying each left-half plane pole with a mirror-image zero in the right-half plane (Fig. 3-12). Thus the AP filter is always a nonminimum-phase network. Then a phasor drawn from each pole and zero to a point on the $j\omega$ axis has the same length; hence the magnitude response is unity for all values of ω (see 2.5-49).

Only first-order and second-order AP functions are considered here, since any AP characteristic can be realized as a product of these simpler forms. To appreciate this fact, consider the nth-order AP transfer function with real coefficients of powers of s,

$$H(s) = \frac{s^n + b_1 s^{n-1} + \cdots + a_n}{s^n + a_1 s^{n-1} + \cdots + a_n} \qquad (5.2\text{-}1)$$

Since $H(s)$ is AP, the poles and zeros must occur in mirror-image pairs, and a polynomial with real coefficients has roots that must be real or must occur in complex-conjugate pairs. Suppose all poles and zeros of $H(s)$ are complex except for one real pole and one real zero. Then $H(s)$, after factoring numerator and denominator and pairing mirror-image poles and zeros, can be expressed as

$$H(s) = \frac{s - \alpha_1}{s + \alpha_1} \times \frac{(s - \alpha_2)^2 + \beta_2^2}{(s + \alpha_2)^2 + \beta_2^2} \times \cdots \times \frac{(s - \alpha_{(n+1)/2})^2 + \beta_{(n+1)/2}^2}{(s + \alpha_{(n+1)/2})^2 + \beta_{(n+1)/2}^2} \qquad (5.2\text{-}2)$$

which is now a product of only first- and second-order AP functions. This decomposition can always be accomplished for AP transfer functions with real coefficients of powers of s.

In practice, $H(s)$ is realized by cascading first- and second-order AP networks. The total delay is then the sum of the individual section delays. These networks can be cascaded without any interaction because they possess the constant-resistance property, that is, the filter input impedance is R ohms at all frequencies when the network is designed for and terminated with R ohms. Sections of order higher than 2 are seldom used because there is no saving in components, and component-value spread, sensitivity, and tuning difficulty are all increased.

5.2.1 First-Order All-Pass Function

The first-order AP transfer function, having one real pole at $-\gamma$ and one real zero at $+\gamma$, is

$$H_1(s) = \frac{\gamma - s}{\gamma + s} \qquad (\gamma > 0) \qquad (5.2\text{-}3)$$

The phase response, from (2.5-52), is

$$\theta_1(\omega) = -2 \tan^{-1} \frac{\omega}{\gamma} \qquad (5.2\text{-}4)$$

shown in Fig. 5-3a. The phase shift is zero degrees at $\omega = 0$ and monotonically decreases, approaching $-180°$ as ω approaches infinity. At $\omega = \gamma$ the phase shift is 90°. The group delay is

$$D_1(\omega) = -\frac{d\theta_1}{d\omega} = \frac{2}{\gamma} \frac{1}{1 + (\omega/\gamma)^2} \qquad (5.2\text{-}5)$$

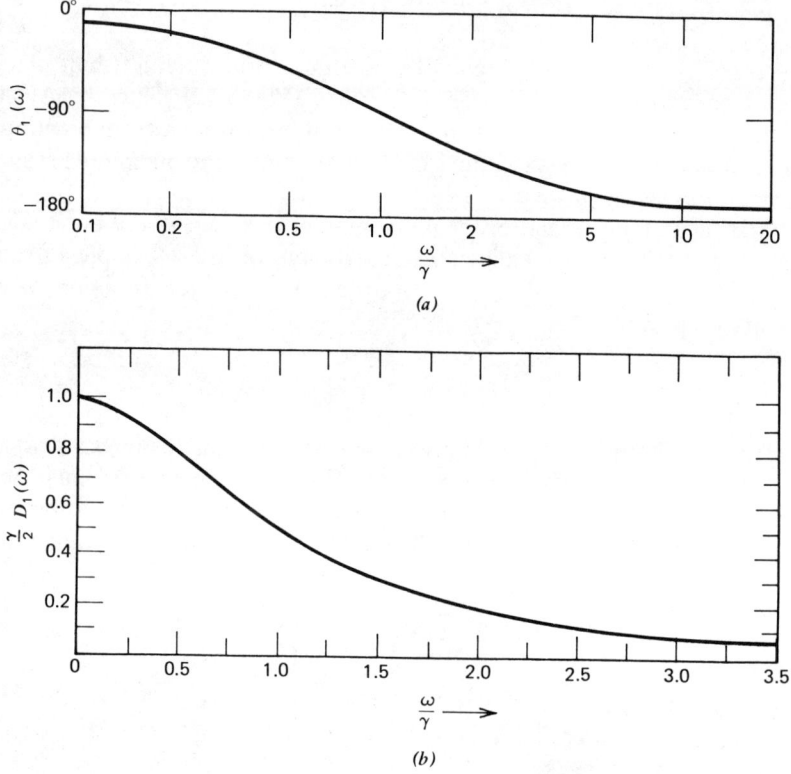

Figure 5-3. First-order all-pass (*a*) phase function and (*b*) group delay function.

shown in Fig. 5-3*b*. Because the group delay is a decreasing function, it is usually only useful for equalizing LP systems whose delay increases as the cutoff frequency is approached.

The impulse response is obtained by first expressing $H_1(s)$ as the proper fraction,

$$H_1(s) = -1 + \frac{2\gamma}{s + \gamma} \qquad (5.2\text{-}6)$$

and then, from Table 2-1,

$$h_1(t) = [-\delta(t) + 2\gamma e^{-\gamma t}]u_{-1}(t) \qquad (5.2\text{-}7)$$

shown in Fig. 5-4*a*. A negative impulse function occurs at $t = 0$, reflecting the fact that abrupt changes of the input signal immediately appear at the output reversed in sign. This is similar to the HP impulse response and occurs because the bandwidth is theoretically infinite, thus allowing all high-frequency components, which are necessary to achieve the abrupt change, to pass. In practice, parasitic elements prevent the magnitude response from being all pass, and discontinuous inputs do not immediately appear at the output.

The step response $g_1(t)$ (Fig. 5-4*b*) is

$$g_1(t) = \mathscr{L}^{-1}\left[\frac{1}{s}H_1(s)\right] = \mathscr{L}^{-1}\left[\frac{1}{s} - \frac{2}{s+\gamma}\right] = [1 - 2e^{-\gamma t}]u_{-1}(t) \qquad (5.2\text{-}8)$$

All-Pass Transfer Function

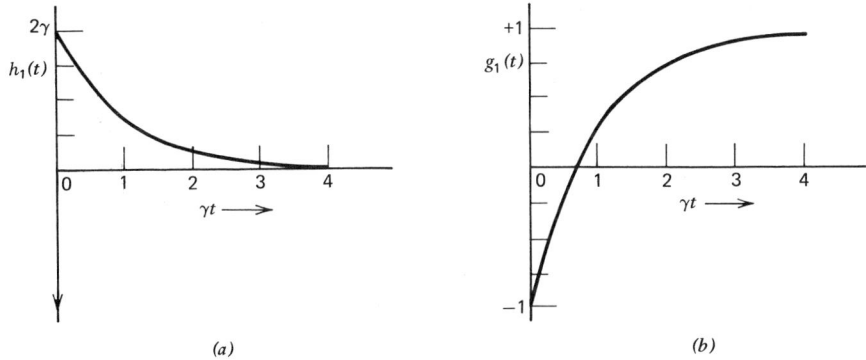

Figure 5-4. (a) Impulse response $h_1(t)$ and (b) step response $g_1(t)$ of first-order all-pass filter.

At $t = 0$ the abrupt change of the input immediately appears at the output reversed in sign but then exponentially approaches the input constant value of unity.

5.2.2 Example of First-Order All-Pass Delay Calculation

The phase and group delay of a first-order AP filter at 1 MHz is now determined when it is known that its phase at 333.333 kHz is $-90°$. Since we know from Fig. 5-3a that a phase shift of $-90°$ occurs at $\omega = \gamma$, we have $\gamma = 2\pi \times 333.333 \times 10^3$ rads/sec, and 1 MHz denormalized is

$$\frac{\omega}{\gamma} = \frac{2\pi \times 1 \times 10^6}{2\pi \times 333.333 \times 10^3} = 3 \qquad (5.2\text{-}9)$$

Enter Fig. 5-3a at $\omega/\gamma = 3$ to find a phase shift of $-143°$, and from Fig. 5-3b $(\gamma/2)D_1(\omega) = 0.1$ or

$$D_1 = \text{delay at 1 MHz} = \frac{2}{\gamma} \times 0.1 = 95.5 \text{ nsec} \qquad (5.2\text{-}10)$$

5.2.3 Butterworth Delay Equalization with a First-Order All-Pass Function

The group delay of the Butterworth family ($n = 2$ to $n = 10$) is now equalized with a first-order AP function by choosing the value of γ to ensure that the overall delay is as flat as possible at $\omega = 0$.

From (3.4-5), the Butterworth group delay is

$$D_B(\omega) = \frac{M_1 + M_3\omega^2 + M_5\omega^4 + \cdots + M_{2n-1}\omega^{2(n-1)}}{1 + \omega^{2n}} \qquad (5.2\text{-}11)$$

where

$$M_{2l+1} = \frac{1}{\sin\dfrac{(2l+1)\pi}{2n}} \qquad (5.2\text{-}12)$$

and the Taylor series, obtained by long division, is

$$D_B(\omega) = M_1 + \sum_{l=1}^{\infty} N_{2l}\omega^{2l} \qquad (5.2\text{-}13)$$

where the value of each N is plus or minus a specific value of M in (5.2-11). We know from Section 3.2.1 that the coefficients of ω^{2l} are proportional to the delay derivatives at $\omega = 0$, and since the coefficients of odd powers of ω are zero the odd derivatives at $\omega = 0$ are always zero. This is characteristic of all filter delay functions because they are even functions, hence contain no odd powers of ω. Furthermore, (5.2-13) allows explicit description of the even derivatives of $D_B(\omega)$ at $\omega = 0$, from (3.2-5), as

$$\left. \frac{d^l D_B(\omega)}{d\omega^l} \right|_{\omega=0} = \begin{cases} (l)! N_l & \text{for } l \text{ even} \\ 0 & \text{for } l \text{ odd} \end{cases} \quad (5.2\text{-}14)$$

The AP delay from (5.2-5) is likewise expanded as the Taylor series

$$D_1(\omega) = \frac{2}{\gamma} - \frac{2}{\gamma^3} \omega^2 + \frac{2}{\gamma^5} \omega^4 + \cdots \quad (5.2\text{-}15)$$

and the total delay $D_t(\omega)$ is then the sum of the filter delay from (5.2-13) and the equalizer delay from (5.2-15). For $n > 1$; $N_2 = M_3$, and

$$D_t(\omega) = D_B(\omega) + D_1(\omega) = M_1 + \frac{2}{\gamma} + \left(M_3 - \frac{2}{\gamma^3}\right)\omega^2 + \left(N_4 + \frac{2}{\gamma^5}\right)\omega^4 + \cdots \quad (5.2\text{-}16)$$

We select the one adjustable parameter γ so that the lowest nonzero delay derivative at $\omega = 0$ vanishes. This is achieved by setting the coefficient of ω^2 to zero, yielding

$$M_3 = \frac{2}{\gamma^3} \quad (5.2\text{-}17)$$

and from (5.2-12) γ is explicitly obtained as

$$\gamma = \sqrt[3]{2 \sin \frac{3\pi}{2n}} \quad (5.2\text{-}18)$$

For this value of γ, given in Table 5-1 for $n = 2$ to $n = 10$, the first three derivatives of $D_t(\omega)$ at $\omega = 0$ are now zero.

Table 5-1 Values of γ from (5.2-18) and $D_t(0)$ from (5.2-16) for Flat Butterworth Delay at $\omega = 0$, Using One First-Order AP Function

n	γ	$D_t(0)$
2	1.1225	3.196
3	1.2599	3.587
4	1.2271	4.243
5	1.1740	4.940
6	1.1225	5.646
7	1.0763	6.352
8	1.0358	7.057
9	1.0000	7.759
10	0.9683	8.458

All-Pass Transfer Function

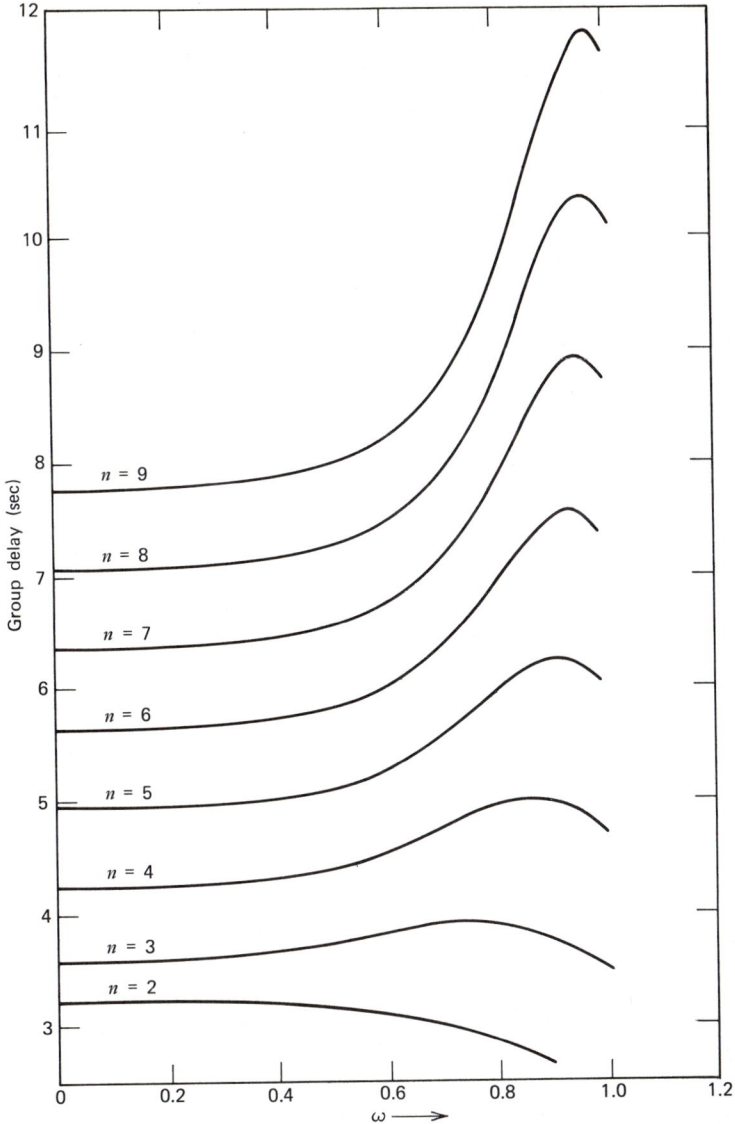

Figure 5-5. Butterworth group delays equalized with one first-order all-pass function for flat delay at $\omega = 0$.

Comparison of equalized delays (Fig. 5-5) to the unequalized delays in Fig. 3-25 shows a significant improvement in delay flatness about $\omega = 0$, and a reduction of passband delay variation. Consider the fourth-order response, for example. Unequalized, the delay variation is approximately 1.3 sec, whereas the equalized variation is reduced to 0.75 sec.

5.2.4 Second-Order All-Pass Function

The second-order AP transfer function, containing a pair of complex poles $-\alpha \pm j\beta$ and mirror-image zeros $\alpha \pm j\beta$, is

$$H_2(s) = \frac{(s-\alpha)^2 + \beta^2}{(s+\alpha)^2 + \beta^2} \qquad (\alpha > 0) \tag{5.2-19}$$

The phase response is

$$\theta_2(\omega) = -2\tan^{-1}\frac{2\alpha\omega}{\alpha^2 + \beta^2 - \omega^2} \tag{5.2-20}$$

and the group delay function is

$$D_2(\omega) = -\frac{d\theta_2}{d\omega} = \frac{4\alpha(\omega^2 + \alpha^2 + \beta^2)}{(\alpha^2 + \beta^2 - \omega^2)^2 + 4\alpha^2\omega^2} \tag{5.2-21}$$

Comparison of these responses for various values of α and β is more meaningful with the introduction of the new variables

$$\bar{\omega}_0^2 = \alpha^2 + \beta^2$$
$$k = \frac{2\alpha}{\bar{\omega}_0},$$
$$\mu = \frac{\omega}{\bar{\omega}_0}, \tag{5.2-22}$$
$$\alpha = \frac{k\bar{\omega}_0}{2}$$
$$\beta = \bar{\omega}_0\sqrt{1 - \frac{k^2}{4}}$$

The phase response is now

$$\theta_2(\mu) = -2\tan^{-1}\frac{k\mu}{1 - \mu^2} \tag{5.2-23}$$

shown in Fig. 5-6 for $k = 0.1, 0.5, 1$, and 5. At $\mu = 0$ ($\omega = 0$), the phase shift is zero degrees, and it approaches $-360°$ as the frequency becomes infinite. The phase is $-180°$ at $\mu = 1$ ($\omega = \bar{\omega}_0$) for all values of k. The second-order phase function offers a variety of shapes, depending on the value of k.

With the new variables, the group delay is

$$\bar{\omega}_0 D_2(\mu) = \frac{2k(1 + \mu^2)}{(1 - \mu^2)^2 + k^2\mu^2} \tag{5.2-24}$$

shown in Fig. 5-7 for $k = 0.1, 0.5, 1$, and $\sqrt{3}$. Small k values give narrow BP delay functions with relatively large peak values. As k increases, the peaks decrease and the function broadens. Finally at $k = \sqrt{3}$ the peak disappears and the function is maximally flat at $\omega = 0$. The delay for $k = 2$ (not shown) corresponds to $\beta = 0$; that is, two real poles occur at $s = -\alpha$ and two real zeros occur at $s = +\alpha$. Therefore the delay is just twice the delay of the first-order function. For $k > 2$, the poles are real and simple, and the second-order function is equivalent to the product of two first-order functions. Because the delay function can have a BP shape, it is useful for equalizing both LP and BP systems. The delay of selective

All-Pass Transfer Function

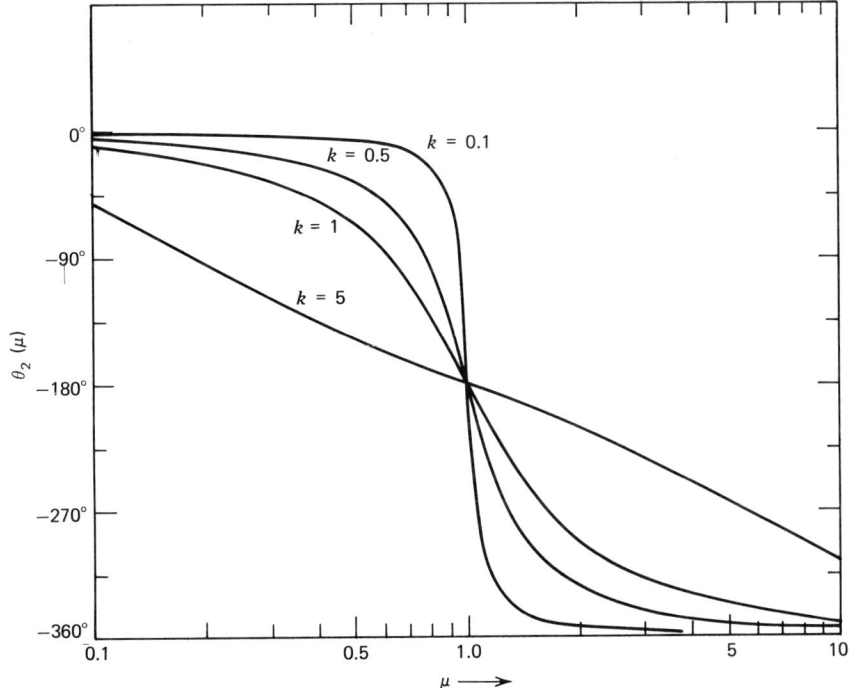

Figure 5-6. Second-order all-pass phase functions.

BP systems increases as each cutoff frequency is approached; thus the desired correction is a delay function with $k < \sqrt{3}$.

The maximum value of delay occurs at $\mu = \mu_p$ and is an important design parameter. To find this value, equate the derivative of $D_2(\mu)$ with respect to μ to zero and solve for μ_p as

$$\mu_p = \sqrt{\sqrt{4-k^2}-1} \tag{5.2-25}$$

plotted in Fig. 5-8. For $k > \sqrt{3}$, μ_p is imaginary, indicating that the delay does not peak but rather decreases monotonically. For $k < 0.4$, $\mu_p \approx 1$ and the delay peaks at $\omega \approx \bar{\omega}_0$.

The peak delay D_p (also shown in Fig. 5-8) is obtained by substituting μ_p from (5.2-25) into (5.2-24) to yield

$$\bar{\omega}_0 D_p = \frac{2k}{2\sqrt{4-k^2}+k^2-4} \tag{5.2-26}$$

For $k < 0.5$, a good approximation to D_p is

$$\bar{\omega}_0 D_p \approx \frac{4}{k} \tag{5.2-27}$$

designated by the dashed curve in Fig. 5-8.

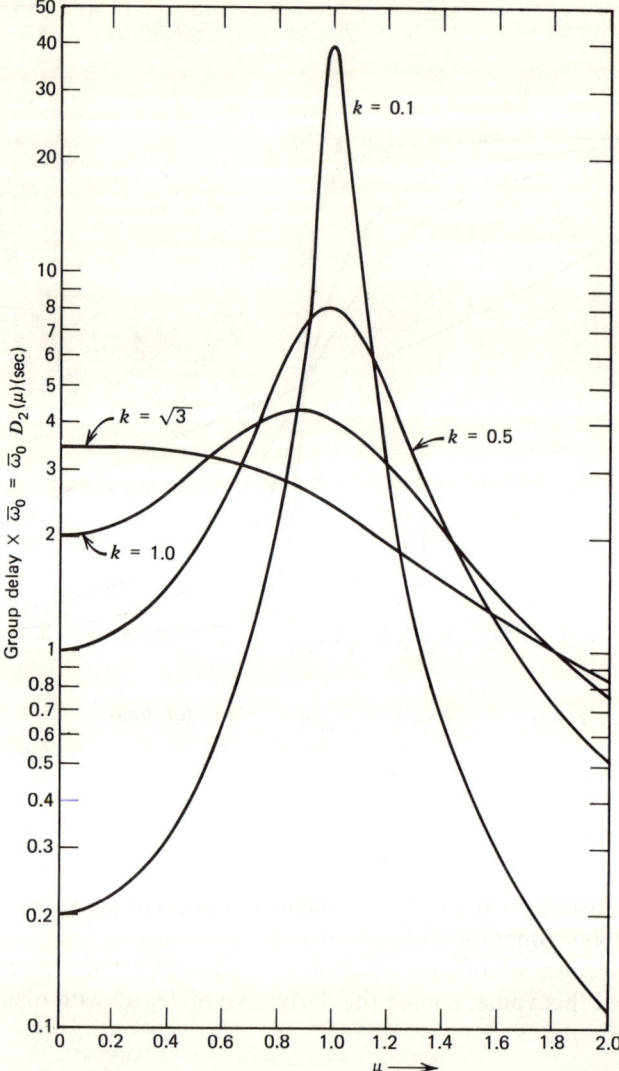

Figure 5-7. Second-order all-pass group delay functions.

5.2.5 Example of Second-Order All-Pass Delay Calculation

To illustrate denormalization of the phase and delay curves for the second-order AP function with $\alpha = \sqrt{3}$ and $\beta = 3$, we determine the phase and delay at $\omega = 4.16$. From (5.2-22) the normalized values are

$$\bar{\omega}_0 = \sqrt{\alpha^2 + \beta^2} = 2\sqrt{3} = 3.464$$

$$k = \frac{2\alpha}{\bar{\omega}_0} = 1 \qquad (5.2\text{-}28)$$

$$\mu = \frac{\omega}{\bar{\omega}_0} = \frac{4.16}{3.464} = 1.2$$

All-Pass Transfer Function

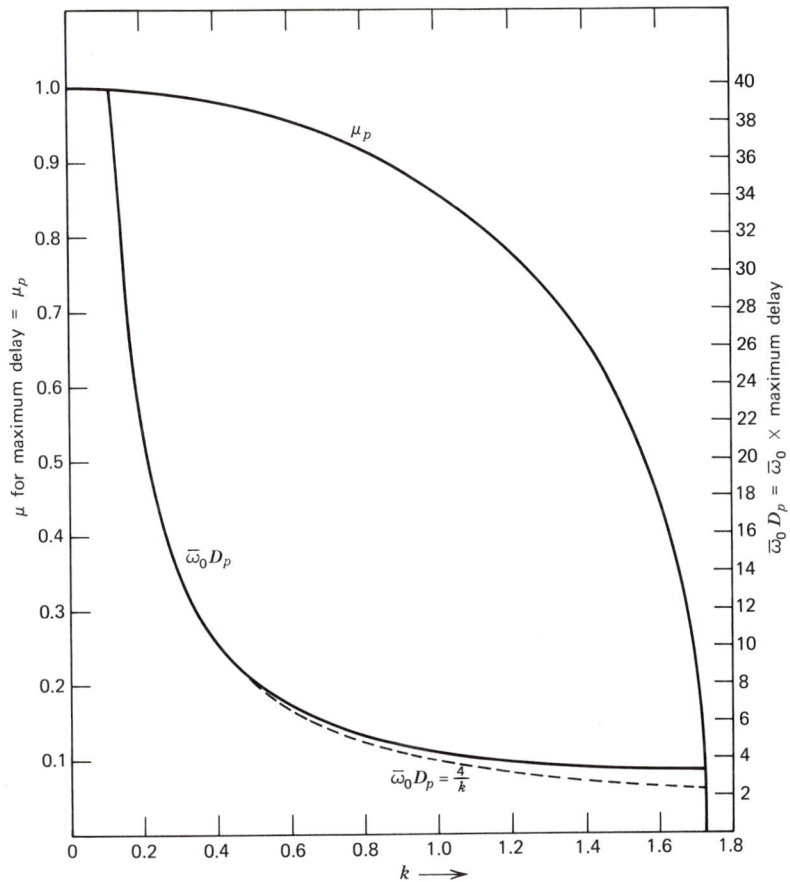

Figure 5-8. Plots of μ_p and $\bar{\omega}_0 D_p$ versus k for second-order all-pass delay function.

Enter Fig. 5-6 at $\mu = 1.2$ and, from the $k = 1$ curve, the phase is $-220°$. Enter Fig. 5-7 at $\mu = 1.2$ and, from the $k = 1$ curve, read 3. Therefore at $\omega = 4.16$, the denormalized delay is

$$D_2 = \frac{3}{\bar{\omega}_0} = \frac{3}{2\sqrt{3}} = 0.866 \text{ sec} \tag{5.2-29}$$

From Fig. 5-8, the peak delay occurs at $\mu_p = 0.855$ and the denormalized frequency is

$$\omega_p = 0.855\bar{\omega}_0 = 2.96 \text{ rads/sec} \tag{5.2-30}$$

The peak delay at ω_p is also obtained from Fig. 5-8 as

$$D_p = \frac{4.3}{\bar{\omega}_0} = \frac{4.3}{3.464} = 1.24 \text{ sec} \tag{5.2-31}$$

5.2.6 Transient Responses of a Second-Order All-Pass System

The impulse response of the second-order AP system is determined by first expressing (5.2-19) as

$$H_2(s) = 1 - \frac{4\alpha s}{(s+\alpha)^2 + \beta^2} \tag{5.2-32}$$

and referring to Table 2-1 to write

$$h_2(t) = \delta(t) - 4\alpha e^{-\alpha t}\left(\cos \beta t - \frac{\alpha}{\beta}\sin \beta t\right)u_{-1}(t) \tag{5.2-33}$$

In terms of k and $\bar{\omega}_0$ from (5.2-22),

$$\frac{\alpha}{\beta} = \frac{k}{\sqrt{4-k^2}} \tag{5.2-34}$$

and

$$h_2(t) = \delta(t) - 2k\bar{\omega}_0 e^{-(k/2)\bar{\omega}_0 t}\left[\cos\sqrt{1-\frac{k^2}{4}}\,\bar{\omega}_0 t - \frac{k}{\sqrt{4-k^2}}\sin\sqrt{1-\frac{k^2}{4}}\,\bar{\omega}_0 t\right]u_{-1}(t) \tag{5.2-35}$$

The input impulse function immediately appears at the output, at $t = 0_+$ the response is $-2k\bar{\omega}_0$, and thereafter, for $k < 2$, it oscillates exponentially to zero. Figure 5-9 gives the normalized responses for $k = 0.1, 0.5, 1,$ and $\sqrt{3}$. The impulse

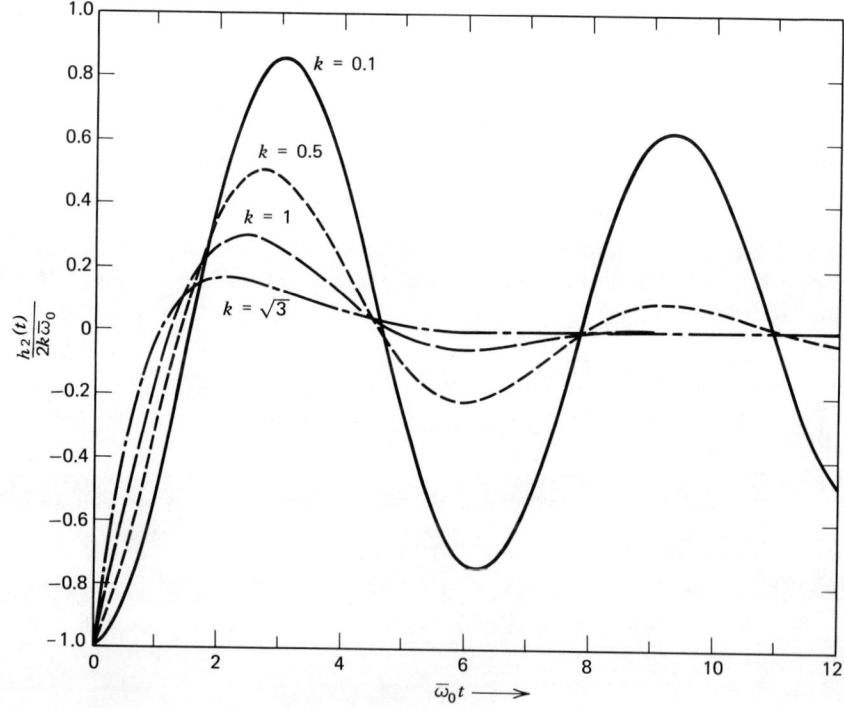

Figure 5-9. Impulse responses of second-order all-pass filter for various values of k. Impulse function at $t = 0$ is not shown.

All-Pass Transfer Function

function at $t = 0$ is not shown. As α decreases, the responses are less damped, and the amplitudes decrease.

The step response $g_2(t)$ is

$$g_2(t) = \mathcal{L}^{-1}\left[\frac{1}{s}H_2(s)\right] = \mathcal{L}^{-1}\left[\frac{1}{s} - \frac{4\alpha}{(s+\alpha)^2+\beta^2}\right] = \left[1 - \frac{4\alpha}{\beta}e^{-\alpha t}\sin\beta t\right]u_{-1}(t) \quad (5.2\text{-}36)$$

and in terms of the normalized variables

$$g_2(t) = \left[1 - \frac{4k}{\sqrt{4-k^2}}e^{-(k/2)\bar{\omega}_0 t}\sin\sqrt{1-\frac{k^2}{4}}\bar{\omega}_0 t\right]u_{-1}(t) \quad (5.2\text{-}37)$$

At $t = 0$ the output is unity, it oscillates (for $k < 2$) and approaches unity as t approaches infinity. Figure 5-10 plots the step responses for the range of k values

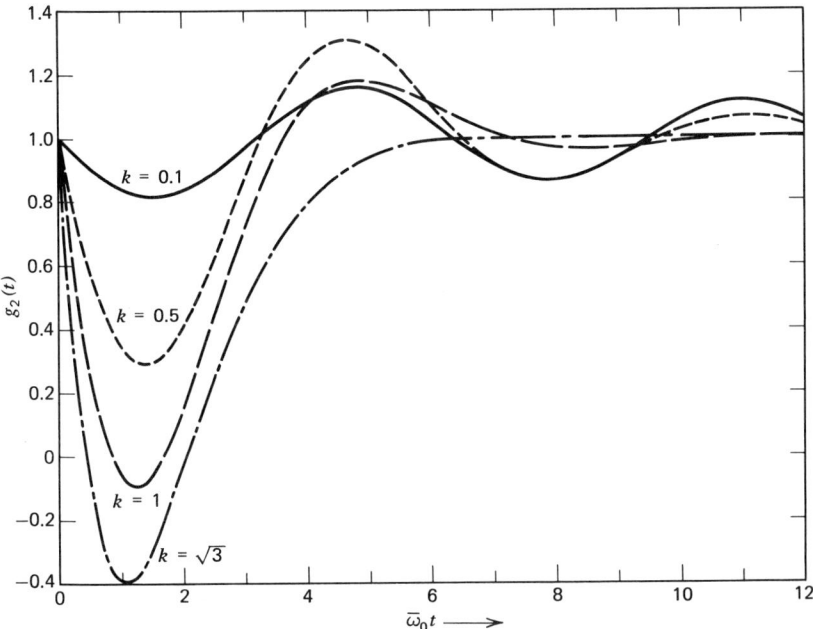

Figure 5-10. Step responses of second-order all-pass filter for various values of k.

that yield delay responses with peaks. The responses are less damped as α decreases, but the initial dip is larger as k increases. The response with $k = \sqrt{3}$, the maximally flat delay case, essentially reaches its final value and remains there in a shorter time ($\bar{\omega}_0 t = 6$) than the responses with smaller values of k.

5.2.7 Butterworth Delay Equalization with a Second-Order All-Pass Function

The Butterworth group delay for $n = 3$ to $n = 10$ is now equalized with a second-order AP function, ensuring that the total delay is as flat as possible at

$\omega = 0$. Similar to the operations in Section 5.2.3, we first replace μ by $\omega/\bar{\omega}_0$ in (5.2-24) to yield

$$D_2(\omega) = \frac{2k}{\bar{\omega}_0} \frac{1+\left(\frac{\omega}{\bar{\omega}_0}\right)^2}{1-\left(\frac{k^2-2}{\bar{\omega}_0^2}\right)\omega^2+\left(\frac{\omega}{\bar{\omega}_0}\right)^4} \qquad (5.2\text{-}38)$$

and then by long division we express $D_2(\omega)$ as the Taylor series,

$$D_2(\omega) = \frac{2k}{\bar{\omega}_0} + \frac{2k(3-k^2)}{\bar{\omega}_0^3}\omega^2 - \frac{2k[1+(k^2-2)(3-k^2)]}{\bar{\omega}_0^5}\omega^4 + \cdots \qquad (5.2\text{-}39)$$

The total delay $D_t(\omega)$ is then the sum of the Butterworth delay in (5.2-13) and the AP delay in (5.2-39),

$$D_t(\omega) = M_1 + \frac{2k}{\bar{\omega}_0} + \left[M_3 + \frac{2k(3-k^2)}{\bar{\omega}_0^3}\right]\omega^2 + \left[M_5 - \frac{2k}{\bar{\omega}_0^5} - \frac{2k(k^2-2)(3-k^2)}{\bar{\omega}_0^5}\right]\omega^4 + \cdots$$
$$(5.2\text{-}40)$$

For this case there are two adjustable parameters (k and $\bar{\omega}_0$), thus the coefficients of ω^2 and ω^4 are each set to zero

$$M_3 + \frac{2k(3-k^2)}{\bar{\omega}_0^3} = 0 \qquad (5.2\text{-}41)$$

$$M_5 - \frac{2k}{\bar{\omega}_0^5} - \frac{2k(k^2-2)(3-k^2)}{\bar{\omega}_0^5} = 0 \qquad (5.2\text{-}42)$$

Substitute $\bar{\omega}_0$ from (5.2-41)

$$\bar{\omega}_0 = \sqrt[3]{\frac{2k(k^2-3)}{M_3}} \qquad (5.2\text{-}43)$$

into (5.2-42) to obtain

$$\frac{4M_5^3}{M_3^5} k^2(3-k^2)^5 = -[1+(k^2-2)(3-k^2)]^3 \qquad (5.2\text{-}44)$$

Let

$$K_n = \frac{4M_5^3}{M_3^5} \quad \text{and} \quad y = k^2 - 3 \qquad (5.2\text{-}45)$$

and expand (5.2-44) to yield the polynomial equation

$$(K_n+1)y^6 + 3(K_n+1)y^5 - 5y^3 + 3y - 1 = 0 \qquad (5.2\text{-}46)$$

whose roots are easily found on a digital computer for each order n. The parameters $\bar{\omega}_0$ in (5.2-43) and α and β in (5.2-22) must each be positive. These restrictions lead to the inequality

$$4 \geq k^2 \geq 3 \qquad (5.2\text{-}47)$$

and from (5.2-45)

$$0 \leq y \leq 1 \qquad (5.2\text{-}48)$$

Therefore acceptable roots of (5.2-46) must lie between 0 and +1. Computations reveal that only one root for each order falls in this range. The important parameters are listed in Table 5-2, and the equalized delay responses for $n = 3$ to $n = 9$ appear in Fig. 5-11.

Table 5-2 **Second-Order AP Parameters that Produce Flat Delay at $\omega = 0$ for Butterworth Responses (Section 5.2.7)**

n	k	$\bar{\omega}_0$	α	β	$D_t(0)$
3	1.81523	1.02321	0.928682	0.429543	5.54812
4	1.84027	1.09546	1.00797	0.428981	5.97294
5	1.84994	1.08123	1.0001	0.410907	6.658
6	1.85571	1.05204	0.976139	0.392342	7.39156
7	1.85977	1.02085	0.949267	0.375522	8.13754
8	1.86286	0.991045	0.92309	0.36066	8.88522
9	1.86535	0.963504	0.898634	0.347559	9.63077
10	1.86741	0.938308	0.876103	0.335957	10.3728

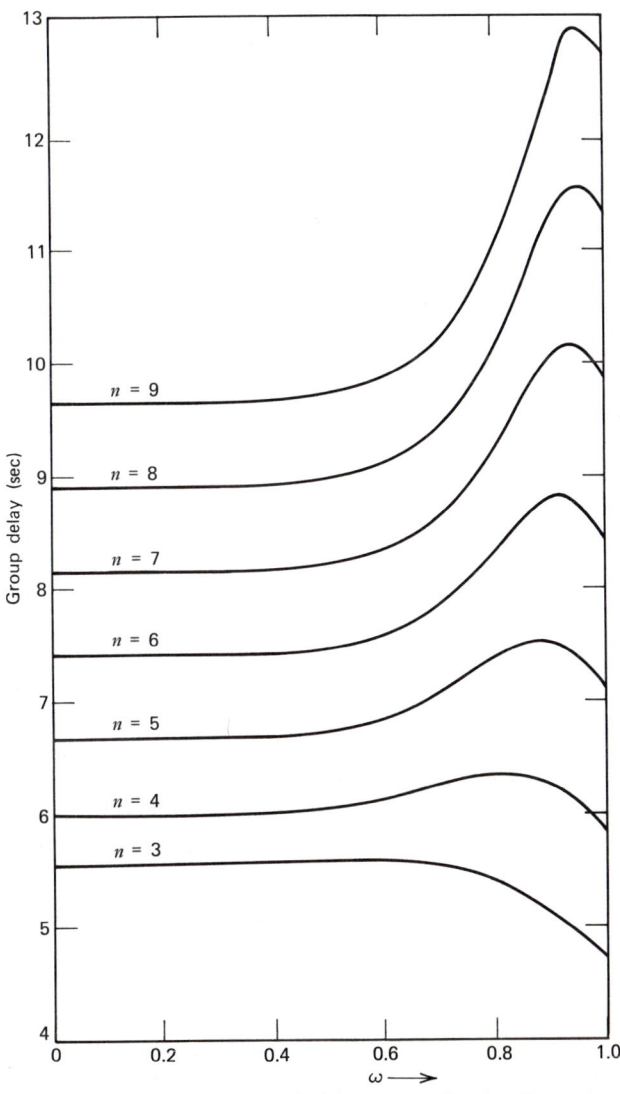

Figure 5-11. Butterworth group delays equalized with one second-order all-pass function for flat delay at $\omega = 0$.

There is considerable improvement in the delay flatness near $\omega = 0$ by using the second-order AP function because now the fourth derivative at $\omega = 0$ is also zero. In fact the delay is essentially constant over the lower half of the passband. The total passband variation for the fourth-order response is 0.35 sec compared to 0.75 sec using a first-order AP function. Table 5-2 shows that the variation of k is less than 3%. These values of k are greater than $\sqrt{3}$, hence produce delay curves that monotonically decrease, although the poles and zeros are complex numbers (the poles and zeros are real for $k \geq 2$). Therefore this delay response cannot be achieved with first-order AP functions because their poles and zeros are real.

5.2.8 Effect of Delay Equalization on Transient Responses

In Section 3.1.7 we mentioned that nonlinear phase (nonconstant delay) increases impulse and step response overshoots over those obtained when the phase is linear. By the same token, linearizing the phase should reduce these overshoots. Furthermore, the impulse and step responses should become more symmetric as the phase linearity improves, for we showed in Section 2.1.3 that linear phase ensures a symmetric impulse response. In fact, the degree of impulse response symmetry is a meaningful measure of the filter phase linearity. To demonstrate this improvement, we now determine these time-domain responses for the fourth-order equalized Butterworth filter.

The transfer function of the equalized filter $H_t(s)$ is the product of $H_B(s)$, the fourth-order Butterworth transfer function from (3.4-3) and $H_2(s)$, the AP transfer function from (5.2-19) with the parameters from Table 5-2, $n = 4$. The impulse response $h_t(t)$ is then

$$h_t(t) = \mathcal{L}^{-1}[H_t(s)] = \mathcal{L}^{-1}[H_B(s)H_2(s)] \qquad (5.2\text{-}49)$$

and the step response $g_t(t)$ is

$$g_t(t) = \mathcal{L}^{-1}\left[\frac{1}{s}H_t(s)\right] = \mathcal{L}^{-1}\left[\frac{1}{s}H_B(s)H_2(s)\right] \qquad (5.2\text{-}50)$$

Figure 5-12 gives $h_t(t)$ along with the fourth-order Butterworth impulse response from Fig. 3-26, and Fig. 5-13 shows $g_t(t)$ and the fourth-order Butterworth step response from Fig. 3-27. These curves substantiate the original claims. In summary:

1. The overshoots of the equalized filter are less than those of the unequalized filter. The first impulse response overshoot is now 18 dB down from the maximum value, whereas it is 15 dB for the unequalized filter. The first step response overshoot is 1.069 compared with 1.108 for the unequalized filter.
2. The symmetry of each response has greatly improved. The AP filter introduces preshoots in both responses, which aid in attaining symmetry.
3. The delay of each response has increased to approximately 5.95 sec, where the delay is the occurrence of the impulse response maximum and the step-response one-half amplitude. This compares favorably with the filter plus equalizer delay at $\omega = 0$; 5.97 sec from Table 5-2. The fact that both times are almost the same is also a good indication of the passband group delay constancy. Remember that the group delay at frequencies of low attenuation has greater

All-Pass Transfer Function

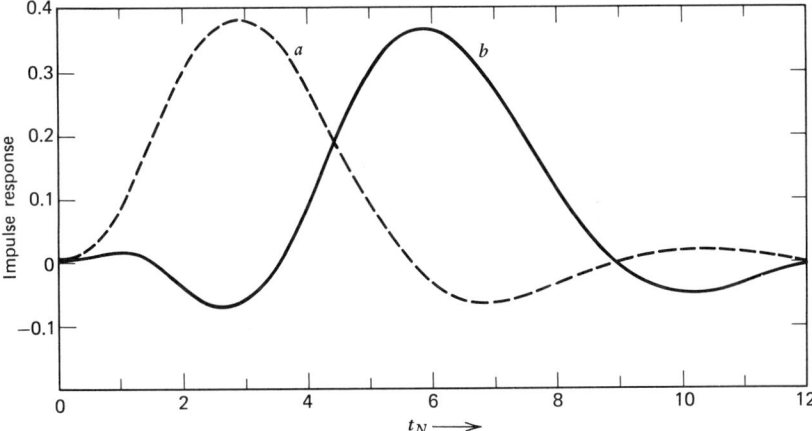

Figure 5-12. Fourth-order Butterworth impulse responses when group delay is (a) unequalized and (b) equalized with one second-order all-pass function for flat delay at $\omega = 0$ (Section 5.2.8).

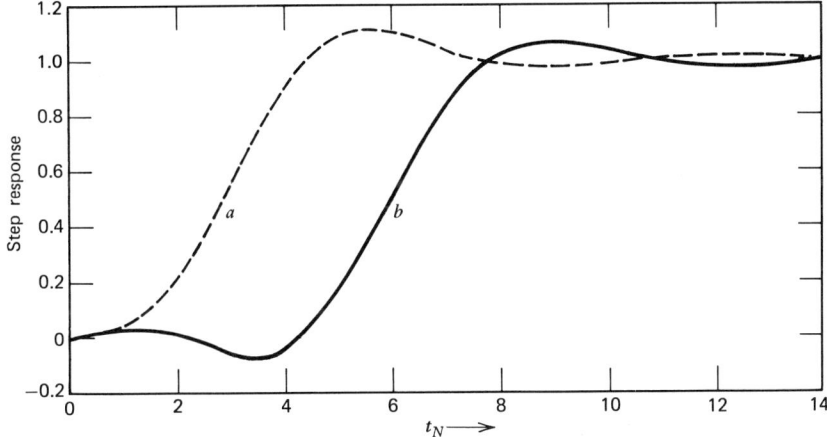

Figure 5-13. Fourth-order Butterworth step responses when group delay is (a) unequalized and (b) equalized with one second-order all-pass function for flat delay at $\omega = 0$ (Section 5.2.8).

influence on the time-domain responses than do the delay characteristics at frequencies of larger attenuation. For this equalized filter, the delay is essentially flat at the lower frequencies, where the attenuation is negligible.

4. Considerable improvement in both the time domain (symmetry and response overshoots) and the frequency domain (phase linearity and constant group delay) has been obtained by the addition of only one second-order AP function, whose parameters were obtained to achieve flat delay at $\omega = 0$. Other error criterions result in further improvements in these responses, as we find in Section 5.3.4.

We have just showed that improvement in passband phase linearity (by now you should realize that linear phase and constant delay are synonymous) has reduced the time-domain overshoots. This raises an interesting question. How much will the time-domain overshoots be reduced if we use an infinite number of

AP functions to obtain a linear phase shift over the infinite frequency range? The answer would tell us when to stop equalizing, for a few equalizers may yield results only slightly inferior to those obtained with linear phase. Furthermore it would tell us what parts the magnitude response and the phase response play in producing the overshoots. We know from the discussion in Section 3.1.2 that even though linear phase is assumed, the first impulse response overshoot is only 13.3 dB down. Likewise, we expect that approximations to this response will produce overshoots even if the phase is linear, but they should be reduced. Let us now determine the impulse response for this idealized phase condition.

Consider the frequency response $H(j\omega)$ represented as the magnitude and phase function

$$H(j\omega) = |H(j\omega)|e^{j\theta(\omega)} \tag{5.2-51}$$

To remove the phase nonlinearity, we assume $\theta(\omega)$ to be proportional to frequency, a condition that introduces a constant delay in the impulse response and step response. However by setting the phase angle to zero, we only remove this delay, and the impulse response is then maximum at $t = 0$. The impulse response is determined from the Fourier inversion integral in (2.1-6), noting that $|H(j\omega)|$ is an even function, as

$$h(t) = \frac{1}{\pi} \int_0^\infty |H(j\omega)| \cos \omega t \, d\omega \tag{5.2-52}$$

Since $\cos \omega t$ is an even function, $h(t)$ is symmetric about $t = 0$, an expected result, since the phase is assumed linear with frequency. Except for the very simplest functions, $h(t)$ cannot be determined analytically, and a digital computer is used.

5.2.9 Transient Responses of Butterworth Filters with Linear Phase

We now determine the impulse response overshoots for the Butterworth family when linear phase is assumed. The magnitude function is

$$|H(j\omega)| = \frac{1}{\sqrt{1 + \omega^{2n}}} \tag{5.2-53}$$

and the impulse response, from (5.2-52) is

$$h(t) = \frac{1}{\pi} \int_0^\infty \frac{\cos \omega t}{\sqrt{1 + \omega^{2n}}} d\omega \tag{5.2-54}$$

By numerical integration on a digital computer, we determined $h(t)$ for $n = 2$ to $n = 10$. Figure 5-14 gives the attenuation of the first impulse response overshoot relative to the main peak. The average reduction for each order is about 6 dB. As n approaches infinity, the Butterworth response approaches the rectangular response of (3.1-1), and the overshoot attenuation approaches 13.3 dB. Figure 5-12 shows that the equalizer with parameters given in Table 5-2 ($n = 4$) reduces the first overshoot from 15 to 18 dB, whereas linear phase, requiring an infinite number of equalizers, only reduces this overshoot by an additional 2.6 dB. This

All-Pass Transfer Function

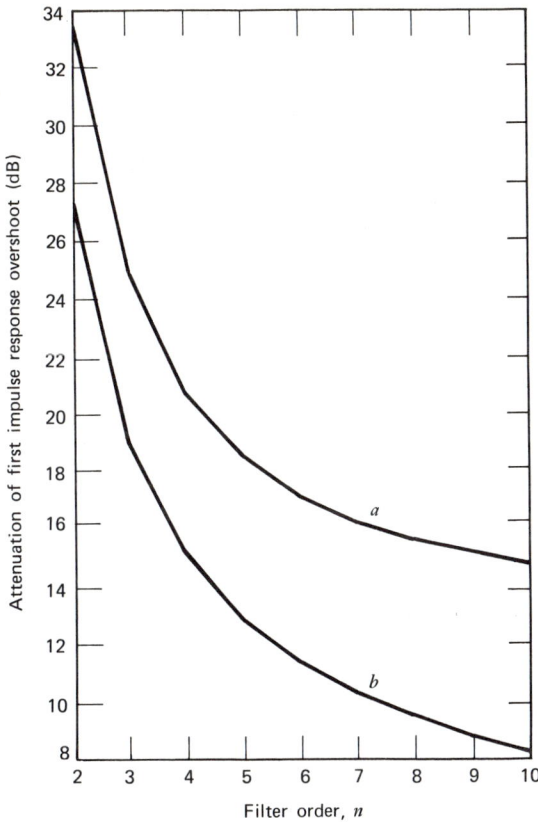

Figure 5-14. Attenuation of first impulse response overshoot for Butterworth family when group delay is (*a*) completely equalized and (*b*) unequalized.

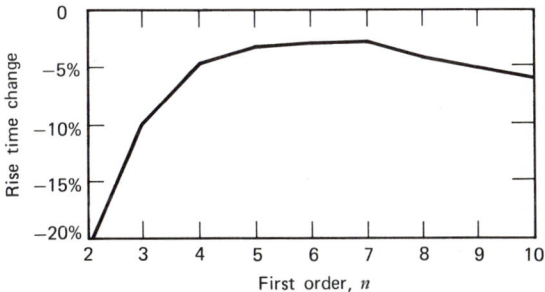

Figure 5-15. Butterworth rise time change for phase-equalized condition compared to Fig. 3-41 values.

illustrates the efficiency of one equalizer in reducing time-domain overshoots.

Another aspect of this investigation is the increase in the impulse response peak value as the phase is linearized. One definition of rise time, from (3.8-1), is the reciprocal of this peak value; thus its increase reflects as a desirable rise time decrease. Figure 5-15 compares the percentage decrease in rise time to the Butterworth rise time in Fig. 3-41. Except for the second- and third-order systems, the reduction in rise time is only 3 to 6%. Thus phase equalization is not the answer for Butterworth rise time reduction; the magnitude response shape is the major contribution to this quantity. Other responses, however, may have a more significant decrease in rise times when the phase is linearized.

5.2.10 Narrow-Band Filter Equalization

Each equalized Butterworth response of Sections 5.2.3 and 5.2.7 can be considered a new LP prototype characteristic, and its narrow-band BP equivalent is obtained by applying the narrow-band transformation of (4.5-7)

$$s_i = \frac{\omega_2 - \omega_1}{2} S_i \pm j\omega_0 \qquad (5.2\text{-}55)$$

The second-order AP function, however, transforms to a fourth-order AP function for the BP case, and we have already stated the desirability of only first- and second-order functions. Therefore this BP equalizer is now realized as two second-order AP functions.

Assume that the LP equalizer zeros are $\alpha_0 \pm j\beta_0$ and its poles are $-\alpha_0 \pm j\beta_0$. Application of the transformation in (5.2-55) results in the BP pole pairs

$$(a) \quad -\frac{\omega_2 - \omega_1}{2} \alpha_0 \pm j\omega_0 \left(1 + \frac{\gamma}{2}\beta_0\right)$$

$$(b) \quad -\frac{\omega_2 - \omega_1}{2} \alpha_0 \pm j\omega_0 \left(1 - \frac{\gamma}{2}\beta_0\right) \qquad (5.2\text{-}56)$$

where

$\omega_0 = $ BP filter center frequency
$\omega_2 - \omega_1 = $ BP filter 3-dB radian bandwidth
$\gamma = \dfrac{\omega_2 - \omega_1}{\omega_0} = $ relative bandwidth

and k and $\bar{\omega}_0$ for each second-order AP function are determined from the real and imaginary parts of these BP poles by applying the definitions in (5.2-22) to each pole pair in (5.2-56). For pole pair (a)

$$\bar{\omega}_{01} = \omega_0 \sqrt{1 + \gamma\beta_0 + \frac{\gamma^2}{4}(\alpha_0^2 + \beta_0^2)}$$

$$k_1 = \frac{(\omega_2 - \omega_1)\alpha_0}{\bar{\omega}_{01}} \qquad (5.2\text{-}57)$$

and for pole-pair (b)

$$\bar{\omega}_{02} = \omega_0 \sqrt{1 - \gamma\beta_0 + \frac{\gamma^2}{4}(\alpha_0^2 + \beta_0^2)}$$

$$k_2 = \frac{(\omega_2 - \omega_1)\alpha_0}{\bar{\omega}_{02}} \qquad (5.2\text{-}58)$$

Equations 5.2-57 and 5.2-58 characterize the two second-order AP functions for BP equalization expressed in terms of the second-order AP function parameters for LP equalization.

5.3 LEAST-SQUARES APPROXIMATION

We have shown that first- and second-order AP functions are sufficient to describe any AP function. Thus the total delay of M first-order and N second-order AP functions is

$$D_a(\omega) = 2 \sum_{i=1}^{M} \frac{\gamma_i}{\omega^2 + \gamma_i^2} + 4 \sum_{j=1}^{N} \frac{\alpha_j(\omega^2 + \alpha_j^2 + \beta_j^2)}{(\alpha_j^2 + \beta_j^2 - \omega^2)^2 + 4\alpha_j^2 \omega^2} \quad (5.3\text{-}1)$$

In any given application, the design task is to determine the M values of γ_i and the N values of α_j and β_j in a meaningful manner. Their selection for a least-squares approximation is now considered.

Application of the least-squares approximation in Section 3.2.3 yields the error number

$$\varepsilon = \int_{\omega_1}^{\omega_2} [D_d(\omega) - D_a(\omega)]^2 \, d\omega \quad (5.3\text{-}2)$$

where $D_d(\omega)$ is the desired delay for $\omega_1 \leq \omega \leq \omega_2$. For minimum ε, the partial derivatives of ε with respect to each variable must be zero; that is,

$$\frac{\partial \varepsilon}{\partial \gamma_i} = 0; \quad \frac{\partial \varepsilon}{\partial \alpha_j} = 0; \quad \frac{\partial \varepsilon}{\partial \beta_j} = 0 \quad (5.3\text{-}3)$$

and the simultaneous solution of these equations yields the optimum parameters. The optimum parameters are those for which the global minimum of ε is achieved. Other parameters may satisfy (5.3-3) but these result in local minima. Unfortunately the nonlinear relationships among the unknowns prevent these equations from being solved analytically; hence a digital computer must be used. Rather than solve the system of nonlinear equations in (5.3-3) on the computer, we find it more efficient to minimize ε directly, for either problem requires a computer algorithm for minimizing a function of several variables.

5.3.1 Numerical Integration

The digital computer integrates numerically, and many integration schemes exist. A simple and accurate algorithm, provided the function is reasonably well behaved, is to replace the integral by equivalent rectangles, that is, (5.3-2) becomes

$$\varepsilon = \Delta\omega \sum_{k=1}^{P} (D_{d_k} - D_{a_k})^2 \quad (5.3\text{-}4)$$

This function is computed at P frequencies in the interval $\omega_1 \leq \omega \leq \omega_2$, and $\Delta\omega$ is the interval between frequencies. The first frequency of computation is $\omega_1 + \Delta\omega/2$, the second is $\omega_1 + 3\Delta\omega/2$, and so on; the last frequency of computation is $\omega_2 - \Delta\omega/2$. For this sequence

$$\Delta\omega = \frac{\omega_2 - \omega_1}{P} \quad (5.3\text{-}5)$$

The desired delay is entered as discrete values; hence no initial approximation of $D_d(\omega)$ is necessary. Figure 5-16 presents the partitioning of the function for numerical integration by this method. More sophisticated numerical integration

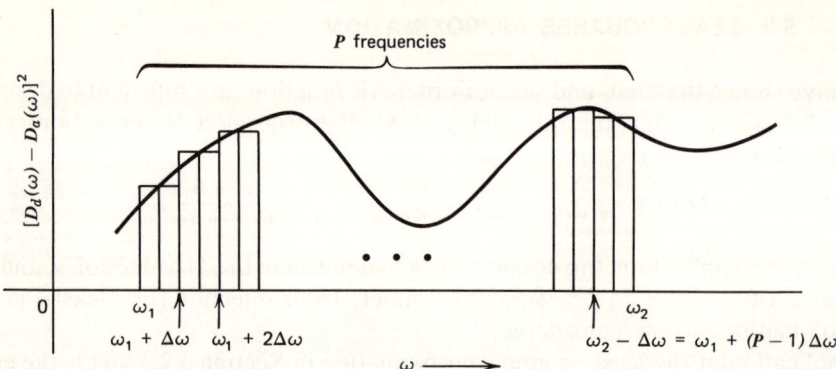

Figure 5-16. Numerical integration by the method of equivalent rectangles.

schemes exist, but the method of equivalent rectangles additionally illustrates how a problem that is not analytically solvable, is easily formulated for a computer solution.

5.3.2 Equalization of Filter Group Delay

Consider the commonly encountered situation of equalizing a filter's group delay $D_f(\omega)$ between frequencies ω_1 and ω_2 such that the sum of $D_f(\omega)$ and the AP delay $D_a(\omega)$ approximates the constant delay D_0. Then $D_d(\omega) = D_0 - D_f(\omega)$ and (5.3-2) becomes

$$\varepsilon = \int_{\omega_1}^{\omega_2} [D_0 - D_f(\omega) - D_a(\omega)]^2 \, d\omega \tag{5.3-6}$$

The value of D_0 that minimizes ε satisfies

$$\frac{\partial \varepsilon}{\partial D_0} = 2 \int_{\omega_1}^{\omega_2} [D_0 - D_f(\omega) - D_a(\omega)] \, d\omega = 0 \tag{5.3-7}$$

or

$$D_0 = \frac{1}{\omega_2 - \omega_1} \int_{\omega_1}^{\omega_2} [D_f(\omega) + D_a(\omega)] \, d\omega \tag{5.3-8}$$

Therefore D_0, for minimum ε, is the average value of the filter's delay plus the AP delay. Because the group delay is the negative of the derivative of phase with respect to frequency, (5.3-8) can be rewritten in terms of the phase functions as

$$D_0 = \frac{-1}{\omega_2 - \omega_1} \int_{\omega_1}^{\omega_2} \left[\frac{d\theta_f(\omega)}{d\omega} + \frac{d\theta_a(\omega)}{d\omega} \right] d\omega$$

$$= \frac{1}{\omega_2 - \omega_1} [\theta_f(\omega_1) - \theta_f(\omega_2) + \theta_a(\omega_1) - \theta_a(\omega_2)] \tag{5.3-9}$$

The AP phase shift is obtained from (5.2-4) and (5.2-20).

The error number ε in (5.3-6) is now shown to be expressible as the area under the square of the filter delay plus equalizer delay minus the area under the square

Least-Squares Approximation

of the constant delay. First square the integrand of (5.3-6) to give

$$\varepsilon = \int_{\omega_1}^{\omega_2} D_0^2 \, d\omega - 2D_0 \int_{\omega_1}^{\omega_2} [D_f(\omega) + D_a(\omega)] \, d\omega + \int_{\omega_1}^{\omega_2} [D_f(\omega) + D_a(\omega)]^2 \, d\omega \qquad (5.3\text{-}10)$$

and then substitute from (5.3-8) into (5.3-10) to give

$$\varepsilon = \int_{\omega_1}^{\omega_2} [D_f(\omega) + D_a(\omega)]^2 \, d\omega - (\omega_2 - \omega_1)D_0^2 \qquad (5.3\text{-}11)$$

Although interesting academically, (5.3-11) presents a computational problem, for as ε approaches its minimum value, ε becomes the difference of large, almost-equal numbers and is very susceptible to round-off errors. For this reason (5.3-4) is preferred, with D_0 obtained from (5.3-8) or (5.3-9).

5.3.3 Starting Values for Computation

Schmidt [4] investigated delay equalization with second-order filters and arrived at some empirical formulas concerning the number of computation frequencies and starting values for the minimization process. To avoid overlooking any of the extrema of the desired curve while simultaneously minimizing the cost of computation, he chooses P in (5.3-4) to be

$$P = 4N + 5 \qquad (5.3\text{-}12)$$

where N is the number of second-order filters used. This value is intended as a guideline only. The larger the value of P, the more accurately the summation in (5.3-4) approximates the integral in (5.3-2). Schmidt also states that the following starting values for α_j and β_j usually give good results:

$$\begin{aligned} \alpha_j &= \frac{\omega_2 - \omega_1}{N} \\ \beta_j &= \omega_1 + \frac{(2j-1)(\omega_2 - \omega_1)}{2N} \end{aligned} \qquad j = 1, 2, \ldots, N \qquad (5.3\text{-}13)$$

5.3.4 Butterworth Delay Equalization with a Second-Order All-Pass Function

The Butterworth group delay responses are now equalized with one second-order AP function; thus the overall group delay approximates a constant delay in a least-squares sense from $\omega = 0$ to $\omega = 1$, the filter passband. The Butterworth delay, $D_B(\omega)$, from (3.4-5), and the AP delay, $D_2(\omega)$ from (5.2-21), are substituted into (5.3-6) to give

$$\varepsilon = \int_0^1 [D_0 - D_B(\omega) - D_2(\omega)]^2 \, d\omega \qquad (5.3\text{-}14)$$

and from (5.3-9)

$$D_0 = -[\theta_B(1) + \theta_2(1)] \qquad (5.3\text{-}15)$$

The AP phase is obtained from (5.2-20) and the phase shift of the nth-order Butterworth filter at $\omega = 1$ is $-n\pi/4$. Then D_0 becomes

$$D_0 = \frac{n\pi}{4} + 2 \tan^{-1} \frac{2\alpha}{\alpha^2 + \beta^2 - 1} \qquad (5.3\text{-}16)$$

Equation 5.3-4 was implemented on the digital computer for $n = 4$ to $n = 8$, and α and β were adjusted until ε was minimized. The optimum parameters are listed in

Table 5-3 **Second-Order AP Parameters that Produce a Least Squares Approximation to a Constant Passband Delay for Butterworth Responses**

n	k	$\bar{\omega}_0$	α	β	D_0
4	1.905	1.0621	1.0119	0.32268	6.15627
5	1.724	0.79787	0.68784	0.40432	7.58446
6	1.643	0.69671	0.57241	0.39719	8.69814
7	1.590	0.63754	0.50691	0.38666	9.69783
8	1.551	0.59690	0.46300	0.37672	10.6385

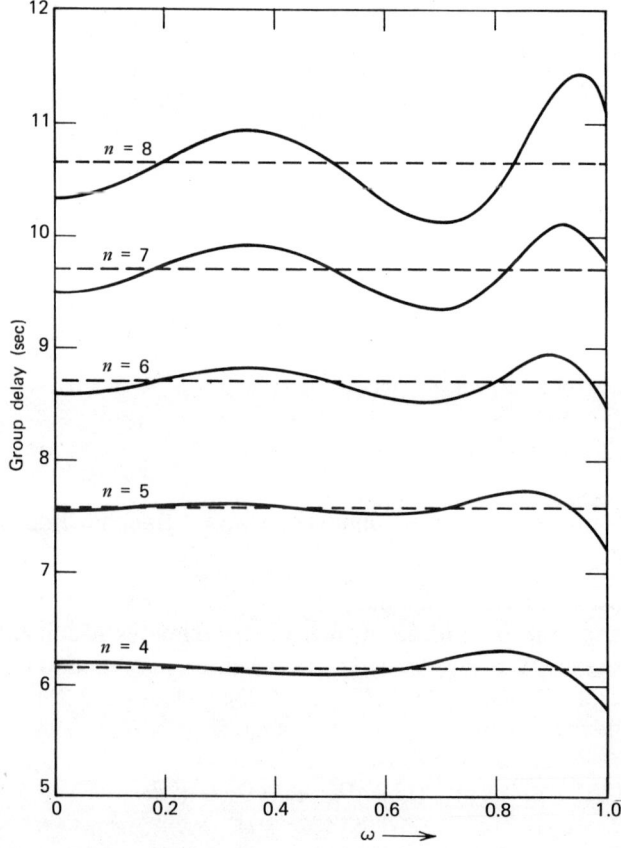

Figure 5-17. Butterworth group delays equalized with one second-order all-pass function for a least-squares approximation to a constant passband delay.

Network Realizations

Table 5-3 and the equalized delay curves appear in Fig. 5-17, where the dashed lines are the respective values of D_0.

As expected, the curves in Fig. 5-17 are more constant over the passband than the curves of Fig. 5-11, since the least-squares approximation considers the entire interval, whereas the flat approximations in Fig. 5-11 consider only the frequency $\omega = 0$. The delays of Fig. 5-17 exhibit increasing deviation from the constant value as $\omega = 1$ is approached, a feature of the least-squares approximation. The values of D_0 in Table 5-3 were numerically computed from (5.3-8), and they agree very well with the analytically determined values from (5.3-16), verifying the accuracy of the numerical integration algorithm previously discussed.

To determine the passband delay's effect on the time-domain responses, we computed the impulse response of the fourth-order equalized Butterworth filter (Fig. 5-18). The first overshoot is now 19.2 dB down from the maximum value compared with 18 dB for the flat delay equalization in Fig. 5-12. The peak response occurs at $t = 5.93$ sec, consistent with the constant delay from Table 5-3, $D_0 = 6.16$ sec.

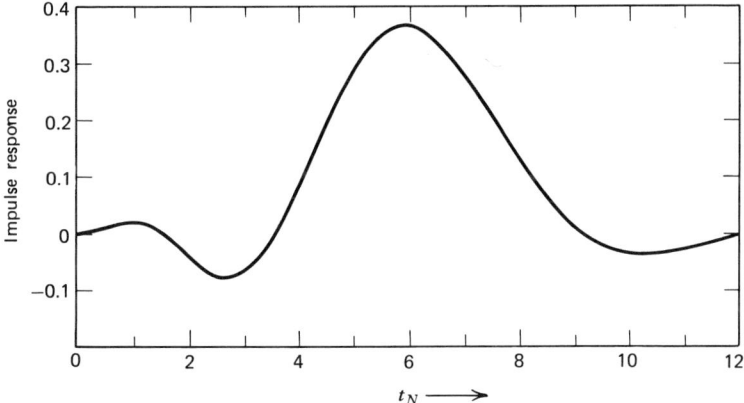

Figure 5-18. Fourth-order Butterworth impulse response when group delay is equalized with one second-order all-pass function for a least-squares approximation to a constant passband delay.

5.4 NETWORK REALIZATIONS

Because the AP filter is a nonminimum-phase network, it cannot be realized as a ladder network, but it can be realized by the lattice network in Fig. 5-19. This lattice must possess the constant-resistance property, permitting the individual sections to be cascaded without any interaction. This condition is achieved when

$$Z_a Z_b = R^2 \tag{5.4-1}$$

Then the lattice transfer function is

$$H(s) = \frac{R - Z_a}{R + Z_a} \tag{5.4-2}$$

and so long as Z_a is a reactance function $jX(\omega)$, the filter is all pass, for then

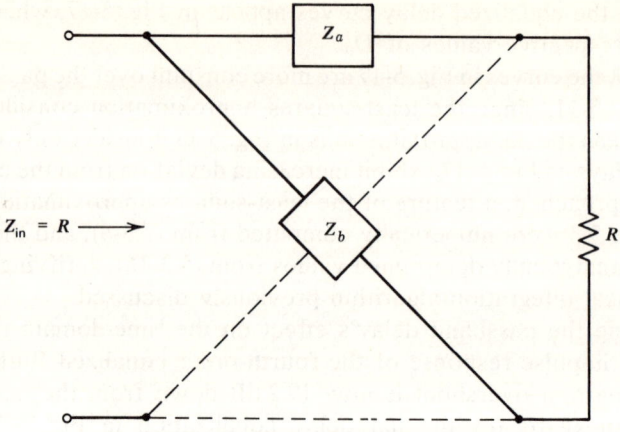

Figure 5-19. Constant-resistance lattice network.

$$H(j\omega) = \frac{R - jX}{R + jX} \quad (5.4\text{-}3)$$

and

$$|H(j\omega)| = \frac{\sqrt{R^2 + X^2}}{\sqrt{R^2 + X^2}} = 1 \quad (5.4\text{-}4)$$

5.4.1 First-Order Lattice Section

For the first-order section, $Z_a = sL$, which substituted into (5.4-2) gives

$$H(s) = \frac{R/L - s}{R/L + s} \quad (5.4\text{-}5)$$

Equate $H(s)$ to $H_1(s)$ in (5.2-3) and determine the value of the inductor as

$$L = \frac{R}{\gamma} \quad (5.4\text{-}6)$$

and from (5.4-1)

$$Z_b = \frac{R^2}{sL} = \frac{1}{sC} \quad (5.4\text{-}7)$$

the reactance of the capacitor whose value is

$$C = \frac{L}{R^2} = \frac{1}{\gamma R} \quad (5.4\text{-}8)$$

Thus L and C resonate at γ.

5.4.2 Second-Order Lattice Section

For the second-order section, Z_a is the impedance of an antiresonant circuit,

$$Z_a = \frac{Ls}{LCs^2 + 1} \quad (5.4\text{-}9)$$

Network Realizations

which substituted into (5.4-2) gives

$$H(s) = \frac{s^2 - (1/RC)s + 1/LC}{s^2 + (1/RC)s + 1/LC} \tag{5.4-10}$$

Equate $H(s)$ and $H_2(s)$ in (5.2-19) and use the normalized variables in (5.2-22) to obtain

$$\frac{1}{RC} = 2\alpha = k\bar{\omega}_0$$
$$\frac{1}{LC} = \alpha^2 + \beta^2 = \bar{\omega}_0^2 \tag{5.4-11}$$

from which

$$L = \frac{kR}{\bar{\omega}_0}$$
$$C = \frac{1}{k\bar{\omega}_0 R} = \frac{1}{L\bar{\omega}_0^2} \tag{5.4-12}$$

The parallel resonance of L and C occurs at $\bar{\omega}_0$. From (5.4-1), Z_b is

$$Z_b = \frac{R^2(LCs^2 + 1)}{Ls} = R^2Cs + \frac{1}{(L/R^2)s} = sL_1 + \frac{1}{sC_1} \tag{5.4-13}$$

This is the impedance of a series circuit composed of L_1 and C_1, where

$$L_1 = R^2 C$$
$$C_1 = \frac{L}{R^2} \tag{5.4-14}$$

and the series resonance again occurs at $\bar{\omega}_0$. The design equations for first- and second-order AP lattice networks are summarized in Fig. 5-20.

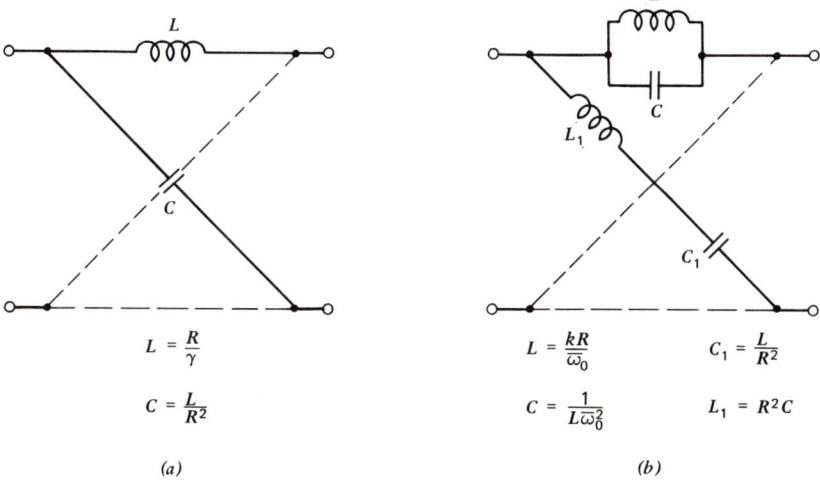

Figure 5-20. Lattice forms of (a) first-order and (b) second-order all-pass networks.

5.4.3 First-Order Bridge Networks

The lattice AP network has the disadvantage that it does not have a common input-output ground connection, and it contains redundant elements. However other equivalent networks offer the advantage of a common input-output ground connection, as well as a reduction in the number of components. One is the bridged-Tee network in Fig. 3-15a, and another is the differential bridge (Fig. 3-15b). Both are obtained by inspection once the lattice form is known. Here we discuss the first-order bridge network; second-order bridge networks are covered in Section 5.4.4.

We decompose the first-order lattice into the ladder containing a negative inductor as in Fig. 5-21a. Use of the equivalence in Fig. 4-27 allows this network to be realized with a unity-coupled transformer, as in Fig. 5-21b. Transformers with prescribed mutual, primary, and secondary inductances can be realized at frequencies below 1 MHz, but they present problems at higher frequencies because of parasitic elements.

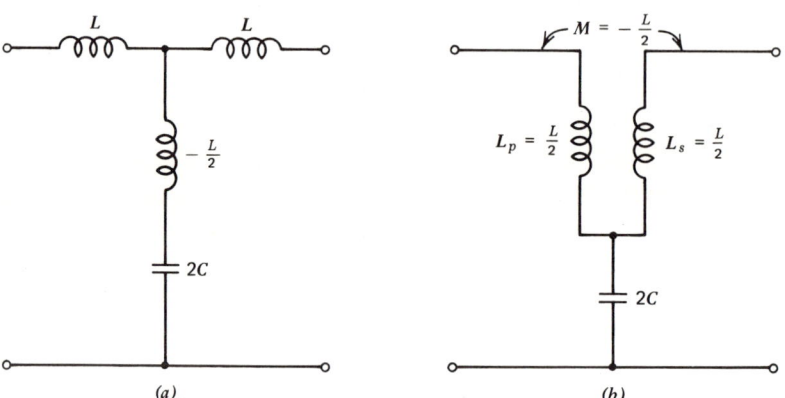

Figure 5-21. Equivalent networks for a first-order all-pass lattice. (a) Ladder equivalent (unrealizable), (b) realizable network.

5.4.4 Second-Order Bridge Networks

The second-order AP lattice is decomposed into the bridge of Fig. 5-22, useful for $k > 1$, where the transformer is not unity coupled. For $k \leq 1$, schematics without mutual inductance are realizable, and Fig. 5-23 presents two canonic forms. Alternate canonic networks are obtained by transforming the capacitor Tee (C, C, C_A) or inductor Tee (L, L, L_A) to a π network. The schematic in Fig. 5-23a is preferred in practice because at high frequencies inductors have few turns; thus the balance in the schematic in Fig. 5-23b cannot be precisely achieved. However, except for small values of capacitance, close-tolerance capacitors can be obtained. At lower frequencies, where inductors become bulky, the schematic of Fig. 5-23a is preferred because it contains one less inductor.

The network of Fig. 5-23a can be modified to absorb the distributed capacitance of the $2L$ inductor by splitting C_A into two capacitors and transforming the remaining T to a π network as illustrated in Fig. 5-24. This technique, also useful for obtaining practical capacitance values, leads to the network in Fig. 5-25.

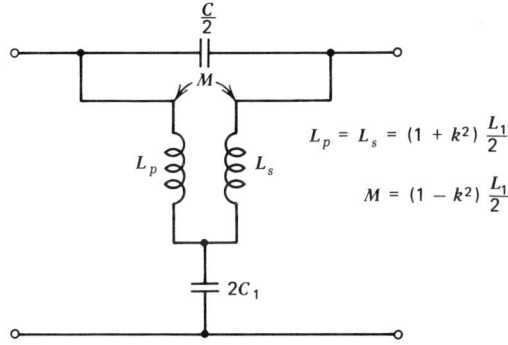

Figure 5-22. Bridged-Tee equivalent of a second-order all-pass lattice.

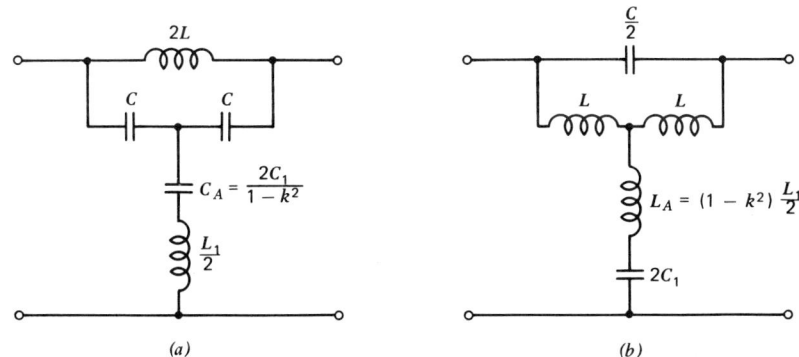

Figure 5-23. Bridged-Tee equivalents of a second-order all-pass lattice for $k \leq 1$.

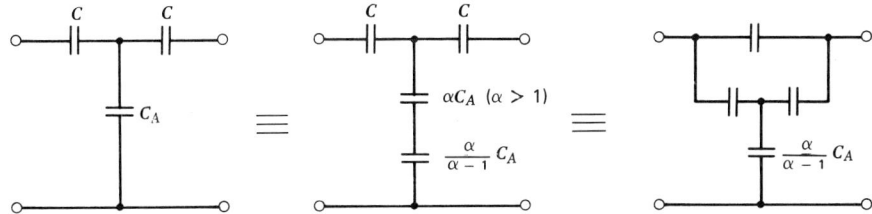

Figure 5-24. Transformation leading to practical bridged-Tee schematic.

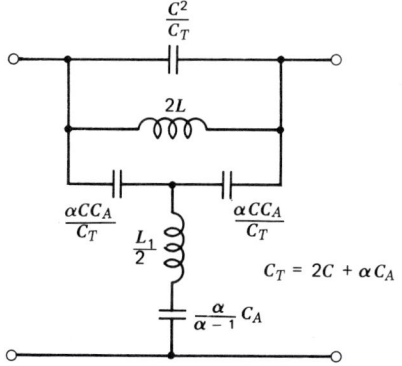

Figure 5-25. Practical bridged-Tee schematic ($k \leq 1$).

241

5.4.5 Effect of Losses in a Second-Order Network

The preceding discussions assume that the AP elements are lossless, but in practice the elements are lossy and attenuation occurs, destroying the AP characteristic. Although high-Q elements result in negligible loss per section, the cascading of many sections can produce unacceptable attenuation. If the element-Q is low, even the loss per section can be appreciable.

The attenuation characteristic of the second-order network with losses is now examined, for these AP filters are the most frequently used in practice. Typically, losses change the AP response to a bandstop response, illustrated in Fig. 5-26 for

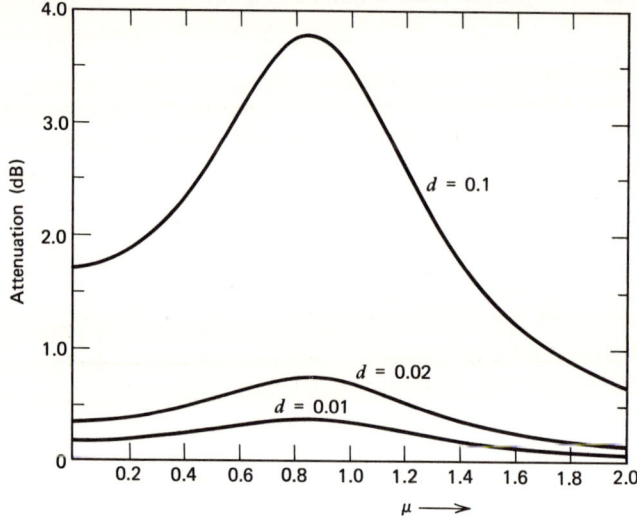

Figure 5-26. Attenuation of second-order all-pass filter with various amounts of loss ($k = \bar{\omega}_0 = 1$).

$k = 1$. The dissipation factor d is defined as

$$d = \frac{\bar{\omega}_0}{Q} \tag{5.4-15}$$

where Q is the quality factor of each element at the frequency $\bar{\omega}_0$. It can be shown that the maximum attenuation occurs at the frequency

$$\omega_a = \sqrt{2\sqrt{(\alpha^2 + \beta^2)(\beta^2 + d^2)} - (\alpha^2 + \beta^2 + d^2)} \tag{5.4-16}$$

which, after use of (5.2-22) and (5.4-15), corresponds to the normalized frequency

$$\mu_a = \frac{\omega_a}{\bar{\omega}_0} = \sqrt{\sqrt{4\left(1 + \frac{1}{Q^2}\right) - k^2} - \left(1 + \frac{1}{Q^2}\right)} \tag{5.4-17}$$

For $Q > 10$, μ_a is approximately equal to μ_p, the frequency of maximum delay given by (5.2-25). As Q approaches infinity (no losses), μ_a approaches μ_p. Table 5-4 compares μ_a and μ_p for the values of d in Fig. 5-26, and the negligible difference is apparent.

Network Realizations

Table 5-4 Values of μ_p and μ_a for $k = 1$ ($\bar{\omega}_0 = 1$)

d	Q	μ_a	$\mu_p = 0.855600$
0.1	10	0.856480	
0.02	50	0.855636	
0.01	100	0.855609	

The maximum attenuation at ω_a is

$$A_M = 20 \log_{10} \frac{\alpha\sqrt{\beta^2 + d^2} + d\sqrt{\alpha^2 + \beta^2}}{\beta(\alpha - d)} \text{ decibels} \quad (5.4\text{-}18)$$

and in terms of k and Q,

$$A_M = 20 \log_{10} \frac{\frac{k}{2}\sqrt{1 + \frac{1}{Q^2} - \frac{k^2}{4} + \frac{1}{Q}}}{\sqrt{1 - \frac{k^2}{4}\left(\frac{k}{2} - \frac{1}{Q}\right)}} \text{ decibels} \quad (5.4\text{-}19)$$

Table 5-5 Peak Attenuation at μ_a

	Attenuation (dB)		
Q	$k = 0.1$	$k = 0.5$	$k = 1$
2	—	—	∞
4	—	∞	10.201
10	—	7.473	3.790
20	∞	3.579	1.877
50	7.364	1.416	0.749
100	3.524	0.707	0.374
200	1.744	0.353	0.187
500	0.696	0.141	0.075

tabulated in Table 5-5 for $k = 0.1$, 0.5, and 1.0 for various values of Q. Table 5-5 shows that the attenuation at μ_a increases with decreasing values of Q until it is infinite when $Q = (2/k)(d = \alpha)$. Substitute $d = \alpha$ into (5.4-16) to determine that this notch occurs at $\omega_\infty = \beta$, and from (5.4-17) with $Q = 2/k$,

$$\mu_\infty = \frac{\omega_\infty}{\bar{\omega}_0} = \sqrt{1 - \frac{k^2}{4}} \quad (5.4\text{-}20)$$

With $k = 1$, $\mu_\infty = \sqrt{3}/2 = 0.866$, which is approximately the values in Table 5-4. Thus sufficient loss changes the AP characteristic to a notch characteristic, with infinite attenuation occurring at $\omega_\infty = \beta$ when $d = \alpha$.

The occurrence of the notch at ω_∞ is also explained by the movement of the

s-plane poles and zeros as losses are introduced. In Section 6.2.1 we show that equal loss in each element causes the transfer-function poles and zeros to shift to the left by the amount d. Thus infinite attenuation occurs at $\omega = \beta$ because the zeros at $\alpha \pm j\beta$ shift to the $j\omega$ axis when $d = \alpha$. Therefore networks with small k values are more sensitive to losses than those with larger k values because their transfer-function zeros are closer to the $j\omega$ axis, and a small change in their position has a greater effect on the attenuation function. For example, consider two AP functions, one with $\alpha = 0.05$ when $k = 0.1$ and the other with $\alpha = 0.5$ when $k = 1$ ($\bar{\omega}_0 = 1$). Table 5-5 indicates that elements with $Q = 100$ cause a peak attenuation of 3.524 dB for the first network, whereas the second network experiences an attenuation of only 0.374 dB.

In addition to the losses in the elements, which may appreciably degrade the theoretical response, the element-value tolerances influence a practical design. Suppose the delay ripple associated with a specified design is too large. Theoretically, additional cascaded AP networks, appropriately designed, will reduce this ripple, but in practice a limit is eventually reached. An increase in the number of AP filters also increases the mean delay and reduces the ripples. Finally the variations caused by element tolerances are greater than the design ripple and the design becomes impractical. A good rule of thumb for practical equalizer design is that the ratio of mean delay to design ripple should not exceed 200 for element tolerances of 1% [4].

5.4.6 Example of Bandpass Delay Equalization

The delay of a 70-MHz IF filter for satellite ground equipment (Fig. 5-27) is now equalized to ensure that the overall delay approximates a constant delay in the least-squares sense between 50 and 90 MHz. Since the unequalized delay is approximately parabolic, only second-order AP functions are used. For computation purposes $f_1 = 48$ MHz, $f_2 = 92$ MHz, and $\Delta f = 1$ MHz (see Fig. 5-16); P, the number of computation frequencies from (5.3-4), is 44. In comparison, values of P given by (5.3-12) are 17, 21, and 25 for $N = 3$, 4, and 5, respectively. The error function in (5.3-4) was minimized on the computer for $N = 3$, 4, and 5, and the equalized delay deviations are shown in Fig. 5-28. For $N = 5$, the ratio of mean

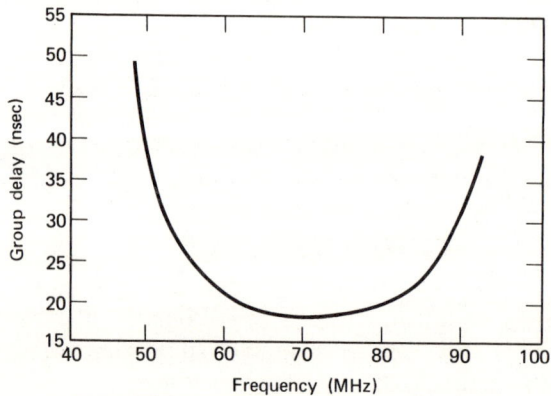

Figure 5-27. Unequalized group delay of filter in Section 5.4.6.

Network Realizations

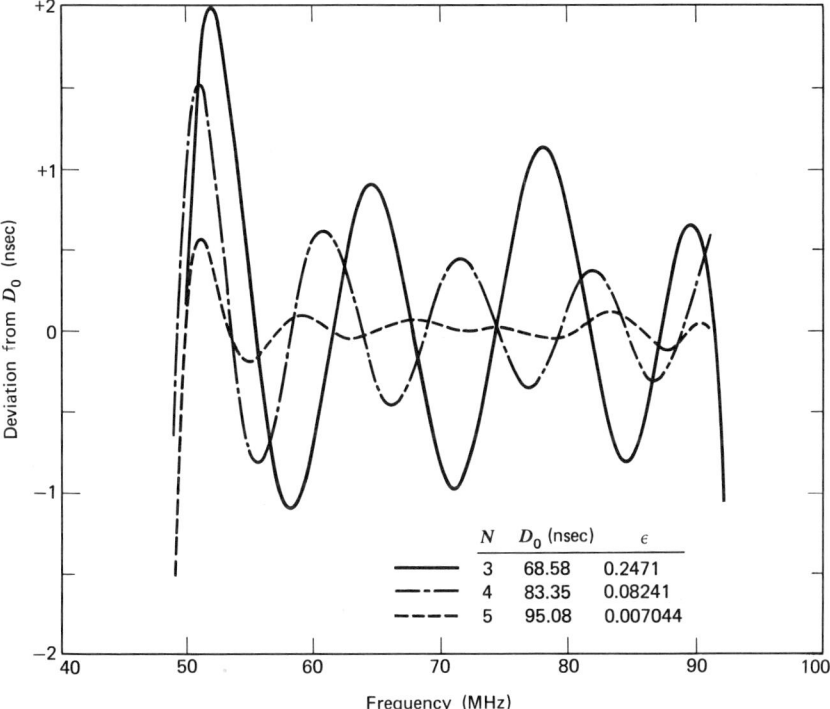

Figure 5-28. Delay deviations for equalized responses in Section 5.4.6.

delay to peak ripple is $95/0.5 = 190$. This is near the critical value for this ratio, and if 1% elements are used, the computed ripple may not be realized because of element tolerances.

Table 5-6 lists the final normalized values of α_j and β_j and the corresponding k values for the $N = 3$ case. The denormalized values of α_j and β_j are obtained by multiplying each value by $2\pi \times 92 \times 10^6$. Since the three values of k are all less than unity, the AP filters are realized in the unbalanced form of Fig. 5-23a, and these are given in Fig. 5-29 for source and load equal to 100 ohms.

Delay equalization with AP filters extends from the very low frequencies to the hundreds of megahertz, the upper limit occurring when the lumped elements can no longer be realized. Delay equalization is still possible beyond this frequency, but the equalizer realization changes to UHF and microwave components [3, 5].

Table 5-6 **Final Normalized Values for $N = 3$ (Section 5.4.6)**

α	β	k
0.11936	0.77206	0.3056
0.11220	0.62177	0.3552
0.11105	0.92617	0.2381

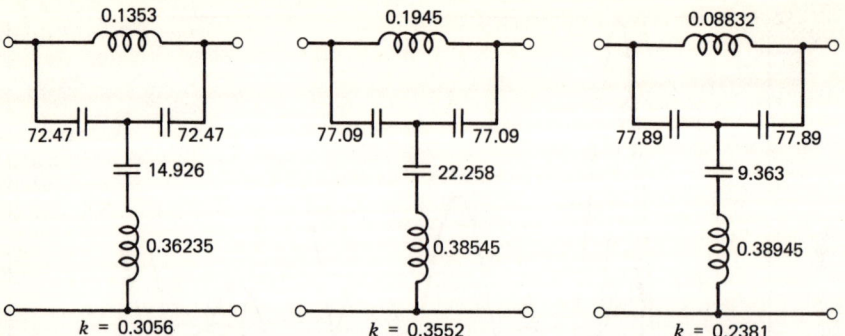

Figure 5-29. Three bridged-Tee all-pass sections that equalize delay of Fig. 5-27 ($R = 100$ ohms, L in microhenries, C in picofarads).

REFERENCES

1. Crane, R. L., "All-Pass Network Synthesis," *IEEE Trans. Circuit Theory*, Vol. CT-15, pp. 474–477, December 1968.
2. Humpherys, D. S., "Rational Function Approximation of Polynomials to Give an Equiripple Error," *IEEE Trans. Circuit Theory*, Vol. CT-11, pp. 479–486, December 1964.
3. Scanlan, S. O., and J. D. Rhodes, "Microwave Allpass Networks—Part I and Part II," *IEEE Trans. Microwave Theory Tech.*, Vol. MTT-16, pp. 62–79, February 1968.
4. Schmidt, C. E., "Delay Equalizers," Presented at Wescon, Los Angeles, Cal., August 1970.
5. Sleven, R. L., "Minimum Distortion Filters for Satellite Communications," *Microwave J.*, pp. 67–72, May 1972.

PROBLEMS

5-1. Two first-order AP sections are cascaded. At $\omega = 0$, the group delay of the first section is 2 sec and the delay of the second section is 1 sec.
  (a) What is the overall delay function expressed as a ratio of polynomials?
  (b) What is the impulse response of the cascade?

5-2. Consider the first-order AP filter. What value of γ ensures that the slope of the delay response at $\omega = 1$ is -1?

5-3. Twenty-five identical first-order AP sections are cascaded. What is the value of γ if the overall delay at $\omega = 10^4$ is 2 msec and the delay at $\omega = 1.5 \times 10^4$ is 1 msec?

5-4. Design a second-order AP filter to ensure that the maximum delay is 0.5 μsec and occurs at 0.9874 MHz. Use the lattice schematic in Fig. 5-20b with $R = 100$ ohms.

5-5. What is the step response of the second-order AP filter with $k = 2$?

5-6. What is the area under the delay curve ($\omega = 0$ to $\omega = \infty$) of
  (a) The first-order AP filter?
  (b) The second-order AP filter?

Problems

5-7. The AP transfer function
$$H(s) = \frac{s^2 - 3s + 16}{s^2 + 3s + 16}$$
is to be realized using a lattice network terminated in 1 ohm.
  (a) What are the element values of the network?
  (b) What is the delay at $\omega = 0$?
  (c) What is the impulse response of this network?

5-8. Show that the response of a first-order AP filter with the transfer function of (5.2-3), excited by the function $u_{-1}(t)f(t)$, can be expressed as
$$u(t) = -f(t) + 2\gamma e^{-\gamma t} \int_0^t e^{\gamma \tau} f(\tau)\, d\tau$$

Note that at $t = 0$, $u(0) = -f(0)$, that is, the input is instantly inverted.

5-9. A first-order AP filter is cascaded with a second-order Butterworth filter ($\omega_c = 1$). If $h_B(t)$ is the impulse response of the Butterworth filter alone, show that $h(t)$, the impulse response of the cascade, can be expressed as
$$h(t) = c\left[\frac{\gamma - c}{c\gamma} h_B(t) - \frac{d}{dt} h_B(t) + e^{-\gamma t}\right] u_{-1}(t)$$
where $c = \dfrac{2\gamma}{\gamma^2 - \gamma\sqrt{2} + 1}$.

5-10. Consider the second-order AP section in Fig. 5-23a with $R = 5000$ ohms. The maximum delay occurs at $\mu_p = 0.96773$ and is 8.131 μsec there.
  (a) What are k and $\bar{\omega}_0$?
  (b) What are the network element values?

Chapter Six

Finite-Q Elements and Predistortion

Except for a brief discussion in Section 5.4.5, we have dealt only with lossless filter elements. In practice losses are always present, however, and they cause the frequency- and time-domain responses to deviate from those predicted by theory. In addition to the response distortion, losses introduce a flat passband loss referred to as insertion loss, which often directly influences system performance. For example, the insertion loss of preselectors in the receiver (front-end filters) is required to be less than 0.5 dB, for such loss adds directly to the system noise figure. Also filters used at the transmitter output for achieving spectral purity are required to have low insertion loss. The seemingly small value of 1-dB insertion loss indicates that 20.6% of the power is dissipated in the filter. This figure is better appreciated by noting that for a 100-kW transmitter, 20.6 kW is dissipated in the filter. For such applications, even 1-dB insertion loss is too large!

In this chapter we determine the effect of losses on the attenuation and transient responses, establish for easy reference the insertion loss of the popular filter types, and introduce the technique of predistorting the responses so that with losses present, the responses "return" to the theoretically predicted ones.

6.1 THE QUALITY FACTOR

Filter losses can be represented by a resistor in series with each inductor and a resistor in parallel with each capacitor, but as is commonly done, we represent the element loss by its quality factor (Q). Thus lossless elements have infinite Q and as Q decreases, the difference between the lossless response and the lossy response increases.

Furthermore Q is a convenient description, for it is invariant under simple frequency scaling. For example, consider the LP inductor L with loss represented as a resistor r in series. Then at the frequency ω_a, the quality factor q_a is

$$q_a = \frac{\omega_a L}{r} \tag{6.1-1}$$

To standardize the frequency at which q is computed, we choose the cutoff frequency for LP and HP filters and the band center for BP and BS filters. We designate the LP quality factor by q, as in Section 4.5.8, and the transformed BP or BS quality factor by Q. If ω_a in (6.1-1) is the cutoff frequency, and we wish to transform to a new cutoff frequency ω_b, then at ω_b, after L is frequency scaled,

Lossy-Filter Responses

the quality factor is

$$q_b = \frac{\omega_b}{r}\frac{L}{\omega_b/\omega_a} = \frac{\omega_a L}{r} = q_a \qquad (6.1\text{-}2)$$

Thus q is preserved by frequency scaling.

Of particular interest is the effect of finite-q elements on the attenuation response. Qualitatively, passband ripples, if present, are smeared; the attenuation response near the cutoff frequency is rounded, changing the 3-dB bandwidth, and the infinite attenuation of the stopband peaks is reduced to a finite value. Whenever the losses grow exceedingly large, the passband attenuation response becomes so rounded that the shape becomes Gaussian regardless of its characteristic with lossless elements.

6.2 LOSSY-FILTER RESPONSES

To consider the effect of resistive losses on the filter responses, we now examine the LP filter with equal-q elements. One may insist that inductors generally have a higher quality factor than do capacitors, but remember that each LP element transforms to a BP resonator, when the LP-to-BP transformation is applied. Since the most common filter type in practice is the narrow band BP filter, and since the BP resonators are then essentially identical, the choice of equal-q LP elements is certainly reasonable.

6.2.1 Transfer Function

The impedance of the LP inductor L in series with the loss resistor r is

$$Z(s) = sL + r = L\left(s + \frac{r}{L}\right) = L\left(s + \frac{\omega_a r}{\omega_a L}\right) = L\left(s + \frac{\omega_a}{q}\right) \qquad (6.2\text{-}1)$$

where q is given by (6.1-1). Since q is invariant under frequency scaling, we let ω_a be the unit cutoff frequency, yielding

$$Z(s) = L\left(s + \frac{1}{q}\right) \qquad (6.2\text{-}2)$$

The admittance of the LP capacitor C in parallel with the loss resistor r is

$$Y(s) = Cs + \frac{1}{r} = C\left(s + \frac{1}{rC}\right) = C\left(s + \frac{\omega_a}{q}\right) \qquad (6.2\text{-}3)$$

and with $\omega_a = 1$,

$$Y(s) = C\left(s + \frac{1}{q}\right) \qquad (6.2\text{-}4)$$

Therefore the transfer function of the lossy filter $H_q(s)$ is obtained by replacing the variable s by $s + 1/q$ in the normalized lossless filter transfer function $H_L(s)$, that is,

$$H_q(s) = H_L(s)|_{s=s+1/q} = H_L\left(s + \frac{1}{q}\right) \qquad (6.2\text{-}5)$$

The effect of equal q is to shift the s-plane locations to the left by $1/q$ as in Fig.

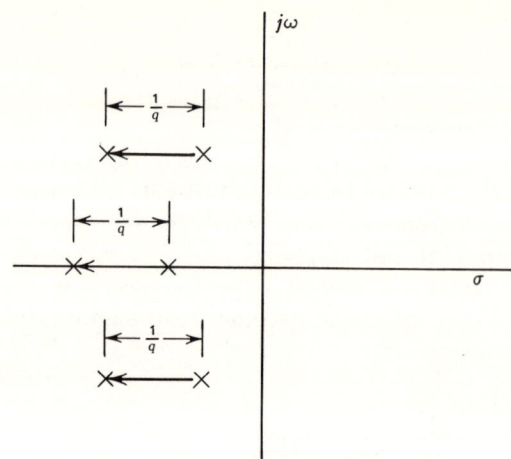

Figure 6-1. Shift of LP poles to the left due to equal-q elements.

6-1. This is better seen by considering a typical denominator factor $(s - s_k)$ in the lossless transfer function, where s_k is a left-half plane pole. Replace s by $s + 1/q$, equate to zero, and determine the new pole s_{k_1} as

$$s_{k_1} = s_k - \frac{1}{q} \qquad (6.2\text{-}6)$$

which is just the lossless pole shifted to the left by $1/q$.

6.2.2 Transient Responses

The lossy-filter impulse response $h_q(t)$ is the inverse Laplace transform of $H_L(s + 1/q)$. From (2.3-1)

$$h_q(t) = \frac{1}{2\pi j} \int_c H_L(s + 1/q) e^{st} \, ds \qquad (6.2\text{-}7)$$

where c is a suitable contour in the s-plane. With the change of variable $x = s + 1/q$,

$$h_q(t) = \frac{1}{2\pi j} \int_{c_1} H_L(x) e^{(x-1/q)t} \, dx = e^{-t/q} \frac{1}{2\pi j} \int_{c_1} H_L(x) e^{xt} \, dx = e^{-t/q} h_L(t) \qquad (6.2\text{-}8)$$

where c_1 is also a suitable contour in the x-plane, and $h_L(t)$, the lossless impulse response, is the inverse Laplace transform of $H_L(s)$. The lossy impulse response is the lossless impulse response, weighted by the decreasing exponential $e^{-t/q}$. This simple relationship allows the lossless impulse responses in catalogs such as Ref. 6 to be easily converted to the lossy responses.

For large values of q, the exponential is essentially unity over the significant time span of $h_L(t)$, and then $h_q(t)$ and $h_L(t)$ are approximately the same. For small values of q, $e^{-t/q}$ appreciably reduces the lossless response at moderate values of t and generally improves the response symmetry. This is an expected

Lossy-Filter Responses

result, for we earlier mentioned that the filter's magnitude and impulse responses approach the symmetric Gaussian response as losses increase.

The lossy-filter step response $g_q(t)$, from (1.8-12), is

$$g_q(t) = \int_0^t h_q(x)\, dx = \int_0^t e^{-x/q} h_L(x)\, dx \qquad (6.2\text{-}9)$$

which can be rewritten as

$$g_q(t) = \int_0^\infty e^{-x/q} h_L(x)\, dx - \int_t^\infty e^{-x/q} h_L(x)\, dx \qquad (6.2\text{-}10)$$

The first integral is the Laplace transform of $h_L(t)$ evaluated at $1/q$. But this is just $H_L(1/q)$, and $g_q(t)$ simplifies to

$$g_q(t) = H_L(1/q) - \int_t^\infty e^{-x/q} h_L(x)\, dx \qquad (6.2\text{-}11)$$

As t approaches ∞, $g_q(t)$ approaches the steady-state value $H_L(1/q)$, which, from (6.2-5), is the lossy magnitude response at $s = 0$.

The difference between the lossless and lossy responses raises the following question. What is the minimum value of q for which the resulting responses are still a respectable approximation to the lossless responses? Obviously the answer is not unique, but we may make an intelligent estimate by noting that since each pole shifts the same amount $1/q$, the pole closest to the $j\omega$ axis shifts the most relatively. Thus it seems logical to limit its shift, say, to a maximum of 10% and thereby arrive at a lower limit for q as

$$q > \frac{10}{|\alpha_c|} \qquad (6.2\text{-}12)$$

where α_c is the real part of the pole closest to the $j\omega$ axis. The lossy responses computed from the lossy poles verify that (6.2-12) is a valid condition. These responses are additionally useful for designing the lossless filter to anticipate the accompanying response change with losses.

6.2.3 Butterworth Filter Analysis

We now examine the change in the Butterworth responses when the filter elements are lossy. The analysis of the nth-order filter follows the procedure now described for the third-order normalized transfer function of (3.4-3),

$$H_L(s) = \frac{1}{s^3 + 2s^2 + 2s + 1} \qquad (6.2\text{-}13)$$

The transfer function of the lossy filter with equal-q elements, from (6.2-5), is

$$H_q(s) = \frac{1}{\left(s + \dfrac{1}{q}\right)^3 + 2\left(s + \dfrac{1}{q}\right)^2 + 2\left(s + \dfrac{1}{q}\right) + 1}$$

$$= \frac{1}{s^3 + \left(2 + \dfrac{3}{q}\right)s^2 + \left(\dfrac{3}{q^2} + \dfrac{4}{q} + 2\right)s + \left(\dfrac{1}{q^3} + \dfrac{2}{q^2} + \dfrac{2}{q} + 1\right)} \qquad (6.2\text{-}14)$$

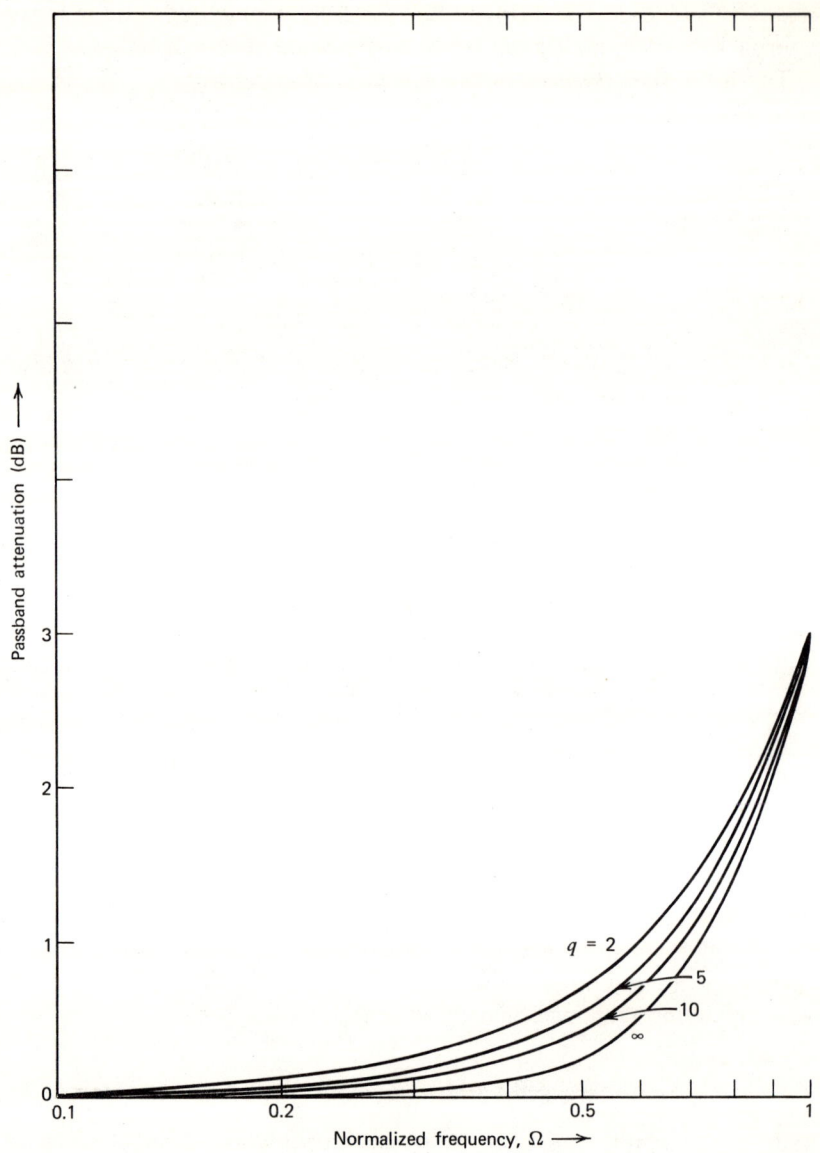

Figure 6-2. Attenuation of the $n = 2$ Butterworth filter with finite q, normalized to unity radian cutoff.

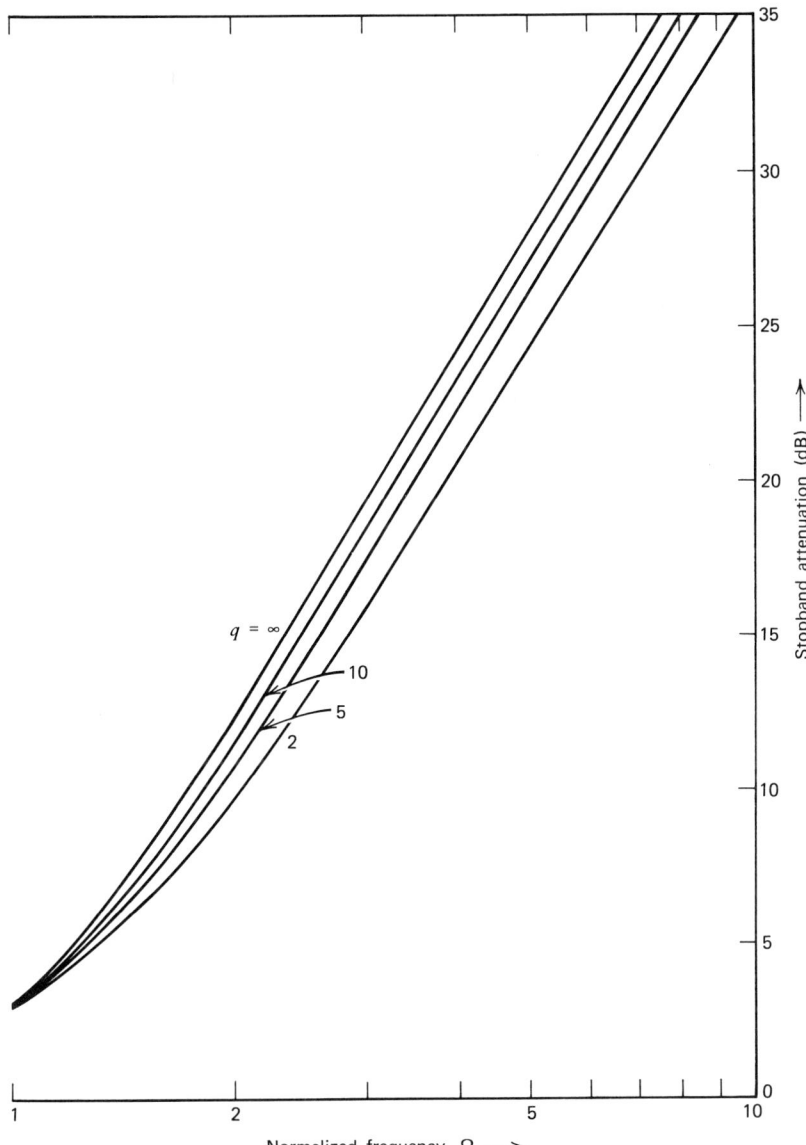

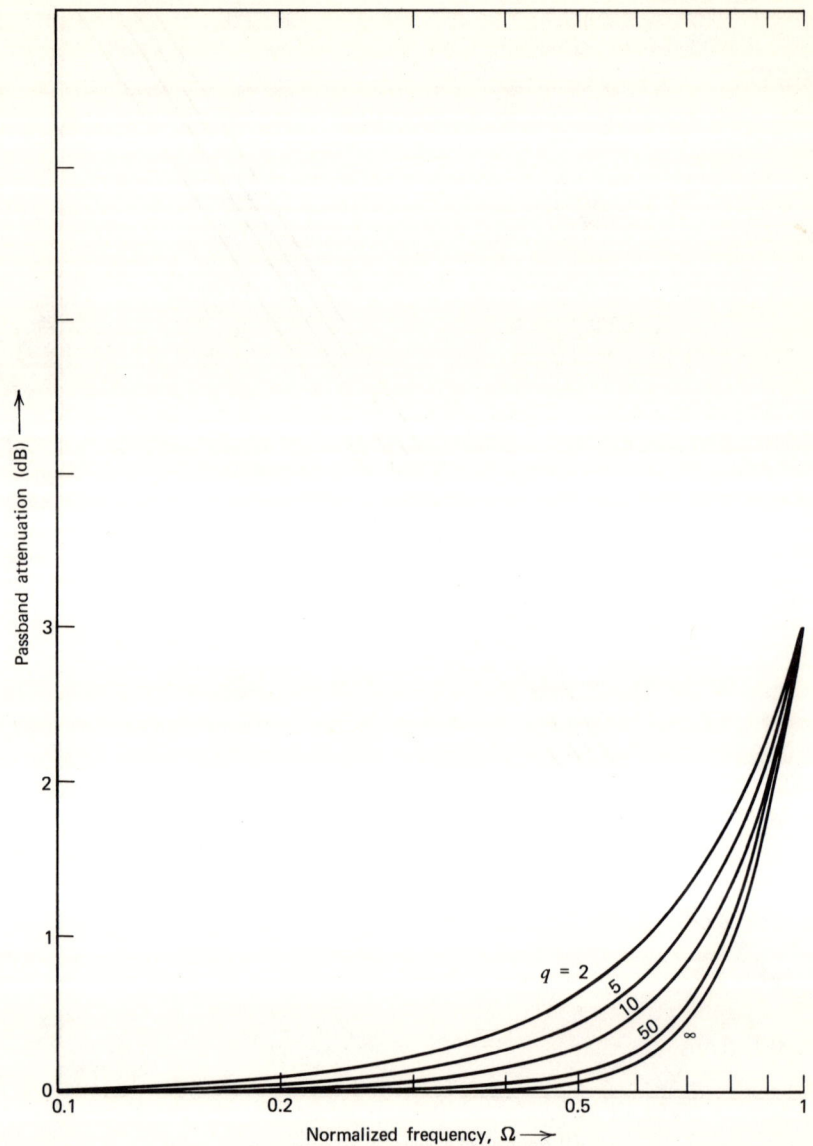

Figure 6-3. Attenuation of the $n = 3$ Butterworth filter with finite q, normalized to unity radian cutoff.

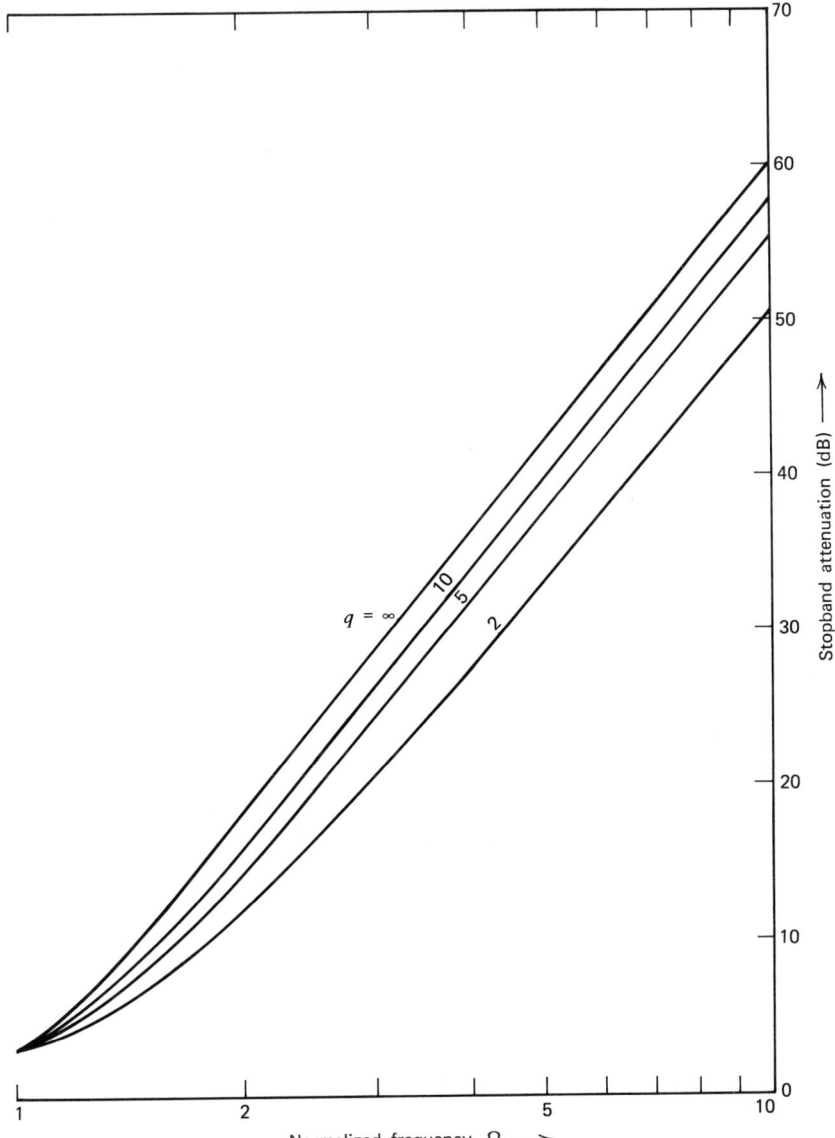

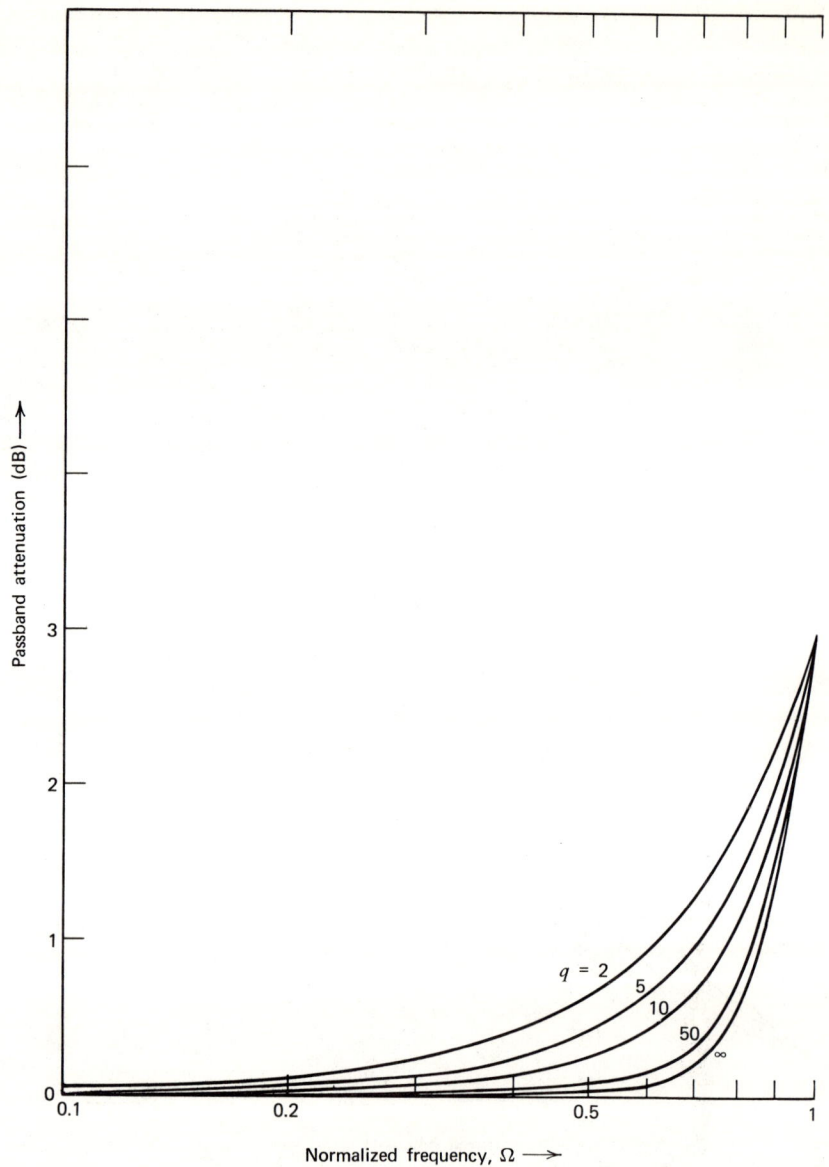

Figure 6-4. Attenuation of the $n = 4$ Butterworth filter with finite q, normalized to unity radian cutoff.

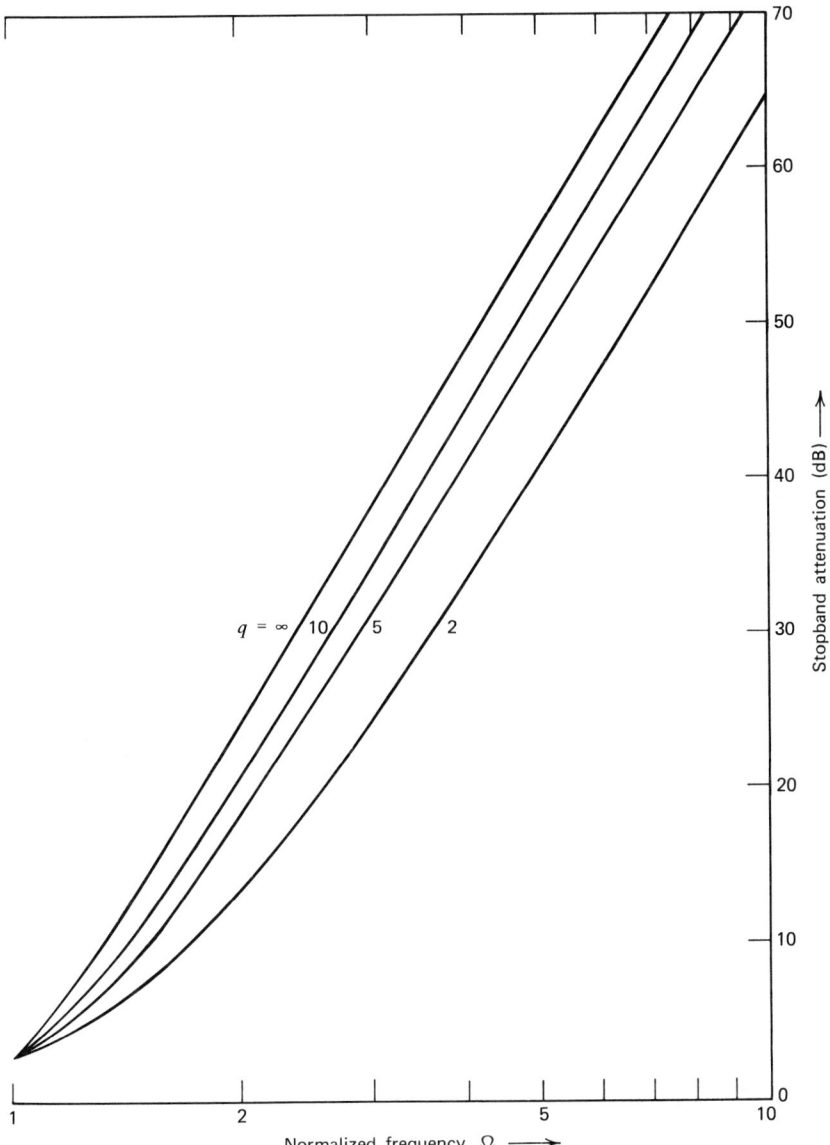

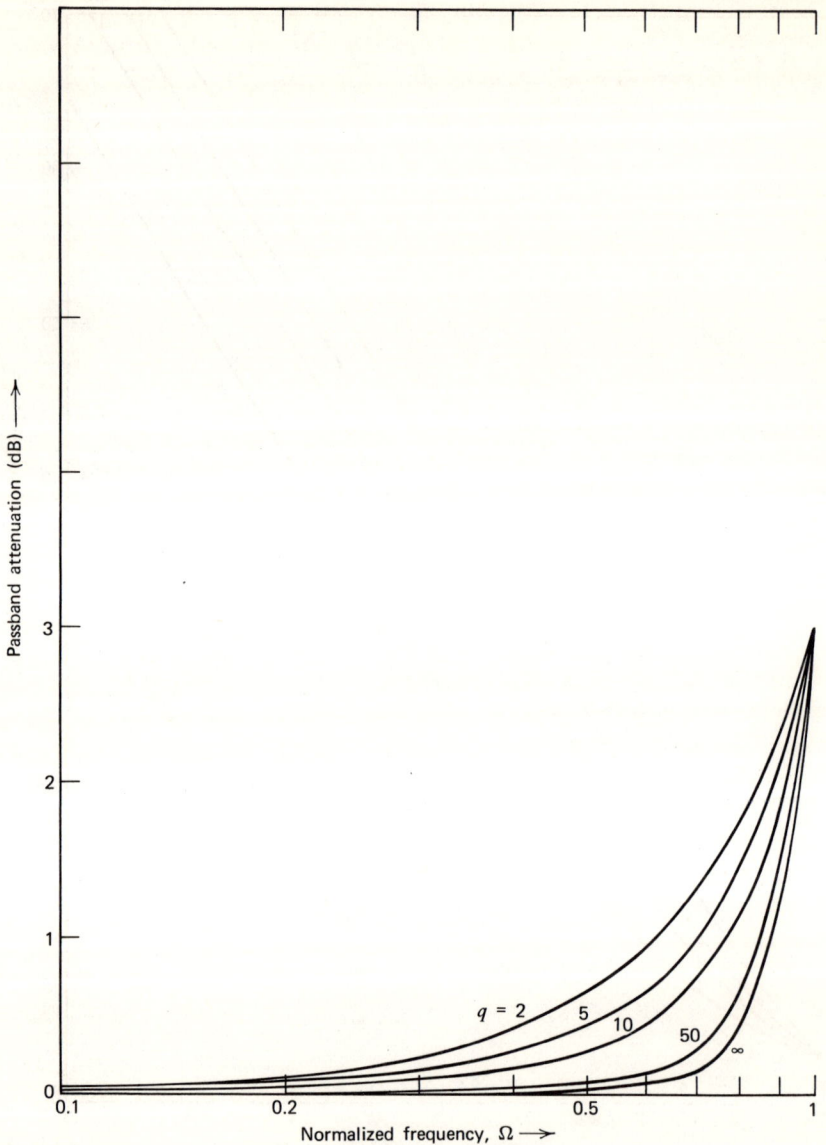

Figure 6-5. Attenuation of the $n = 5$ Butterworth filter with finite q, normalized to unity radian cutoff.

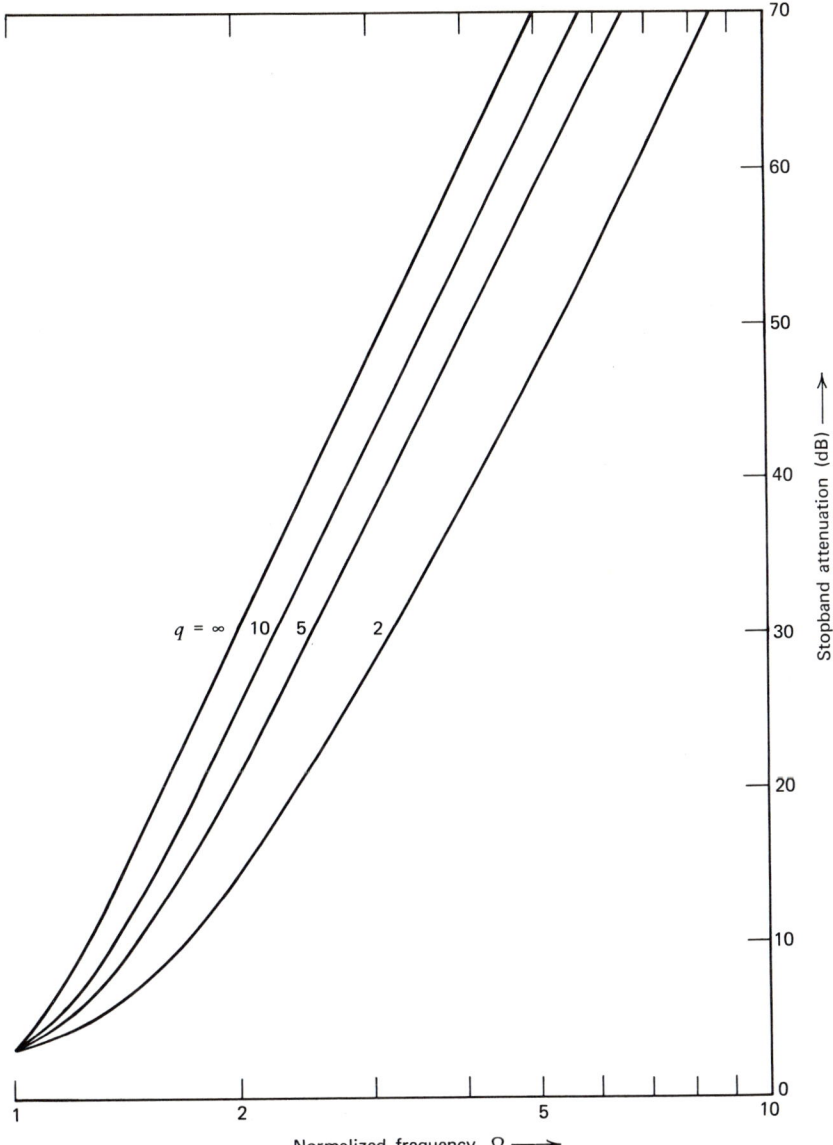

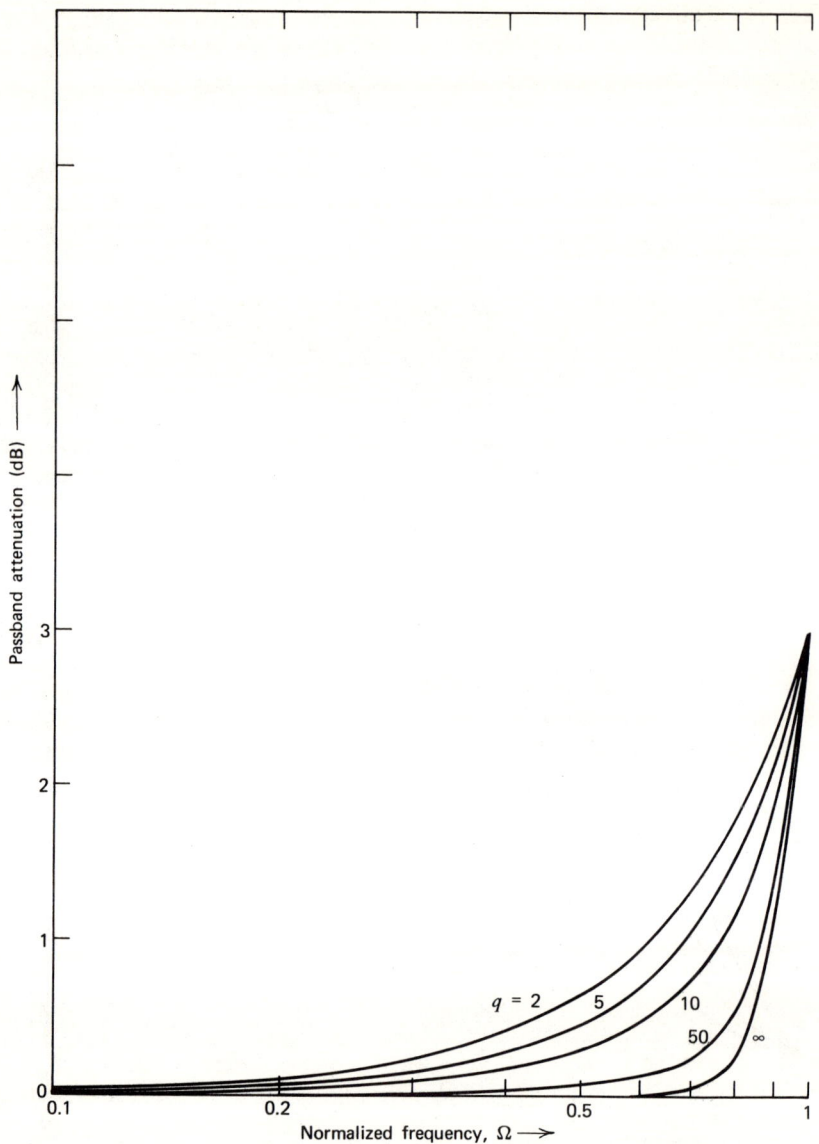

Figure 6-6. Attenuation of the $n = 6$ Butterworth filter with finite q, normalized to unity radian cutoff.

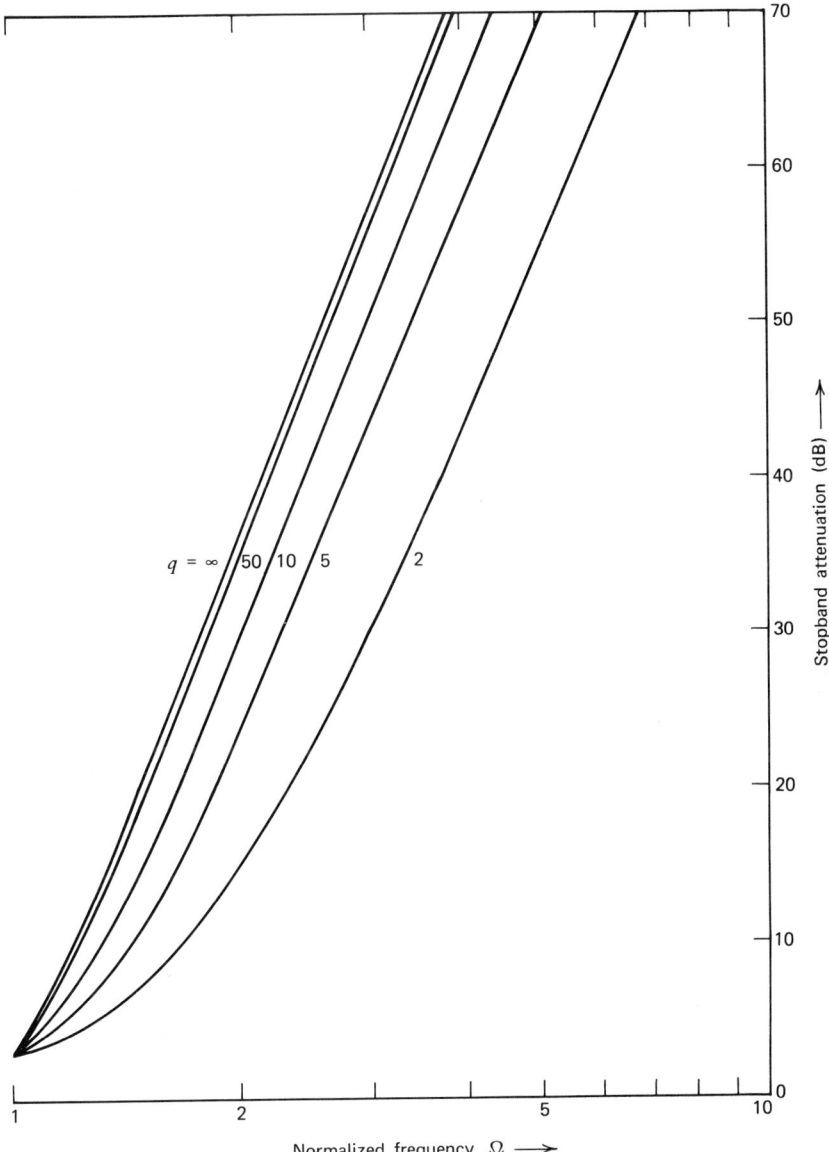

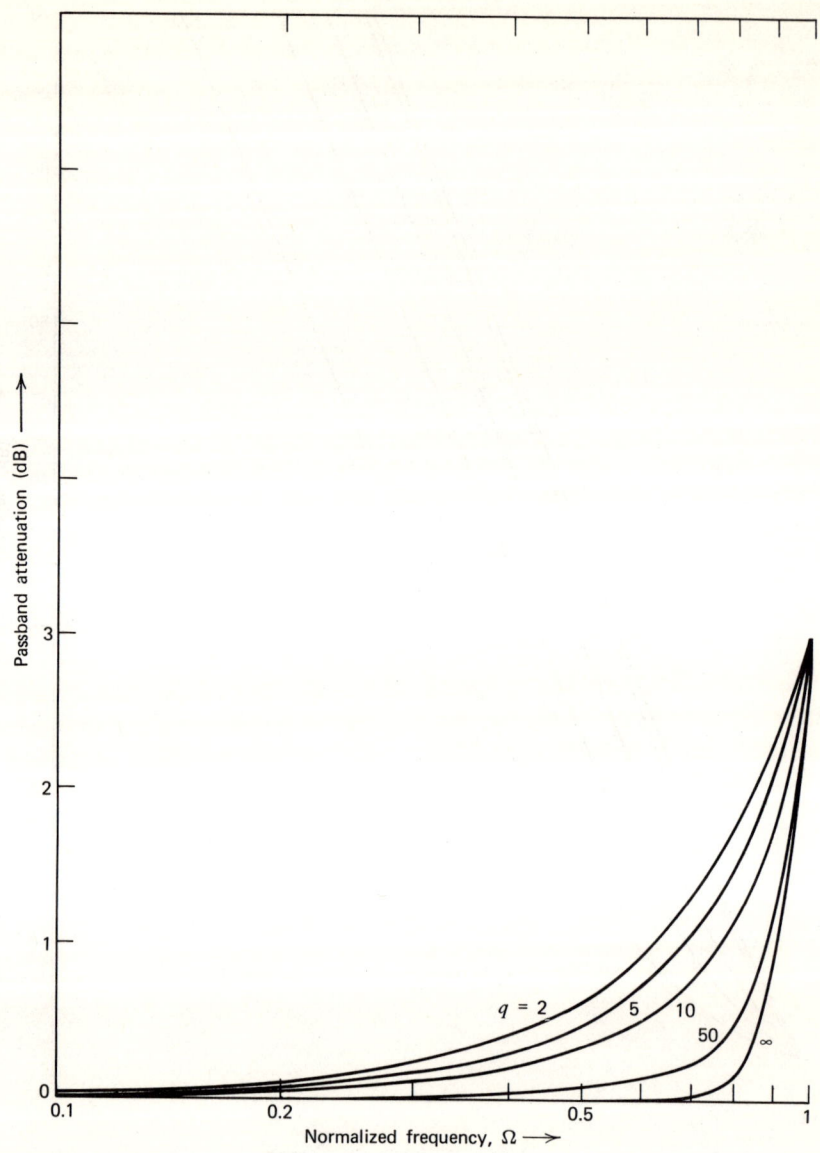

Figure 6-7. Attenuation of the $n = 7$ Butterworth filter with finite q, normalized to unity radian cutoff.

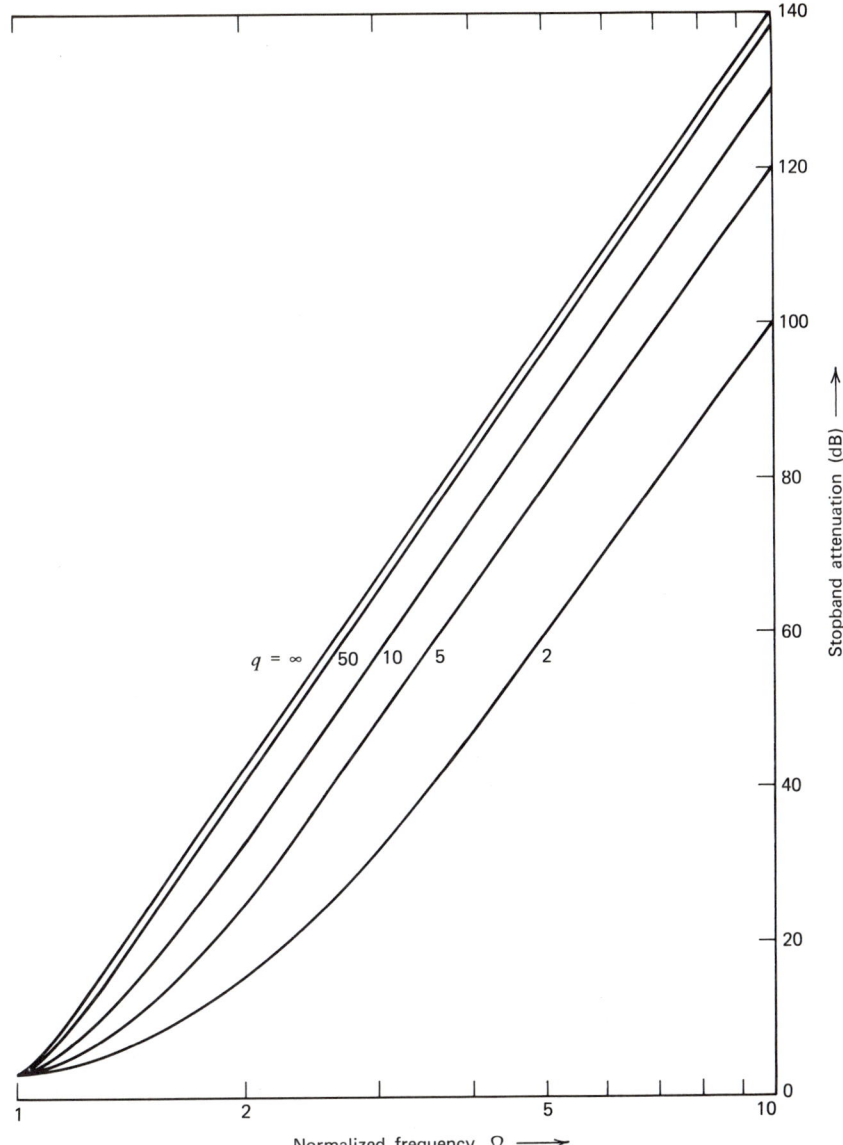

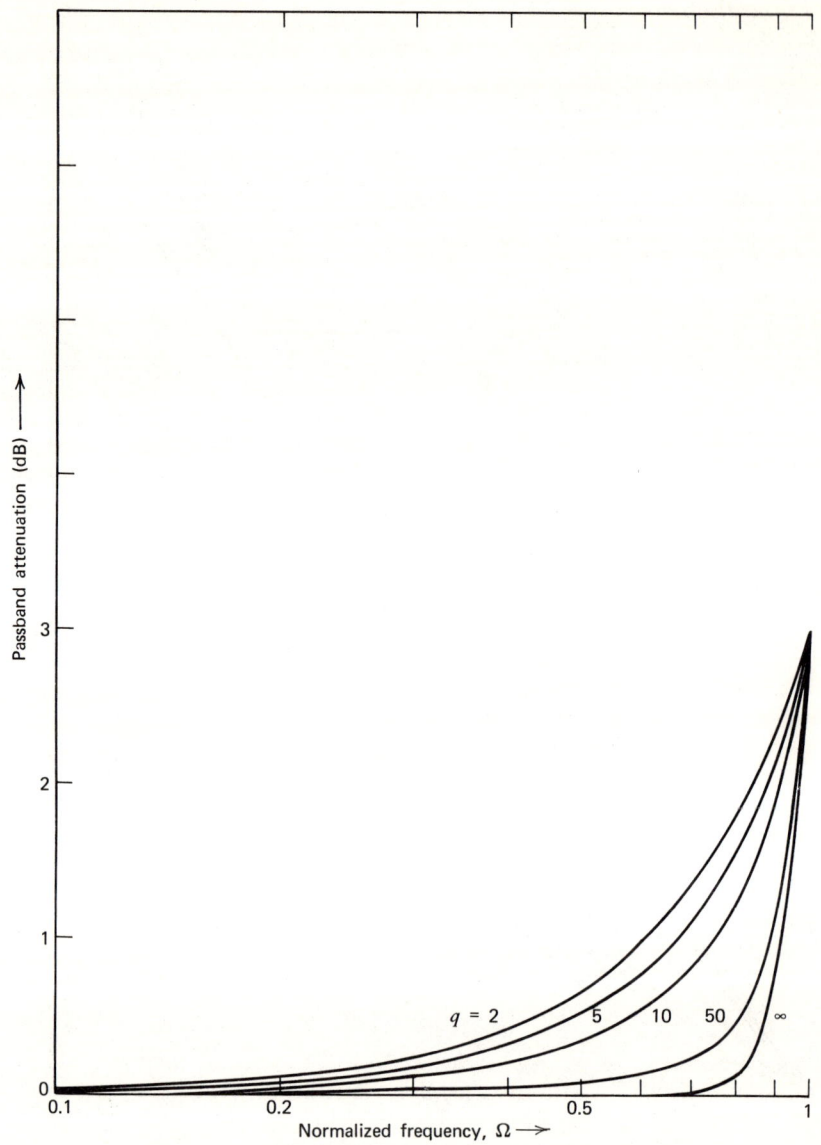

Figure 6-8. Attenuation of the $n = 8$ Butterworth filter with finite q, normalized to unity radian cutoff.

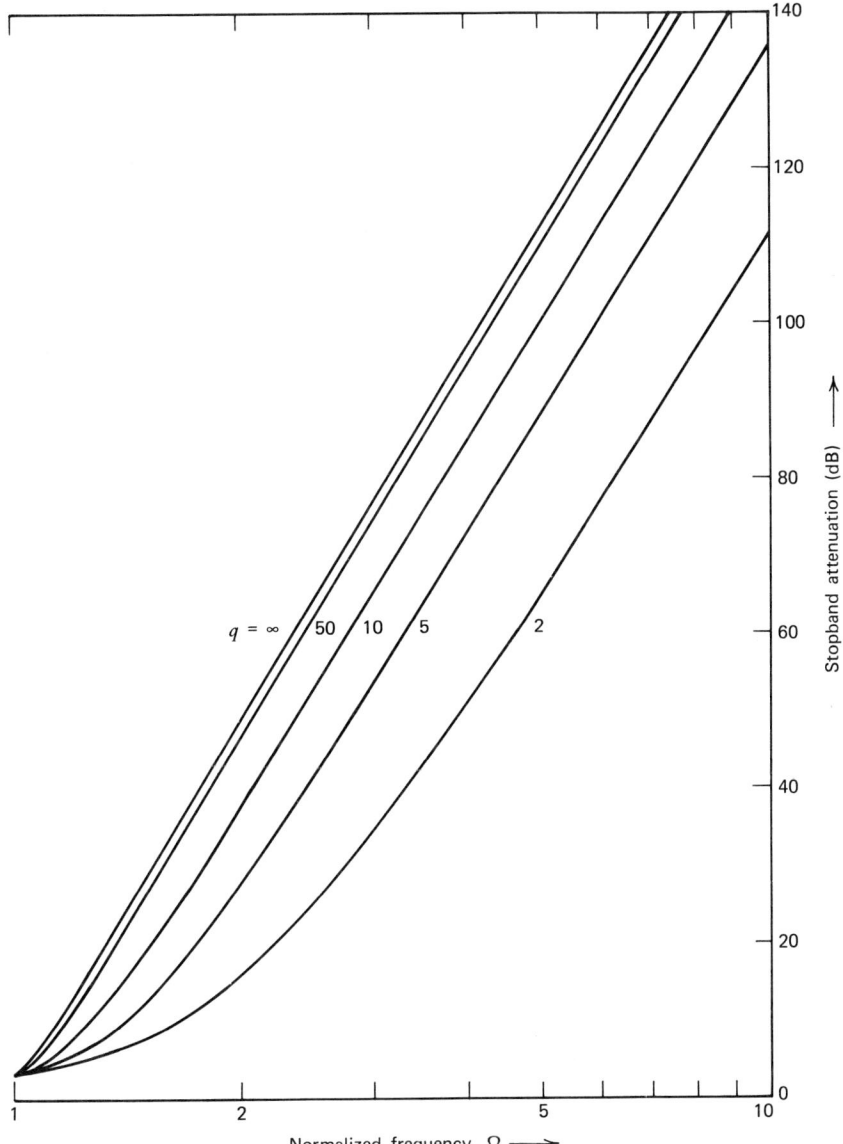

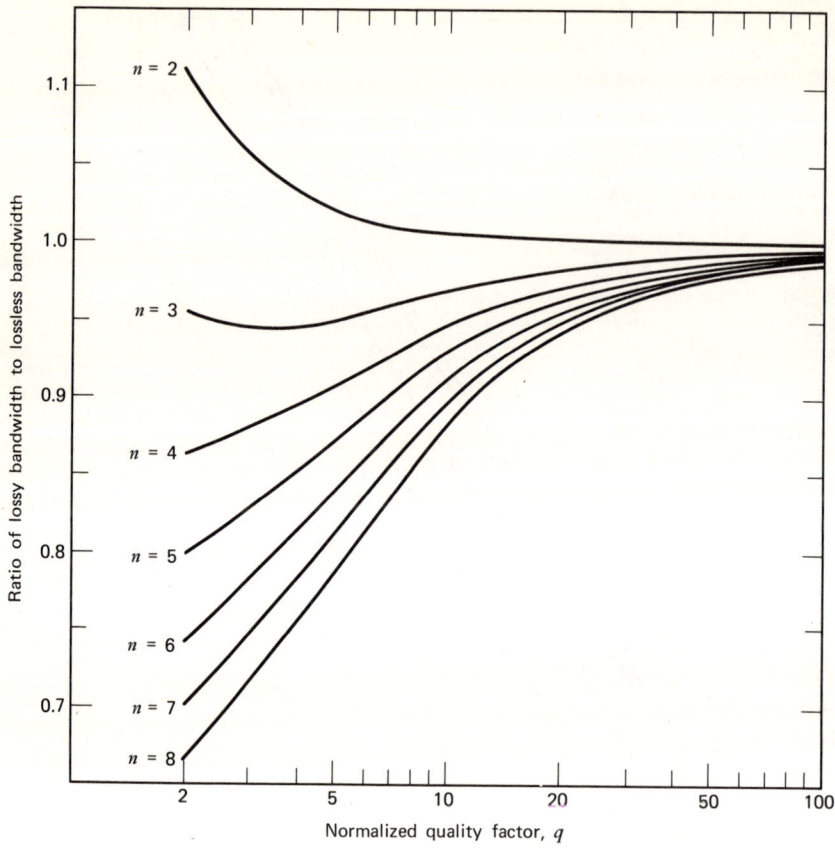

Figure 6-9. Bandwidth change of Butterworth filters due to finite q.

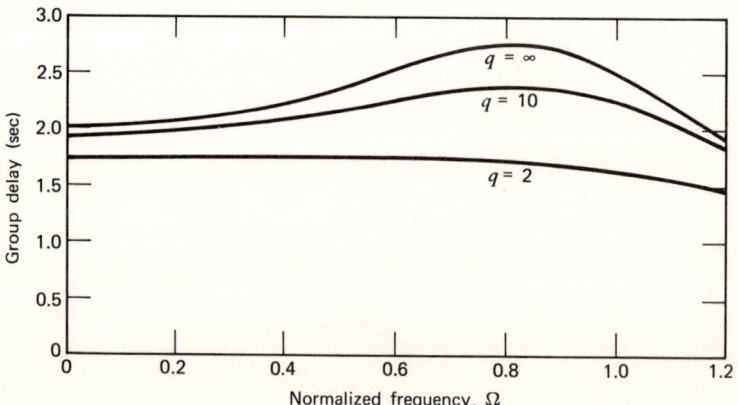

Figure 6-10. The $n = 3$ Butterworth filter group delay as a function of q.

Lossy-Filter Responses

whose poles are the lossless poles shifted to the left by $1/q$,

$$\text{lossy poles} = \begin{cases} -\left(1+\dfrac{1}{q}\right) \\ -\left(\dfrac{1}{2}+\dfrac{1}{q}\right) \pm j\dfrac{\sqrt{3}}{2} \end{cases} \quad (6.2\text{-}15)$$

For specified q values, the attenuation, phase, and other pertinent responses are calculated in the usual manner. Figures 6-2 ($n = 2$) to 6-8 ($n = 8$) give the attenuation for selected values of q, normalized to unity radian cutoff. The maximally flat property has disappeared for the finite-q condition, but for $q > 10$ the resulting response change from the theoretical response over the lower half of the passband is small, hence the responses are acceptable for many practical situations. The asymptotic behavior with losses is still $6n$ dB/octave.

From the bandwidth change due to q in Fig. 6-9, we note that for $n = 2$, the bandwidth increases with loss, contrary to the accepted conclusion that losses cause a bandwidth decrease. Figure 6-9 is valuable for design because the bandwidth of the lossless filter can be designed to anticipate the bandwidth reduction. For example, with $n = 3$ and $q = 10$, the bandwidth reduction with losses is 3%. Thus if the desired bandwidth is 10 kHz, the lossless filter should be designed for a bandwidth of $10^4/0.97 = 10.31$ kHz, and the presence of losses will then shrink the bandwidth to the desired 10 kHz. The attenuation response is obtained from the $q = 10$ curve in Fig. 6-3, where $\Omega = 1$ corresponds to 10 kHz.

The passband group delay in Fig. 6-10 is for $n = 3$ ($q = 2$, 10, and ∞), and we see the typical consequence of finite q. As q decreases, the passband delay variation decreases, and the delay approaches the Gaussian delay response. Thus finite q aids in reducing passband delay variation, but the price is rounding of the attenuation response near cutoff.

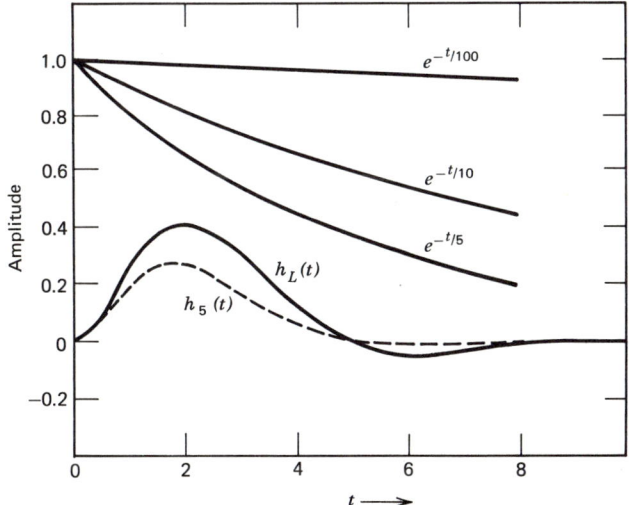

Figure 6-11. Effect of finite q on the $n = 3$ Butterworth filter impulse response.

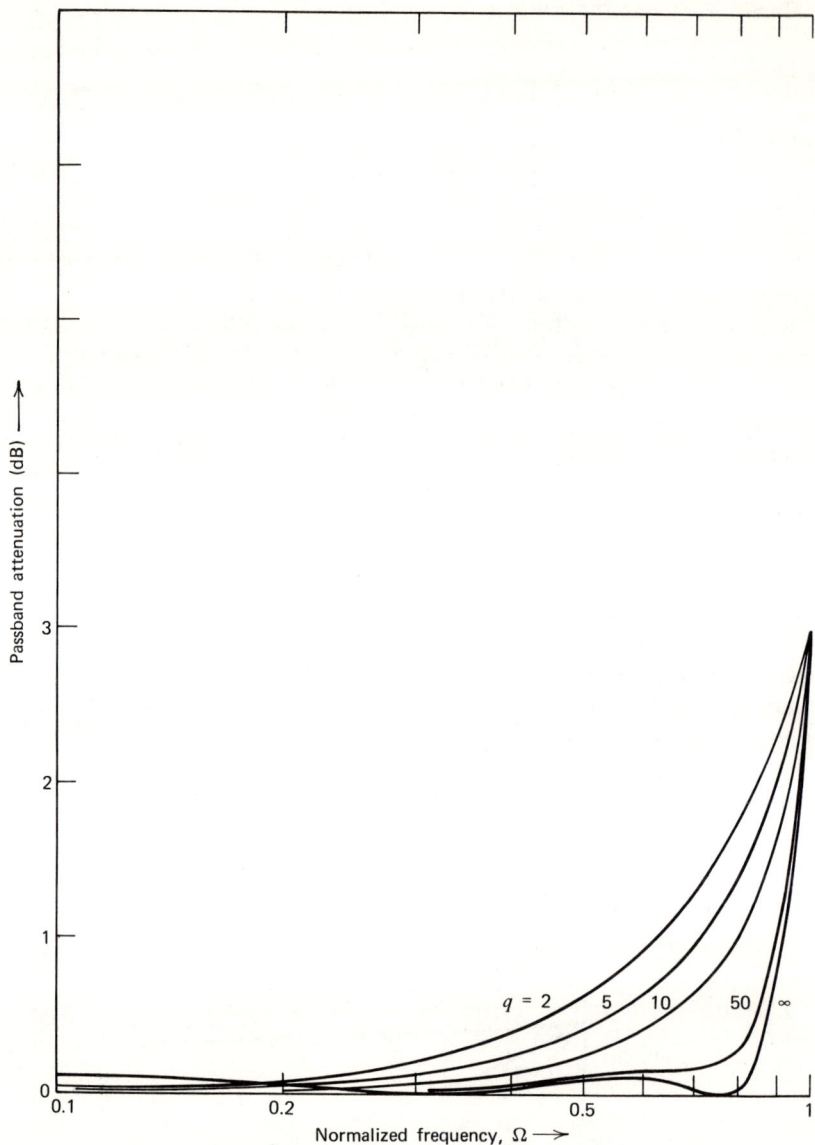

Figure 6-12. Attenuation of the fourth order, 0.1-dB ripple Chebyshev filter with finite q, normalized to unit radian cutoff.

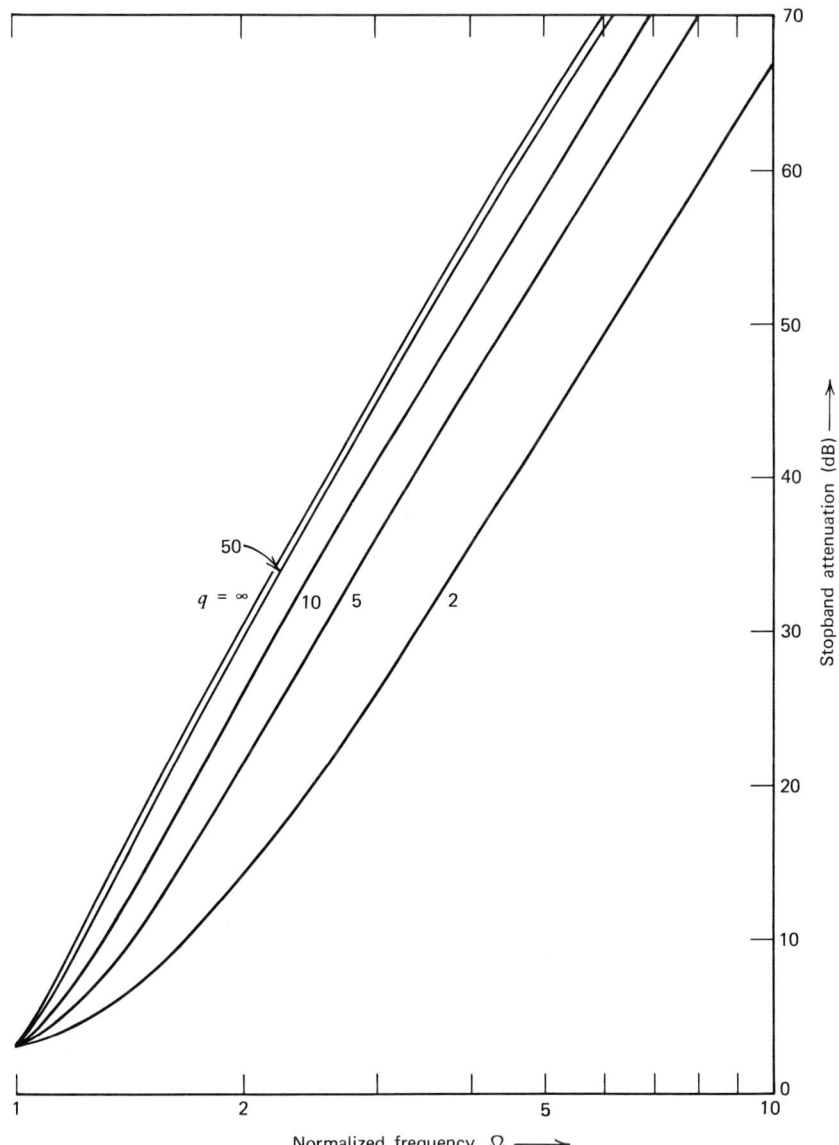

The lossy impulse response is obtained from (6.2-8) as

$$h_q(t) = e^{-t/q}\left[e^{-t} - \frac{2}{\sqrt{3}} e^{-t/2} \cos\left(\frac{\sqrt{3}}{2} t + \frac{\pi}{6}\right)\right] u_{-1}(t) \qquad (6.2\text{-}16)$$

where $h_L(t)$ is given by (3.4-14). Figure 6-11 gives the lossless impulse response and the exponential $e^{-t/q}$ for $q = 5, 10$, and 100. For $q = 100$, $e^{-t/100}$ is essentially unity over the impulse response duration, whereas $e^{-t/5}$ and $e^{-t/10}$ tend to improve the response symmetry. This fact is verified by the dashed curve, which is the lossy impulse response with $q = 5$.

Equation 6.2-12 suggests a minimum value of q for a negligible response change. From (3.4-4), the Butterworth pole closest to the $j\omega$ axis occurs with $i = 1$; hence this value of q is

$$q > \frac{10}{\sin(\pi/2n)} = 10 D_L(0) \qquad (6.2\text{-}17)$$

and is given in Table 6-1 for $n = 2$ to $n = 8$. A glance at the attenuation curves shows that this criterion is indeed a reasonable choice. In fact for some orders a lower value of q still yields an acceptable response, but (6.2-12) offers a good rule of thumb.

Table 6-1 Lower Limit for q for Negligible Response Change

	\multicolumn{7}{c}{n}						
	2	3	4	5	6	7	8
Lower limit for q	14.1	20	26.1	32.4	38.6	44.9	51.3

6.2.4 Chebyshev and Gaussian Filter Analysis

Lossy responses of other filter types are also obtained as shown in Section 6.2.3. Figure 6-12 presents the attenuation of the fourth-order 0.1-dB ripple, Chebyshev filter with $q = 2, 5, 10, 50$, and ∞ to illustrate the degradation of the attenuation response with finite q. As q decreases, the passband ripples are smeared and minimum attenuation then occurs at $\Omega = 0$. For this family, the bandwidth change with element q is shown in Fig. 6-13; again, for $n = 2$, the bandwidth increases. The general conclusions of the lossy Butterworth responses likewise apply to the lossy Chebyshev responses.

For the Gaussian filter, the effect of $q > 2$ on the passband attenuation shape is negligible, after renormalizing the 3-dB cutoff frequency to $\Omega = 1$; hence the lossless response accurately describes these lossy passband responses. The stopband behavior is likewise insensitive to q, and Fig. 6-14 typically shows this behavior for the fifth-order case with $q = 2$ and $q = \infty$. It is not surprising that the Gaussian filter attenuation shapes are not significantly degraded by q because, as previously stated, large amounts of loss cause all attenuation shapes to become

Lossy-Filter Responses

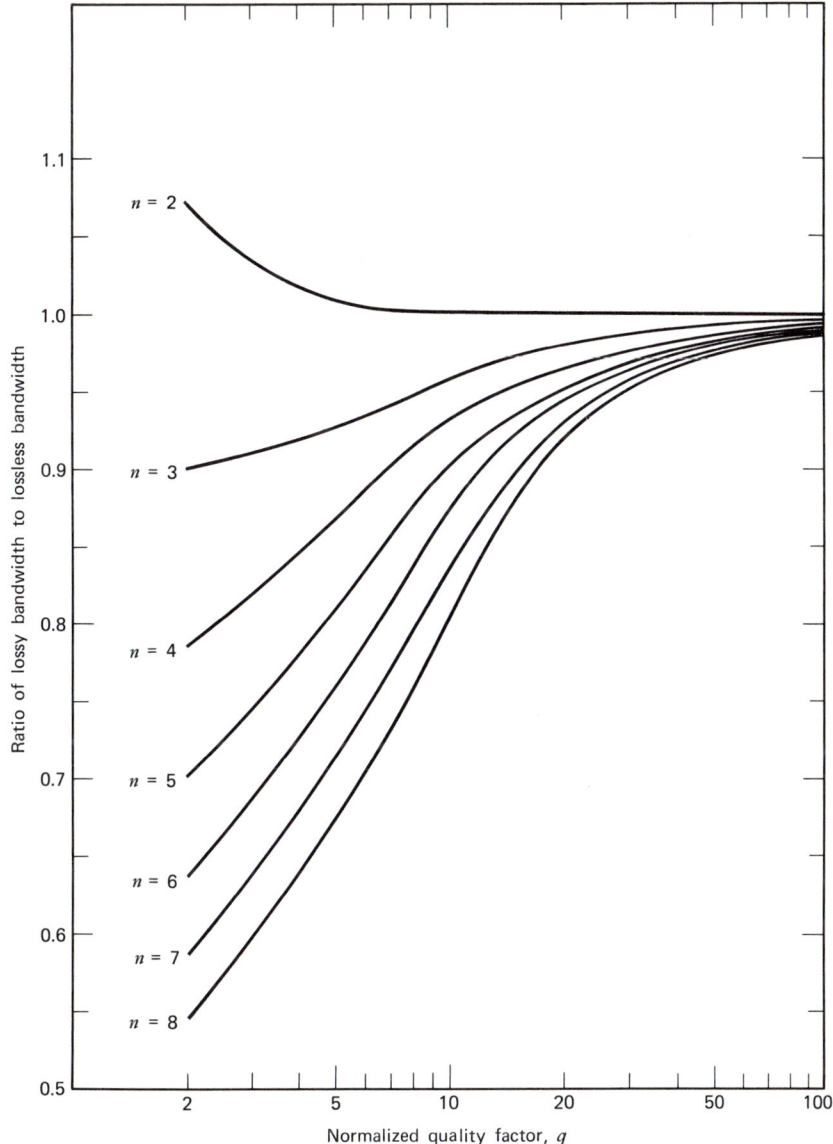

Figure 6-13. Bandwidth change of the 0.1-dB ripple Chebyshev filters due to finite q.

Gaussian, regardless of the initial approximation. The Gaussian filter attenuation; however, is already an approximation to the Gaussian function, hence little change occurs with finite q. The significant change with finite q is the bandwidth, and this function is shown in Fig. 6-15. Contrary to the Butterworth and Chebyshev bandwidth reduction, the Gaussian filter bandwidth increases with finite q.

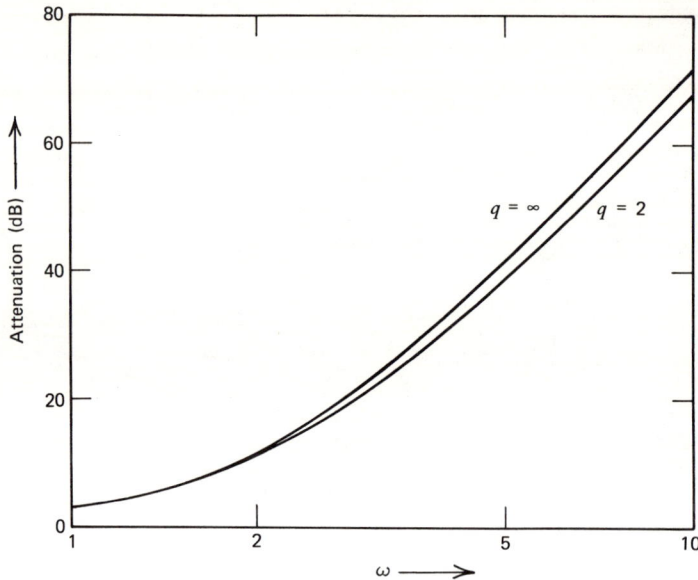

Figure 6-14. Stopband attenuation of the $n = 5$ Gaussian filter with $q = 2$ and $q = \infty$.

6.2.5 Q Requirement for Bandpass Filters

The LP q information is now extended to describe the BP situation. Use the LP to BP transformation to transform the LP inductor L in series with the resistor r as

$$SL + r \rightarrow \frac{L}{\omega_2 - \omega_1} s + \frac{1}{(\omega_2 - \omega_1)s/(L\omega_0^2)} + r \qquad (6.2\text{-}18)$$

where $\omega_2 - \omega_1$ is the 3-dB radian bandwidth and ω_0 is the BP center frequency. Then the Q of the BP inductor at ω_0 is

$$Q = \omega_0 \frac{[L/(\omega_2 - \omega_1)]}{r} = \frac{\omega_0}{\omega_2 - \omega_1} \cdot \frac{L}{r} \qquad (6.2\text{-}19)$$

but from (6.1-2), with $\omega_a = 1$, the BP quality factor reduces to

$$Q = \frac{\omega_0}{\omega_2 - \omega_1} q = \frac{q}{\gamma} \quad \left(\gamma = \frac{\omega_2 - \omega_1}{\omega_0}\right) \qquad (6.2\text{-}20)$$

The capacitor quality factor is also q/γ, and this value is customarily referred to as the resonator quality factor. Thus a BP quality factor $\omega_0/(\omega_2 - \omega_1)$ times greater than the LP quality factor is necessary to maintain the same response degradation. This means that all lossy LP responses with quality factor q apply to the corresponding BP responses with q replaced by γQ.

6.2.6 Illustrative Example

We now illustrate use of the LP curves by considering a BP filter centered at 100 kHz with 3-dB bandwidth of 5 kHz and a minimum 40-dB bandwidth of 20 KHz. The resonator Q is 100, and the response is to be reasonably flat in the

Lossy-Filter Responses

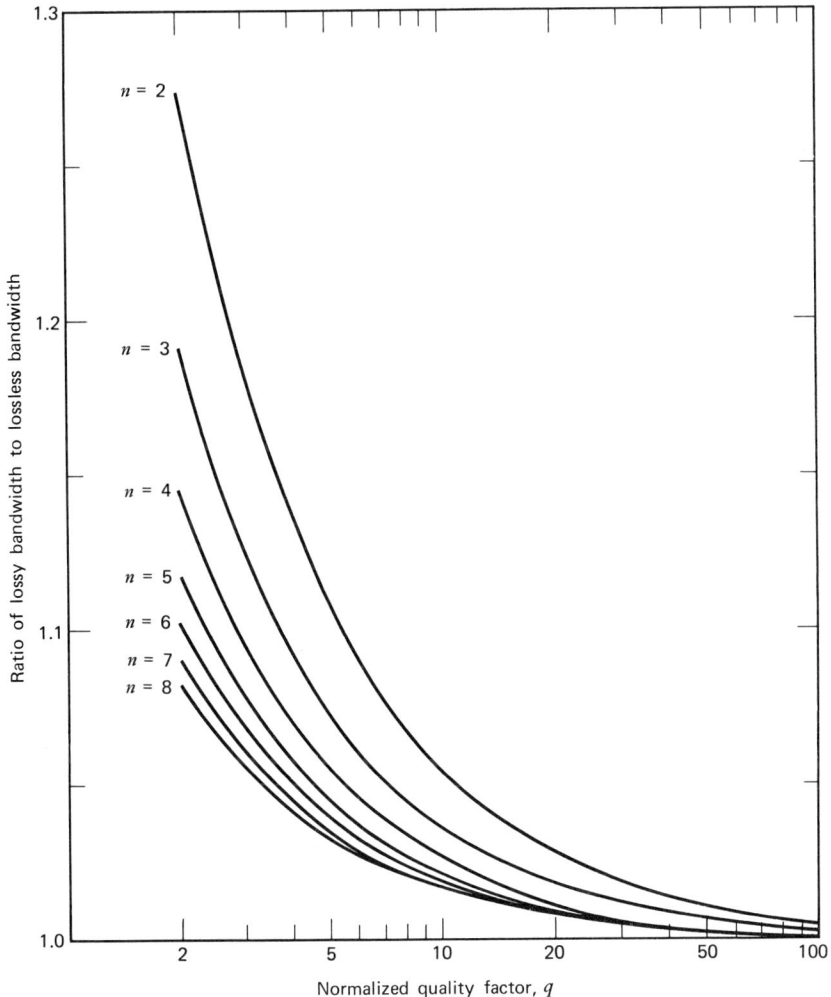

Figure 6-15. Bandwidth change of the Gaussian filters due to finite q.

vicinity of band center. First we compute the equivalent LP q from (6.2-20) as

$$q = \gamma Q = \frac{5}{100} \times 100 = 5 \qquad (6.2\text{-}21)$$

The Butterworth response is selected, for we know from Section 6.2.3 that the low-frequency response is reasonably flat with $q = 5$. Since this filter is considered narrow band, the frequency scale of Figs. 6-2 to 6-8 can be replaced by $\Omega = \Delta\omega/(\omega_2 - \omega_1)$ as given by (4.5-5). The 40-dB reject bandwidth corresponds to $\Omega = 4$, and a search of these attenuation curves with $q = 5$ reveals that $n = 4$ is the minimum order that meets this requirement. Figure 6-9 shows that the lossy bandwidth is 0.908 times the lossless bandwidth; therefore to achieve the desired response, we increase the lossless bandwidth to 5 kHz/0.908 = 5.507 kHz. Losses then shrink the bandwidth to the desired 5 kHz and the resulting attenuation response is obtained from the $q = 5$ curve in Fig. 6-4. Actually the value of q has now increased to

$$q = \frac{5.507}{100} \times 100 = 5.507 \tag{6.2-22}$$

which causes a bandwidth reduction of 0.912 (see Fig. 6-9) rather than 0.908, reflecting a lossy bandwidth of 5.022 kHz. This procedure can be repeated until the correct lossless bandwidth is achieved. Usually, however, one iteration gives results that are accurate enough for practical situations.

6.3 INSERTION LOSS

The insertion loss of the filter in Fig. 6-16, for constant input voltage, is defined as

$$l = 10 \log \frac{P_{max}}{P_{out}} \text{ decibels} \tag{6.3-1}$$

where P_{max} is the maximum power delivered to any load and P_{out} is the power delivered to R_2 when the filter is inserted as shown. This definition is valid at any frequency and therefore includes the attenuation due to the filter's selectivity. Of interest here is the loss at the frequency of minimum attenuation, and henceforth

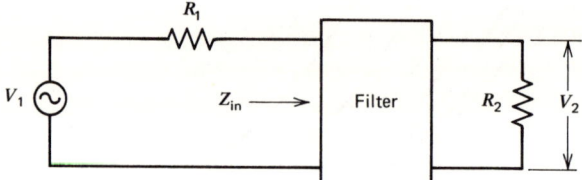

Figure 6-16. Filter under test conditions.

the term "insertion loss" refers to this value, designated l_0. Usually l_0 is measured at $\omega = 0$ for LP filters and at ω_0 for BP filters, since the attenuation is usually minimum there. The most notable exception is the class of even-ordered-LP Chebyshev filters, but even these have minimum attenuation at $\omega = 0$ if enough losses are present.

Insertion loss can also be expressed in terms of the easily measured input and output voltages. From Fig. 6-16, the maximum power delivered to any load occurs when the filter is absent and $R_2 = R_1$, that is,

$$P_{max} = \frac{|V_1|^2}{4R_1} \tag{6.3-2}$$

With the filter inserted, the power delivered to R_2 is

$$P_{out} = \frac{|V_2|^2}{R_2} \tag{6.3-3}$$

and the insertion loss, from (6.3-1), is

$$l_0 = 10 \log \frac{|V_1|^2/4R_1}{|V_2|^2/R_2} = 20 \log \frac{1}{2} \sqrt{\frac{R_2}{R_1}} \left| \frac{V_1}{V_2} \right| \text{ decibels} \tag{6.3-4}$$

Insertion Loss

Although V_1 and V_2 are easily measured, (6.3-4) is not very useful for predicting the insertion loss. Furthermore, (6.3-4) does not reveal that l_0 is expressible as

$$l_0 = l_m + l_d \qquad (6.3\text{-}5)$$

where l_m is the insertion loss due to impedance mismatch between R_1 and Z_{in}, and l_d is the insertion loss due to dissipation in the network elements. These quantities are now discussed, and methods are given for computing l_m and l_d. This allows us to predict the total insertion loss by use of (6.3-5).

6.3.1 Mismatch Loss

The loss due to impedance mismatch is obtained by considering the network in Fig. 6-16 to be lossless; hence it dissipates no power. At the frequency of minimum attenuation, the filter's input impedance is assumed resistive, and we designate it as $Z_{in} = R_i$. Then, since the network is lossless, all power delivered to R_i is delivered to the load R_2. What is the insertion loss as R_i is allowed to vary? To answer this question, first consider the equivalent circuit in Fig. 6-17. The maximum power delivered to any load is still given by (6.3-2), while the power

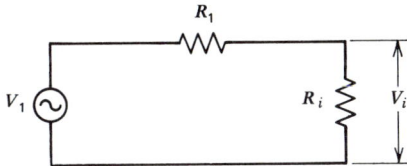

Figure 6-17. Equivalent circuit at frequency of minimum attenuation.

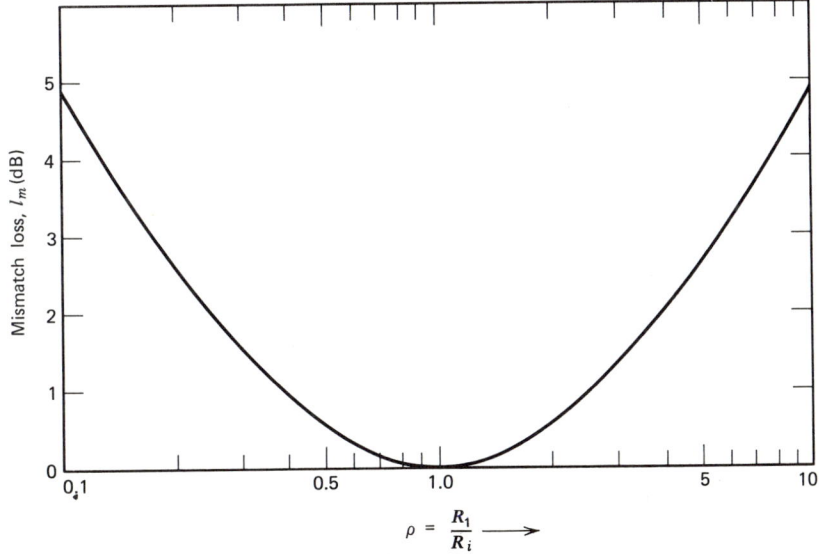

Figure 6-18. Insertion loss due to impedance mismatch.

delivered to R_i is

$$P_{out} = \frac{|V_i|^2}{R_i} = \left(\frac{R_i}{R_1 + R_i}\right)^2 \frac{|V_1|^2}{R_i} \qquad (6.3\text{-}6)$$

Substitute these results into (6.3-1) to obtain

$$l_m = 10 \log \frac{(R_1 + R_i)^2}{4R_1 R_i} = 20 \log \frac{1+\rho}{2\sqrt{\rho}} \text{ decibels} \qquad (6.3\text{-}7)$$

shown in Fig. 6-18, where $\rho = R_1/R_i$ or $\rho = R_i/R_1$. When $R_i = R_1$, $\rho = 1$ and $l_m = 0$ dB, the matched condition. For most practical situations except, of course, the even-ordered Chebyshev filter, R_i is matched to the source resistance, but occasionally it is purposely mismatched.

Suppose we select $R_s = 2\,\Omega$ for the second-order Butterworth filter in Fig. 3-28. Then $\rho = 2$ at $\omega = 0$, and from Fig. 6-18 the mismatch insertion loss is 0.5 dB, even though the filter is lossless.

6.3.2 Resistive Loss

For the matched condition, the insertion loss l_0 is due solely to element resistive losses and is symbolized by l_d. This matched condition, however, can be achieved even though R_1 and R_2 in Fig. 6-16 are not equal, for impedance transformations within BP filters are often used to match the filter's input impedance to R_1. Conversely, equal source and load resistors do not necessarily imply a matched condition.

From a design and application viewpoint it is desirable to predict the insertion loss for a given response and specified q and also to predict the required q for a given insertion loss, for the matched condition. This information is contained in (6.2-5). We compute $H_L(s + 1/q)$ at $s = 0$, since the attenuation is minimum there, to obtain

$$l_d = -20 \log H_L\left(\frac{1}{q}\right) \text{ decibels} \qquad (6.3\text{-}8)$$

Thus $H_L(1/q)$ is the normalized LP transfer function evaluated at $s = 1/q$.

6.3.3 Insertion Loss Curves

The calculation of l_d is now illustrated for the third-order Butterworth response. From (3.4-3), the transfer function is

$$H_L(s) = \frac{1}{s^3 + 2s^2 + 2s + 1} \qquad (6.3\text{-}9)$$

and the insertion loss, from (6.3-8), is

$$l_d = -20 \log \frac{1}{1 + 2/q + 2/q^2 + 1/q^3} \text{ decibels} \qquad (6.3\text{-}10)$$

Evaluation of (6.3-10) for various q values yields the third-order insertion loss curve in Fig. 6-19, which also includes the insertion loss for $n = 2$ to $n = 8$, obtained in a similar manner.

Figures 6-20 to 6-22 likewise show the insertion loss for other filter types.

Insertion Loss

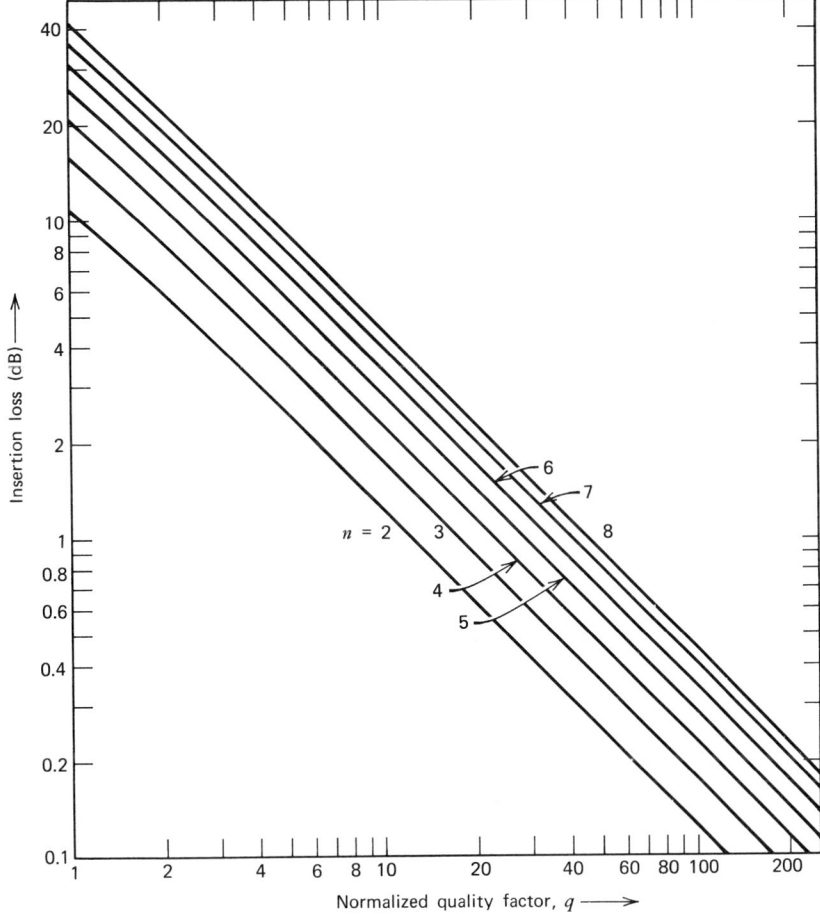

Figure 6-19. Insertion loss of Butterworth filters as a function of q.

Computation of the even-ordered Chebyshev filter insertion loss is modified because the frequencies of minimum attenuation for the lossless and lossy filters are not the same, as Fig. 6-12 indicates. For the lossless filter, minimum attenuation does not occur at $\omega = 0$, but for the lossy filter two cases arise. For the first case the element q is low, the passband ripples disappear, and the minimum attenuation occurs at $\omega = 0$. We evaluate (6.3-8) to obtain the insertion loss, since maximum output of the lossless filter is unity. For the second case, the element q is higher, the ripples do not quite disappear, and minimum attenuation again does not occur at $\omega = 0$. Therefore we compute the insertion loss from (6.3-8) and subtract the lossy-response ripple attenuation at $\omega = 0$, which we obtain from a detailed analysis. Computing insertion loss in this manner removes the effect of mismatch loss, and the resulting loss is due solely to dissipation. The insertion loss for the Gausian and Bessel filters is essentially the same and accurately approximates the loss of other constant-delay type filters.

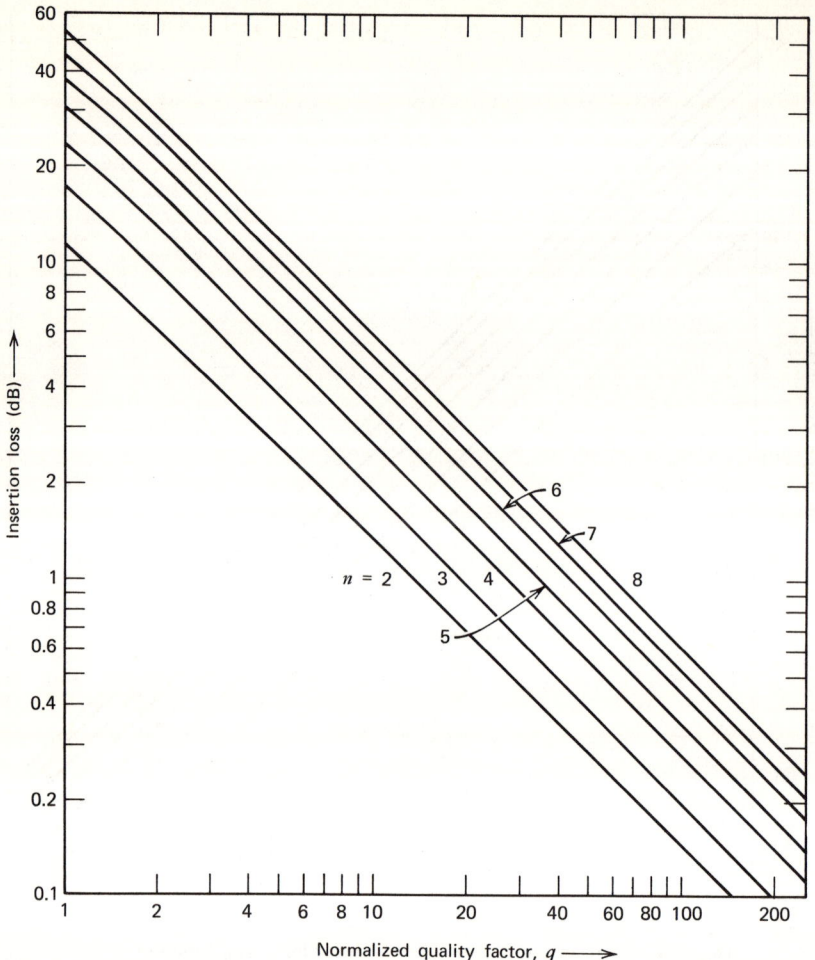

Figure 6-20. Insertion loss of 0.1-dB ripple Chebyshev filters as a function of q.

6.3.4 Approximate Expression for Insertion Loss

For $q > 4$ the insertion loss curves are extremely linear. This explicit dependence on q is revealed by considering the normalized all-pole LP transfer function

$$H_L(s) = \frac{1}{a_0 s^n + a_1 s^{n-1} + \cdots + a_{n-1} s + 1} \quad (6.3\text{-}11)$$

with unity output at $\omega = 0$. Then the insertion loss, from (6.3-8) is

$$l_d = 20 \log \left[a_0 \left(\frac{1}{q}\right)^n + \cdots + a_{n-1} \left(\frac{1}{q}\right) + 1 \right] \text{ decibels} \quad (6.3\text{-}12)$$

We now assume that the contribution of higher-order terms is negligible; thus l_d is approximately

$$l_d \approx 20 \log \left(1 + \frac{a_{n-1}}{q}\right) = 20 \log e \, \ln \left(1 + \frac{a_{n-1}}{q}\right) \quad (6.3\text{-}13)$$

Insertion Loss

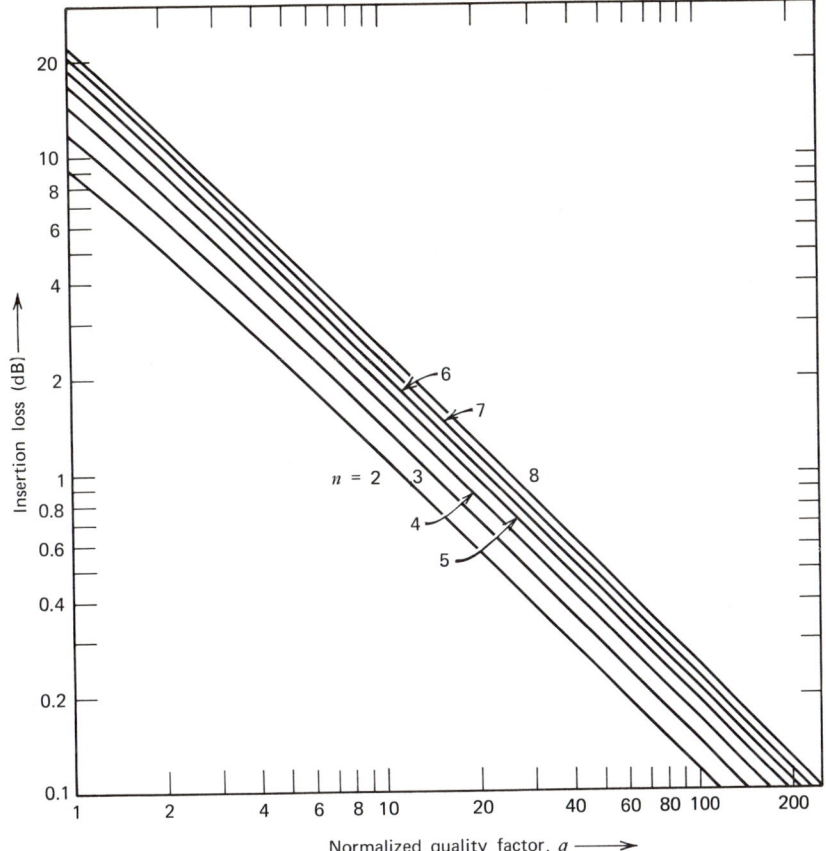

Figure 6-21. Insertion loss of Gaussian filters as a function of q.

Retain the first term of the expansion for $\ln[1+(a_{n-1}/q)]$ to give

$$l_d \approx 20 \left(\frac{a_{n-1}}{q}\right) \log e = 8.686 \frac{a_{n-1}}{q} \text{ decibels} \quad (6.3\text{-}14)$$

exposing the linear relationship between l_d and $1/q$. Use of (3.3-6) and (3.3-9) allows l_d to be expressed as

$$l_d \approx \frac{8.686 D_L(0)}{q} \text{ dB} = \frac{4.343}{q} \sum_{k=1}^{n} x_k \text{ dB} \quad (6.3\text{-}15)$$

We again see a practical use for $D_L(0)$, the normalized delay at $\omega = 0$. Values of x_k, the element value for the equally terminated normalized filter, are given in catalogs such as Ref. 6. The approximation in (6.3-15) is quite accurate for $q > 4$, especially for those responses approximating the rectangular magnitude response. For example, the insertion loss of the fourth-order Butterworth filter with $q = 5$ is 4.515 dB, whereas (6.3-15) yields 4.540 dB.

If the cutoff frequency is ω_a rather than unity, we replace s in (6.3-11) by ω_a/q

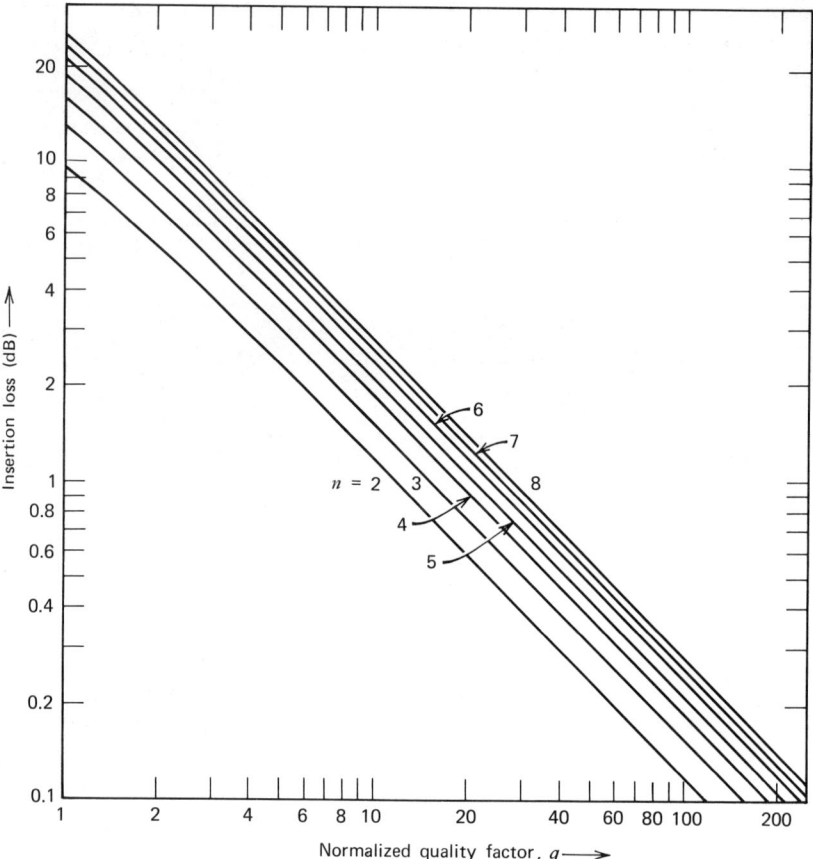

Figure 6-22. Insertion loss of Bessel filters as a function of q.

instead of by $1/q$ [see (6.2-1)]. Then (6.3-15) becomes

$$l_d \approx \frac{8.686\omega_a D_l(0)}{q} \text{ dB} = \frac{4.343\omega_a}{q} \sum_{k=1}^{n} x_k \text{ dB} \qquad (6.3\text{-}16)$$

where $D_l(0)$ and x_k now correspond to the filter with cutoff frequency ω_a. But $\omega_a D_l(0)$ is the normalized group delay at $\omega = 0$ and $\omega_a x_k$ is the normalized element value. Furthermore, since q is invariant under simple frequency scaling, (6.3-15) and (6.3-16) are identical. We conclude that for a given value of q evaluated at the cutoff frequency, the LP insertion loss is independent of the cutoff frequency.

For the BP case the insertion loss *is* dependent on the lossless filter bandwidth, for this bandwidth is used in computing the normalized quality factor in (6.2-20).

Example 6-1. If the bandwidth degradation resulting from finite q is temporarily disregarded, the filter in Section 6.2.6 has, from Fig. 6-19 ($q = 5$), an insertion loss of 4.5 dB. However the lossless bandwidth is increased to 5.507 kHz to ensure a lossy bandwidth of 5 kHz. The corresponding q of the lossless filter is then given by (6.2-22), and from Fig. 6-19 the expected insertion loss is now 4.1 dB.

Insertion Loss

Example 6-2. We now determine the necessary BP resonator Q to ensure that the final 3-dB bandwidth is 1.5 MHz and the insertion loss is 1.5 dB maximum. The center frequency is 30 MHz, and the response is derived from the fourth-order Butterworth LP filter. The solution requires both the fourth-order curves of Figs. 6-9 and 6-19. From Fig. 6-19, the value of q for 1.5 dB insertion loss is 15. From Fig. 6-9, the bandwidth reduction due to losses is 0.966 for $q = 15$, necessitating a lossless bandwidth of $1.5/0.966$ MHz $= 1.553$ MHz. Then, from (6.2-20), the required BP resonator Q is

$$Q = \frac{q}{\gamma} = 15 \times \frac{30}{1.553} = 289.8 \qquad (6.3\text{-}17)$$

6.3.5 Minimum Insertion Loss Filter

Now that we are able to compute the insertion loss for a specified filter response and value of q, we ask whether for a given stopband bandwidth, there exists a set of element values for which the insertion loss is minimized. The answer is yes, and closed-form solutions for n even have been obtained, while for n odd, the solution requires solving a polynomial [4]. These element values, however, are functions of the element q, and consequently the passband attenuation characteristics vary widely. For $q > 10$ this design accurately approximates Cohn's equal-element design [2], for which the insertion loss is nearly minimum.

The attenuation characteristics of these filters (Fig. 6-23) exhibit pronounced ripples, which are even larger for $n = 6, 7$, and 8. Passband ripple as large as 2 dB is generally undesirable, but these ripples are reduced, and often eliminated, as element losses are introduced. For $n = 2$ the equal-element filter is the Butterworth filter.

The equal-element filter leads to a BP filter with identical resonators and coupling reactances, and this has important implications for high-power microwave filtering [1]. The electric field in each microwave cavity is then the same at band center, and this in turn increases the filter's power handling capabilities.

Table 6-2 presents the transfer-function coefficients and element values x_n, and Table 6-3 lists the transfer-function poles for the normalized filter. If all element values are normalized to unity, the transfer-function polynomial coefficients are given by the binomial coefficients [5].

The attenuation characteristics of the lossy filter can be found in Ref. 5, and the insertion loss, computed from (6.3-8) using the data in Table 6-2, is shown in Fig. 6-24. Note that this insertion loss for a specified n and value of q is larger, for example, than the Chebyshev insertion loss of Fig. 6-20. How is this possible when the equal-element filter should yield a lesser insertion loss for the same n and q? The answer lies in the frequency normalization.

The equal-element filter has minimum loss when all responses are normalized for the same stopband bandwidth. Furthermore the quality factor is then not evaluated at the cutoff frequency of each filter, but it is evaluated at a convenient frequency, for example $\omega = 1$. One way to ensure the same stopband behavior for all filters with the same value of n is to normalize all element values so that their product is unity. Then the asymptotic behavior is identical. For this condition, the minimum insertion loss for even-ordered filters is [4]

$$l_d = 4.343\, n\, \sinh^{-1} \frac{1}{q} \text{ decibels} \qquad (6.3\text{-}18)$$

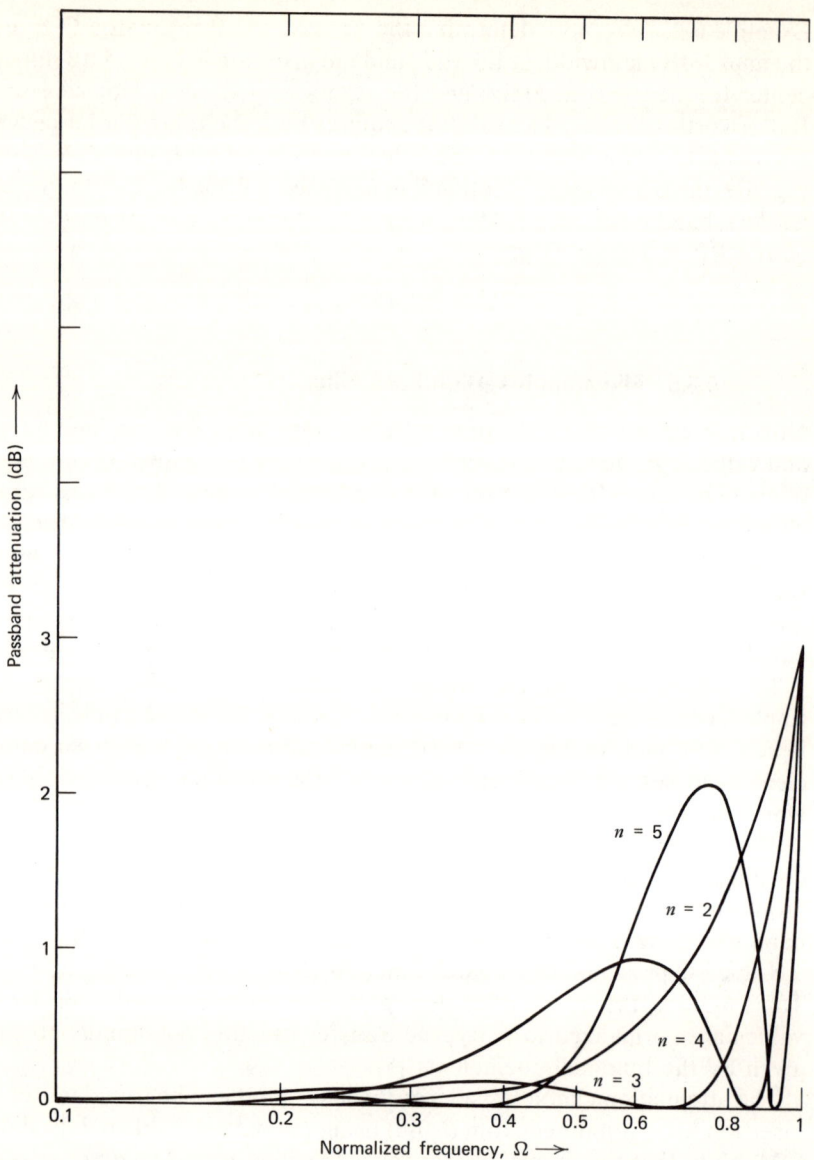

Figure 6-23. Attenuation of the lossless equal-element filter normalized to unity radian cutoff.

6.3.6 Insertion Loss Comparison of Various Filters

Let us now determine the insertion loss of the fourth-order Butterworth, 0.1 dB Chebyshev, Gaussian, and equal-element filter when each is normalized for the same asymptotic stopband attenuation. Assume that the loss q, measured at $\omega = 1$, is 5. To solve this problem using the computed insertion loss curves, we must first determine the effective q value to enter these curves.

Insertion Loss

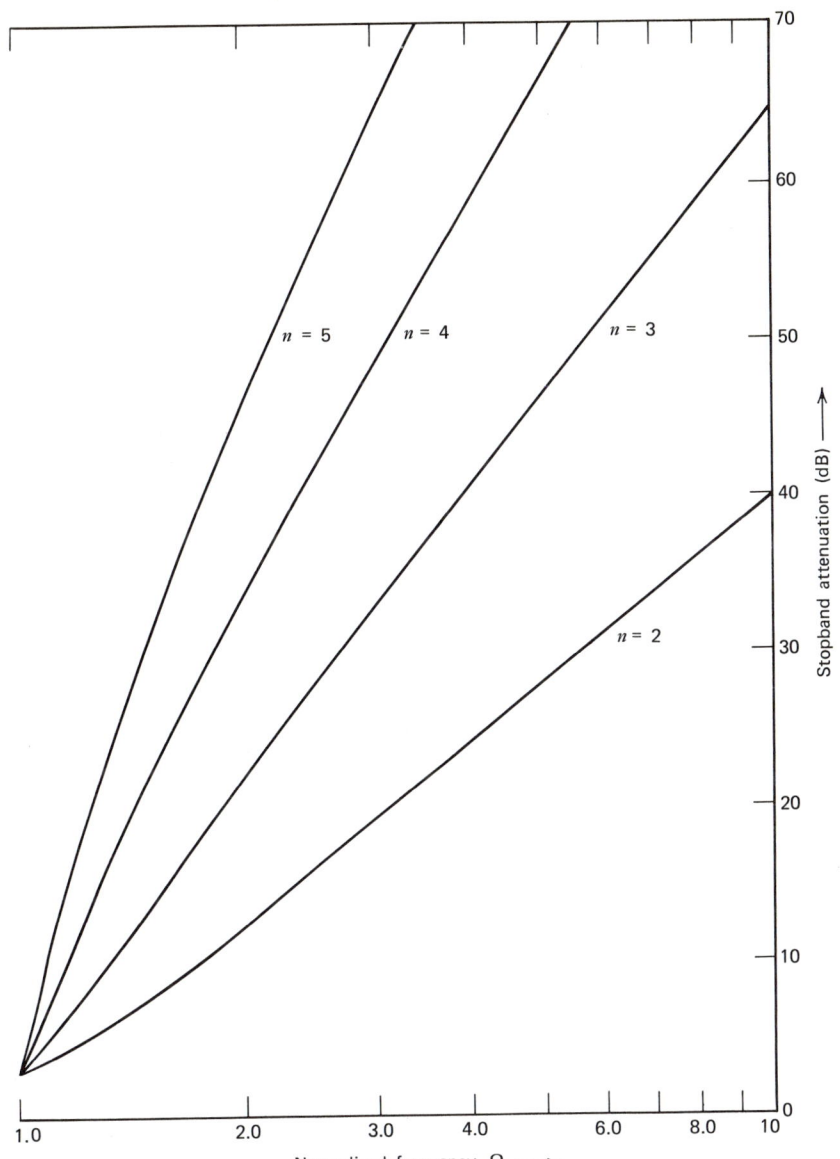

Consider the normalized LP transfer function in (6.3-11). Denormalizing the 3-dB bandwidth to ω_c produces the new transfer function $H_l(s)$,

$$H_l(s) = \frac{1}{\dfrac{a_0}{\omega_c^n} s^n + \dfrac{a_1}{\omega_c^{n-1}} s^{n-1} + \cdots + \dfrac{a_{n-1}}{\omega_c} s + 1} \quad (6.3\text{-}19)$$

and the response is now the same at ω_c as it previously was at $\omega = 1$. With q evaluated at ω_a, as in (6.2-1), we determine the insertion loss from (6.3-8), except

Table 6-2 Polynomial Coefficients of Equal-Element Filter Transfer Function and Filter-Element Values ($\omega_c = 1$)

$$H_L(s) = \frac{a_n}{s^n + a_1 s^{n-1} + \cdots + a_{n-1} s + a_n}$$

n	a_n	a_{n-1}	a_{n-2}	a_{n-3}	a_{n-4}	x_n
2	1.0000	1.4142				1.4142
3	0.5679	1.2961	1.3146			1.5214
4	0.2679	0.8858	1.4641	1.2100		1.6529
5	0.1233	0.5383	1.1276	1.6405	1.1456	1.7458

Table 6-3 Pole Locations of the Equal-Element Filter Transfer Function ($\omega_c = 1$)

$n = 2$	$n = 3$	$n = 4$	$n = 5$
$-0.707107 \pm j0.707107$	-0.657298	$-0.449476 \pm j0.320098$	-0.409686
	$-0.328649 \pm j0.869524$	$-0.155525 \pm j0.925101$	$-0.286403 \pm j0.496065$
			$-0.081560 \pm j0.954382$

$H_l(s)$ is now evaluated at $s = \omega_a/q$ rather than at $s = 1/q$; that is,

$$H_l\left(\frac{\omega_a}{q}\right) = \frac{1}{a_0\left(\frac{\omega_a}{\omega_c q}\right)^n + a_1\left(\frac{\omega_a}{\omega_c q}\right)^{n-1} + \cdots + a_{n-1}\left(\frac{\omega_a}{\omega_c q}\right) + 1} \qquad (6.3\text{-}20)$$

But this is the same value that is obtained by evaluating $H_L(s)$ at

$$s = \frac{1}{q_{\text{eff}}} \qquad (6.3\text{-}21)$$

where the effective quality factor q_{eff}, from (6.3-20), is

$$q_{\text{eff}} = \frac{\omega_c}{\omega_a} q \qquad (6.3\text{-}22)$$

For this problem, since ω_a is unity and $q = 5$, we need only find the cutoff frequency for each filter. We choose the cutoff frequency ω_c; thus the product of the denormalized element values is unity (see Fig. 3-23). That is,

$$\frac{x_1}{\omega_c} \cdot \frac{x_2}{\omega_c} \cdot \frac{x_3}{\omega_c} \cdot \frac{x_4}{\omega_c} = 1 \qquad (6.3\text{-}23)$$

or

$$\omega_c = \sqrt[4]{x_1 x_2 x_3 x_4} \qquad (6.3\text{-}24)$$

We obtain the Butterworth element values from (3.4-7), the Chebyshev element values (0.1-dB ripple) from (3.4-29), the equal-element value from Table 6-2, and the Gaussian element values from Ref. 6. Then ω_c is computed from (6.3-24), q_{eff} is obtained from (6.3-22), and the insertion loss is determined from (6.3-8) with $q = q_{\text{eff}}$. These data are summarized in Table 6-4 along with the minimum insertion loss computed from (6.3-18).

Insertion Loss

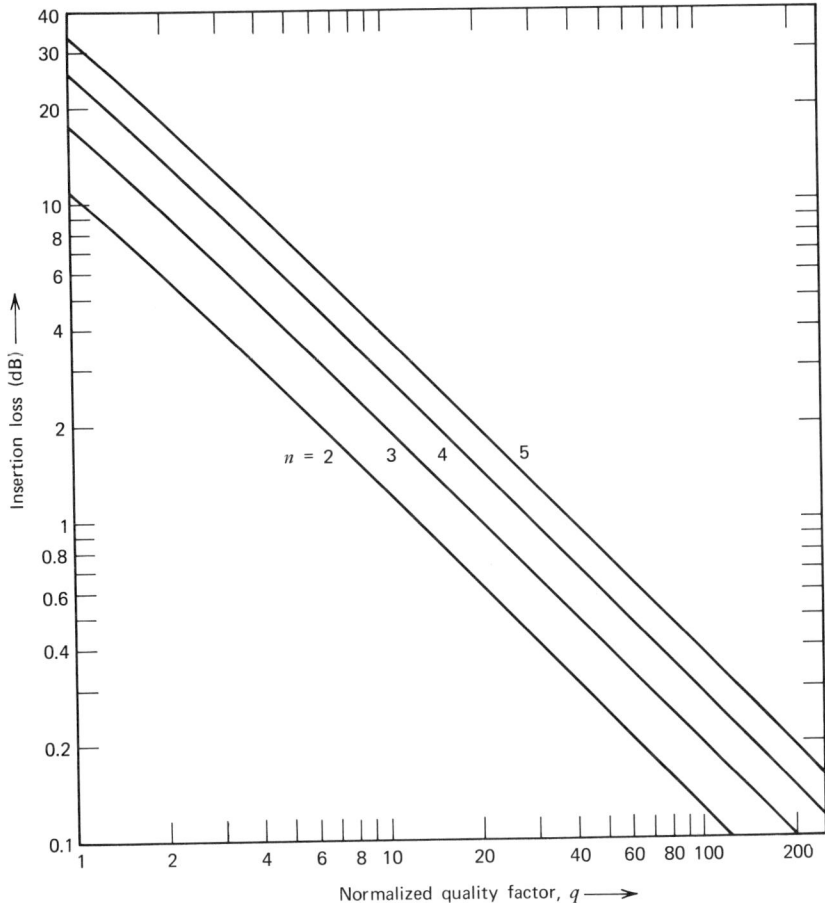

Figure 6-24. Insertion loss of equal-element filters as a function of q.

Table 6-4 **Summary of Data for Insertion Loss Comparison**

Filter Type	x_1	x_2	x_3	x_4	ω_c	q_{eff}	Insertion Loss (dB)
Butterworth	0.7654	1.8478	1.8478	0.7654	1.1892	5.946	3.8
Chebyshev (0.1 dB)	0.9924	2.1476	1.5845	1.3451	1.4599	7.300	3.55
Gaussian	0.1772	0.5302	0.9321	2.2450	0.6659	3.330	4.8
Equal element	1.6529	1.6529	1.6529	1.6529	1.6529	8.265	3.4572
Minimum insertion loss from (6.3-18)							3.4516

As expected, the equal-element filter has the lowest insertion loss when all filters have the same stopband asymptote. Furthermore we now see why this filter is considered to have minimum insertion loss (at least for moderate q values), since the difference between its insertion loss and the theoretic minimum is insignificant.

However the insertion loss of the 0.1-dB ripple Chebyshev filter is comparable to the equal-element filter insertion loss, but the Chebyshev passband ripple is more acceptable when the q is not low enough to smear the ripples. For this reason, a 0.1-dB Chebyshev filter is recommended when the application requires low insertion loss, regular passband behavior, and moderately high stopband attenuation.

6.4 PREDISTORTION

We have seen that equal-q element losses shift the lossless pole-zero arrangement to the *left* by the amount $1/q$. For an all-pole filter, we can anticipate this shift by initially designing a filter with transfer-function poles shifted to the *right* by an amount $1/q$, ensuring that when losses are added these poles will shift to the left into the desired positions. In other words, we predistort the desired function; hence this approach is known as the predistortion technique. We limit the technique to all-pole filters here, although predistortion is possible for any zero locations. However, if zeros on the $j\omega$ axis were moved to the right, they would be in the right-half plane and no longer realizable with the desirable ladder topology.

The minimum value of q for which a predistorted network can be obtained is designated q_{min}. It is referenced to the filter with unity-radian cutoff frequency and is given by

$$q_{min} = \frac{1}{|\alpha_c|} \tag{6.4-1}$$

where α_c is the real part of the pole closest to the $j\omega$ axis. For $q < q_{min}$, the shift to the right then introduces a right-half plane pole, and of course, this is not realizable with passive elements. Equation 6.2-12 gives a good rule of thumb for the minimum q that causes an insignificant change in the filter response. Coupled with this discussion, we find that this quality factor should be 10 times larger than q_{min}; that is,

$$q > 10 q_{min} \tag{6.4-2}$$

6.4.1 Minimum Value of Quality Factor for Various Responses

We now give closed-form expressions for q_{min} characterizing the Butterworth and Chebyshev responses by referring to (3.4-4) and (3.4-28), the respective pole locations. The Butterworth pole closest to the $j\omega$ axis ($i = 1$) has real part $-(\sin \pi/2n)$, hence

$$\text{Butterworth } q_{min} = \frac{1}{\sin \pi/2n} = D_L(0) \tag{6.4-3}$$

The value of q_{min} is the group delay at $\omega = 0$. For the Chebyshev response,

$$\text{Chebyshev } q_{min} = \frac{\cosh[1/n \cosh^{-1} 1/\epsilon]}{\sinh[1/n \sinh^{-1} 1/\epsilon]} \times \frac{1}{\sin \pi/2n} \tag{6.4-4}$$

Table 6-5 lists q_{min} for various filter responses.

Table 6-5 Values of q_{min} for Various Filter Responses ($\omega_c = 1$)

Response	n						
	2	3	4	5	6	7	8
Butterworth	1.414	2.000	2.613	3.236	3.864	4.494	5.126
0.1-dB Chebyshev	1.638	2.866	4.592	6.814	9.529	12.738	16.441
0.5-dB Chebyshev	1.950	3.727	6.234	9.461	13.407	18.071	23.453
Gaussian	0.779	0.728	0.705	0.692	0.682	0.675	0.669
Bessel	0.908	0.955	1.005	1.044	1.075	1.099	1.120

6.4.2 Insertion Loss

We now give a physical interpretation of the predistortion process. Predistortion is usually applied to the Butterworth and Chebyshev responses because the presence of element losses causes an undesirable rounding of the attenuation response in the vicinity of the cutoff frequency. This rounding is equalized by properly introducing impedance mismatch loss (by shifting the poles) about $\omega = 0$ (ω_0 for the BP case) for an exact overall maximally flat or equiripple response. This results in a relatively high insertion loss, especially for the Chebyshev responses, for it requires considerable mismatch loss to achieve passband ripples in the presence of dissipation.

Zverev [6] gives the exact insertion loss for the classical responses with various q values and includes the predistorted k and q values for these same responses. Fubini and Guillemin [3] present an accurate approximate formula for determining the insertion loss of equal-terminated predistorted Butterworth and Chebyshev narrow-band BP filters as

$$l_p = 20c \log \frac{u}{u-1} \text{ decibels} \tag{6.4-5}$$

where $u = q/q_{min}$ and c is a correction factor depending on n, given in Table 6-6.

Equation (6.4-5) reveals that the insertion loss approaches infinity as the element q approaches q_{min}, but this is expected, since the amount of introduced mismatch loss increases rapidly as q approaches q_{min}.

Table 6-6 Correction Factor c for Insertion Loss of Predistorted Filter

	n						
	2	3	4	5	6	7	8
c	1	1.08	1.14	1.19	1.23	1.27	1.32

6.4.3 Example of Insertion Loss Calculation

We now compare the insertion loss of the fourth-order filter in Example 6-1 with $q = 5.507$ and the insertion loss of the predistorted filter. After increasing the bandwidth to allow for the shrinkage with losses, we obtained an insertion loss of 4.1 dB. The insertion loss resulting from predistortion is obtained from (6.4-5) with q_{min} obtained from Table 6-5. Since the theoretical response is preserved by predistortion, $q = 5$ and

$$u = \frac{q}{q_{min}} = \frac{5}{2.613} = 1.9135 \tag{6.4-6}$$

and with the correction factor from Table 6-6,

$$l_p = 20 \times 1.14 \log \frac{1.9135}{0.9135} = 7.32 \text{ dB} \tag{6.4-7}$$

Thus one of the prices we pay to achieve the exact Butterworth response is an insertion loss of 7.32 dB compared to the 4.1 dB of the Butterworth filter with losses.

6.4.4 Example of Predistorted Filter Design

To obtain the maximally flat response with $q = 5$, we now predistort the third-order Butterworth LP filter. The predistorted lossless poles are the Butterworth poles shifted to the right by $1/q = 1/5$,

$$\begin{aligned} s_1 &= -1 + \frac{1}{5} = -0.8 \\ s_{2,3} &= -\frac{1}{2} + \frac{1}{5} \pm j\frac{\sqrt{3}}{2} = -0.3 \pm j0.866 \end{aligned} \tag{6.4-8}$$

and the corresponding attenuation and group delay appear in Fig. 6-25. The attenuation characteristic now has a passband ripple and the group delay peak has increased to a 4-sec maximum compared to the Butterworth maximum of 2.74 sec. With the addition of losses corresponding to $q = 5$ at $\omega = 1$, the poles shift to the left and the Butterworth response is obtained. The incurred insertion loss is given

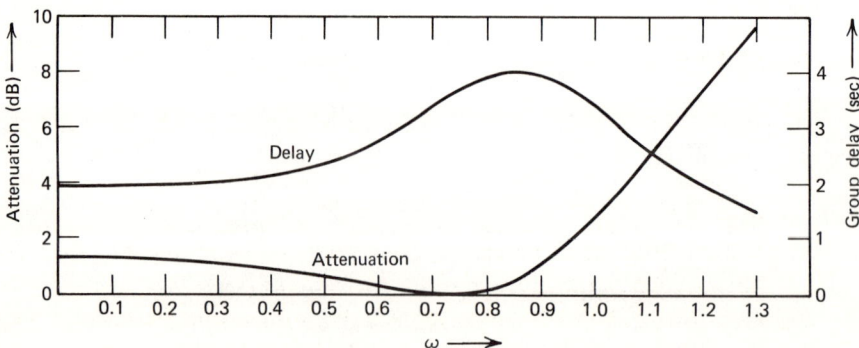

Figure 6-25. Attenuation and group delay of lossless predistorted filter of Section 6.4.4.

by (6.4-5), with $q_{min} = 2$ from Table 6-5, $u = 2.5$, and

$$l_p = 20 \times 1.08 \log \frac{2.5}{1.5} = 4.79 \text{ dB} \qquad (6.4\text{-}9)$$

The exact value for insertion loss is 4.742 dB, verifying the accuracy of (6.4-5). It is understood that this predistorted filter is likewise the LP prototype for the narrow-band BP filter with resonator $Q = 5 \times \omega_0/(\omega_2 - \omega_1)$.

6.4.5 Summary

In addition to increased insertion loss, other undesirable features of predistorted networks are a wide range of element values, increased sensitivity to small element-value changes, increased mismatch between the filter's input impedance and the source resistance, and the appearance of passband ripples when the element q is small and differs from the design value. Consequently predistortion should only be used when q is relatively high and the exact response is desired. Otherwise, a more practical solution is the adjustment of the filter's bandwidth to accommodate losses as explained in Section 6.2.3. In most cases the exact response is not a necessity; the lossy response (e.g., as in Figs. 6-2–6-8) is satisfactory, and the disadvantages associated with the predistortion method are then absent.

REFERENCES

1. Cohn, S. B., "Design Considerations for High-Power Microwave Filters," *IRE Trans. Microwave Theory and Techniques*, Vol. MTT-7, pp. 149–153, January 1959.
2. Cohn, S. B., "Dissipation Loss in Multiple-Coupled-Resonator Filters," *Proc. IRE*, Vol. 47, pp. 1342–1348, August 1959.
3. Fubini, E. G., and E. A. Guillemin, "Minimum Insertion Loss Filters," *Proc. IRE*, Vol. 47, pp. 37–41, January 1959.
4. Humpherys, D. S., "Minimum Insertion Loss Filters of Even Order," *IEEE Trans. Circuit Theory*, Vol. CT-11, pp. 414–416, September 1964.
5. Taub, J. J., "Design of Minimum Loss Band-Pass Filters," *Microwave J.*, pp. 67–76, November 1963.
6. Zverev, A. I., *Handbook of Filter Synthesis*, Wiley, New York, 1967.

PROBLEMS

6-1. Consider a third-order equally terminated Butterworth LP filter with unity-radian cutoff frequency.
    (a) If all elements have $q = 4$, what is the lossy transfer function of the filter?
    (b) What is the insertion loss? Compare the values obtained from (6.3-15) and the lossy transfer function evaluated at $\omega = 0$.

6-2. We wish to realize a BP filter (from a fifth-order Butterworth LP filter) centered at $\omega_0 = 10^6$ and with a 3-dB bandwidth $\Delta \omega = 5 \times 10^4$. What is the Q of the BP resonators if the insertion loss is to be 2.5 dB?

6-3. For the filter below, $|H(j\omega)|$ is maximally flat at $\omega_0 = 10^7$. Each inductor is 20 μH and the lossless 3-dB bandwidth is 2×10^5 rads/sec.

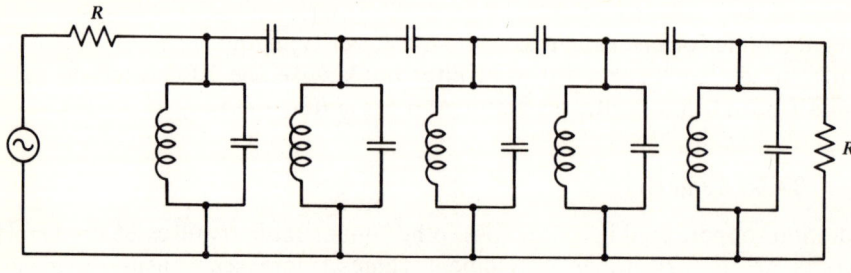

Figure P.6-3.

(a) Losses are represented by a 60,000-ohm resistor across each inductor. What is the filter's insertion loss?
(b) What is Q_{min} for this response?
(c) What is the lossy-filter bandwidth?

6-4. Consider the third-order filter below, operating between the unequal resistances shown.

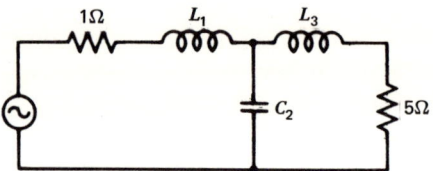

Figure P.6-4.

(a) What is the insertion loss according to (6.3-4)?
(b) What is the insertion loss due to impedance mismatch?
(c) What is the insertion loss due to resistive losses?

6-5. Assume that the filter in Problem 6-4 exhibits the Butterworth response ($\omega_c = 1$) with element values $L_1 = 7.9102$ H, $C_2 = 0.2842$ F, $L_3 = 2.6687$ H, and each element q is 5. Reanswer the questions in Problem 6-4. Note that (6.3-5) is satisfied.

6-6. Reanswer the questions in Problem 6-4 for the equally terminated case; $L_1 = L_3 = 1$ H, $C_2 = 2$ F, and $q = 5$. Again (6.3-5) is satisfied.

6-7. A fifth-order LP Chebyshev filter (0.1-dB ripple) is to have a 3-dB bandwidth of 100 kHz. The element Q is 10. Allowing for bandwidth reduction due to losses, what should the lossless bandwidth be?

6-8. It is desired to predistort a fourth-order Butterworth LP filter with unity-radian cutoff frequency ($q = 4$).
(a) What are the poles of the predistorted lossless filter?
(b) What is the insertion loss of the predistorted filter?

6-9. We wish to predistort a sixth-order Chebyshev BP filter (0.1-dB ripple) with $\omega_0 = 10^6$ and $\Delta\omega = 10^5$. If $Q = 200$, what is the insertion loss?

6-10. The filter below, with equal-q elements, is to exhibit the Butterworth response. Assume a unity-radian cutoff frequency.

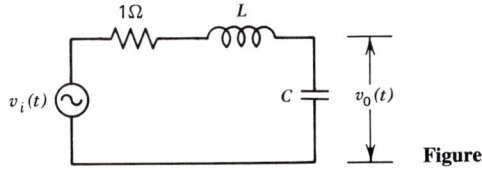

Figure P.6-10.

(a) Determine L and C as a function of q.
(b) At $\omega = 0$, what is the ratio of the output voltage to the input voltage as a function of q?

Chapter Seven

Optimum Linear Filtering

A primary function of electronic communication systems, including radar, is the transmission and reception of signals that contain the desired message. A message has broad meaning and is not restricted to electronic signals, although obviously this must be converted to such signals to be processed electronically. Examples of messages are spoken and written words, the liquid flow rate in a pipe, temperature changes in an industrial process, the pulse rate of an astronaut in space, and traffic flow on a highway.

A message that is completely predictable contains no additional information, for then one need only send a code number to identify the function and thereby disclose all its details. Thus a sine wave contains no further information once its amplitude, phase, and frequency are known, for it is now completely predictable over the infinite time interval.

This situation does not occur in practice, however, because associated with each desired message is some unwanted message, and this disturbance is designated "noise." An example of electrical noise is the familiar thermal noise. Furthermore, noise in one transmission channel may be the desired message in another channel, and vice versa. For example, a radio receiver tuned to station A may also receive audible music transmitted by station B. The music, in this case, is considered noise, whereas it is the desired message when the receiver is tuned to station B.

The basic signal processing problem is to determine the necessary operations on the received signal (message plus noise) to extract specific information about the message signal. Then follows the practical implementation of these operations. This major problem can be subdivided into the following two problems.

Signal detection—the determination of whether a message signal is present in the noisy environment.

Signal extraction—the measurement or recovery of the important parameters contained in the message signal corrupted by noise, such as frequency, amplitude, phase, or epoch.

Both problems arise in radar receiver design. Consider the transmitted pulse reflected from a target. The signal detection problem is to determine this reflected signal's presence at the receiver, whereas the arrival-time determination is the signal extraction problem. For this case, determination of the arrival time solves both problems. Suppose also that the pulse is reflected from a moving target. Then the received carrier frequency differs from the transmitted carrier frequency by the Doppler frequency, and a signal extraction problem is then to determine these shifts and thereby obtain the target's radial velocity.

There is no one "best" solution to any of these problems, for there are many criteria for system optimization and many different classes of problems contained in these broad categories. And, of course, the mere postulate of a criterion does not necessarily admit of a solution that can be easily obtained and implemented. Furthermore, the allowance of nonlinear systems as possible solutions introduces additional mathematical and practical difficulties.

Solutions, if they can be found, are generally a mixture of disciplines from decision theory and filtering theory, and it is often difficult to separate them. The filtering aspect of signal detection and extraction is lightly touched on in the usual treatments of electronic filtering, even for the cases in which the filter is the most important portion. Matched filtering is one such case, and here we discuss it within the framework of signals and filters. We hope to show that it is not a mystical area and that it belongs in the domain of the filter engineer.

The majority of this chapter is devoted to the matched filter because of its importance in modern communication and radar systems. It is the optimum filter under a wide variety of criteria when the input noise is additive, white, and Gaussian. Even if this optimum filter cannot be exactly realized, it remains as the standard against which other realizations can be compared.

The study of the matched filter here includes its derivation, its time- and frequency-domain characteristics, its synthesis, its sensitivity to waveform changes, its relationship to cross-correlation, and its characterization when the input noise spectrum is nonconstant. Also discussed is pulse compression, a topic of both theoretic and practical importance. For large time-bandwidth products, the pulse compression system is essentially a matched filter system, thus an optimum detection system. The two methods of pulse compression considered here use the linear FM signal and the Barker sequences. The compressed waveforms include undesirable sidelobes, and we discuss methods of reducing them by appropriate filtering. Again, these filters are not included in the usual treatments of electronic filtering.

Before beginning our presentation of matched filtering and pulse compression, we introduce the important aspects of the autocorrelation, cross-correlation, and the power density functions. Knowledge of these subjects is necessary in any discussion of signal filtering in the presence of noise.

Another important area of filtering is the linear estimation problem, whose solution was first proposed by Wiener and Kolmogoroff in the early 1940s using the least-squares criterion. The applications of this filter were limited because the problem of practical synthesis was not solved, but Kalman, in 1961, introduced an adaptive filter algorithm that produced the best signal estimate in the least-squares sense. The topic is briefly discussed to acquaint the reader with this type of filtering.

7.1 AUTOCORRELATION, CROSS-CORRELATION, AND POWER DENSITY FUNCTIONS

The study of optimum linear filtering includes deterministic functions, which can be described analytically, and nondeterministic (random) functions, which cannot be described analytically. Random functions are therefore not Fourier transformable, and another tool for handling these functions is used.

One such characterization is the autocorrelation function, which is a measure of the regularity of the specified function. The autocorrelation function is analytic, and its Fourier transform has an important physical interpretation. Another useful function is the cross-correlation function, which describes the relationship between two different functions. If the two functions are the same, the cross-correlation function reduces to the autocorrelation function. Here we discuss the important properties of the aforementioned functions as related to both deterministic and nondeterministic functions.

The definition of $\phi_f(\tau)$, the autocorrelation function of $f(t)$, depends on the function class to which $f(t)$ belongs. Accordingly, we classify the possible functions as (a) periodic functions, (b) aperiodic (transient) functions, and (c) random functions. Unless otherwise noted, $f(t)$ is assumed real. The Fourier transform of the autocorrelation function is

$$\Phi_f(\omega) = \int_{-\infty}^{\infty} \phi_f(\tau) e^{-j\omega\tau} \, d\tau \tag{7.1-1}$$

and the inversion formula is

$$\phi_f(\tau) = \frac{1}{2\pi} \int_{-\infty}^{\infty} \Phi_f(\omega) e^{j\omega\tau} \, d\omega \tag{7.1-2}$$

The transform-pair relationship of (7.1-1) and (7.1-2), known as the Wiener-Khintchine theorem, is one of the most important results in the analysis of time waveforms. It permits $\Phi_f(\omega)$ to be interpreted as the power density spectrum of functions belonging to classes a and c and the energy density spectrum of functions belonging to class b. Where power and energy are introduced, $f(t)$ is assumed to be a voltage across a one-ohm resistor or a current through a one-ohm resistor. Then this power or energy is dissipated in the one-ohm resistor.

We now briefly consider each function class and give the important properties of $\phi_f(\tau)$ and $\Phi_f(\omega)$.

7.1.1 Periodic Function Analysis

The autocorrelation function of a periodic function $f(t)$ of period T_1, expressed as the Fourier series

$$f(t) = \sum_{n=-\infty}^{\infty} F(n) e^{jn\omega_1 t} \quad \left(\omega_1 = \frac{2\pi}{T_1}\right) \tag{7.1-3}$$

where

$$F(n) = \frac{1}{T_1} \int_{-T_1/2}^{T_1/2} f(t) e^{-jn\omega_1 t} \, dt = |F(n)| e^{j\theta(n)} \tag{7.1-4}$$

is defined as

$$\phi_f(\tau) = \frac{1}{T_1} \int_{-T_1/2}^{T_1/2} f(t) f(t+\tau) \, dt \tag{7.1-5}$$

Autocorrelation, Cross-Correlation, Power Density

To determine the explicit behavior of $\phi_f(\tau)$, we substitute from (7.1-3) into (7.1-5) to yield

$$\phi_f(\tau) = \frac{1}{T_1} \int_{-T_1/2}^{T_1/2} f(t) \sum_{n=-\infty}^{\infty} F(n) e^{jn\omega_1(t+\tau)} \, dt$$

$$= \sum_{n=-\infty}^{\infty} F(n) e^{jn\omega_1\tau} \frac{1}{T_1} \int_{-T_1/2}^{T_1/2} f(t) e^{jn\omega_1 t} \, dt$$

$$= \sum_{n=-\infty}^{\infty} F(n) e^{jn\omega_1\tau} F^*(n) = \sum_{n=-\infty}^{\infty} |F(n)|^2 e^{jn\omega_1\tau} \qquad (7.1\text{-}6)$$

which can be rewritten as

$$\phi_f(\tau) = |F(0)|^2 + 2 \sum_{n=1}^{\infty} |F(n)|^2 \cos n\omega_1\tau \qquad (7.1\text{-}7)$$

Thus the autocorrelation function of a periodic function is also periodic and is completely characterized by the Fourier coefficients $F(n)$.

The power density spectrum $\Phi_f(\omega)$, in watts per hertz, is obtained from (7.1-1) as

$$\Phi_f(\omega) = \int_{-\infty}^{\infty} \sum_{n=-\infty}^{\infty} |F(n)|^2 e^{jn\omega_1\tau} e^{-j\omega\tau} \, d\tau$$

$$= \sum_{n=-\infty}^{\infty} |F(n)|^2 \int_{-\infty}^{\infty} e^{j(n\omega_1 - \omega)\tau} \, d\tau \qquad (7.1\text{-}8)$$

and using the result of (2.1-46),

$$\Phi_f(\omega) = 2\pi \sum_{n=-\infty}^{\infty} |F(n)|^2 \delta(\omega - n\omega_1) \qquad (7.1\text{-}9)$$

The power density spectrum is a discrete spectrum located only at the frequencies $\omega = n\omega_1$. Note that for each frequency (except zero), one-half of the power is assigned to the negative frequency and one-half is assigned to the corresponding positive frequency.

The total power in $f(t)$ is the sum of the power due to each frequency component, obtained by evaluating (7.1-2) at $\tau = 0$ as

$$\phi_f(0) = \frac{1}{2\pi} \int_{-\infty}^{\infty} \Phi_f(\omega) \, d\omega \qquad (7.1\text{-}10)$$

Thus the value of the autocorrelation function at $\tau = 0$ is the total power in $f(t)$, in watts. From (7.1-6)

$$\phi_f(0) = \sum_{n=-\infty}^{\infty} |F(n)|^2 \qquad (7.1\text{-}11)$$

the same result obtained by substituting from (7.1-9) into (7.1-10). The total power as expected is the sum of the power at each frequency $n\omega_1$. Substituting $\phi_f(0)$

from (7.1-5) into (7.1-11) establishes Parseval's relationship for periodic functions

$$\frac{1}{T} \int_{-T_1/2}^{T_1/2} f^2(t)\, dt = \sum_{n=-\infty}^{\infty} |F(n)|^2 \qquad (7.1\text{-}12)$$

Equations 7.1-10 and 7.1-12 suggest a method for measuring the power density spectrum at each frequency contained in the periodic signal $f(t)$. First construct a parallel bank of narrow-band filters, each center frequency ω_0 tuned to a different harmonic of the fundamental frequency, each with bandwidth $\Delta\omega$, as in Fig. 7-1. Then square and average the output of each filter to yield the power in that bandwidth. This value of power includes the contribution from the negative frequencies (see Section 2.1.1).

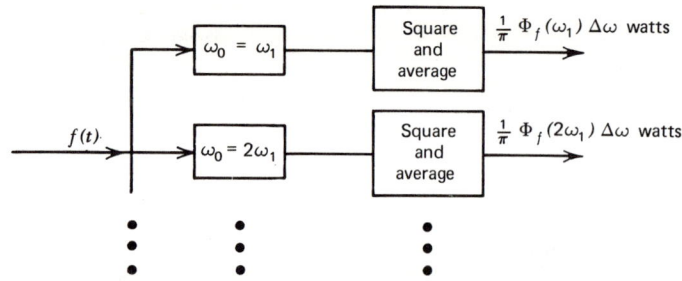

Figure 7-1. Measurement of power density spectrum.

Example 7-1. We illustrate the preceding development by considering the sine wave

$$f(t) = A_0 \sin(\omega_1 t + \theta) \qquad (7.1\text{-}13)$$

Express $f(t)$ as the sum of exponentials

$$f(t) = \frac{A_0 e^{j\theta}}{2j} e^{j\omega_1 t} - \frac{A_0 e^{-j\theta}}{2j} e^{-j\omega_1 t} \qquad (7.1\text{-}14)$$

and compare to (7.1-3). Then $F(-1) = -A_0 e^{-j\theta}/2j$, $F(1) = A_0 e^{j\theta}/2j$, and all other coefficients are zero. The autocorrelation function, from (7.1-7), is

$$\phi_f(\tau) = 2 \times \frac{A_0^2}{4} \cos \omega_1 \tau = \frac{A_0^2}{2} \cos \omega_1 \tau \qquad (7.1\text{-}15)$$

which is periodic and independent of θ. The power spectrum, from (7.1-9), is shown in Fig. 7-2, where each component is $2\pi |F(n)|^2$. One-half of the power is at

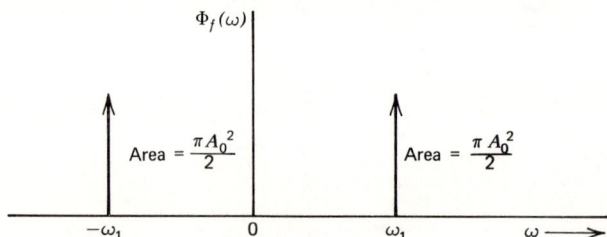

Figure 7-2. Power density spectrum for $f(t) = A_0 \sin(\omega_1 t + \theta)$.

Autocorrelation, Cross-Correlation, Power Density

$-\omega_1$ and one-half the power is at $+\omega_1$. The total power in $f(t)$, from (7.1-11) is then

$$\phi_f(0) = \frac{A_0^2}{4} + \frac{A_0^2}{4} = \frac{A_0^2}{2} \text{ watts} \tag{7.1-16}$$

7.1.2 Aperiodic (Transient) Function Analysis

The class of aperiodic functions contains those functions $f(t)$ for which

$$\int_{-\infty}^{\infty} |f(t)| \, dt < \infty \tag{7.1-17}$$

the same condition for which the classical Fourier transform exists [see (2.1-8)]. Thus the pair of (2.1-1) and (2.1-6) relating $f(t)$ and $F(\omega)$ is valid for this class of functions. Then the autocorrelation function is defined as†

$$\phi_f(\tau) = \int_{-\infty}^{\infty} f(t) f(t+\tau) \, dt \tag{7.1-18}$$

and $\Phi_f(\omega)$ in (7.1-1) is now interpreted as an energy density spectrum with units of joules per hertz.

Similar to the periodic-function development, substitute from (2.1-6) into (7.1-18) to yield

$$\phi_f(\tau) = \int_{-\infty}^{\infty} f(t) \frac{1}{2\pi} \int_{-\infty}^{\infty} F(\omega) e^{j\omega(t+\tau)} \, d\omega \, dt$$

$$= \frac{1}{2\pi} \int_{-\infty}^{\infty} F(\omega) e^{j\omega\tau} \left[\int_{-\infty}^{\infty} f(t) e^{j\omega t} \, dt \right] d\omega$$

$$= \frac{1}{2\pi} \int_{-\infty}^{\infty} F(\omega) e^{j\omega\tau} F^*(\omega) \, d\omega = \frac{1}{2\pi} \int_{-\infty}^{\infty} |F(\omega)|^2 e^{j\omega\tau} \, d\omega \tag{7.1-19}$$

Thus the autocorrelation function is the Fourier transform of $|F(\omega)|^2$. But (7.1-2) tells us that $\phi_f(\tau)$ is the Fourier transform of the energy density spectrum $\Phi_f(\omega)$, and we conclude that

$$\Phi_f(\omega) = |F(\omega)|^2 \tag{7.1-20}$$

Evaluate $\phi_f(\tau)$ in (7.1-19) at $\tau = 0$ to obtain the total energy in $f(t)$,

$$\phi_f(0) = \frac{1}{2\pi} \int_{-\infty}^{\infty} |F(\omega)|^2 \, d\omega \tag{7.1-21}$$

which is the total area under the energy density curve, and from (7.1-18)

$$\phi_f(0) = \int_{-\infty}^{\infty} f^2(t) \, dt \tag{7.1-22}$$

†If $f(t)$ is complex, then $\phi_f(\tau)$ is defined as $\phi_f(\tau) = \int_{-\infty}^{\infty} f^*(t) f(t+\tau) \, dt$.

Equating (7.1-21) and (7.1-22) yields Parseval's relationship in (2.1-31),

$$\int_{-\infty}^{\infty} f^2(t)\, dt = \frac{1}{2\pi} \int_{-\infty}^{\infty} |F(\omega)|^2\, d\omega \qquad (7.1\text{-}23)$$

It is now shown that $\phi_f(0)$ is the maximum value of the autocorrelation function. Apply the inequality

$$\int_a^b p(\omega)q(\omega)\, d\omega \leq \int_a^b |p(\omega)||q(\omega)|\, d\omega \qquad (7.1\text{-}24)$$

to (7.1-19). Then

$$\phi_f(\tau) \leq \frac{1}{2\pi} \int_{-\infty}^{\infty} |F(\omega)|^2 |e^{j\omega\tau}|\, d\omega = \frac{1}{2\pi} \int_{-\infty}^{\infty} |F(\omega)|^2\, d\omega = \phi_f(0) \qquad (7.1\text{-}25)$$

giving the desired result,

$$\phi_f(\tau) \leq \phi_f(0) \qquad (7.1\text{-}26)$$

7.1.3 Example of Aperiodic Function Calculation

Consider the triangular function in Fig. 7-3. Figure 7-4 illustrates the autocorrelation process of (7.1-18). The $f(t)$ remains fixed in each sequence, and τ is the varying time interval between the functions. In Fig. 7-4a, $f(t + \tau)$ is shifted 2 seconds to the right. For each second increase in τ, $f(t + \tau)$ slides one second to the left, beginning at $\tau = -2$ in Fig. 7-4a through $\tau = 2$ in Fig. 7-4e. The integral of the product of $f(t)$ and $f(t + \tau)$ is $\phi_f(\tau)$, given in Fig. 7-4f. Since there is no overlap in Fig. 7-4a and 7-4e, $\phi_f(\tau)$ has a nonzero value only when τ is greater than -2 and less than 2. The greatest value of $\phi_f(\tau)$ occurs when $\tau = 0$, at which time the two signals are coincident, as in Fig. 7-4c. We now see why the maximum value of $\phi_f(\tau)$ always occurs at $\tau = 0$.

Using the graphical display in Fig. 7-4 as a guide, we obtain the proper limits on the integral of (7.1-18) as

$$\phi_f(\tau) = \begin{cases} 4 \displaystyle\int_{-(1+\tau)}^{1} (1-t)(1-t-\tau)\, dt & -2 \leq \tau \leq 0 \\ 4 \displaystyle\int_{-1}^{1-\tau} (1-t)(1-t-\tau)\, dt & 0 \leq \tau \leq 2 \\ 0 & \text{elsewhere} \end{cases} \qquad (7.1\text{-}27)$$

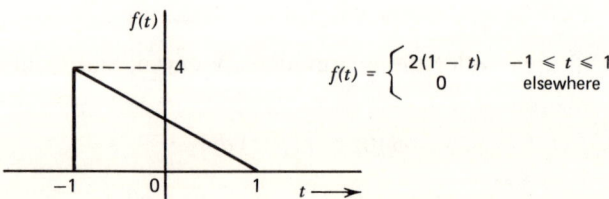

$$f(t) = \begin{cases} 2(1-t) & -1 \leq t \leq 1 \\ 0 & \text{elsewhere} \end{cases}$$

Figure 7-3. Triangular function in Section 7.1.3.

Autocorrelation, Cross-Correlation, Power Density

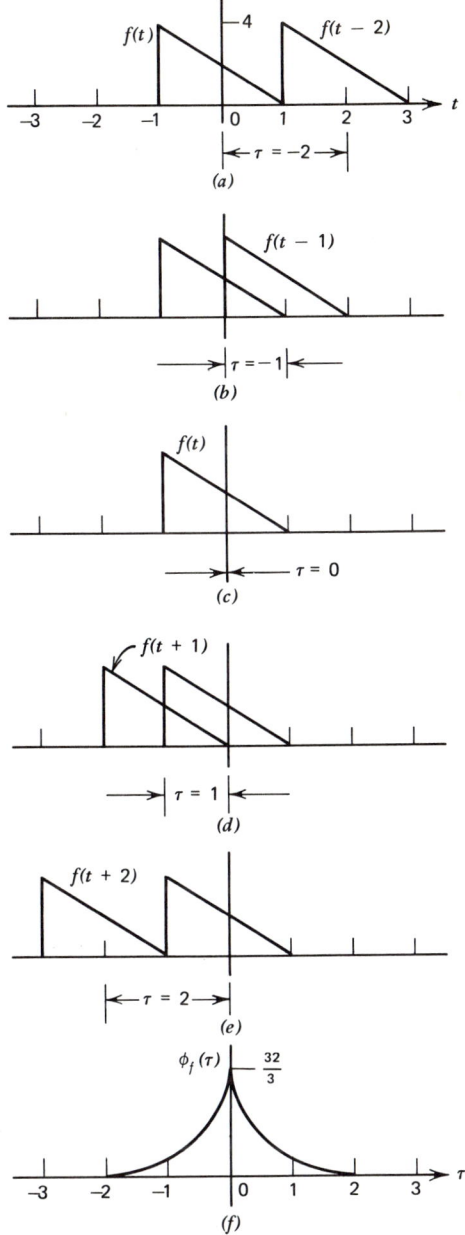

Figure 7-4. Graphical interpretation of the autocorrelation process (Section 7.1.3).

Evaluation of these integrals yields

$$\phi_f(\tau) = \begin{cases} \dfrac{32}{3} + 8\tau - \dfrac{2}{3}\tau^3 & -2 \leq \tau \leq 0 \\ \dfrac{32}{3} - 8\tau + \dfrac{2}{3}\tau^3 & 0 \leq \tau \leq 2 \\ 0 & \text{elsewhere} \end{cases} \quad (7.1\text{-}28)$$

But $\phi_f(\tau)$ is an even function; hence it can be expressed as

$$\phi_f(\tau) = \begin{cases} \frac{32}{3} - 8|\tau| + \frac{2}{3}|\tau|^3 & 0 \le |\tau| \le 2 \\ 0 & \text{elsewhere} \end{cases} \quad (7.1\text{-}29)$$

The energy density spectrum is obtained from (7.1-20), where $F(\omega)$, from (2.1-1), is

$$F(\omega) = 2 \int_{-1}^{1} (1-t)e^{-j\omega t}\, dt = 2 \int_{-1}^{1} (1-t)(\cos \omega t - j \sin \omega t)\, dt$$

$$= 2 \int_{-1}^{1} \cos \omega t\, dt - j2 \int_{-1}^{1} \sin \omega t\, dt - 2 \int_{-1}^{1} t \cos \omega t\, dt + j2 \int_{-1}^{1} t \sin \omega t\, dt$$

(7.1-30)

The second and third integrals are zero because the integrands are odd functions, hence

$$F(\omega) = 4 \int_{0}^{1} \cos \omega t\, dt + j4 \int_{0}^{1} t \sin \omega t\, dt = \frac{4 \sin \omega}{\omega} + j4\left[\frac{\sin \omega}{\omega^2} - \frac{\cos \omega}{\omega}\right] \quad (7.1\text{-}31)$$

Then, from (7.1-20),

$$\Phi_f(\omega) = |F(\omega)|^2 = 16\left[\frac{1}{\omega^2} + \frac{\sin^2 \omega}{\omega^4} - \frac{\sin 2\omega}{\omega^3}\right] \quad (7.1\text{-}32)$$

shown in Fig. 7-5 for $\omega \ge 0$. Since $\Phi_f(\omega)$ is an even function, the response for $\omega < 0$ is not shown.

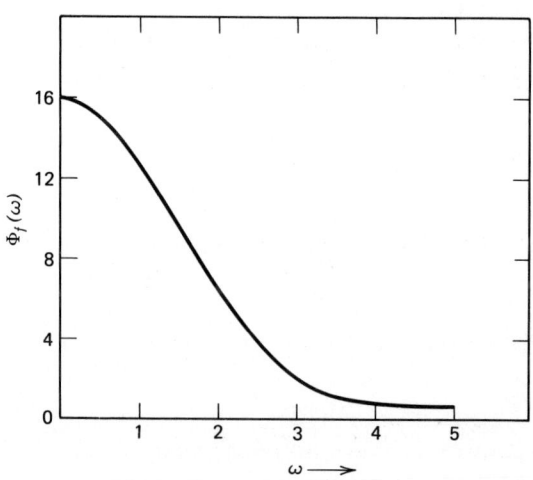

Figure 7-5. Energy density spectrum for the triangular waveform in Fig. 7-3.

7.1.4 Random Function Analysis

For this discussion we consider $f(t)$ to be any member of an ensemble of random functions comprising a random process.† This process is assumed to be ergodic, since many of the important physical processes encountered in scientific work possess this property, and the mathematical difficulty is considerably reduced. Furthermore, ergodicity implies that the process is strictly stationary.

An ergodic process is one for which the time average for any member of the ensemble is equal to the average value of all ensemble members at a given instant of time. The latter is a statistical average. Mathematically, this equivalence is given by

$$\lim_{T \to \infty} \frac{1}{2T} \int_{-T}^{T} f(t) \, dt = \int_{-\infty}^{\infty} x P_\xi(x) \, dx \quad (7.1\text{-}33)$$

where $P_\xi(x)$ is the probability density of the random variable ξ. A strictly stationary process is one whose statistics are not affected by a shift in the time origin, consequently all probability densities are invariant to a time-axis shift.

With the assumption of ergodicity, the autocorrelation function of $f(t)$ is defined as

$$\phi_f(\tau) = \lim_{T \to \infty} \frac{1}{2T} \int_{-T}^{T} f(t) f(t + \tau) \, dt \quad (7.1\text{-}34)$$

It is emphasized that $\phi_f(\tau)$ is the same for every member function of the ensemble; thus it is a characteristic of the ensemble.

The power density spectrum in watts per hertz is given by (7.1-1), and again we stress the importance of its interpretation. Characteristic of density functions, its value at a single frequency is zero, since the frequency increment is zero. The power P in an incremental bandwidth $\Delta \omega$ about ω_0 is

$$P = \frac{\Delta \omega}{2\pi} \Phi_f(\omega_0) \text{ watts} \quad (7.1\text{-}35)$$

Figure 7-6a shows a typical random function $f(t)$ and $f(t + \tau_1)$; the autocorrelation function of $f(t)$ for different values of τ appears in Fig. 7-6b.

The total power in $f(t)$ is the integral of $\Phi_f(\omega)$ over all ω, obtained from (7.1-2) with $\tau = 0$ as

$$\phi_f(0) = \frac{1}{2\pi} \int_{-\infty}^{\infty} \Phi_f(\omega) \, d\omega \quad (7.1\text{-}36)$$

The $\phi_f(0)$ is also the mean square value of $f(t)$, obtained from (7.1-34). This leads to the equality

$$\lim_{T \to \infty} \frac{1}{2T} \int_{-T}^{T} f^2(t) \, dt = \frac{1}{2\pi} \int_{-\infty}^{\infty} \Phi_f(\omega) \, d\omega \quad (7.1\text{-}37)$$

Assuming that $f(t)$ has no periodicities, the value of $\phi_f(\tau)$ at $\tau = \infty$ is the DC power in the waveform.

---

†A random process that is a function of time is sometimes called a stochastic process.

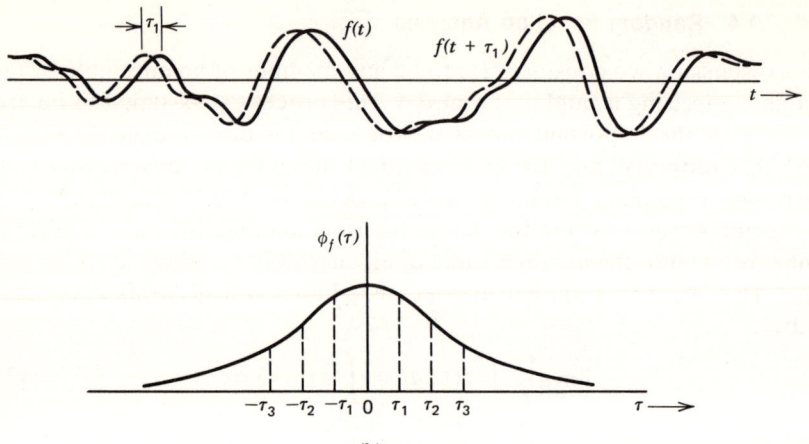

Figure 7-6. (a) Typical random time function $f(t)$ and the same function advanced by τ_1 seconds; (b) the autocorrelation function of $f(t)$.

The random telegraph wave and its autocorrelation function are presented in Fig. 7-7. The total power in the wave is $\phi_f(0) = 0.5$ watt and the DC power in the waveform is $\phi_f(\infty) = 0.25$ watt.

Figure 7-8 illustrates some representative time functions, autocorrelation functions, and power density spectra from each of the three classes discussed. Both $\phi_f(\tau)$ and $\Phi_f(\omega)$ are even functions; hence they are shown only for positive values of their arguments. The time functions in Figs. 7-8a to 7-8e are periodic but can be truncated at any time to yield an aperiodic function. Then $\phi_f(\tau)$ is no longer periodic and $\Phi_f(\omega)$ is no longer discrete. Some random functions that arise in practical situations appear in Figs. 7-8f to 7-8j. Note that wideband Gaussian noise has a constant power density spectrum, and $\phi_f(\tau)$ approaches an impulse function. For this reason it is often used to simulate an impulse function in

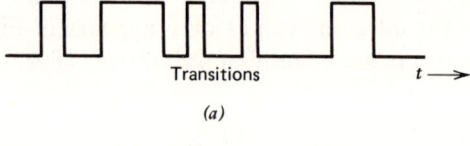

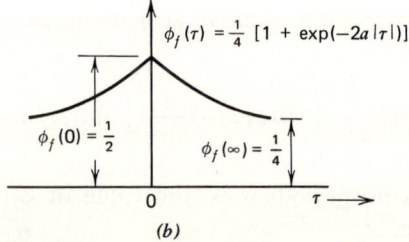

Figure 7-7. (a) Random telegraph wave (Poisson distributed) and (b) its autocorrelation function, where a = average number of transitions per unit time.

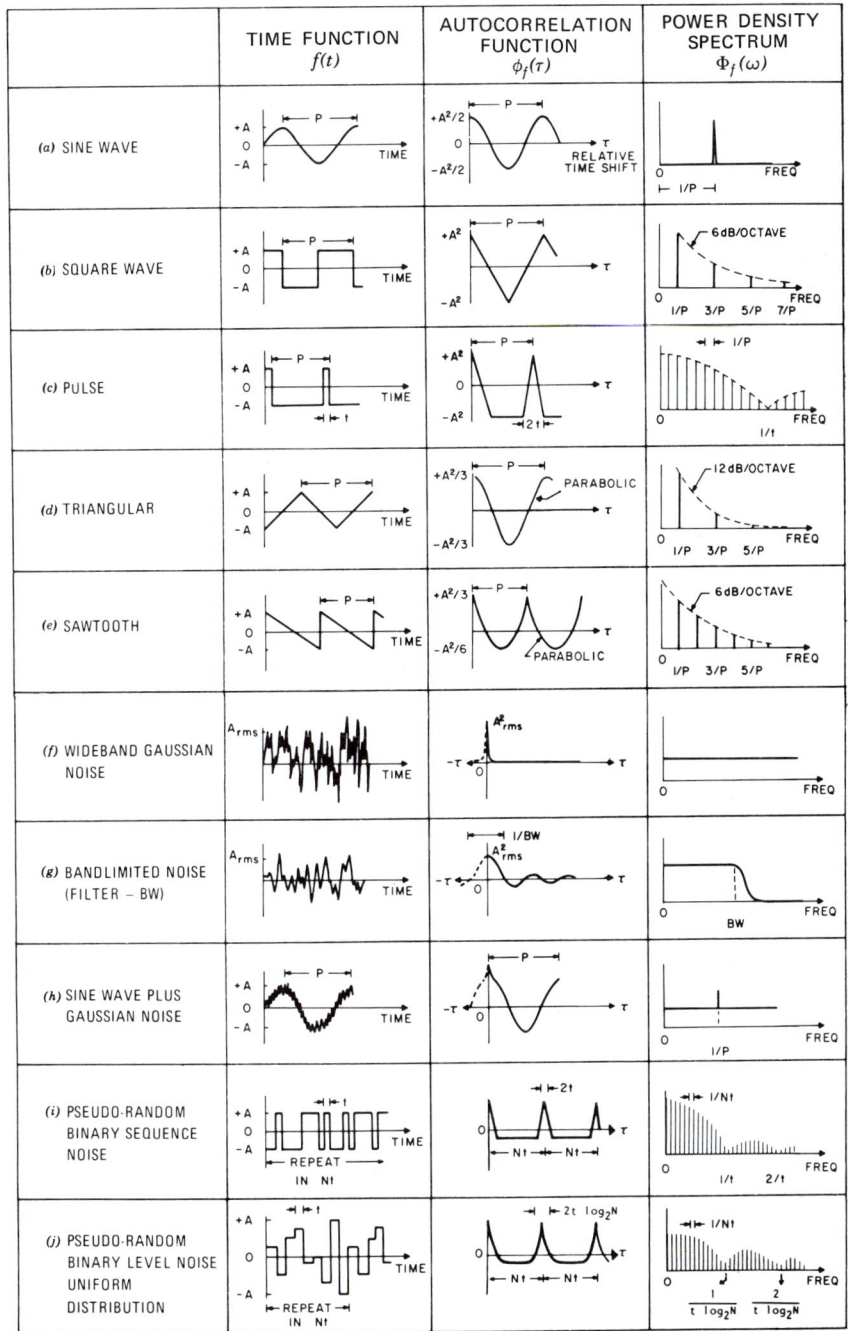

Figure 7-8. Representative time functions, autocorrelation functions, and power density spectra (courtesy of Nicolet Scientific Corporation).

practice. The last three random functions are also periodic; thus $\phi_f(\tau)$ is periodic and $\Phi_f(\omega)$ is discrete.

7.1.5 Properties of the Autocorrelation Function

We now list some interesting and useful properties of the autocorrelation function, which are applicable to the three classes of functions just discussed [7, 9].

1. If $\phi_f(\tau)$ is continuous at $\tau = 0$, it is continuous everywhere.
2. The phase function of periodic and aperiodic functions disappears in the autocorrelation function. Hence $\phi_f(\tau)$ is not unique, and many functions may have the same autocorrelation function. This property is useful in analyzing random functions because they possess no phase relationships. For this reason $\phi_f(\tau)$ is the important characteristic in the harmonic analysis of random functions.
3. The function $\phi_f(\tau)$ is even; hence it is symmetric about $\tau = 0$, that is,

$$\phi_f(-\tau) = \phi_f(\tau) \tag{7.1-38}$$

4. The maximum value of $\phi_f(\tau)$ occurs at $\tau = 0$, that is

$$\phi_f(\tau) \leq \phi_f(0) \tag{7.1-39}$$

5. The value of $\phi_f(\tau)$ at $\tau = 0$ has a different meaning depending on the class of functions. This is summarized in Table 7-1, along with the units for $\Phi_f(\omega)$.
6. The Fourier transform of the autocorrelation function is $\Phi_f(\omega)$. Conversely, the inverse transform of $\Phi_f(\omega)$ is the autocorrelation function.

Table 7-1 Summary of $\phi_f(0)$ and Units of $\Phi_f(\omega)$ for the Three Function Classes

Function	$\phi_f(0)$	Units of $\Phi_f(\omega)$
Periodic	$\dfrac{1}{T_1} \displaystyle\int_{-T_1/2}^{T_1/2} f^2(t)\, dt$ = average power	watts/hertz
Aperiodic	$\displaystyle\int_{-\infty}^{\infty} f^2(t)\, dt$ = total energy	joules/hertz
Random	$\displaystyle\lim_{T\to\infty} \dfrac{1}{2T} \int_{-T}^{T} f^2(t)\, dt$ = average power	watts/hertz

7.1.6 Cross-Correlation Function

The cross-correlation function plays an important role in signal processing, for this function is used to distinguish incoming signals and to measure the signal parameters. In Section 7.3.7 we take up the relationship between the matched filter and the cross-correlator.

Autocorrelation, Cross-Correlation, Power Density

The cross-correlation function between two different functions $f(t)$ and $g(t)$ is designated $\phi_{fg}(\tau)$. Table 7-2 summarizes the definitions of $\phi_{fg}(\tau)$ when $f(t)$ and $g(t)$ are periodic, aperiodic, or random functions. The function $\phi_{fg}(\tau)$ is a quantitative measure of the similarity between $f(t)$ and $g(t)$. If $f(t) = g(t)$, the cross-correlation function reduces to the autocorrelation function.

Table 7-2 Cross-Correlation Definitions for the Three Function Classes

Function	$\phi_{fg}(\tau)$	Notes
Periodic	$\dfrac{1}{T_1} \displaystyle\int_{-T_1/2}^{T_1/2} f(t)g(t+\tau)\,dt$	$f(t)$ and $g(t)$ have the same fundamental frequency
Aperiodic	$\displaystyle\int_{-\infty}^{\infty} f(t)g(t+\tau)\,dt$	
Random	$\displaystyle\lim_{T\to\infty} \dfrac{1}{2T} \int_{-T}^{T} f(t)g(t+\tau)\,dt$	$f(t)$ is a member of one ensemble and $g(t)$ is the corresponding member of another ensemble

The order of performing cross-correlation results in different functions. The correlation between $f(t)$ and $g(t)$ is related to the correlation between $g(t)$ and $f(t)$ by the simple formula

$$\phi_{fg}(\tau) = \phi_{gf}(-\tau) \tag{7.1-40}$$

Thus $\phi_{fg}(\tau)$ is the mirror image of $\phi_{gf}(\tau)$ with respect to the ordinate at $\tau = 0$.

The mechanics of cross-correlation and of autocorrelation are similar except that two different functions are involved. Figure 7-9 illustrates this process for two periodic waveforms. Figure 7-9c shows the relative positions when $g(t)$ is shifted by $\tau = \tau_1$, and Fig. 7-9d gives the periodic cross-correlation function.

7.1.7 Relationship Between Convolution and Cross-Correlation

Notice that the aperiodic cross-correlation function in Table 7-2 resembles the general convolution integral of (1.5-1), namely,

$$p * q = \int_{-\infty}^{\infty} p(t-\tau)q(\tau)\,d\tau \tag{7.1-41}$$

This similarity allows both functions to be obtained graphically by the same displacement, multiplication, and integration operations. However convolution differs from cross-correlation in that the displaced function $p(\tau)$ is reflected about the point $\tau = t$, whereas this reflection is not required for cross-correlation. The reason for this reflection stems from the physical considerations discussed in Section 1.8.

Because the reflection operation is necessary in convolution, it is clear that the correlation of two functions, one of which has first been folded back, is in fact a

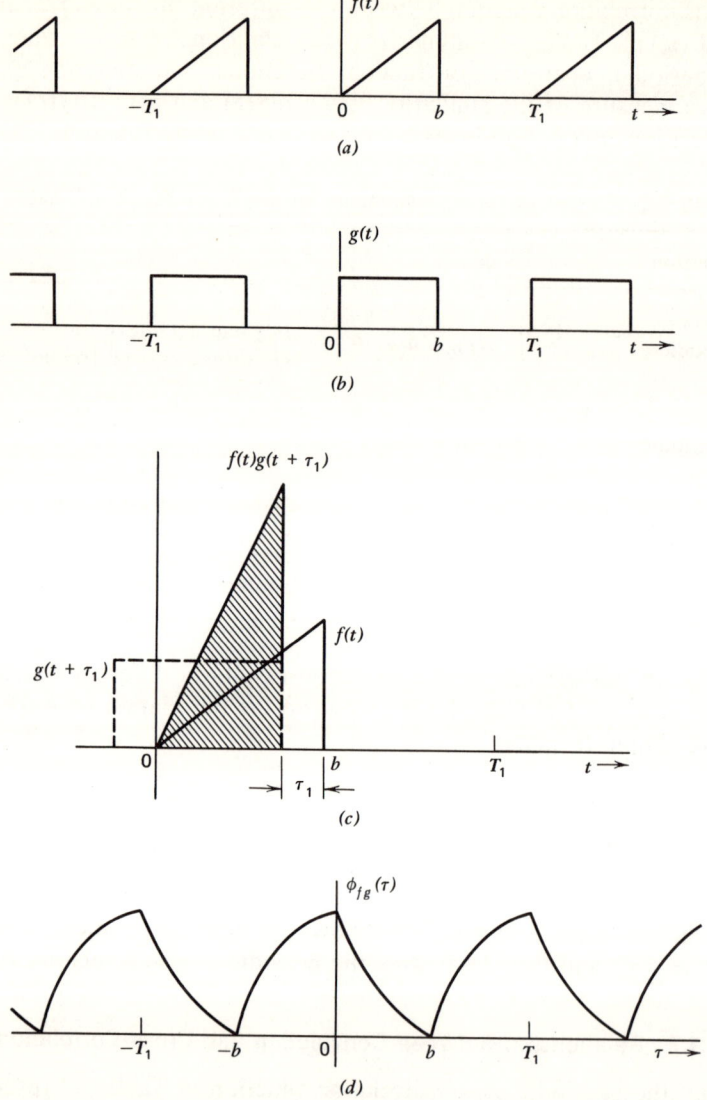

Figure 7-9. Cross-correlation process for periodic triangular and rectangular pulses. (*a*) Periodic triangular pulses, (*b*) periodic rectangular pulses, (*c*) cross-correlation for $\tau = \tau_1$ [shaded area is $\phi_{fg}(\tau_1)$], (*d*) cross-correlation function of $f(t)$ and $g(t)$.

convolution, and conversely, the convolution of two functions, one of which takes the folded form prior to convolution, is actually a correlation. For example, the convolution of $p(t)$ and $q(t)$ in (7.1-41) can be rewritten, with $x = -\tau$, as

$$p * q = \int_{-\infty}^{\infty} p(x+t)q(-x)\,dx \tag{7.1-42}$$

Now substitute t for x ($dt = dx$) and τ for t and compare $p * q$ with $\phi_{fg}(\tau)$ for aperiodic functions (Table 7-2). Thus the convolution of $q(t)$ and $p(t)$ is the

Autocorrelation, Cross-Correlation, Power Density

cross-correlation between $q(-t)$ and $p(t)$. Similarly, the cross-correlation between $f(t)$ and $g(t)$, with $x = -t$, is

$$\phi_{fg}(\tau) = \int_{-\infty}^{\infty} f(-x)g(\tau - x)\, dx \qquad (7.1\text{-}43)$$

Now substitute τ for x ($d\tau = dx$) and t for τ and compare with (7.1-41). The cross-correlation between $f(t)$ and $g(t)$ is the convolution of $f(-t)$ and $g(t)$.

7.1.8 Output Autocorrelation Function of a Linear System

If $\phi_f(\tau)$ is the autocorrelation function of the input signal, the output autocorrelation function $\phi_u(\tau)$ of a linear time-invariant system with transfer function $H(s)$ and real impulse response $h(t)$ is the double convolution [9]

$$\phi_u(\tau) = \phi_f(\tau) * h(-\tau) * h(\tau) \qquad (7.1\text{-}44)$$

Furthermore the cross-correlation function between the input and output function is

$$\phi_{fu}(\tau) = \phi_f(\tau) * h(-\tau) \qquad (7.1\text{-}45)$$

thus (7.1-44) can be expressed as

$$\phi_u(\tau) = \phi_{fu}(\tau) * h(\tau) = \int_{-\infty}^{\infty} h(\tau - x)\phi_{fu}(x)\, dx \qquad (7.1\text{-}46)$$

A system interpretation is presented in Fig. 7-10. In Fig. 7-10a $\phi_{fu}(\tau)$ is the response of a system with impulse response $h(-\tau)$ excited by $\phi_f(\tau)$. Then $\phi_{fu}(\tau)$ excites a system with impulse response $h(\tau)$, yielding the output $\phi_u(\tau)$. Alternatively, as in Fig. 7-10b, $\phi_u(\tau)$ can be considered to be the response of a system with impulse response $h(-\tau) * h(\tau)$ excited by the input $\phi_f(\tau)$.

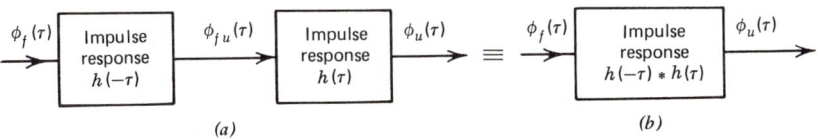

Figure 7-10. System interpretation of (7.1-44) to (7.1-46).

For the special case when the input is white noise of constant power density N_0 (Fig. 7-8f), its autocorrelation function is $\phi_f(\tau) = N_0 \delta(\tau)$. Then, from (7.1-45),

$$\phi_{fu}(\tau) = \int_{-\infty}^{\infty} h(-x)\phi_f(\tau - x)\, dx = \int_{-\infty}^{\infty} h(-x) N_0 \delta(\tau - x)\, dx = N_0 h(-\tau) \qquad (7.1\text{-}47)$$

and from (7.1-46)

$$\phi_u(\tau) = N_0 \int_{-\infty}^{\infty} h(\tau - x) h(-x)\, dx = N_0 \int_{-\infty}^{\infty} h(x) h(x - \tau)\, dx = N_0 \phi_h(-\tau) \qquad (7.1\text{-}48)$$

But from (7.1-38) $\phi_h(\tau)$ is symmetric about $\tau = 0$; thus $\phi_u(\tau) = N_0 \phi_h(\tau)$. For a white noise input, therefore, the output autocorrelation function is the autocorrelation function of the system impulse response scaled by the power density N_0.

7.1.9 Output Power Density Spectrum of a Linear System

The Fourier transform relationship between $\phi(\tau)$ and $\Phi(\omega)$ allows us to obtain the output power density spectrum from (7.1-44) as

$$\Phi_u(\omega) = \Phi_f(\omega) H^*(j\omega) H(j\omega) = |H(j\omega)|^2 \Phi_f(\omega) \tag{7.1-49}$$

remembering that convolution in the time domain corresponds to multiplication in the frequency domain, and $H^*(j\omega)$ is the Fourier transform of $h(-t)$. Thus the output power density spectrum $\Phi_u(\omega)$ is the product of the input power density spectrum $\Phi_f(\omega)$ and $|H(j\omega)|^2$.

Example 7-2. Consider an LP RC filter with transfer function

$$H(s) = \frac{\omega_c}{s + \omega_c} \tag{7.1-50}$$

and impulse response

$$h(t) = \omega_c e^{-\omega_c t} u_{-1}(t) \tag{7.1-51}$$

excited by white noise of power density spectrum N_0 watts per hertz. Then

$$|H(j\omega)|^2 = \frac{1}{1 + (\omega/\omega_c)^2} \tag{7.1-52}$$

and from (7.1-49), the output power density spectrum is

$$\Phi_u(\omega) = \frac{N_0}{1 + (\omega/\omega_c)^2} \tag{7.1-53}$$

shown in Fig. 7-11a. The output autocorrelation function is obtained from (7.1-48) and (7.1-51). The relative positions of $h(x)$ and $h(x - \tau)$ for $\tau < 0$ and $\tau > 0$ appear

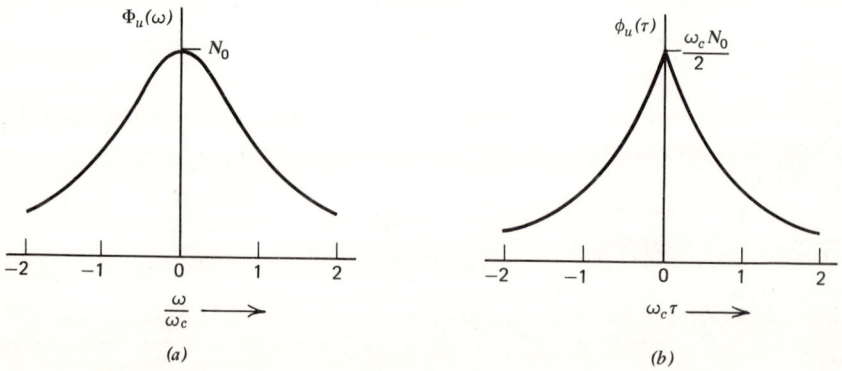

Figure 7-11. (a) Power density spectrum and (b) autocorrelation function at RC filter output for white noise input.

Linear Mean-Square Estimation

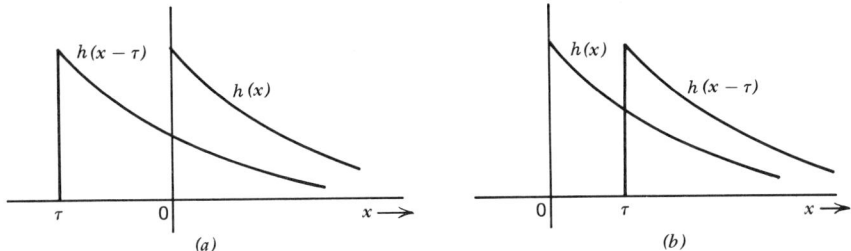

Figure 7-12. Relative positions of impulse responses for autocorrelation of Example 7-2. (a) $\tau < 0$, (b) $\tau > 0$.

in Fig. 7-12. Then, with the designated limits,

$$\phi_u(\tau) = \begin{cases} \omega_c^2 N_0 \int_\tau^\infty e^{-\omega_c x} e^{-\omega_c(x-\tau)} \, dx = \dfrac{\omega_c N_0}{2} e^{-\omega_c \tau} & \tau \geq 0 \\ \omega_c^2 N_0 \int_0^\infty e^{-\omega_c x} e^{-\omega_c(x-\tau)} \, dx = \dfrac{\omega_c N_0}{2} e^{\omega_c \tau} & \tau \leq 0 \end{cases}$$

$$= \frac{\omega_c N_0}{2} e^{-\omega_c |\tau|} \tag{7.1-54}$$

shown in Fig. 7-11b.

7.2 LINEAR MEAN-SQUARE ESTIMATION

An important problem arising in many areas of physical analysis is the estimation $\hat{x}(t)$ of a stochastic process $x(t)$ from the available data, given by the stochastic process $z(t)$. We can think of $\hat{x}(t)$ as being the estimate of a signal, or a function of the signal, and $z(t)$ as the signal plus noise. The general analytic solution to this problem has not yet been found, but significant results have been obtained for specified signal and noise models, and filtering operations on the input data [1, 13].

There are two major milestones in this area. The first is the initial work of Kolmogoroff (1941) and Wiener (1942), which laid the foundation for the subsequent contributions of Kalman (1960), and Kalman and Bucy (1961). Each assumed that the filter was linear and used the mean-squared error as the measure of filter performance. Consequently the resulting procedure is known as linear mean-square estimation.

7.2.1 Wiener–Kolmogoroff Filtering

Kolmogoroff and Wiener considered the filter to be time-invariant in addition to being linear, and they independently solved the prediction problem [12]. Wiener considered $x(t)$ and $z(t)$ to be continuous-time functions, hence his scheme is known as continuous-time filtering, whereas Kolmogoroff considered the data to be discrete-time values, and his scheme is referred to as discrete-time filtering.

Wiener also tried to estimate functions of $x(t)$ in a noisy environment for which prediction is a special case. The applications of the Wiener–Kolmogoroff theory were few, however, because the problem of practical synthesis was not solved. Traditionally Wiener's approach, rather than Kolmogoroff's, is used to illustrate mean-square filtering. Accordingly we now examine Wiener filtering in more detail.

In the linear filter problem we are given a message signal that is corrupted by noise, and it is desired to extract a function of this signal by a linear system, with minimum error in accordance with a chosen performance measure. The desired output signal may be the input signal, the input delayed, the input advanced (prediction), or some linear transformation of the input.

The presence of noise does not allow complete recovery of the desired output unless the situation is exceptional, such as nonoverlapping input signal and noise spectra. However, we can consider the ideal output. Therefore we let $u(t)$ be the desired output signal, $v(t)$ be the actual output signal, and the instantaneous error be

$$e(t) = v(t) - u(t) \tag{7.2-1}$$

For a measure of system performance, Wiener [7] chose the mean-square error

$$\varepsilon = \lim_{T \to \infty} \frac{1}{2T} \int_{-T}^{T} e^2(t)\, dt \tag{7.2-2}$$

which results in comparatively simple mathematics and furthermore provides a meaningful measure for many practical problems. The linear time-invariant system that results from the minimization of ε is referred to as the Wiener filter, although many references call it the optimum linear filter. The design problem is to determine $h(t)$, the filter impulse response, or equivalently $H(s)$, the filter transfer function.

The input signal $f(t)$ is in general some function of a message signal $m(t)$ and a noise signal $n(t)$, but it is commonly assumed that $f(t)$ is the sum of the message and noise as

$$f(t) = m(t) + n(t) \tag{7.2-3}$$

Both $m(t)$ and $n(t)$ are assumed to be stationary random processes. The impulse response that minimizes ε in (7.2-2) is determined from the integral equation [7]

$$\phi_{fu}(\tau) = \int_{-\infty}^{\infty} h_{\text{opt}}(x) \phi_f(\tau - x)\, dx \quad \text{for} \quad \tau \geq 0 \tag{7.2-4}$$

where

$\phi_{fu}(\tau)$ = cross-correlation function between $f(t)$ and $u(t)$
$\phi_f(\tau)$ = autocorrelation function of $f(t)$
$h_{\text{opt}}(x)$ = impulse response that satisfies (7.2-4)

It is emphasized that (7.2-4) holds only for $\tau \geq 0$. This condition is a necessary one.

Equation 7.2-4, known as the Wiener–Hopf equation, arose from earlier unrelated work by Wiener and Hopf. The optimum impulse response cannot be explicitly determined except in certain cases; consequently special methods are necessary for finding its transform [7].

Qualitatively, the Wiener filter, can be viewed as follows in the frequency domain. Suppose that the desired output is $m(t)$. If there is no noise, the optimum filter has constant magnitude and linear phase responses over the frequency spectrum of $m(t)$. If noise is present, the filter bandwidth is decreased to reduce the filter output noise, but at the same time the output signal is also degraded. Therefore, there exists some bandwidth beyond which the output signal is degraded more than the output noise is reduced. This bandwidth is then the optimum value.

Note that the mean-square-error filter is insensitive to the phase of the input signal, since $h_{\text{opt}}(x)$ in (7.2-4) is a function of the correlation functions. We know from Section 7.1 that the phase information is lost when the correlation functions are computed; therefore many different functions have the same correlation function. We should begin suspecting that another approach that distinguishes different phases may be more advantageous in certain situations. One such criterion is the maximization of the peak signal-to-mean-noise power ratio, which is useful for signal detection. The optimum filter for this criterion is the famous matched filter, discussed in Section 7.3.

7.2.2 Kalman Filtering

Many contributions followed the initial discovery of Wiener and Kolmogoroff, but a major advance was made by Kalman [4], who introduced recursive mean-square discrete-time filtering, and Kalman and Bucy [5], who provided the explicit synthesis of the continuous version of the mean-square filter equations. As used here, the term "recursive" implies an implementation that uses a linear combination of previous estimates and the present data.†

Kalman changed the standard formulation of the estimation problem by postulating a state-space model for the signal and noise rather than the conventional signal and noise correlation functions. He showed that the best estimate of the signal process satisfies a linear differential equation driven by the observed process $z(t)$. Furthermore, the error covariance matrix of the optimal estimate satisfies a matrix Riccati equation, which is relatively easy to solve on a digital or analog computer. This technique replaces the more computationally difficult problem of solving the Wiener–Hopf integral equation.

The Kalman filter is the linear, possibly time-varying discrete-time filter that provides a least mean-square error estimate of a discrete-time signal based on noisy observations. It is not normally characterized by an explicit expression for its impulse response but is given as an adaptive (time-varying) algorithm suitable for direct evaluation by a digital computer. The estimate of the process state is weighted between previous estimates and the present data. The essential attributes of the Kalman filter are that it is derived from a time-domain statistical criterion and that it is, in general, time-varying.

Kalman's recursive solution provides an estimate that is equivalent to the estimate obtained by batch-processing the data but offers the advantage of

---

†In Chapter 9 we discuss recursive digital filters, so named because the present output sample is given as a weighted sum of past output samples as well as past and present input samples.

reducing the data-handling requirements [12]. The Kalman filter therefore can be considered to be an efficient computational solution of the least-squares problem. If the Kalman filter is restricted to be time-invariant, it becomes the Wiener filter.

7.2.3 Summary

The Wiener–Kolmogoroff filter theory is the cornerstone of least-squares signal estimation in the presence of noise, providing the foundation for the subsequent development of the Kalman filter theory. The Kalman filter, because of its adaptability for computer simulation and the usefulness of the state-space approach, has had a dramatic impact on linear estimation. Its applications include satellite-orbit determination, submarine detection, and fire control. Succeeding studies have extensively developed linear estimation techniques, and today's important research is in the area of nonlinear filtering [1].

In addition to introducing the important area of mean-square filtering, we have presented the idea that a computer program algorithm can be considered a filter, and this is precisely what the Kalman filter is. In Chapter 9 we find that the algorithm for a digital filter response can likewise be implemented on a digital computer and furthermore can be realized as a special-purpose computer.

7.3 MATCHED FILTERING

One of the most important discoveries arising from World War II investigations of radar systems is the matched filter, sometimes called the North filter, after its discoverer D. O. North [8]. The matched filter is the optimum filter when the signal is corrupted by additive white Gaussian noise under a wide variety of criteria, namely, the signal-to-noise criterion, the likelihood ratio criterion, and the inverse probability criterion [2]. All three yield the same filter response. Here, however, we consider the matched filter with respect to the signal-to-noise criterion only.

The matched filter is essentially a decoder; that is, it recognizes a specified signal shape in the presence of noise (white or colored) and accordingly yields a higher output peak signal-to-mean-noise power ratio (SNR) for this signal shape than for any other signal shape with the same energy. If the noise is not Gaussian, the matched filter is the optimum linear filter. If the noise is also Gaussian, this filter is the optimum of all time-invariant filters, linear or nonlinear, and provides an optimum solution to the signal detection problem. Unlike the Wiener filter, the matched filter makes full use of the signal phase information and uses this characteristic to aid in distinguishing signal shapes.

We first discuss the matched filter when the noise power density spectrum is constant (white). This filter's impulse response $h_0(t)$ is the surprisingly simple expression

$$h_0(t) = Kf(t_1 - t) \qquad (7.3\text{-}1)$$

where $f(t)$ is the input signal and t_1 is a constant delay. For this case, $h_0(t)$ has the same shape as the signal but reversed in time, and is delayed by t_1 for physical realizability. We then say that the filter is *matched* to the input signal, hence the name.

Matched Filtering

The important properties of this matched filter include its spectral characteristics, its synthesis, its relationship to cross-correlators, and its sensitivity to a change in the input signal shape or a change in its own impulse response. Finally, this development can be used to determine the matched filter when the noise spectrum is nonconstant (colored).

7.3.1 Matched Filter Derivation

We assume that the white noise that accompanies the input signal is stationary, but it need not have a Gaussian distribution. Its power density spectrum is the constant value N_0, the input noise power per unit bandwidth. Since in practice the noise spectrum is seldom known precisely, it is commonly assumed to be white.

We now derive the filter that maximizes

$$\text{SNR} = \frac{|u(t)|^2_{\max}}{\overline{u_n^2}} \qquad (7.3\text{-}2)$$

where $u(t)$ is the voltage across a one-ohm load resistor and $\overline{u_n^2}$ is the total mean output noise power dissipated in this same resistor (Fig. 7-13). The input message

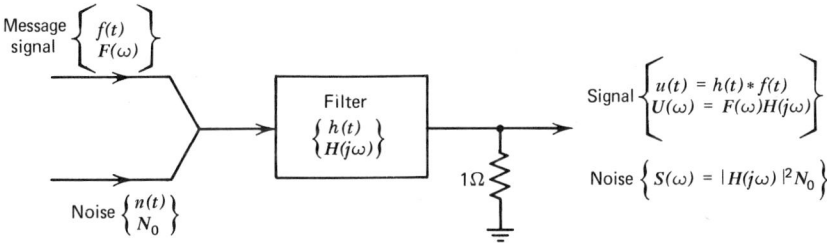

Figure 7-13. Input–output relationship of filter with input signal $f(t)$ plus white noise $n(t)$.

signal of finite-time duration is $f(t)$ with known Fourier transform $F(\omega)$, and the input noise is $n(t)$ with power density spectrum N_0 watts per hertz. The filter is characterized by the impulse response $h(t)$ and transfer function $H(j\omega)$; the output signal is $u(t)$. The derivation can be performed in either the time or the frequency domain. We choose the former in hope that it will provide more insight into the physical process.

The autocorrelation function of the white noise at the filter output, from (7.1-48), is

$$\phi_n(\tau) = \frac{N_0}{2} \int_{-\infty}^{\infty} h(t)h(t+\tau)\,dt \qquad (7.3\text{-}3)$$

The factor of one-half is introduced because N_0 is usually defined from $\omega = 0$ to $\omega = \infty$; thus over the infinite frequency interval ($-\infty \leq \omega \leq \infty$), the average value is $N_0/2$. The total mean output noise power, from (7.1-36), is the value of $\phi_n(\tau)$ at $\tau = 0$,

$$\overline{u_n^2} = \phi_n(0) = \frac{N_0}{2} \int_{-\infty}^{\infty} h^2(t)\,dt \qquad (7.3\text{-}4)$$

From the convolution integral

$$|u(t)|^2 = \left| \int_{-\infty}^{\infty} f(t-\tau) h(\tau) \, d\tau \right|^2 \tag{7.3-5}$$

Since $h(t)$ is causal, the integral upper limit is changed from t to ∞ without changing the value of the integral. Substitute from (7.3-4) and (7.3-5) into (7.3-2) and assume that the maximum value of $|u(t)|^2$ occurs at $t = t_1$. Then

$$\text{SNR} = \frac{\left| \int_{-\infty}^{\infty} f(t_1 - \tau) h(\tau) \, d\tau \right|^2}{\frac{N_0}{2} \int_{-\infty}^{\infty} h^2(t) \, dt} \tag{7.3-6}$$

We now digress to derive a necessary relationship known as the Schwarz inequality. Let P and Q be two complex time functions and λ be a constant. Then, because the integral of a positive function is never negative,

$$\int_{-\infty}^{\infty} |P - \lambda Q|^2 \, dt = \int_{-\infty}^{\infty} (P - \lambda Q)(P^* - \lambda^* Q^*) \, dt$$

$$= \int_{-\infty}^{\infty} |P|^2 \, dt - \lambda \int_{-\infty}^{\infty} QP^* \, dt - \lambda^* \int_{-\infty}^{\infty} PQ^* \, dt + |\lambda|^2 \int_{-\infty}^{\infty} |Q|^2 \, dt \geq 0 \tag{7.3-7}$$

where P^* is the complex conjugate of P and Q^* is the complex conjugate of Q.

Let

$$\lambda = \frac{\int_{-\infty}^{\infty} Q^* P \, dt}{\int_{-\infty}^{\infty} |Q|^2 \, dt} \tag{7.3-8}$$

Then

$$\int_{-\infty}^{\infty} |P|^2 \, dt - \frac{\int_{-\infty}^{\infty} Q^* P \, dt \int_{-\infty}^{\infty} QP^* \, dt}{\int_{-\infty}^{\infty} |Q|^2 \, dt} - \frac{\int_{-\infty}^{\infty} QP^* \, dt \int_{-\infty}^{\infty} PQ^* \, dt}{\int_{-\infty}^{\infty} |Q|^2 \, dt} + \frac{\int_{-\infty}^{\infty} Q^* P \, dt \int_{-\infty}^{\infty} P^* Q \, dt}{\int_{-\infty}^{\infty} |Q|^2 \, dt} \geq 0 \tag{7.3-9}$$

but

$$\int_{-\infty}^{\infty} QP^* \, dt \int_{-\infty}^{\infty} PQ^* \, dt = \left| \int_{-\infty}^{\infty} Q^* P \, dt \right|^2 \tag{7.3-10}$$

Matched Filtering

Therefore

$$\int_{-\infty}^{\infty} |P|^2 \, dt \int_{-\infty}^{\infty} |Q|^2 \, dt \geq \left| \int_{-\infty}^{\infty} Q^*P \, dt \right|^2 \qquad (7.3\text{-}11)$$

the Schwarz inequality. The equality sign applies when

$$P = GQ \quad (G \text{ is a constant}) \qquad (7.3\text{-}12)$$

We now apply the Schwarz inequality to the numerator of (7.3-6) with the identification

$$P = h(\tau) \quad Q^* = f(t_1 - \tau) \qquad (7.3\text{-}13)$$

Then the upper bound on the SNR is

$$\text{SNR} \leq \frac{2}{N_0} \frac{\int_{-\infty}^{\infty} h^2(\tau) \, d\tau \int_{-\infty}^{\infty} f^2(t_1 - \tau) \, d\tau}{\int_{-\infty}^{\infty} h^2(t) \, dt} = \frac{2}{N_0} \int_{-\infty}^{\infty} f^2(t_1 - \tau) \, d\tau \qquad (7.3\text{-}14)$$

But, by Parseval's theorem, the input signal energy is

$$E = \int_{-\infty}^{\infty} f^2(t) \, dt \qquad (7.3\text{-}15)$$

Thus (7.3-14) reduces to

$$\text{SNR} \leq \frac{2E}{N_0} \qquad (7.3\text{-}16)$$

The maximum value of the SNR, designated $(\text{SNR})_M$, is

$$(\text{SNR})_M = \frac{2E}{N_0} \qquad (7.3\text{-}17)$$

and is attained when (7.3-12) is satisfied. Then, because $f(t)$ is real in (7.3-13), the optimum impulse response is

$$h_0(t) = Gf(t_1 - t) \qquad (7.3\text{-}18)$$

where G is the filter gain constant.

The optimum impulse response $h_0(t)$ has the remarkable property of possessing the same shape as the input signal, but it is reversed in time and delayed by t_1 to assure physical realizability. This impulse response is therefore the mirror image of the signal. In this manner the shape of the input signal is "built" into the filter, thus allowing the filter to be shape selective.

Consider the input signal in Fig. 7-14a in the presence of white noise. The matched filter impulse response (Fig. 7-14b), is obtained according to (7.3-18). The function $f(t)$ is reversed in time and delayed by T ($t_1 = T$) seconds; thus no response exists prior to $t = 0$, the time at which the impulse is applied. If the delay T is not introduced, the time reversal creates an impulse response that exists only for $t \leq 0$, hence is nonrealizable.

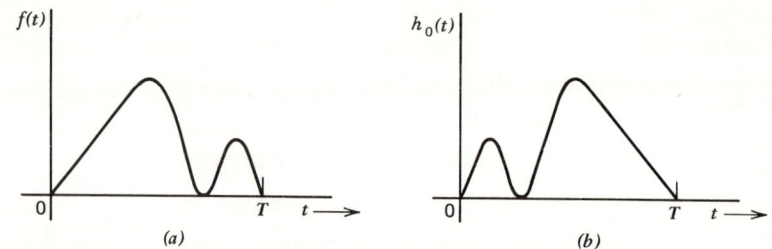

Figure 7-14. (a) Signal $f(t)$ and (b) matched filter impulse response $h_0(t) = f(T-t)$.

For matched filter applications, the signal is usually a repetitive pulse of duration T. Thus the minimum introduced delay for a realizable filter is this signal time T, as illustrated in Fig. 7-14. Values of delay $t_1 > T$ are acceptable, but the output signal is then delayed unnecessarily. Furthermore the filter design is simplified if $t_1 = T$.

If the signal is symmetric about its center, $h_0(t)$ has the same shape as the signal, since the mirror image of a symmetric function is that same function.

Two interesting features of the matched filter deserve special mention. First, the filter design is independent of the white noise level N_0. Once the white noise assumption is made, the filter impulse response is given by (7.3-18). Second, the maximum SNR in (7.3-17) is independent of the signal shape. It is a function of only the signal energy and noise power per bandwidth.

7.3.2 Filtering the Rectangular Pulse

A basic pulse in radar applications is the rectangular pulse in Fig. 7-15a. Because the pulse is symmetric about $T/2$, the matched filter impulse response is identical to

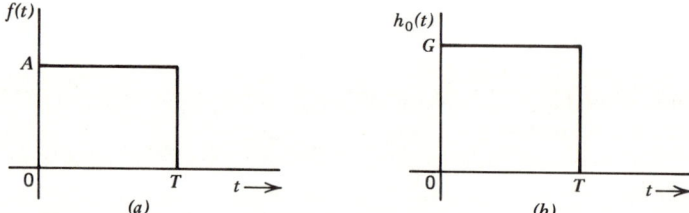

Figure 7-15. (a) Rectangular pulse $f(t)$ and (b) matched filter impulse response $h_0(t) = Gf(T-t)$.

$f(t)$, in Fig. 7-15b. The filter output is given by the convolution integral

$$u(t) = \int_{-\infty}^{t} h_0(t-\tau)f(\tau)\,d\tau \qquad (7.3\text{-}19)$$

The geometric interpretation of the convolution integral in Section 1.5 establishes two possibilities for a nonzero output (Fig. 7-16). For $t < 0$ and $t > 2T$, the output is zero. The shaded area is the overlap region of $f(\tau)$ and $h_0(t-\tau)$. Then the filter

Matched Filtering

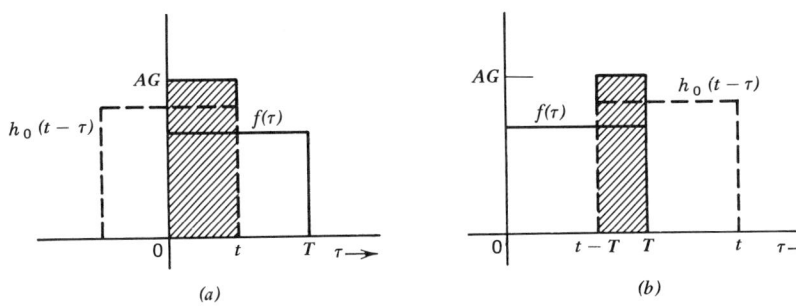

Figure 7-16. Relative positions of $f(\tau)$ and $h_0(t-\tau)$ for non zero output. (a) $0 \leq t \leq T$, (b) $T \leq t \leq 2T$.

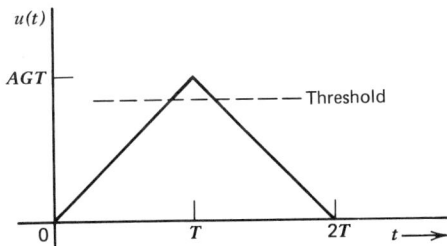

Figure 7-17. Output of filter matched to rectangular pulse of Fig. 7-15a.

output (Fig. 7-17) is

$$u(t) = \begin{cases} AG \int_0^t dt = AGt & 0 \leq t \leq T \\ AG \int_{t-T}^T dt = AG(2T-t) & T \leq t \leq 2T \\ 0 & \text{elsewhere} \end{cases} \quad (7.3\text{-}20)$$

The maximum output occurs at the delay time T, which indicates the completion of the arrival of the input signal. Thus we not only know that the signal is present, we know the exact time of its occurrence. Figure 7-16 shows that at $t = T$, $f(\tau)$ and $h_0(t-\tau)$ are perfectly overlapped, hence the output is maximum. This phenomenon is characteristic of matched filters and is further discussed in Section 7.3.7. Note also that the output is proportional to T. Therefore an increased signal energy yields a larger output.

This analysis has exposed an important attribute of the matched filter; namely, the signal arrival time is unambiguously determined. The peak output is then used as a trigger to indicate the presence of the specific signal. For example, a threshold voltage may be established, such as the dashed line in Fig. 7-17. Voltages above this level indicate the presence of a signal and consequently alert other circuits to this fact.

Suppose that the pulse of Fig. 7-15a has a longer duration, say $3T$, but the same amplitude A. We now examine the effect on the filter response for the filter matched to the pulse of duration T. The geometric picture of the convolution

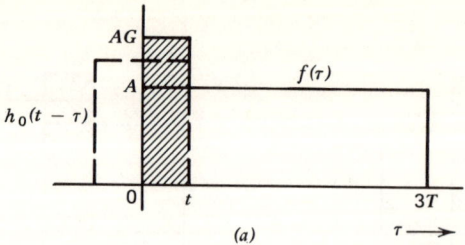

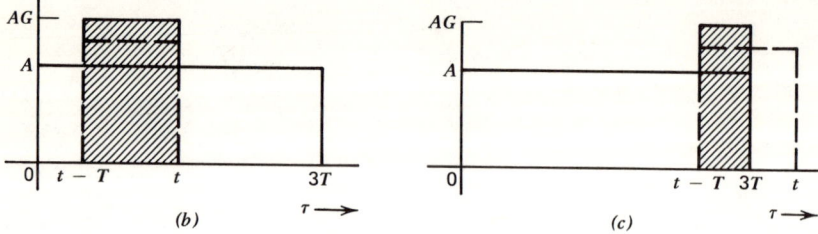

Figure 7-18. Relative positions of $f(\tau)$ and $h_0(t-\tau)$ for nonzero output. (a) $0 \leq t \leq T$, (b) $T \leq t \leq 3T$, (c) $3T \leq t \leq 4T$.

integral now yields three possibilities for a nonzero output, as illustrated in Fig. 7-18. Then the filter output (Fig. 7-19) is

$$u(t) = \begin{cases} AG \int_0^t dt = AGt & 0 \leq t \leq T \\ AG \int_{t-T}^t dt = AGT & T \leq t \leq 3T \\ AG \int_{t-T}^{3T} dt = AG(4T-t) & 3T \leq t \leq 4T \\ 0 & \text{elsewhere} \end{cases} \quad (7.3\text{-}21)$$

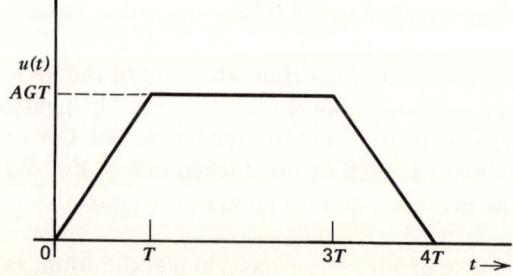

Figure 7-19. Filter output for input rectangular pulse of duration $3T$ and rectangular impulse response of Fig. 7-15b.

Matched Filtering

The maximum output voltage is the same value as that obtained when $f(t)$ is T seconds in duration (see Fig. 7-17). Thus as a signal detector, this filter performs optimally. However this maximum voltage remains there for $2T$ seconds, the same time span for which this pulse duration exceeds the optimum duration. This filter is therefore not as useful as the true matched filter because the signal arrival time is ambiguous, possibly occurring anywhere from $t = T$ to $t = 3T$.

If a filter is matched to a signal $f(t)$, the same filter is still matched if the amplitude and/or the starting time of $f(t)$ is changed. So long as the shape is preserved, the matched filter is invariant to the signal amplitude and its starting time. This can be seen by referring to Fig. 7-15a. Suppose the amplitude of $f(t)$ is changed from A to μA and the starting time is $t = T_1$ instead of $t = 0$. Then the filter response in Fig. 7-17 will have a peak output of μAGT, and this peak will occur at $t = T + T_1$.

For the RF case, where the signal is a pulsed carrier frequency, the matched filter theory is still valid. The filter, however, is now centered at the carrier frequency ω_0. An initial carrier phase change of θ corresponds to the additional delay $(-\theta/\omega_0)$ in the output signal. The amplitudes of the input and output signals, however, remain the same, regardless of the initial phase angle.

If the initial signal phase changes randomly, the time of the peak output is also random. Consequently the optimum receiver for a signal with random initial phase is a device whose output voltage is independent of the random initial phase. Such a device is an amplitude detector, which preserves the amplitude information and disregards the signal phase information. Figure 7-20 is the block diagram of a possible receiver incorporating this detector.

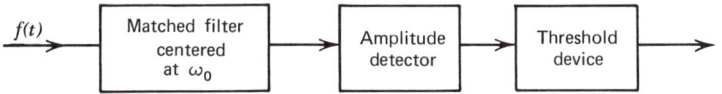

Figure 7-20. Receiver for pulsed carrier at ω_0 with random initial phase.

In radar applications, the initial phase value is usually random, provides no useful information, and most important, does not change the matched filter structure.

7.3.3 Frequency-Domain Characterization

We obtain $H_0(j\omega)$, the matched filter transfer function, by substituting $h_0(t)$ from (7.3-18) into the Fourier transform relationship of (2.1-1) as

$$H_0(j\omega) = \int_{-\infty}^{\infty} h_0(t)e^{-j\omega t}\, dt = G \int_{-\infty}^{\infty} f(t_1 - t)e^{-j\omega t}\, dt \qquad (7.3\text{-}22)$$

Since H_0 represents a transfer function, we let the argument be $j\omega$ instead of the customary ω used for Fourier transforms. With the change of variable $x = t_1 - t$,

$$H_0(j\omega) = Ge^{-j\omega t_1} \int_{-\infty}^{\infty} f(x)e^{j\omega x}\, dx \qquad (7.3\text{-}23)$$

The integral is $F^*(\omega)$, where $F(\omega)$ is the Fourier transform of $f(t)$; thus

$$H_o(j\omega) = GF^*(\omega)e^{-j\omega t_1} \quad (7.3\text{-}24)$$

The matched filter transfer function is the complex conjugate of the signal transform with a linear phase shift introduced for physical realizability. However, the filter magnitude response differs by only the constant G from the amplitude density spectrum of the input signal.

The amplitude density spectrum of the output signal $|U(\omega)|$ also differs from the energy spectrum of the output noise by a constant. The former, using (7.3-24) and the multiplication property of transforms, is

$$|U(\omega)| = |H_o(j\omega)| \, |F(\omega)| = G|F^*(\omega)| \, |F(\omega)| = G|F(\omega)|^2 \quad (7.3\text{-}25)$$

whereas from (7.1-49) and (7.3-24),

$$\text{output noise spectrum} = |H_o(j\omega)|^2 \frac{N_0}{2} = \frac{G^2 N_0}{2}|F(\omega)|^2 \quad (7.3\text{-}26)$$

Therefore

$$|U(\omega)| = \frac{2}{GN_0} \times \text{output noise spectrum} \quad (7.3\text{-}27)$$

Consequently the ratio of $|U(\omega)|$ to the output noise spectrum, at any frequency, is a constant.

The conjugation of $F(\omega)$ in (7.3-24) means that the filter phase function (neglecting the linear term $-\omega t_1$) differs from the signal phase function by only a minus sign. Thus the signal phase is perfectly compensated by the filter, and all frequency components add in phase, producing the maximum output at $t = t_1$. The output frequency distribution due to the signal $f(t)$, using (7.3-24), is

$$U(\omega) = H_o(j\omega)F(\omega) = GF^*(\omega)F(\omega)e^{-j\omega t_1} = G|F(\omega)|^2 e^{-j\omega t_1} \quad (7.3\text{-}28)$$

and is independent of the signal and filter phase function. The output phase function is linear.

The filter phase shifts the noise components but does not change its random character; therefore the summing of these noise components remains random. The probability that the noise components at any given instant will sum in phase and produce a large output is minute.

Example 7-3. We now give a frequency-domain explanation for the filtering process in Section 7.3.2. The transform of the rectangular pulse in Fig. 7-15a is ($A = 1$)

$$F(\omega) = \int_0^T e^{-j\omega t}\, dt = T\frac{\sin(\omega T/2)}{\omega T/2} e^{-j(\omega T/2)} \quad (7.3\text{-}29)$$

and

$$|F(\omega)| = T\left|\frac{\sin(\omega T/2)}{\omega T/2}\right| \quad (7.3\text{-}30)$$

Aided by (7.3-24), (7.3-25), and (7.3-26), we present the normalized input and output spectra of the signal and noise in Figs. 7-21a and b, respectively. Because the spectra are symmetric about $\omega = 0$, only the positive portions are shown. The filter attenuation is greater where the signal energy is lower, since these

Matched Filtering

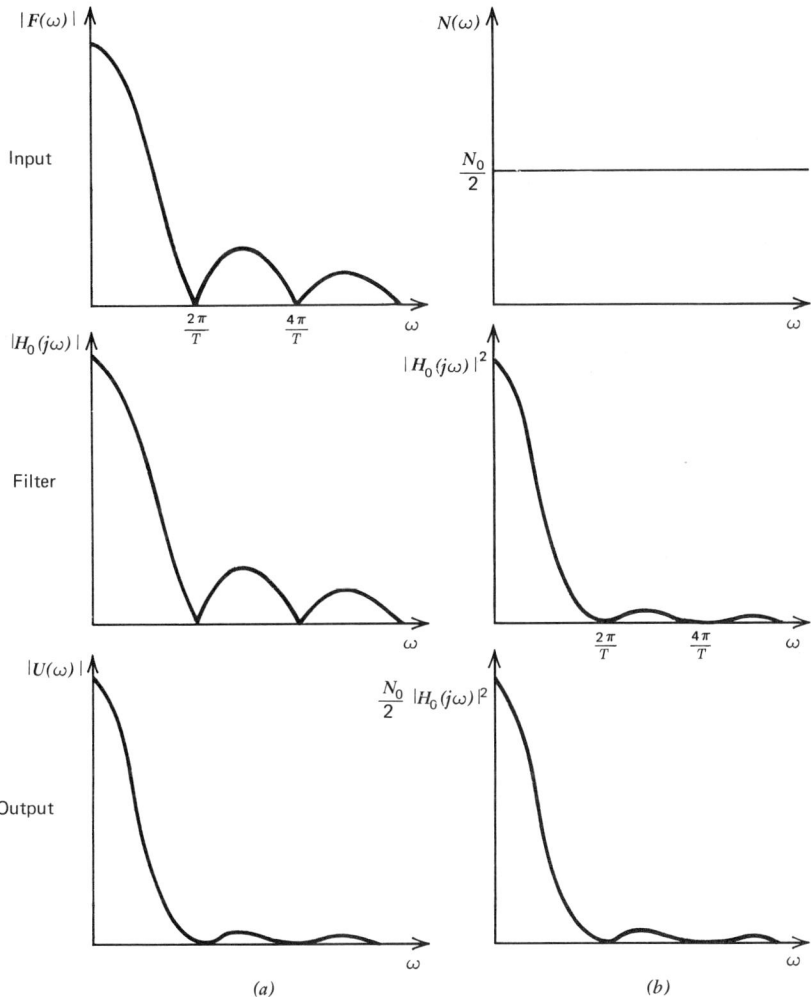

Figure 7-21. Transformation of (*a*) signal spectrum and (*b*) white noise spectrum by the matched filter in Section 7.3.2.

frequencies play a lesser role in creating the peak value of the output signal. The noise is attenuated at all frequencies except at $\omega = 0$, where the signal energy is maximum. Relative to their values at $\omega = 0$, the output signal and noise spectra are the same, as (7.3-27) predicts.

7.3.4 Matched Filter Synthesis

Determination of the optimum impulse response has now been completed, and the next task is to realize this response with practical components. For systems characterized by rational transfer functions, such as LC filters, an intermediate approximation step is necessary because their impulse responses are not of finite duration, whereas the optimum impulse response is. This topic is reserved for Chapter 8.

Here we consider a realization technique that uses four basic components—integrators, delays, multipliers, and adders—and yields the exact time-domain response for many practical pulse shapes. This realization eliminates the intermediate approximation step. Of course imperfect components may cause the response to deviate from the theoretical response, in which case the final result is still an approximation. However, this possibility exists regardless of the realization technique.

This procedure is based on the fact that successive integrations of the impulse function generate the necessary constant (step function), and the linear function (ramp function). From (1.7-3), the step function is the integral of the impulse function, that is,

$$u_{-1}(t) = \int_0^t \delta(x)\, dx \qquad (7.3\text{-}31)$$

Likewise the ramp function $r(t)$, in Fig. 7-22, is the integral of the step function, or the double integral of the impulse function

$$r(t) = K \int_0^t \int_0^x \delta(\xi)\, d\xi\, dx = K \int_0^t u_{-1}(x)\, dx = K t u_{-1}(t) \qquad (7.3\text{-}32)$$

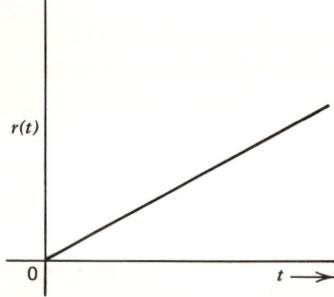

Figure 7-22. The ramp function $r(t)$.

Note that K, the slope of the ramp function, is the amplitude of the step function. Higher-order functions such as the quadratic function are similarly obtained. The desired impulse response is then synthesized by appropriately scaling these basic components, delaying them when necessary, and then adding them. This procedure is illustrated in the next two subsections.

7.3.5 Synthesis of the Rectangular Impulse Response

We now realize the unit-amplitude rectangular impulse response of Fig. 7-15b. In Section 1.7 the synthesis of pulses by superposition of step functions is discussed. Thus the arrangement in Fig. 7-23 yields the desired impulse response. The function at point a is the step function in Fig. 7-24a, obtained by integrating the input impulse function. At point b we have the step function delayed by T units, as in Fig. 7-24b, which is then scaled by -1 and added to $u_{-1}(t)$ to yield the desired rectangular impulse response in Fig. 7-24c. This filter is now matched to the rectangular pulse of duration T in Fig. 7-15a.

Matched Filtering

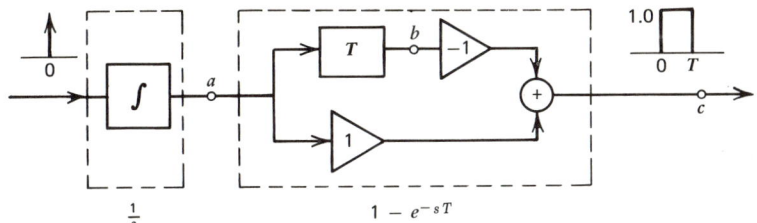

Figure 7-23. Filter with rectangular impulse response.

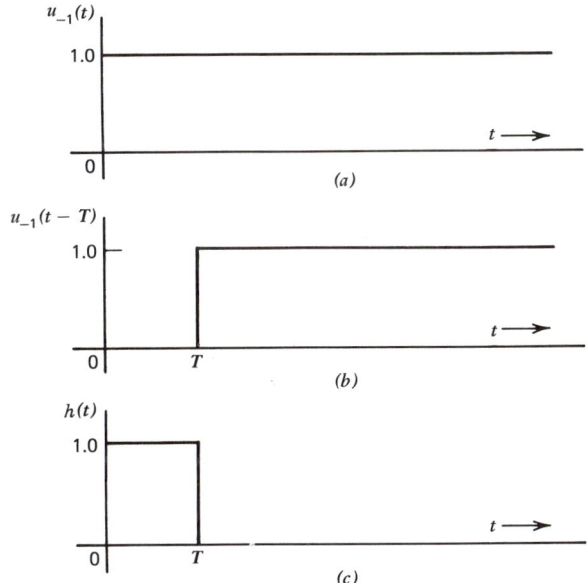

Figure 7-24. Waveforms in synthesis of rectangular pulse. (*a*) Step function, (*b*) delayed step function, (*c*) rectangular pulse.

The frequency domain also reveals the operations necessary to synthesize the pulse. The Laplace transform of the rectangular pulse is

$$H(s) = \frac{1}{s}(1 - e^{-sT}) \tag{7.3-33}$$

which can be realized as the cascade of two simpler systems. The factor $1/s$ is the transfer function of an integrator, and e^{-sT} is the transfer function of a delay T. The factor $(1 - e^{-sT})$ is therefore the transfer function of a system that subtracts a delayed version of the input from the input. The input in this case is the step function at point a. In this cascade arrangement (Fig. 7-23), the dashed lines in the diagram enclose each simpler system.

7.3.6 Synthesis of the Trapezoidal Impulse Response

The filter impulse response matched to the symmetric trapezoidal pulse is shown in Fig. 7-25. Because this pulse has linear variations with time, the ramp function

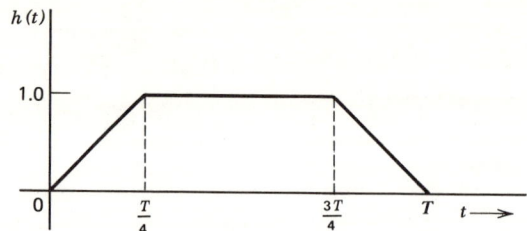

Figure 7-25. Trapezoidal impulse response.

of (7.3-32) is required. The trapezoid is the superposition of the four ramp functions in Fig. 7-26. Each is the same basic component function appropriately delayed and sign-changed. Because the slope of each is not unity, a multiplier of value $4/T$ is necessary. The impulse response function is then represented as

$$h(t) = r_1(t) - r_1\left(t - \frac{T}{4}\right) - r_1\left(t - \frac{3T}{4}\right) + r_1(t - T) \tag{7.3-34}$$

where

$$r_1(t) = \frac{4t}{T} \tag{7.3-35}$$

For example, $r_1(t - T/4)$ is zero for $t < T/4$. Explicitly

$$h(t) = \frac{4}{T}\left[tu_{-1}(t) - \left(t - \frac{T}{4}\right)u_{-1}\left(t - \frac{T}{4}\right) - \left(t - \frac{3T}{4}\right)u_{-1}\left(t - \frac{3T}{4}\right) + (t - T)u_{-1}(t - T)\right] \tag{7.3-36}$$

This equation is verified by evaluating $h(t)$ in the following four intervals.

$0 \leq t \leq \frac{T}{4}$; $h(t) = \frac{4}{T}t$, the first term

$\frac{T}{4} \leq t \leq \frac{3T}{4}$; $h(t) = \frac{4}{T}\left[t - \left(t - \frac{T}{4}\right)\right] = 1$, the sum of first two terms

$\frac{3T}{4} \leq t \leq T$; $h(t) = \frac{4}{T}\left[t - \left(t - \frac{T}{4}\right) - \left(t - \frac{3T}{4}\right)\right] = \frac{4}{T}(T - t)$, the sum of first three terms

$t \geq T$; $h(t) = \frac{4}{T}\left[t - \left(t - \frac{T}{4}\right) - \left(t - \frac{3T}{4}\right) + (t - T)\right] = 0$, the sum of first four terms

One realization of this response is illustrated in Fig. 7-27. The input impulse function is integrated twice and multiplied by $4/T$ to yield $r_1(t)$ at point a. The remainder of the system, enclosed in dashed lines, is known as a tapped delay line or a transversal filter. The output after each equal-delay element is scaled by a weighting factor, commonly called the tap weight. These outputs are then added to yield the desired trapezoid pulse. Note that the center tap weight for this case is zero. The realization of $1 - e^{-sT}$ in Fig. 7-23 is likewise a transversal filter with one delay element.

Direct implementation of (7.3-34) yields the parallel arrangement in Fig. 7-28. Here, however, larger delays are necessary.

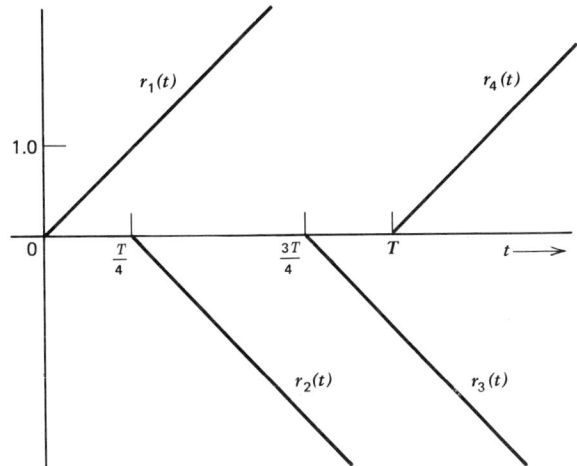

Figure 7-26. Four ramp functions that sum to yield trapezoid pulse of Fig. 7-25.

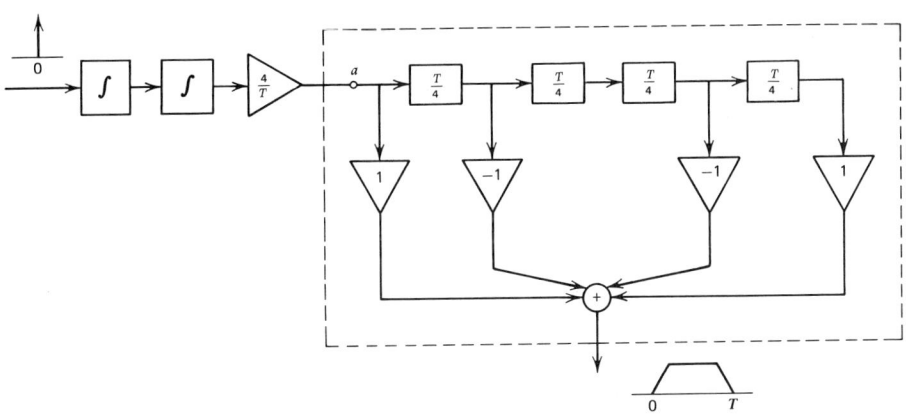

Figure 7-27. Tapped delay line realization of the trapezoidal impulse response.

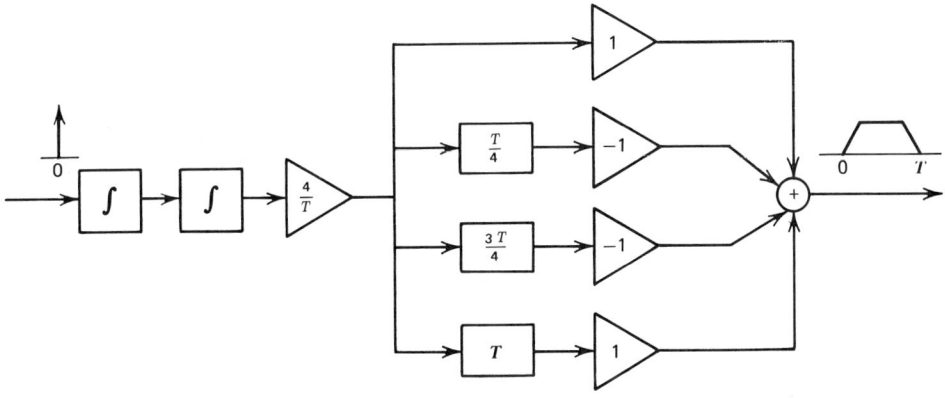

Figure 7-28. Direct implementation of (7.3-34).

The frequency-domain characterization of the trapezoid gives us yet another realization. From Table 2-1, entry 9, the transform of $r_1(t)$ is

$$R_1(s) = \frac{4}{T}\frac{1}{s^2} \tag{7.3-37}$$

and the transform of $h(t)$ in (7.3-36), from Table 2-1, entry 12, is then

$$H(s) = \frac{4}{Ts^2}[1 - e^{-sT/4} - e^{-s3T/4} + e^{-sT}] = \frac{4}{Ts^2}(1 - e^{-sT/4})(1 - e^{-s3T/4}) \tag{7.3-38}$$

Reference to (7.3-33) shows that $H(s)$ is the cascade of two rectangular pulse transfer functions, where the pulses have different time durations. This is not surprising, since the convolution of two such pulses is a trapezoidal pulse. This realization (Fig. 7-29) has the two integrators in series. However, the product form of (7.3-38) allows us to realize each factor in any convenient sequence. Practical considerations usually dictate the choice of one sequence over another.

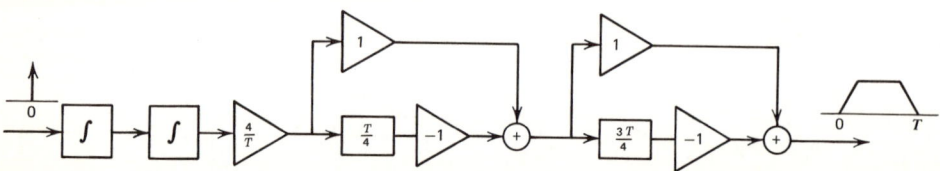

Figure 7-29. Filter with trapezoidal impulse response obtained from frequency-domain characterization.

The advances in digital processing have made it popular and competitive to analog signal processing. Consequently these analog tapped delay lines have an equivalent digital representation. This topic is discussed in Chapter 9, and there we find that the digital filter with a finite impulse response is easily achieved. The tap weights there are the sampled values of the impulse response. These digital filters are likewise called transversal filters or nonrecursive filters.

7.3.7 Relationship of Matched Filtering to Cross-Correlation

As shown in Section 7.1.7 the convolution of the input signal $f_1(t)$ and the impulse response $h(t)$ is equivalent to the cross-correlation function between $h(-t)$ and $f_1(t)$. But for a filter matched to the signal $f(t)$ the impulse response is given by (7.3-18) and

$$h(-t) = Gf(t + t_1) \tag{7.3-39}$$

Therefore, the matched filter output for the input $f_1(t)$ is proportional to the cross-correlation between $f(t)$, the signal to which the filter is matched, and $f_1(t)$, that is,

$$u(t) = G\int_{-\infty}^{\infty} f(x)f_1(x + t - t_1)\,dx = G\phi_{ff_1}(t - t_1) \tag{7.3-40}$$

Matched Filtering

When the input signal is $f(t)$, the filter response is

$$u(t) = G \int_{-\infty}^{\infty} f(x)f(x + t - t_1)\,dx = G\phi_f(t - t_1) \qquad (7.3\text{-}41)$$

which is the scaled autocorrelation function of $f(t)$, delayed by t_1 seconds. The maximum output occurs when $\phi_f(t - t_1)$ is maximum, which, we have shown, is $\phi_f(0)$. Therefore at $t = t_1$,

$$u(t_1) = G \int_{-\infty}^{\infty} f^2(x)\,dx = GE \qquad (7.3\text{-}42)$$

revealing that the maximum response is proportional to E, the energy in the pulse $f(t)$.

Figure 7-17 displays the features just discussed. The triangular response $u(t)$ is the autocorrelation function of the rectangular pulse in Fig. 7-15a, and the maximum response of AGT is proportional to the pulse energy (G is a constant times A). This maximum occurs at $t = T$, the minimum value of t_1 for which the filter remains causal.

Interpreting the matched filter as a correlator more clearly reveals its decoding properties. Each input signal is correlated with the filter's time-reversed impulse response, which is the same form as the desired input signal. When this signal is present, the filter output is then this signal's autocorrelation function and the maximum response value is used as a trigger to indicate the signal occurrence. For signal detection the matched filter output need not resemble the input signal, and indeed it does not. In the manner just described, the matched filter optimally distinguishes a particular signal shape from all others with the same energy, in the presence of white noise.

Since the matched filter response is proportional to the cross-correlation function between the desired input signal and all other input signals, the cross-correlation receiver of Fig. 7-30 is theoretically an equivalent realization.

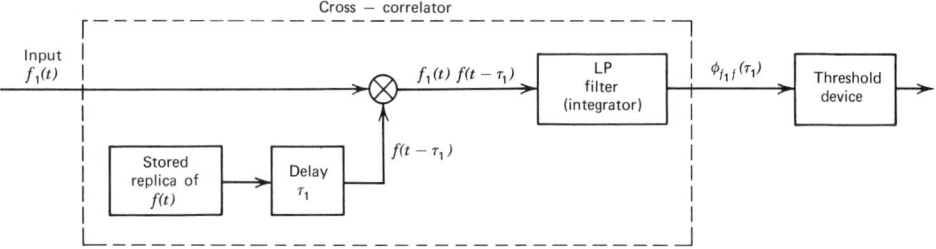

Figure 7-30. Cross-correlation receiver.

The desired pulsed carrier signal $f(t)$ is stored as an internal reference signal. The input signal $f_1(t)$ is multiplied by $f(t - \tau_1)$, a delayed version of $f(t)$, and the product is passed through an LP filter that essentially averages the product. The output is $\phi_{f_1 f}(\tau_1)$ the cross-correlation function for the shift τ_1. Because the LP filter is not a perfect integrator, the output is only approximate, but usually it is good enough for most applications. The output is then passed through the threshold device.

Threshold detection is a common radar technique for deciding whether a signal is present. The signal is considered to be present if the input signal level exceeds a prescribed threshold (Fig. 7-17), which may be fixed by system design or set by an input parameter measurement (e.g., the average input power). The existence of an output voltage from the threshold device indicates the presence of the signal. When the output voltage does not exist, the received signal is assumed to be noise.

This detection scheme, however, has two shortcomings. First, the noise input may be so large that the output of the correlator is higher than the threshold level. Consequently the threshold device indicates the presence of the signal, thus creating a false alarm. Second, the combination of input signal and noise may yield a correlator output voltage that is less than the threshold level, and the signal is then undetected. The setting of the threshold level is therefore a compromise between these two conditions, and the relative weight of each error depends on the application.

In a radar receiver, the delay time τ_1 corresponds to a target at the range R, where

$$R = \frac{c\tau_1}{2} \quad (7.3\text{-}43)$$

and c is the speed of light. Thus this receiver will indicate the presence of a target only at the distance corresponding to τ_1, provided the output exceeds the threshold. Obviously the system must have the capability to detect targets at other ranges. One way to achieve this is to vary τ_1, but this requires a longer search time. Alternatively, the search time is reduced if we parallel a number of the receivers in Fig. 7-30, each with a different value of delay, as in Fig. 7-31. Then each LP filter output yields the cross-correlation function for a different value of

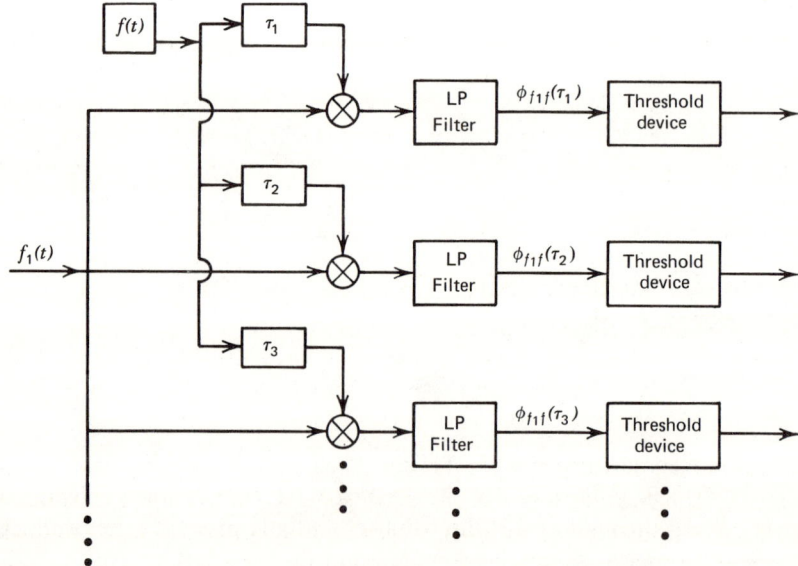

Figure 7-31. Cross-correlation receiver with variable delays.

Matched Filtering

τ. When the LP filter output of the channel with delay τ_k exceeds the prescribed threshold, the threshold device indicates the presence of the target at the range given by (7.3-43) with $\tau_1 = \tau_k$.

Cross-correlation is a fundamental process in the detection of signals in noise. In addition to their use in search-radar optimum receivers, cross-correlators are also used to

- Distinguish one signal from another signal.
- Distinguish each signal in an orthogonal set.
- Distinguish signals characterized by many parameter values.
- Measure signal amplitude.
- Measure signal time delay, hence target distance.
- Measure signal frequency (Doppler shift), consequently target radial velocity.

The choice between the matched filter and cross-correlator depends on the application, for both are mathematically equivalent to maximizing the SNR. If long integration times are necessary to extract weak signals from noise (such as signal returns from distant celestial bodies), it is usually more efficient to cross-correlate digitally than to use an analog matched filter.

7.3.8 Nonoptimal Conditions

Of interest is the degradation in the SNR when the filter is not matched to the incoming signal. Theoretically we can always realize the optimum impulse response (assuming physical realizability) regardless of the input signal form. Practically, however, this may cause increased design and manufacturing cost, as well as filter complexity. It is therefore worthwhile to know the consequences of not optimizing the filter for the specified signal, and conversely, the effect of a signal change, in which case the filter is no longer optimum. For the former situation the output noise is different, dependent on the filter characteristics, but for the latter situation the output filter noise remains the same. These two situations are examined here by examples. (In Section 7.3.2 we computed the response of a filter whose rectangular impulse response is shorter than the rectangular input pulse. The maximum output voltage is then the same as that for the optimum condition, but longer in duration, thus creating ambiguity in the signal arrival time.)

Example 7-4. Consider the rectangular pulse of Fig. 7-15a exciting the first-order filter with transfer function

$$H(s) = \frac{\omega_c}{s + \omega_c} \tag{7.3-44}$$

and impulse response

$$h(t) = u_{-1}(t)\omega_c e^{-\omega_c t} \tag{7.3-45}$$

The white input noise has a power density spectrum of N_0 watts per hertz. The bandwidth ω_c is to be chosen to ensure that the SNR of (7.3-2) is maximized. We then compare this value with the optimum value given by (7.3-17) to determine the resulting degradation due to nonoptimization.

The filter output to the pulse input is given by (1.5-7), and Fig. 1-5 shows that the peak output occurs at $t = T$. Therefore for the pulse of amplitude A

$$|u(t)|^2_{max} = A^2(1 - e^{-\omega_c T})^2 \qquad (7.3\text{-}46)$$

This peak value is not very pronounced, thus the signal arrival time is not clearly defined in a practical situation. The total mean output noise power is obtained from (7.3-4) and (7.3-45) as

$$\overline{u_n^2} = \frac{N_0}{2} \int_0^\infty \omega_c^2 e^{-2\omega_c t}\, dt = \frac{N_0 \omega_c}{4} \qquad (7.3\text{-}47)$$

Then, from (7.3-2),

$$\text{SNR} = \frac{A^2(1 - e^{-\omega_c T})^2}{N_0 \omega_c /4} \qquad (7.3\text{-}48)$$

Since the optimum value of SNR for the rectangular pulse of duration T, from (7.3-17), is

$$(\text{SNR})_M = \frac{2A^2 T}{N_0} \qquad (7.3\text{-}49)$$

the ratio $\text{SNR}/(\text{SNR})_M$ is

$$\frac{\text{SNR}}{(\text{SNR})_M} = \frac{2(1 - e^{-\omega_c T})^2}{\omega_c T} \qquad (7.3\text{-}50)$$

shown in Fig. 7-32 as a function of $\omega_c T$. The maximum value of $(\text{SNR})/(\text{SNR})_M$ is

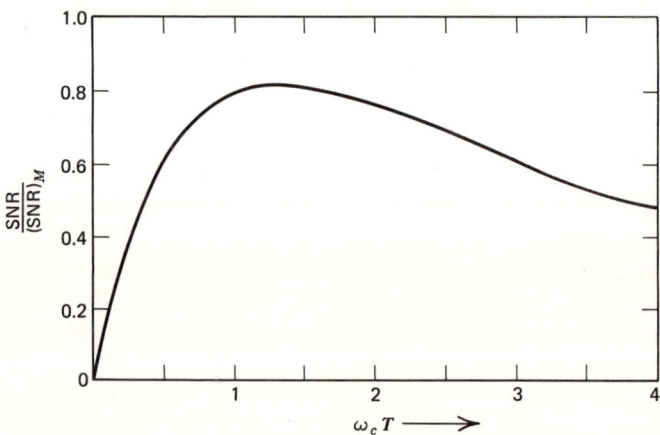

Figure 7-32. Efficiency of a first-order filter relative to the matched filter for the rectangular pulse of Fig. 7-15a.

0.815 and occurs when $\omega_c T = 1.256$. This value corresponds to an SNR degradation of 0.89 dB compared with the true matched filter. For the BP case, ω_c is replaced by $(\omega_2 - \omega_1)/2$, so the BP bandwidth is

$$\omega_2 - \omega_1 = \frac{2.512}{T} \text{ rads./sec} \qquad (7.3\text{-}51)$$

Example 7-5. In this example we assume that the filter is matched to the unit-amplitude rectangular pulse of duration T and energy T. We wish to design

Matched Filtering

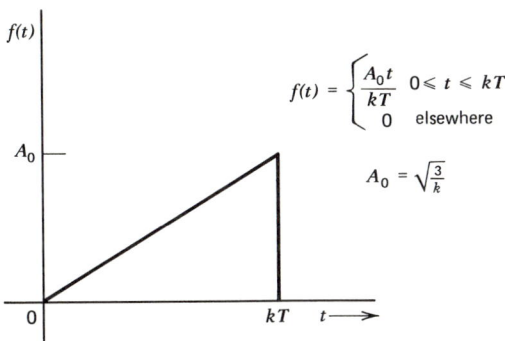

Figure 7-33. Triangular pulse with energy T exciting matched filter with rectangular impulse response.

the triangular pulse in Fig. 7-33, to optimize the SNR in (7.3-6). To compare the triangular pulse response and the rectangular pulse response, we normalize the triangular pulse energy to T. Then

$$\text{energy} = \int_0^{kT} f^2(t)\, dt = \frac{A_0^2}{k^2 T^2} \times \frac{k^3 T^3}{3} = \frac{kT}{3} A_0^2 \qquad (7.3\text{-}52)$$

and $A_0 = \sqrt{3/k}$. For this comparison, the output noise is the same for both cases, since the filter is unchanged. Therefore we need only compute the maximum output voltage and compare it to T, the maximum rectangular pulse response. Since k can take on any positive value, we have the six possible situations, depicted in Fig. 7-34. For $k \leq 1$ it is clear that the maximum output voltage occurs for the situation in Fig. 7-34b. Then, at $t = kT$,

$$u(kT) = \frac{A_0}{kT} \int_0^{kT} (kT - \tau)\, d\tau = \frac{A_0 kT}{2} = \frac{\sqrt{3k}\, T}{2} \qquad (7.3\text{-}53)$$

which is maximum for $k = 1$, and

$$u(kT)|_{\max} = \frac{\sqrt{3}\, T}{2} \qquad 0 \leq k \leq 1 \qquad (7.3\text{-}54)$$

For $k \geq 1$, the maximum output voltage occurs for the situation in Fig. 7-34e. Then

$$u(t) = \int_0^T \frac{A_0}{kT}(t - \tau)\, d\tau = \frac{A_0}{kT}\left[tT - \frac{T^2}{2}\right] = \frac{A_0}{k}\left(t - \frac{T}{2}\right) \qquad (7.3\text{-}55)$$

which is maximum when $t = kT$, at which time

$$u(kT) = \frac{\sqrt{3}\, T}{2} \frac{2k - 1}{k^{3/2}} \qquad (7.3\text{-}56)$$

This output voltage is further maximized when $k = \tfrac{3}{2}$, yielding

$$u(kT)|_{\max} = \frac{2\sqrt{2}}{3} T \qquad k \geq 1 \qquad (7.3\text{-}57)$$

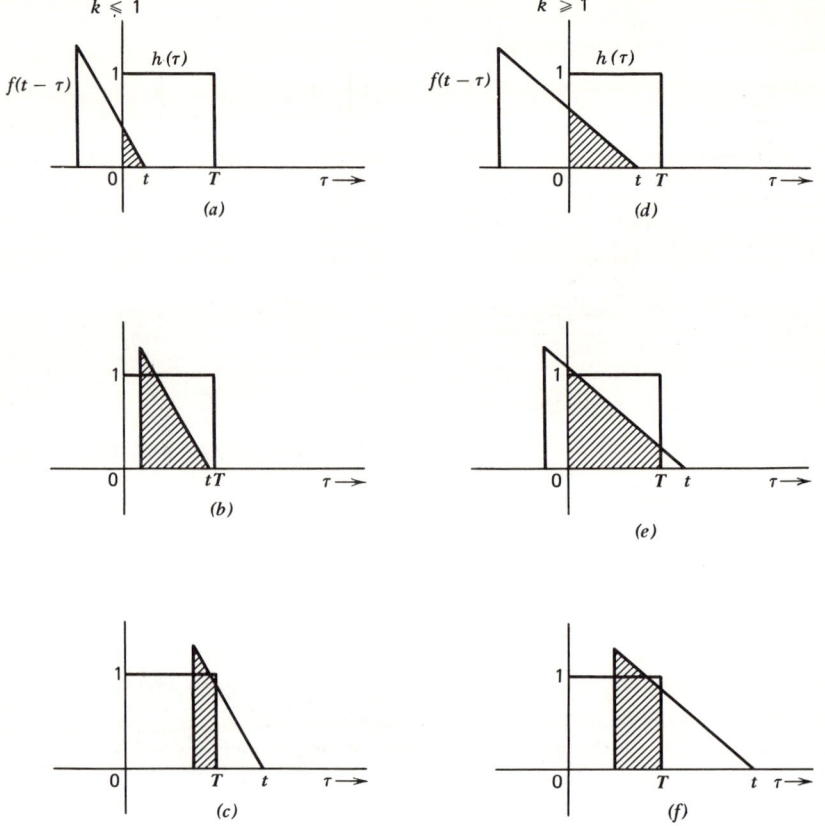

Figure 7-34. Geometric interpretation of the convolution integral for the six possibilities in Example 7-5. (a) $0 \leq t \leq kT$, (b) $kT \leq t \leq T$, (c) $T \leq t \leq (k+1)T$, (d) $0 \leq t \leq T$, (e) $T \leq t \leq kT$, (f) $kT \leq t \leq (k+1)T$.

which is greater than the value in (7.3-54). The output SNR is greatest when the triangular pulse is of duration $\frac{3}{2}T$ and amplitude $\sqrt{2}$. The overlap for maximum output voltage and the filter response to this signal are represented in Figs. 7-35a and b, respectively. Comparing this output to the optimum signal output of T, we obtain the efficiency

$$\text{eff} = \frac{(2\sqrt{2}/3)T}{T} = 0.9428 \tag{7.3-58}$$

which corresponds to a loss of 0.256 dB. Thus a properly designed triangular pulse degrades the SNR by 0.256 dB compared to the matched filter response.

Examples 7-4 and 7-5 suggest that the optimum filter is relatively insensitive to the signal form. With respect to the peak output, this is true. Insensitivity arises because the criterion used for maximizing the SNR is integral. Thus moderate differences in the signal shape and impulse response shape are smoothed out by the integration operation in convolution, as illustrated in Fig. 7-35a. The triangular pulse is vastly different from the rectangular pulse, yet the maximum response in Fig. 7-35b is almost unity, the optimum output. Only the shaded portion contributes to the maximum output, but note that this shaded area (obtained by

Matched Filtering

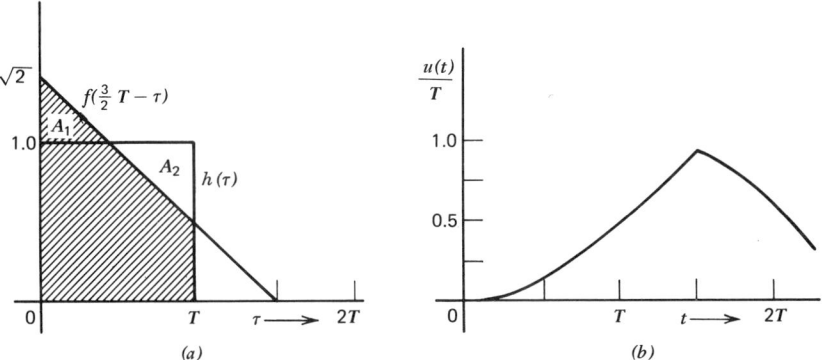

Figure 7-35. (a) Optimum triangular input signal and rectangular impulse response position for maximum output, (b) filter response to $f(t)$.

integration) is almost equal to the area of the rectangle. The area A_1 is only 5.7% less than area A_2.

Peak output, however, is not the entire story. The more important factor is the false-alarm probability P_{fa}, which magnifies the apparently small SNR degradation from the optimum value. For example, suppose the probability of detection of a single pulse by a threshold device is set at 0.9. Then, with a linear rectifier detector, P_{fa} decreases approximately an order of magnitude for every decibel increase in the SNR in the vicinity of SNR = 12 dB [11, p. 2-19]. If SNR = 11.8 dB, $P_{fa} = 10^{-4}$, but when SNR = 12.6 dB, P_{fa} decreases to 10^{-5}.

Usually the system designer must decide whether the expense and effort necessary to realize an improved approximation to the optimum filter to increase the SNR are worthwhile. Often other system components can be more easily improved to obtain the desired value of P_{fa}. Ultimately it becomes a matter of ascertaining where the cost per extra decibel of SNR is minimum.

The optimum filter need not always be designed to include every small signal detail. For this reason, many matched filter approximations are relatively simple networks, because the required SNR has already been achieved and a more sophisticated filter is not necessary. Example 7-4 shows the extreme case of filter design simplification, and even this filter is acceptable in many situations. However, there are always cases in which the additional 0.2 to 0.3 dB of SNR provided by the optimum filter is of utmost importance for system performance.

7.3.9 Nonwhite Input Noise Spectrum

We now consider input noise known as nonwhite or colored noise, whose power density spectrum $N(\omega)$ is nonconstant. For the following analysis, the noise need not have a Gaussian distribution. We can show that it is possible to obtain the matched filter for this case using the theory for the white noise case.

First we convert $N(\omega)$ to the constant power density spectrum $N_0/2$ by passing it through the filter with transfer function

$$H_1(j\omega) = \sqrt{\frac{N_0}{2}} \frac{1}{\sqrt{N(\omega)}} e^{j\theta(\omega)} \tag{7.3-59}$$

where $\theta(\omega)$ is an arbitrary phase function. Then the output noise spectrum, from (7.1-49), is

$$\text{output noise spectrum} = |H_1(j\omega)|^2 N(\omega) = \frac{N_0}{2} \qquad (7.3\text{-}60)$$

and the output signal in the frequency domain is

$$\text{output signal} = H_1(j\omega)F(\omega) = \sqrt{\frac{N_0}{2}} \frac{F(\omega)}{\sqrt{N(\omega)}} e^{j\theta(\omega)} \qquad (7.3\text{-}61)$$

The filter with transfer function $H_1(j\omega)$ is known as a prewhitening filter because it transforms the power density spectrum $N(\omega)$ to white noise.

We now have the previously discussed problem of detecting the signal in (7.3-61) immersed in white noise. The optimum filter transfer function is then given by (7.3-24) as

$$H_2(j\omega) = G\sqrt{\frac{N_0}{2}} \frac{F^*(\omega)e^{-j\theta(\omega)}}{\sqrt{N(\omega)}} e^{-j\omega t_1} \qquad (7.3\text{-}62)$$

The overall transfer function is then the cascade of $H_1(j\omega)$ and $H_2(j\omega)$ as shown in Fig. 7-36, or

$$H_0(j\omega) = H_1(j\omega)H_2(j\omega) = G_1 \frac{F^*(\omega)}{N(\omega)} e^{-j\omega t_1} \qquad \left(G_1 = \frac{GN_0}{2}\right) \qquad (7.3\text{-}63)$$

Therefore the transfer function of the optimum filter for the signal corrupted by frequency-dependent noise is proportional to the complex conjugate of $F(\omega)$ and inversely proportional to the input noise power density spectrum. Unlike the white

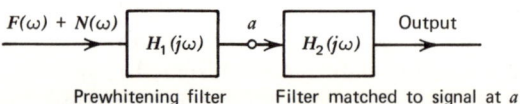

Figure 7-36. Matched filter for nonwhite noise using a prewhitening filter.

noise case, the filter design depends on the input noise spectrum. Since $N(\omega)$ is real, the filter phase function is the negative of the input signal phase function plus the linear phase $-\omega t_1$.

The optimum impulse response $h_0(t)$ is the inverse Fourier transform of $H_0(j\omega)$,

$$h_0(t) = \frac{GN_0}{4\pi} \int_{-\infty}^{\infty} \frac{F^*(\omega)}{N(\omega)} e^{j\omega(t-t_1)} d\omega \qquad (7.3\text{-}64)$$

We can rearrange (7.3-63) as

$$H_0(j\omega)N(\omega) = \frac{GN_0}{2} F^*(\omega)e^{-j\omega t_1} \qquad (7.3\text{-}65)$$

The inverse transform of the right side of (7.3-65) is $(GN_0/2)f(t_1-t)$, while the inverse transform of the left side is the convolution of the respective time functions. Therefore we arrive at the following integral equation for $h_0(t)$,

$$\int_{-\infty}^{\infty} h_0(\tau)\phi_n(t-\tau)\,d\tau = \frac{GN_0}{2} f(t_1 - t) \tag{7.3-66}$$

where $\phi_n(\tau)$ is the autocorrelation function of the noise with power density spectrum $N(\omega)$. This is the nonwhite noise expression analogous to (7.3-18). When the noise is white, $\phi_n(\tau) = (N_0/2)\delta(\tau)$, and (7.3-66) reduces to (7.3-18).

7.4 PULSE COMPRESSION USING LINEAR FREQUENCY MODULATION

Important considerations in radar design are the system's resolution, range accuracy, and detection capabilities. If the transmitted pulse contains enough energy to detect targets at long range, the pulse duration is long and/or the peak power is large. However, if the pulse duration is narrow for good range accuracy or resolution, then the pulse peak power must be large to retain the system detection capability. Unfortunately the peak power cannot always be increased to the desired value because of component breakdown in the transmitter section. Most notable is saturation of the output power tube.

The peak power limitation requires transmission of a long pulse, whereas the resolution capability requires a short pulse, or equivalently, one with a wide frequency spectrum. Therefore the transmitted signal should have a large time-bandwidth product to satisfy both requirements. This feature of the signal appears to be contradictory, since a long pulse, from transform theory, implies a narrow spectrum. However, careful selection of the transmitted signal and the use of the appropriate matched filter allow the advantages of both a long and a short pulse to be obtained through the mechanism of pulse compression.

Pulse compression systems can operate satisfactorily with almost any form of frequency modulation (FM), provided the receiver filter is matched to the transmitted signal. Because of its simplicity, however, the linear FM signal has been extensively used in operational pulse compression systems, and it is this form and its matched filter that we consider here.

7.4.1 The Linear FM Signal

The transmitted waveform is the linear FM signal of duration T and unit amplitude

$$v(t) = \begin{cases} \cos\left(\omega_c t + \frac{\Delta\omega}{2T} t^2\right) & -\frac{T}{2} \leq t \leq \frac{T}{2} \\ 0 & \text{elsewhere} \end{cases} \tag{7.4-1}$$

where

ω_c = carrier frequency in rads/sec

$\phi(t)$ = instantaneous phase angle = $\omega_c t + \frac{\Delta\omega}{2T} t^2 \qquad -\frac{T}{2} \leq t \leq \frac{T}{2}$

$\omega(t)$ = instantaneous frequency = $\frac{d\phi(t)}{dt} = \omega_c + \frac{\Delta\omega}{T} t \qquad -\frac{T}{2} \leq t \leq \frac{T}{2}$

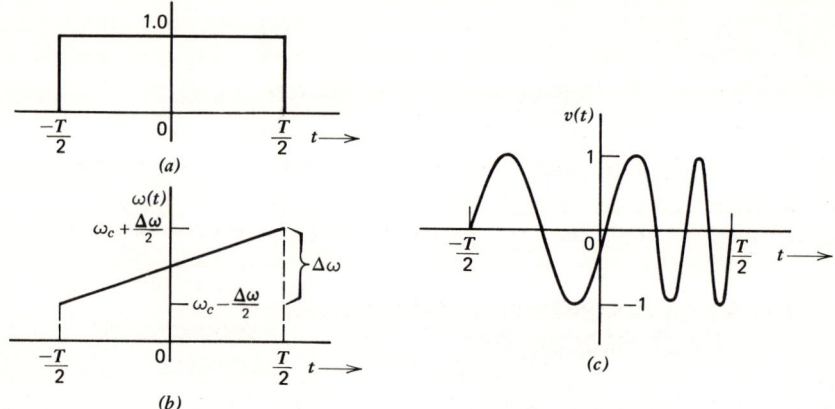

Figure 7-37. (a) Amplitude modulation, (b) frequency modulation, and (c) composite signal as expressed by (7.4-1).

The rectangular amplitude modulation of $v(t)$, the linear instantaneous frequency $\omega(t)$, and the composite signal are shown in Fig. 7-37.

The frequency-domain representation of $v(t)$ is obtained by the Fourier transform relationship of (2.1-1),

$$V(\omega) = \int_{-T/2}^{T/2} \cos\left(\omega_c t + \frac{\Delta\omega}{2T} t^2\right) e^{-j\omega t} \, dt$$

$$= \frac{1}{2} \int_{-T/2}^{T/2} \exp j\left[(\omega_c - \omega)t + \frac{\Delta\omega}{2T} t^2\right] dt + \frac{1}{2} \int_{-T/2}^{T/2} \exp\left\{-j\left[(\omega_c + \omega)t + \frac{\Delta\omega}{2T} t^2\right]\right\} dt \quad (7.4\text{-}2)$$

The first integral essentially defines $V(\omega)$ about the positive carrier frequency ω_c, while the second integral defines $V(\omega)$ about the negative carrier frequency $-\omega_c$. Each response, however, exists for all values of ω, but if $\omega_c/\Delta\omega$ is large, as it is in many practical situations, the contribution of the second integral at positive frequencies is negligible. With this assumption, $V(\omega)$ is given by the first integral, which can be represented as [2]

$$V(\omega) = |V(\omega)| e^{j[\theta_1(\omega) + \theta_2(\omega)]} \quad (7.4\text{-}3)$$

where

$$|V(\omega)| = \frac{T}{2\sqrt{2D}} \{[C(X_1) + C(X_2)]^2 + [S(X_1) + S(X_2)]^2\}^{1/2}$$

$$\theta_1(\omega) = -\pi D \left(\frac{\omega_c - \omega}{\Delta\omega}\right)^2$$

$$\theta_2(\omega) = \tan^{-1}\left[\frac{S(X_1) + S(X_2)}{C(X_1) + C(X_2)}\right]$$

$$X_1 = \sqrt{\frac{D}{2}}\left[1 - 2\frac{\omega_c - \omega}{\Delta\omega}\right]; \quad X_2 = \sqrt{\frac{D}{2}}\left[1 + 2\frac{\omega_c - \omega}{\Delta\omega}\right]$$

Pulse Compression Using Linear Frequency Modulation

$$C(X) = \int_0^X \cos\frac{\pi y^2}{2}\, dy$$
$$S(X) = \int_0^X \sin\frac{\pi y^2}{2}\, dy \quad \bigg\} \text{Fresnel integrals}$$

$$D = \Delta f T = \frac{\Delta \omega T}{2\pi} = \text{time-bandwidth product}$$

Both the spectrum $|V(\omega)|$ and the phase angle $\theta_2(\omega)$ are presented in Fig. 7-38 for $D = 13$ and for $D = 130$. As D increases, both the spectrum and $\theta_2(\omega)$ become

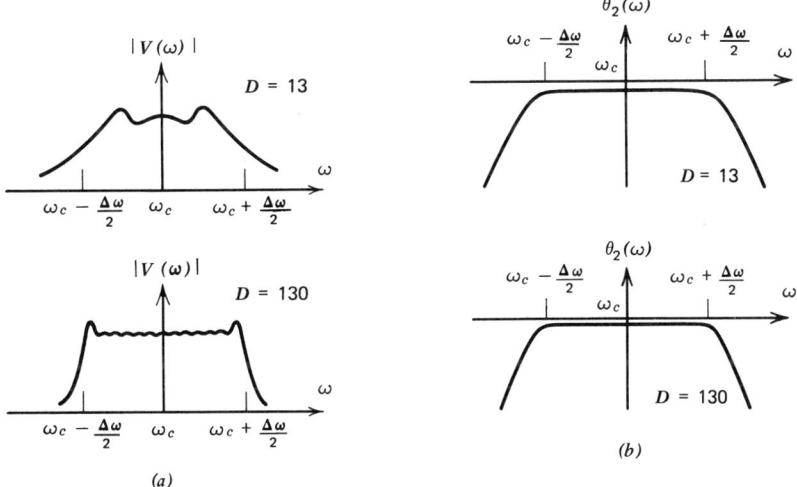

Figure 7-38. (a) Spectrum and (b) phase angle $\theta_2(\omega)$ for linear FM pulse with $D = 13$ and $D = 130$.

more uniform over the frequency band $\Delta\omega$ and change more abruptly at the band edges. In fact, for values of D greater than about 25, $|V(\omega)|$ is generally assumed to be rectangular as

$$|V(\omega)| = \begin{cases} \dfrac{T}{2\sqrt{D}} & \omega_c - \dfrac{\Delta\omega}{2} \leq \omega \leq \omega_c + \dfrac{\Delta\omega}{2} \\ 0 & \text{elsewhere} \end{cases} \qquad (7.4\text{-}4)$$

and the residual phase shift $\theta_2(\omega)$ is usually neglected because it can be approximated by a constant phase angle over the frequency band $\Delta\omega$. Calculations show that even with $D = 10$, almost 95% of the signal energy is within the band $\Delta\omega$, and with $D = 100$ this figure approaches 98%. The total phase spectrum of $v(t)$ over the frequency band $\Delta\omega$ for large values of D is then the quadratic phase

$$\theta(\omega) = \theta_0 + \theta_1(\omega) = \theta_0 - \pi D\left(\frac{\omega_c - \omega}{\Delta\omega}\right)^2 \qquad (7.4\text{-}5)$$

where θ_0 is the constant value of $\theta_2(\omega)$.

Both $|V(\omega)|$ and $\theta(\omega)$ depend on the value of D, which we soon show is the amount that the pulse is compressed after passing through the matched filter.

7.4.2 The Linear FM Matched Filter

From (7.3-18), the impulse response of the filter matched to $v(t)$, the linear FM signal in (7.4-1), is

$$h(t) = \begin{cases} \dfrac{2\sqrt{D}}{T} \cos\left(\omega_c t - \dfrac{\Delta\omega}{2T} t^2\right) & -\dfrac{T}{2} \leq t \leq \dfrac{T}{2} \\ 0 & \text{elsewhere} \end{cases} \quad (7.4\text{-}6)$$

where the constant G is chosen for unity filter gain, and the delay t_1 is neglected. Thus $h(t)$ in (7.4-6) characterizes an anticipatory system because this response exists prior to the application of the impulse function at $t = 0$. However this positioning of $h(t)$ eases the analysis, for the filter response to $v(t)$ is then symmetric about $t = 0$. Of course, the filter realized in practice has a causal impulse response, achieved by delaying $h(t)$ by an appropriate amount t_1.

The matched filter response to $v(t)$ in (7.4-1) is given by the convolution integral

$$u(t) = \frac{2\sqrt{D}}{T} \int_{-\infty}^{\infty} \cos\left(\omega_c \tau + \frac{\Delta\omega \tau^2}{2T}\right) \cos\left[\omega_c(t-\tau) - \frac{\Delta\omega}{2T}(t-\tau)^2\right] d\tau \quad (7.4\text{-}7)$$

The integral limits are obtained by referring to the graphical interpretation of the convolution integral, remembering that $h(t)$ is anticipatory by $T/2$ seconds. Then [2]

$$u(t) = \begin{cases} \dfrac{2\sqrt{D}}{T} \displaystyle\int_{-T/2}^{t+T/2} \cos\left(\omega_c \tau + \dfrac{\Delta\omega \tau^2}{2T}\right) \cos\left[\omega_c(t-\tau) - \dfrac{\Delta\omega}{2T}(t-\tau)^2\right] d\tau & -T \leq t \leq 0 \\[2ex] \dfrac{2\sqrt{D}}{T} \displaystyle\int_{t-T/2}^{T/2} \cos\left(\omega_c \tau + \dfrac{\Delta\omega \tau^2}{2T}\right) \cos\left[\omega_c(t-\tau) - \dfrac{\Delta\omega}{2T}(t-\tau)^2\right] d\tau & 0 \leq t \leq T \end{cases} \quad (7.4\text{-}8)$$

Use of the identity

$$2 \cos\alpha \cos\beta = \cos(\alpha + \beta) + \cos(\alpha - \beta) \quad (7.4\text{-}9)$$

allows (7.4-8) to be written as

$$u(t) = \begin{cases} \dfrac{\sqrt{D}}{T} \displaystyle\int_{-T/2}^{t+T/2} P_1(t, \tau) \, d\tau + \dfrac{\sqrt{D}}{T} \displaystyle\int_{-T/2}^{t+T/2} P_2(t, \tau) \, d\tau & t \leq 0 \\[2ex] \dfrac{\sqrt{D}}{T} \displaystyle\int_{t-T/2}^{T/2} P_1(t, \tau) \, d\tau + \dfrac{\sqrt{D}}{T} \displaystyle\int_{t-T/2}^{T/2} P_2(t, \tau) \, d\tau & t \geq 0 \end{cases} \quad (7.4\text{-}10)$$

where

$$P_1(t, \tau) = \cos\left[\omega_c t - \frac{\Delta\omega}{2T} t^2 + \frac{\Delta\omega}{T} t\tau\right]$$

$$P_2(t, \tau) = \cos\left[\omega_c(2\tau - t) + \frac{\Delta\omega}{T}\tau(\tau - t) + \frac{\Delta\omega}{2T} t^2\right]$$

Pulse Compression Using Linear Frequency Modulation

The first integral in (7.4-10) is evaluated as

$$\int_a^b \cos\left[\omega_c t - \frac{\Delta\omega}{2T}t^2 + \frac{\Delta\omega}{T}t\tau\right] d\tau = \left.\frac{\sin\left[\omega_c t - (\Delta\omega/2T)t^2 + (\Delta\omega/T)t\tau\right]}{(\Delta\omega/T)t}\right|_{\tau=a}^{\tau=b} \quad (7.4\text{-}11)$$

The second integral depends on higher frequency terms, and for practical values of ω_c and $\Delta\omega$ its contribution to the total response is negligible; hence it is neglected. Therefore for $t \geq 0$

$$u(t) = \frac{\sqrt{D}}{\Delta\omega t}\left\{\sin\left[\omega_c t - \frac{\Delta\omega t^2}{2T} + \frac{\Delta\omega}{2}t\right] - \sin\left[\omega_c t - \frac{\Delta\omega t^2}{2T} + \frac{\Delta\omega t}{T}\left(t - \frac{T}{2}\right)\right]\right\} \quad (7.4\text{-}12)$$

which has the form

$$\sin(\alpha + \beta) - \sin(\alpha - \beta) = 2\cos\alpha\sin\beta \quad (7.4\text{-}13)$$

where

$$\alpha = \omega_c t \quad \text{and} \quad \beta = \frac{\Delta\omega t}{2T}(T - t)$$

Evaluation of the first integral in (7.4-10) for $t < 0$ yields a similar expression. Combining these two results yields the output of the matched filter as

$$u(t) = \sqrt{D}\,\frac{\sin \pi D \frac{t}{T}\left(1 - \frac{|t|}{T}\right)}{\pi D \frac{t}{T}} \cos\omega_c t \qquad -T < t < T \quad (7.4\text{-}14)$$

which is also the autocorrelation function of $v(t)$.

The filter output is an amplitude-modulated signal whose carrier frequency is ω_c. The frequency modulation of $v(t)$ has disappeared because the matched filter's impulse response is a time reversal of the input signal.

Of major concern is the modulating function $m_D(t)$, for it contains the information pertaining to the peak power increase and the amount of pulse compression. The shape of function $m_D(t)$ is approximately $(\sin x)/x$, the approximation improving as D increases. This fact is better appreciated by remembering that the spectrum of $v(t)$ is essentially rectangular for large values of D, consequently the product of $|V(\omega)|$ and the filter magnitude response is also rectangular. The quadratic phase angles cancel [see (7.3-25)]; thus the cardinal function, which is the inverse Fourier transform of the rectangular function, is the filter response envelope.

Figure 7-39 shows $[m_D(t)]/\sqrt{D}$ for $D = 13, 25, 50$, and 100. Only values for $t \geq 0$ are given because $m_D(t)$ is an even function, which is zero outside the interval $-T < t < T$. The first sidelobe is attenuated approximately 14 dB relative to the main peak.

Since the cardinal function is the limiting form for $m_D(t)$, we show the function

$$m_\infty(t) = \sqrt{D}\,\frac{\sin \pi \Delta f(t - t_1)}{\pi \Delta f(t - t_1)} \quad (7.4\text{-}15)$$

in Fig. 7-40 to illustrate the important properties of the filter output. The delay necessary for physical realizability t_1, is introduced for generality. The maximum output is $\sqrt{D}$ and occurs at $t = t_1$. The compressed pulse width is measured between the times at which the amplitude is $(2/\pi)\sqrt{D}$ (about 4 dB down from the

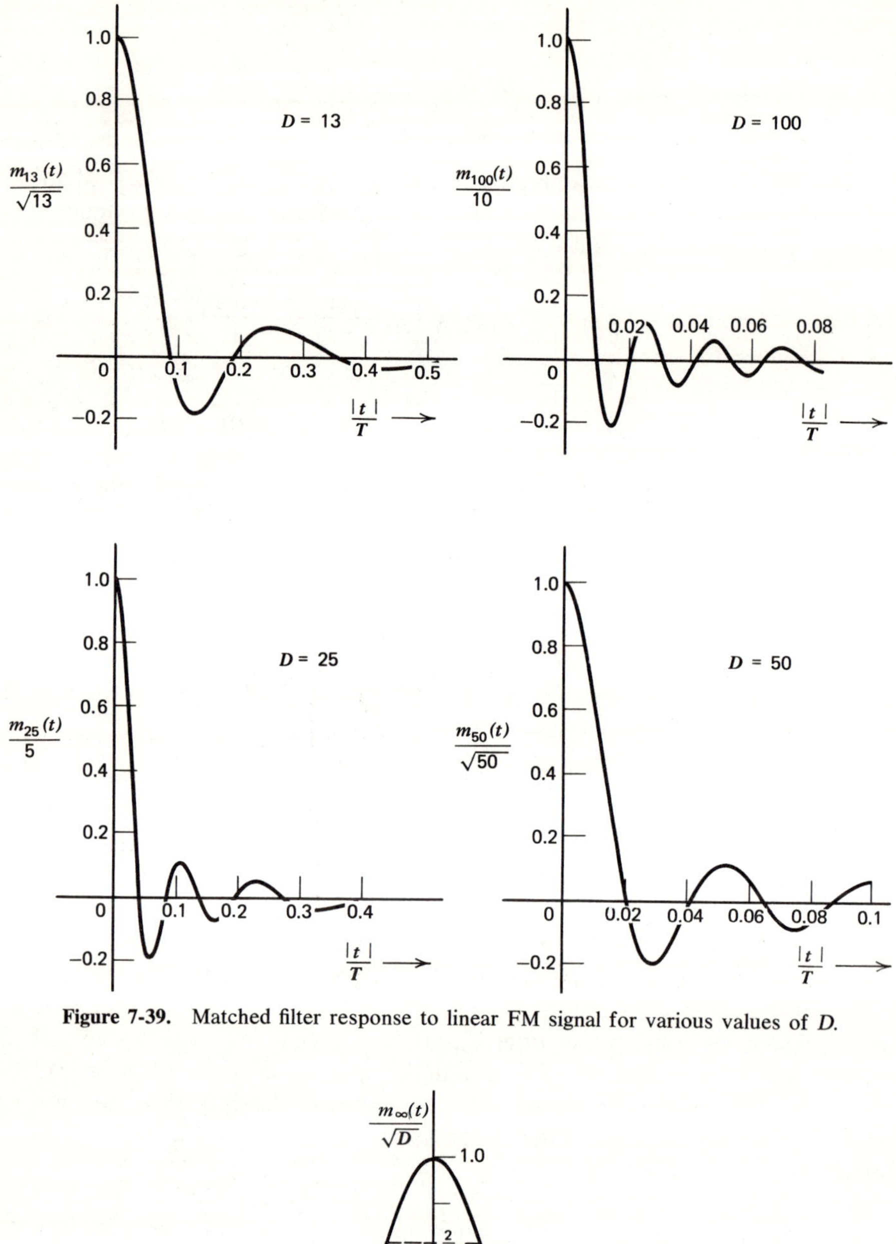

Figure 7-39. Matched filter response to linear FM signal for various values of D.

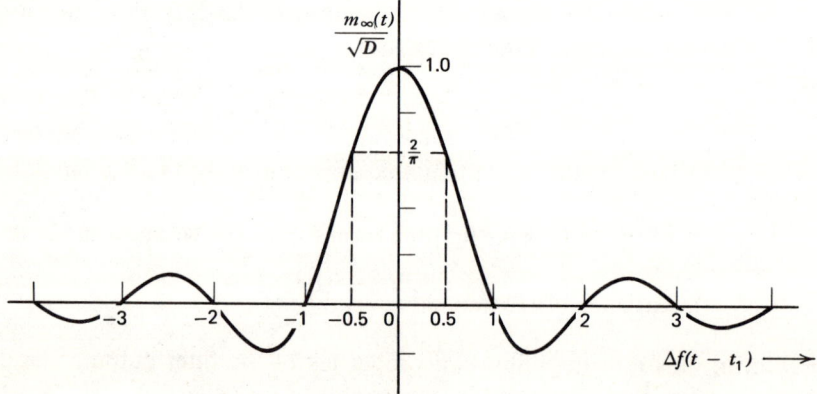

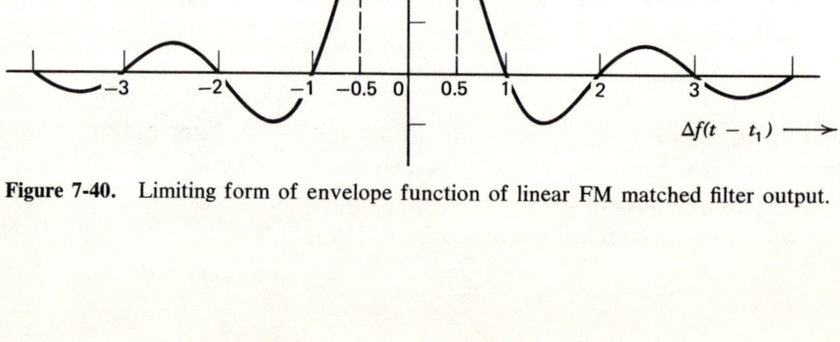

Figure 7-40. Limiting form of envelope function of linear FM matched filter output.

Pulse Compression Using Linear Frequency Modulation

peak value). These times occur at $t = t_1 \pm 1/(2\Delta f)$, hence the total pulse width is

$$\Delta t = t_1 + \frac{1}{2\Delta f} - \left(t_1 - \frac{1}{2\Delta f}\right) = \frac{1}{\Delta f} \tag{7.4-16}$$

which is the reciprocal of the rectangular spectrum width. Since the input pulse duration, from Fig. 7-37a, is T, the ratio of this pulse duration to the output pulse duration is

$$\frac{\text{input pulse duration}}{\text{output pulse duration}} = \frac{T}{1/\Delta f} = \Delta f T = D \tag{7.4-17}$$

Thus the filter compresses the input pulse by D; hence D is also known as the pulse compression ratio.

The peak output signal, interpreted as a voltage across a one-ohm resistor, is $\sqrt{D}$; thus the peak power is D. Comparing this to the peak input power of unity, we conclude that the peak power of the compressed pulse has increased by the factor D.

We have just shown how the matched filter principle, applied to the linear FM signal, achieves the advantages of both a long pulse and a short pulse. Transmitting a long pulse preserves the detection capabilities at the longer ranges, while the pulse compression properties of the filter allow the resolution capability to be retained.

The transfer function of the matched filter $H(j\omega)$ is obtained from (7.3-24) and (7.4-2). However, for the practical cases of interest ($D > 25$), $H(j\omega)$ is expressible as

$$H(j\omega) = \begin{cases} \exp j\left[\pi D\left(\frac{\omega_c - \omega}{\Delta \omega}\right)^2 - \omega t_1 - \theta_0\right] & \omega_c - \frac{\Delta \omega}{2} \leq \omega \leq \omega_c + \frac{\Delta \omega}{2} \\ 0 & \text{elsewhere} \end{cases} \tag{7.4-18}$$

Throughout the band $\Delta \omega$, the magnitude response is unity, the quadratic phase response is

$$\theta(\omega) = \pi D\left(\frac{\omega_c - \omega}{\Delta \omega}\right)^2 - \omega t_1 - \theta_0 \tag{7.4-19}$$

shown in Fig. 7-41a, and the linear group delay is

$$D(\omega) = -\frac{d\theta(\omega)}{d\omega} = t_1 + T\left(\frac{\omega_c - \omega}{\Delta \omega}\right) \tag{7.4-20}$$

shown in Fig. 7-41b. The variation of delay with frequency is called dispersion, and a delay device with this property is called a dispersive delay device.

The linear delay varies in an opposite manner to the instantaneous frequency of the input signal (Fig. 7-37b). Had $\omega(t)$ been assumed to decrease linearly with time, the filter group delay would linearly increase with frequency. This opposite relationship is necessary to compress the incoming pulse and produce the increased output pulse amplitude.

7.4.3 Compression Mechanism

The input instantaneous frequency and the linear filter delay functions provide enough information to yield a qualitative explanation for the sudden output voltage buildup at $t = t_1$. Consider $v(t)$, the input signal pulse in (7.4-1), as being

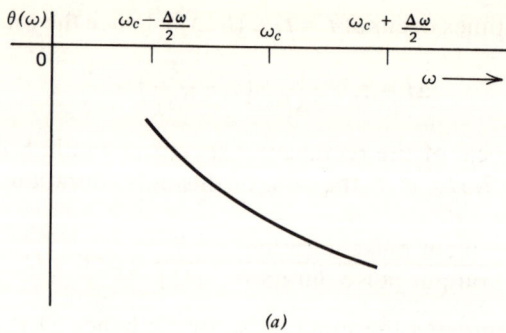

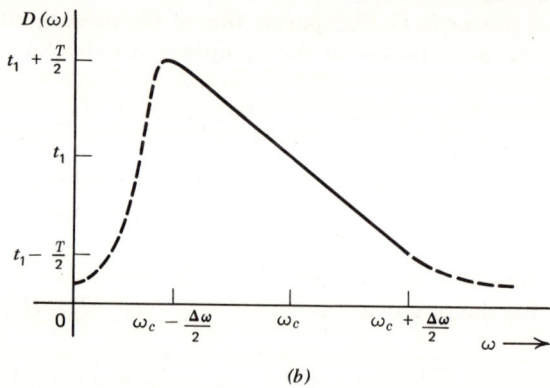

Figure 7-41. (a) Phase response and (b) group delay of linear FM matched filter.

composed of an infinite number of sinusoids, whose individual frequencies are given by $\omega(t)$ at each instant of time. For example, at $t = t_a$, one particular sinusoid has a frequency $\omega(t_a) = \omega_c + (\Delta\omega/T)t_a$. This sinusoid is then delayed by the optimum filter an amount $D(\omega)$ in (7.4-20), where $\omega = \omega(t_a)$, the frequency of the input sinusoid. Thus the sinusoid arrives at the filter output at the time $t_a + D(\omega)|_{\omega=\omega(t_a)}$. We now compute this arrival time at the output for an arbitrary sinusoid comprising the input signal $v(t)$, that is,

$$t + D(\omega)|_{\omega=\omega(t)} = t + t_1 + \frac{T}{\Delta\omega}(\omega_c - \omega)|_{\omega=\omega(t)} = t + t_1 + \frac{T}{\Delta\omega}\left[\omega_c - \left(\omega_c + \frac{\Delta\omega}{T}t\right)\right] = t_1$$

(7.4-21)

Therefore all spectral components of the input signal, independent of their frequency, reach the filter output at the same time t_1. Being in phase at this time, they add constructively, producing the increased output signal at $t = t_1$.

The increase in the output signal amplitude by the amount $\sqrt{D}$ can be explained as follows. Approximate the FM pulse in (7.4-1) by $v_1(t)$, the train of N equal pulses of unit amplitude (Fig. 7-42). Each pulse width is then T/N. The carrier frequency of each subpulse is constant, approximating the linear frequency function of (7.4-1) over the subpulse interval, as in Fig. 7-43. Thus the frequency

Pulse Compression Using Linear Frequency Modulation

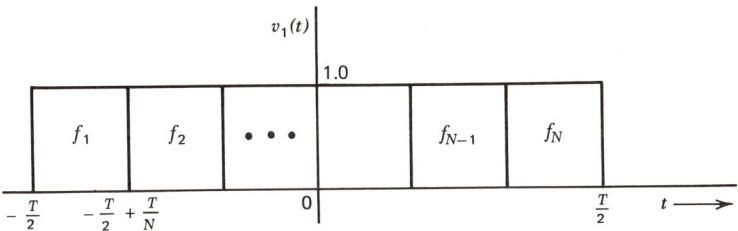

Figure 7-42. Approximation of the linear FM pulse by N pulses having different constant carrier frequencies.

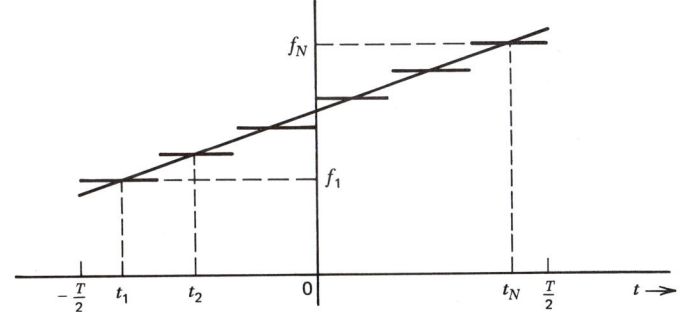

Figure 7-43. Step-approximation of linear FM.

of the kth pulse is

$$f_k = f_c - \frac{\Delta f}{2} + \frac{(2k-1)\Delta f}{2N} \qquad (k = 1, 2, \ldots, N) \qquad (7.4\text{-}22)$$

which coincides with the FM pulse frequency at the instant

$$t_k = -\frac{T}{2} + \frac{(2k-1)T}{2N} \qquad (7.4\text{-}23)$$

We now determine the value of N so that $v_1(t)$ is a meaningful approximation to $v(t)$ in (7.4-1). A good approximation to each subpulse spectrum is the familiar cardinal function given in Fig. 7-44. Overlapping the spectra of the N subpulses so that all zeros crossings coincide, as in Fig. 7-45, then summing them, results in a

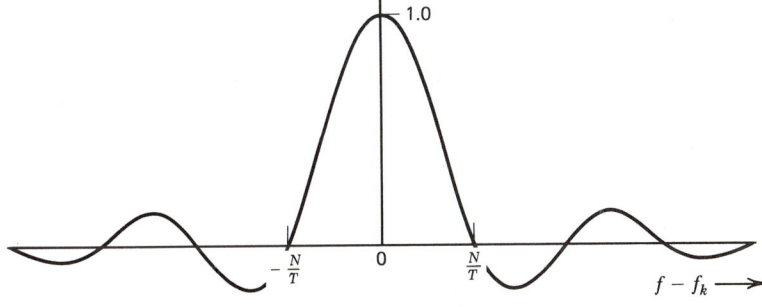

Figure 7-44. Spectrum of each subpulse of $v_1(t)$.

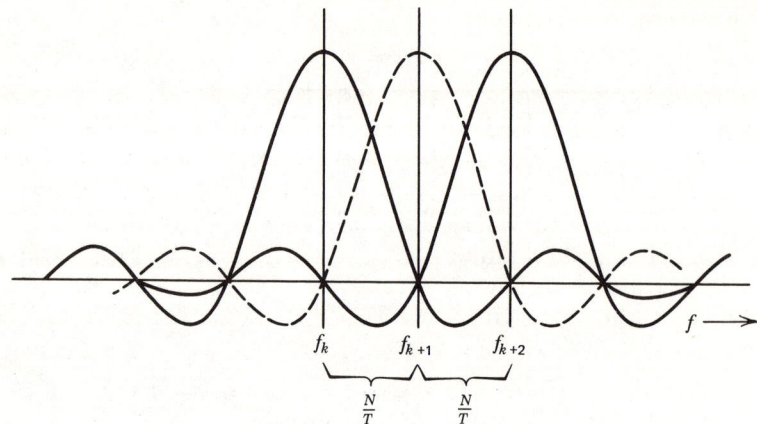

Figure 7-45. Spectrum spacing for overall rectangular spectrum.

spectrum that approximates the rectangular spectrum of the linear FM signal. Overlapping in this manner leaves the unit amplitude at f_k unchanged because all other spectra are zero there. From Fig. 7-45, this overlap condition is obtained when

$$f_{k+1} - f_k = \frac{N}{T} \tag{7.4-24}$$

From (7.4-22)

$$f_{k+1} - f_k = f_c - \frac{\Delta f}{2} + \frac{(2k+1)\Delta f}{2N} - \left[f_c - \frac{\Delta f}{2} + \frac{(2k-1)\Delta f}{2N} \right] = \frac{\Delta f}{N} \tag{7.4-25}$$

which inserted into (7.4-24) yields

$$N = \sqrt{\Delta f T} = \sqrt{D} \tag{7.4-26}$$

A first-order approximation to the linear FM pulse is therefore $\sqrt{D}$ subpulses of time duration $T/\sqrt{D}$. The carrier frequency of the kth pulse is constant and is given by (7.4-22). If $\sqrt{D}$ is not an integer, $v_1(t)$ is not practically realizable, but it is still a valid mathematical model.

Now consider the response of the unity-gain filter matched to $v_1(t)$. We assume that N is large, as it must be for the rectangular spectrum approximation to be valid. Then the maximum value of this filter's output at $t = t_1$ is N times the peak output due to one pulse, since (7.4-21) shows that all sinusoids arrive simultaneously at the filter output at $t = t_1$. Quantitatively the peak filter response to one input pulse is the autocorrelation function of the pulse evaluated at $\tau = 0$,

$$v_k(0) = \frac{N}{T} \int_0^{T/N} \cos^2 \omega_k t \, dt = \frac{N}{T} \times \frac{T}{2N} \left[1 + \frac{\sin(2\omega_k T/N)}{2\omega_k T/N} \right] \tag{7.4-27}$$

where the constant N/T is necessary for the unity filter gain.

But we have assumed that N is large, therefore $v_k(0) \approx 1$. The peak response due to N pulses is then N. But $N = \sqrt{D}$, the same output value obtained for the linear FM matched filter response. By representing $v(t)$ by a train of narrow RF pulses, for which analytic procedures are available, we have shown that the output increase is $\sqrt{D}$.

7.4.4 Sidelobe Reduction

In a multiple-target environment, the sidelobes in the matched filter response (the smaller subsidiary peaks in Fig. 7-40) can mask smaller adjacent targets, thus limiting the dynamic range and/or the range resolution of the radar system. In many applications, therefore, the efficient reduction of these sidelobe levels can lead to a desirable improvement in target detectability. However any change that we make to achieve this goal will also reduce the peak signal-to-mean noise power ratio (SNR), since the matched filter property is then also changed. Intuitively, the SNR should not be drastically affected, since we are reducing output values already attenuated at least 13 dB. And indeed this turns out to be correct.

Assuming the input FM signal is not changed, the sidelobes can be reduced by shaping (weighting) either the filter output frequency spectrum or the output time-function envelope. We know from the discussion in Section 3.1 that the sidelobes arise because of the rapid change of the rectangular output spectrum in the vicinity of the cutoff frequency. Thus whether we attack this problem in the time or frequency domain, the solution will yield an output spectrum having a more gradual transition region.

This result can be obtained in two ways. The first is to replace the matched filter with one whose magnitude response shapes the rectangular input spectrum in the appropriate manner. The second approach is to cascade this appropriate shaping filter with the matched filter. The latter technique, although it may require more hardware, is advantageous in that the shaping filter can be switched in for improved sidelobe reduction or switched out for maximum SNR.

To illustrate frequency shaping, we replace the matched filter by a BP filter whose magnitude response is the Gaussian function

$$|H(j\omega)| = \exp\left[-1.386\left(\frac{f-f_c}{\Delta F}\right)^2\right] \tag{7.4-28}$$

where ΔF is the filter 3-dB bandwidth and f_c is the filter center frequency. The filter phase response is the optimum parabolic phase response of (7.4-19) and is assumed to remain the same, independent of the filter bandwidth. A necessary design parameter is

$$n = \frac{\Delta F}{\Delta f}$$

where Δf is the frequency shift of the input FM signal in (7.4-1) and also the bandwidth of the rectangular signal spectrum. Figure 7-46 shows the output envelope function $m_D(t)/\sqrt{D}$ for $n = 0.4, 0.6$, and ∞. The curve for $n = \infty$ reflects the absence of the shaping filter and is the matched filter response in Fig. 7-40 for which the minimum sidelobe suppression is 13.3 dB. As the filter bandwidth is narrowed, the first sidelobe is attenuated 19.3 dB with $n = 0.6$ and 44.9 dB with $n = 0.4$. However this bandwidth reduction also lowers the SNR, and analysis reveals that minimum degradation occurs at $n = 0.62$. For this value the SNR loss compared to the optimum filter ($n = \infty$) is 1.1 dB. For values of n between 0.38 and 0.96, the maximum SNR loss does not exceed 3.1 dB.

Superior results are obtained by selecting other frequency-shaping functions. Frequency-weighting functions with desirable properties have been the subject of investigation for many years, and the more famous ones are discussed and

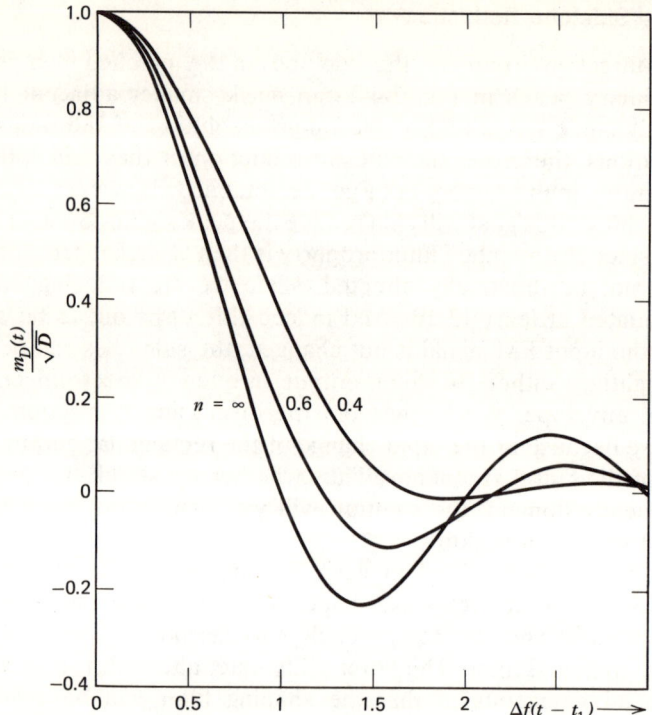

Figure 7-46. Gaussian filter envelope response to linear FM signal of (7.4-1).

summarized in Refs. 2 and 11. In Section 9.6.10 we again consider these functions, using them as weighting functions in the time domain to reduce sidelobes in digital filter magnitude responses.

Briefly, some of these frequency-weighting functions are as follows.

1. Dolph–Chebyshev—originally obtained for antenna design, it is optimum in the sense that it produces the narrowest main lobe for a specified sidelobe level. This function is physically unrealizable, but it provides an optimum measure against which other functions can be compared.

2. Taylor—expressed as a finite periodic series, it approximates the Dolph–Chebyshev function.

3. Hamming—yields the lowest sidelobe levels for the class of cosine-squared weighting functions, it is expressed as

$$W(\omega) = 0.08 + 0.92 \cos^2 \pi \frac{\omega_c - \omega}{\Delta \omega} \qquad (7.4\text{-}29)$$

4. General cosine-power weighting—a more general function than the Hamming function, it is given by

$$W(\omega) = k + (1-k) \cos^n \pi \frac{\omega_c - \omega}{\Delta \omega} \qquad (7.4\text{-}30)$$

Pulse Compression Using Linear Frequency Modulation

7.4.5 The Tapped Delay Line

The periodicity of the weighting functions just given suggests the tapped delay line, often called a transversal filter, as one possible realization. This filter type was introduced and its operation was explained in Section 7.3.6, but we again briefly examine it because of its usefulness as a matched filter and sidelobe reduction filter. Its transfer function is given by

$$H(j\omega) = \sum_{k=-N}^{N} a_k \exp\left[-j\frac{2\pi k(\omega - \omega_c)}{\Delta\omega}\right] \qquad (7.4\text{-}31)$$

and if the a_k's are real and symmetric about a_0, then $H(j\omega)$ is expressible as

$$H(j\omega) = a_0 + 2\sum_{k=1}^{N} a_k \cos 2\pi k \frac{(\omega - \omega_c)}{\Delta\omega} \qquad (7.4\text{-}32)$$

plus a linear phase shift for physical realizability. This linear phase shift is the significant feature of the symmetric transversal filter† for then the cascade of this filter with the optimum filter does not upset the phase characteristics of the latter. Associated with tapped delay lines are periodic frequency-domain functions, illustrated by the magnitude response in Fig. 7-47. The periodicity does not affect the optimum filter output spectrum, which is essentially rectangular, therefore negligible outside the region indicated in Fig. 7-47.

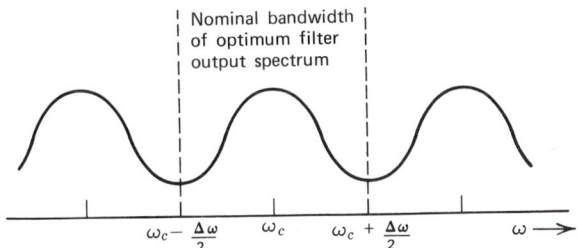

Figure 7-47. Tapped delay line frequency response.

Also characteristic of a tapped delay line filter is the finite duration of its impulse response, which is composed of a train of weighted impulse functions. This property is discussed in Section 9.6.5, where we consider the transversal filter in more detail. There, a class of digital filters is realized as transversal filters, but the basic filter operation remains the same. The essential differences are that the delay is realized by a shift register rather than by an analog delay line, and the inputs are numbers rather than analog functions.

In a typical transversal filter diagram (Fig. 7-48), each rectangle represents a section of a nondispersive delay line of delay T, although in the more general case, the time delays between the taps need not be the same. The triangles represent weighting coefficients of value $a_0, a_1, a_2, \ldots$, and are often called tap weights. The loading of the taps on the transmission line is assumed to be very small, ensuring that the input signal travels the length of the line with negligible reflection or

---

†This property is discussed in Section 9.6.7.

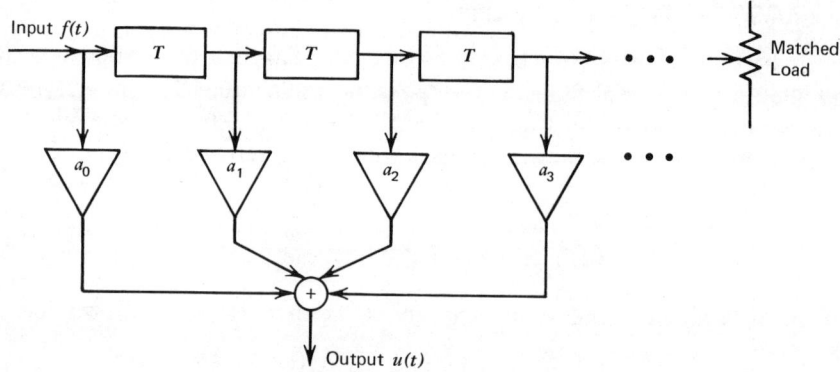

Figure 7-48. Typical transversal filter (tapped delay line).

attenuation, and it is absorbed in the matched load. After each delay the input signal is appropriately scaled by the tap weight, and these individual functions are added to yield the output. Mathematically, the output is synthesized as

$$u(t) = a_0 f(t) + a_1 f(t - T) + a_2 f(t - 2T) + a_3 f(t - 3T) + \cdots \quad (7.4\text{-}33)$$

Figure 7-49 illustrates a symmetric tapped delay line for sidelobe reduction. The notation is changed to accommodate this symmetry: $T = 1/\Delta f$, which is the reciprocal of the spectrum rectangular width, and the total delay of the

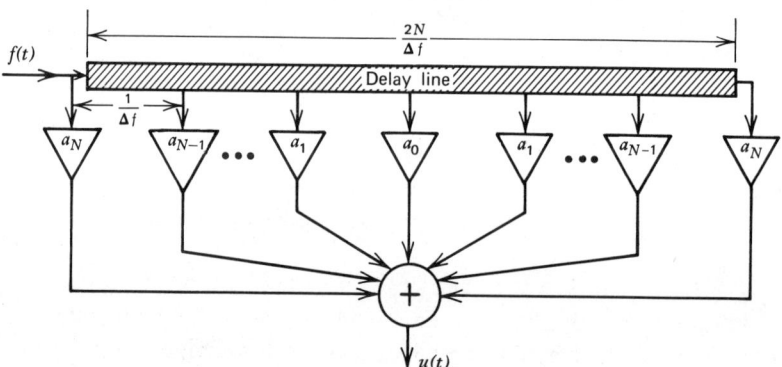

Figure 7-49. Symmetric tapped delay line.

nondispersive delay line is $2N/\Delta f$. Figure 7-50 shows the compressed pulse envelope at the output of a sixth-order transversal filter designed to exhibit the Taylor weighting function. The sidelobes are now attenuated at least 40 dB, compared to the 13.3 dB at the output of the matched filter (Fig. 7-40). The main pulse width has slightly increased. Careful shaping of the rectangular spectrum does indeed reduce the sidelobes, thereby improving target detectability.

7.4.6 Optimum Filter Realizations

The optimum filter transfer function in (7.4-18) is required only over the bandwidth $\Delta \omega$ because the signal spectrum is confined to this bandwidth. The

Pulse Compression Using Linear Frequency Modulation

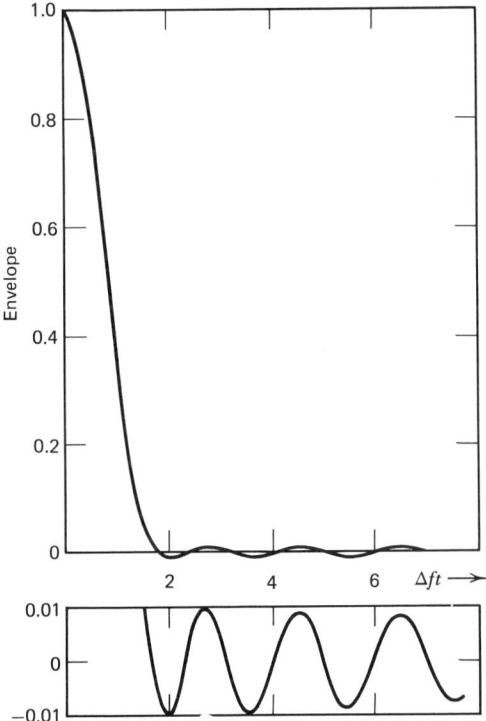

Figure 7-50. Compressed pulse output envelope of the sixth-order Taylor weighting filter.

filter characteristics outside this band do not affect the response to the linear FM signal.

Several devices are available for achieving the required linear delay and constant magnitude response. One approach, and one of the first used in practice, is a cascade of the appropriately designed all-pass sections discussed in Chapter 5. The composite filter, however, may contain hundreds of sections, consequently hundreds of coils and capacitors. The equivalent bridge realizations halve the number of components, but the resulting filter is still cumbersome. As a historical note, the optimum filter for the SPS-37 and SPS-43, two of the earliest operational shipboard radars using pulse compression, was realized by a cascade of 240 identical second-order AP lattice sections. These radar systems are still in operation.

The bulk-wave delay line is often used for optimum filtering of narrow frequency-band but long time-duration signals. These lines are made from aluminum or steel strips, and a piezoelectric transducer at each end transforms electrical energy into mechanical energy, and vice versa. As the longitudinal ultrasonic waves propagate through the device, the medium exhibits the desired linear delay characteristics over the necessary bandwidth.

New filter techniques rely on surface wave propagation and charge transfer. These devices, briefly discussed in Section 7.6, are realized as transversal filters. Surface wave technology has developed to an advanced stage, and many such devices serve as the matched filter in radar and communication systems. Charge-coupled equipment on the other hand, is in the development stage, although prototypes have been produced.

In practice, the input signal is often sampled prior to filtering. The sampled input is then a series of narrow pulses whose amplitudes represent the input function at that instant. Figure 7-51 illustrates this sampling process for a sample time T.

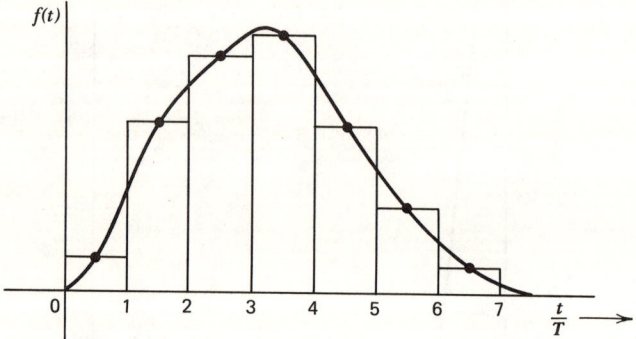

Figure 7-51. Approximation of $f(t)$ by a series of pulses.

Example 7-6. We now determine the response of the third-order transversal filter in Fig. 7-48 to the sampled input signal in Fig. 7-52. The three tap weight values are $a_0 = -1$, $a_1 = +1$, and $a_2 = +1$. The output is determined from (7.4-33) by displaying each term and then summing them all, as in Fig. 7-53. Note that we chose the tap weights to ensure that the transversal filter is the matched filter for this input, that is, the tap weights are the amplitudes of the pulse amplitudes, reversed in time. Consequently the output signal is proportional to the autocorrelation function of $f(t)$, and the peak signal is three times the value of the input pulses. This corresponds to a pulse compression of 3.

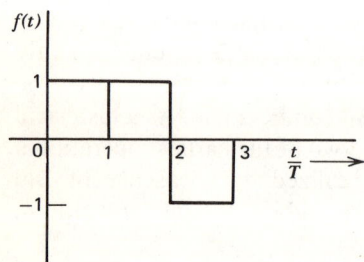

Figure 7-52. Transversal filter input signal.

7.5 PHASE-CODED WAVEFORMS

A phase-coded waveform can also be used to achieve the benefits of a frequency-modulated pulse. The basic pulse is subdivided into short subpulses of equal time duration T, and each subpulse is transmitted with a particular phase, selected in accordance with a specific code. The most popular type of phase coding is binary; that is, the pulse amplitude is either $+1$ or -1. This corresponds to a transmitted carrier phase that is either 0° or 180° (Fig. 7-54). The coded signal

Phase-Coded Waveforms

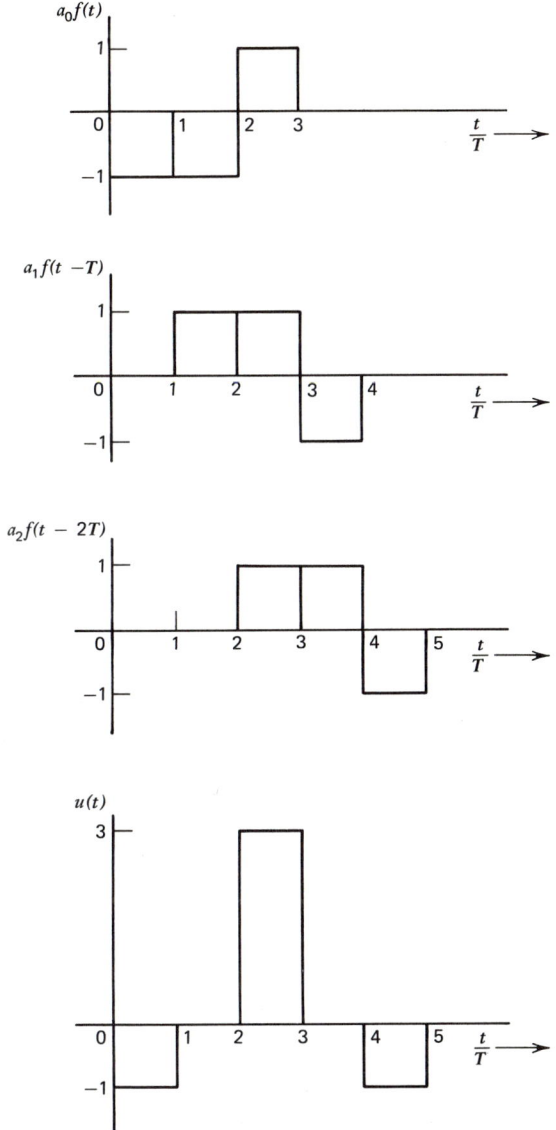

Figure 7-53. Graphical determination of transversal filter output.

may be discontinuous at the phase reversal point because the transmitted frequency is not necessarily a multiple of the reciprocal of the subpulse width.

The pulse compression at the receiver is obtained by either matched filtering or correlator processing. The compressed pulse width at the half-amplitude point is approximately equal to the subpulse width; hence the range resolution is proportional to this value.

Example 7-7 we now illustrate pulse compression by phase coding a modulating pulse of $5T$ duration. Figure 7-55a shows the received pulse and the matched filter impulse response, and the resulting filter output $u(t)$, which is $10T$ in duration

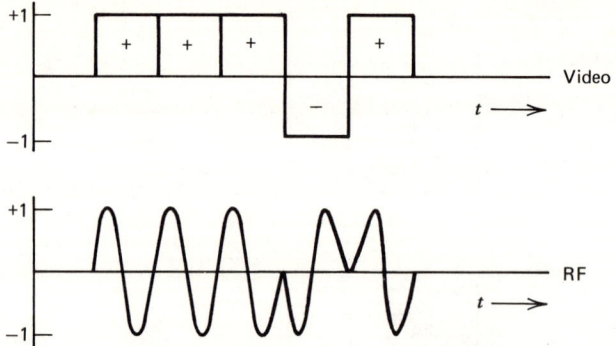

Figure 7-54. Binary phase-coded signal.

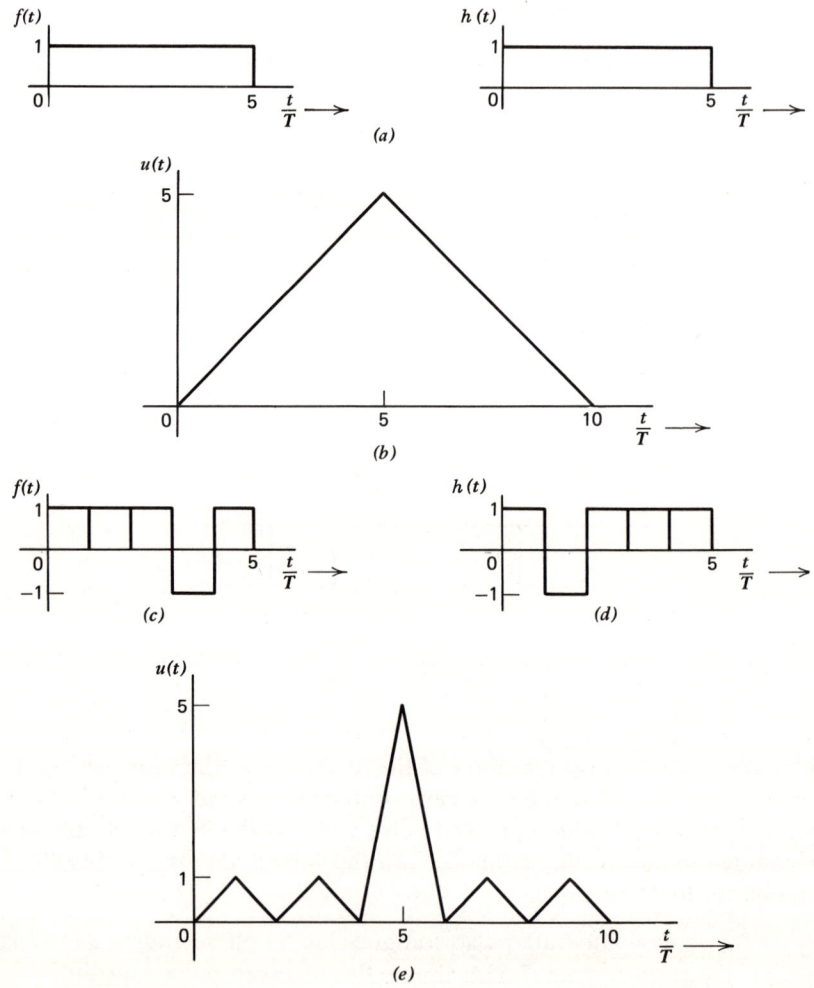

Figure 7-55. Pulse compression achieved by phase coding (Example 7-7).

Phase-Coded Waveforms

appears in Fig. 7-55b. To improve the range resolution, that is, to reduce the processed pulse width, we partition the pulse into five subpulses, and change the sign of the fourth subpulse to -1, as in Fig. 7-55c. The matched-filter impulse response for this phase-coded signal is shown in Fig. 7-55d. The filter output is the convolution of these two functions, and the result is given in Fig. 7-55e. The total duration remains $10T$, but the duration of its principal peak, measured at the half-amplitude point, is now T. Comparing to the original pulse duration of $5T$, we conclude that the pulse compression ratio is 5, which is also equal to the number of subpulses in the waveform. The sequence in Fig. 7-55c is the fifth-order Barker code, which has optimal properties.

7.5.1 The Barker Codes

The most widely used binary codes are the Barker sequences. They are optimum in the sense that the autocorrelation function peak is N and the sidelobe level falls between $+1$ and -1, where N is the number of subpulses (elements). Thus each additional element increases the compression ratio by one. Unfortunately no Barker code with more than 13 elements has been found; furthermore, Barker codes do not even exist for all N less than 13. Table 7-3 lists all possible codes and the corresponding peak-to-sidelobe ratio. The plus sign refers to a pulse amplitude of unity; the minus designates a pulse amplitude of -1.

Table 7-3 The Known Barker Codes

Length of code, N	Code Elements	Peak-to-Sidelobe ratio, $20 \log N$ (dB)
2	+ −, (+ +)	6.0
3	+ + −	9.5
4	+ + − +, (+ + + −)	12.0
5	+ + + − +	14.0
7	+ + + − − + −	16.9
11	+ + + − − − + − − + −	20.8
13	+ + + + + − − + + − + − +	22.3

The autocorrelation functions for the codes $N = 4, 7, 11,$ and 13 in Fig. 7-56 are normalized for unit output at $\tau = 0$. Note that two possibilities exist for $N = 4$. The autocorrelation function for $N = 5$ appears in Fig. 7-55e and is discussed in Example 7-7.

The Barker coded waveforms indeed possess the desirable property for pulse compression. Radar systems (TPS43, FPS27) that are now operational use the thirteenth-order Barker code to achieve this pulse compression. The heart of the signal processor for these radars is a surface wave matched filter (correlator), often referred to as the decoder.

The calculation of the autocorrelation function for these binary coded waveforms at first may seem to be computationally difficult, however, it is easily accomplished by simple multiplication. We now illustrate this procedure for the

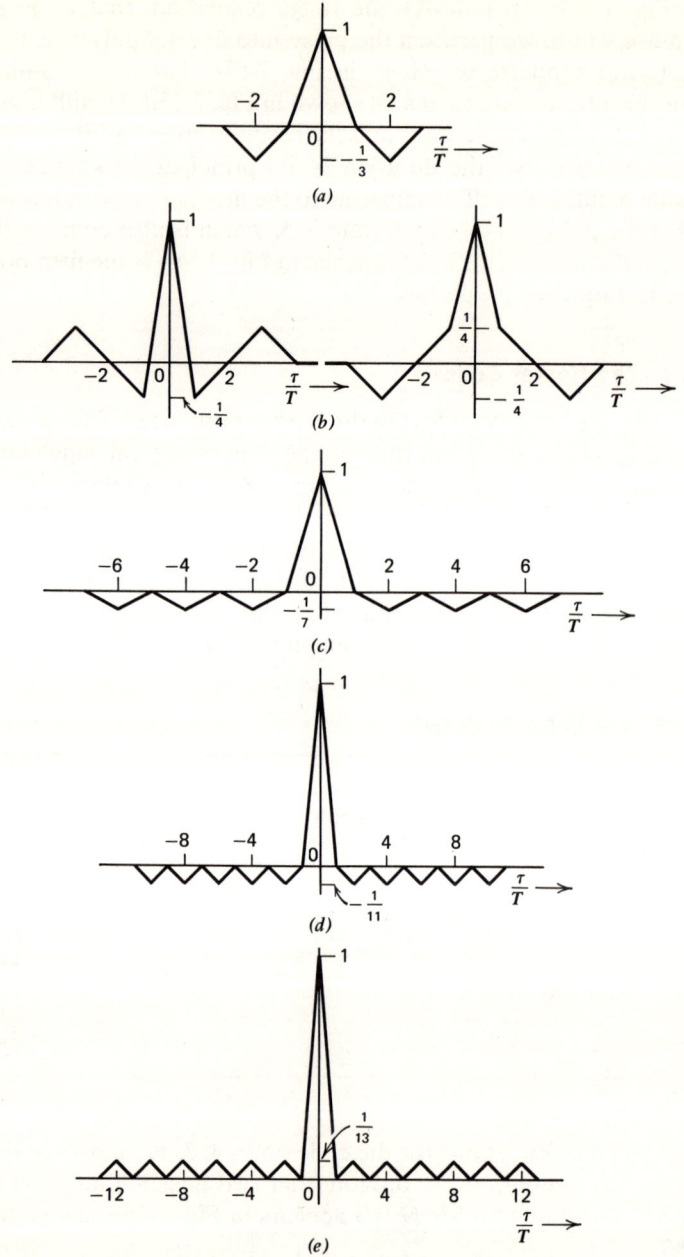

Figure 7-56. Autocorrelation functions for Barker codes. (*a*) $N = 3$, (*b*) $N = 4$, (*c*) $N = 7$, (*d*) $N = 11$, (*e*) $N = 13$.

Phase-Coded Waveforms

fifth-order Barker code (Fig. 7-55c). Its sequence, from Table 7-2, is $+1, +1, +1, -1, +1$, which corresponds to the pulse amplitudes in Fig. 7-55c. The matched filter impulse response is the sequence in Fig. 7-55d, which also corresponds to the tap weights of a transversal filter. The filter response to the Barker sequence is then the autocorrrelation function of this sequence and is obtained by multiplication of the two sequences. Each line in standard multiplication is automatically shifted, and this corresponds to a delay of one pulse width, a necessary operation in autocorrelation computations. The integration of the resulting product is unity, since each element has unit amplitude and is normalized to unit pulse width.

The computation of the filter output is as follows:

```
                        1   1   1  -1   1          Input sequence
      ×                 1  -1   1   1   1          Impulse response
      ─────────────────────────────────────
                        1   1   1  -1   1
                    1   1   1  -1   1
                1   1   1  -1   1
         -1  -1  -1   1  -1
     1   1   1  -1   1
      ─────────────────────────────────────
     1   0   1   0   5   0   1   0   1          Output sequence
     1   2   3   4   5   6   7   8   9          Delay variable ($\tau/T$)
```

The output sequence is then connected by straight lines to yield the correct autocorrelation function, as in Fig. 7-55e. This technique allows the transversal filter response to be determined by arithmetic operations and eliminates tedious integration procedures.

7.5.2 Amplitude Spectra of Pulse Sequences

The frequency representation of a train of video pulses is of considerable interest, for these pulses are often used to modulate a carrier frequency in radar and communication systems. An analytic expression for this representation is now derived, with the Barker representation presented as a special case.

Consider $f(t)$, the train of N video pulses in Fig. 7-57. The time duration of each pulse is T, and its amplitude, indicated by a_k, can be positive or negative. Mathematically $f(t)$ is expressible as

$$f(t) = \sum_{k=0}^{N-1} a_k p(t - kT) \qquad (7.5\text{-}1)$$

where $p(t - kT)$ is the rectangular pulse of unit amplitude delayed by kT (Fig. 7-58).

The frequency domain representation of $f(t)$, from Table 2-1, entry 12, is

$$F(s) = \sum_{k=0}^{N-1} a_k e^{-skT} P(s) \qquad (7.5\text{-}2)$$

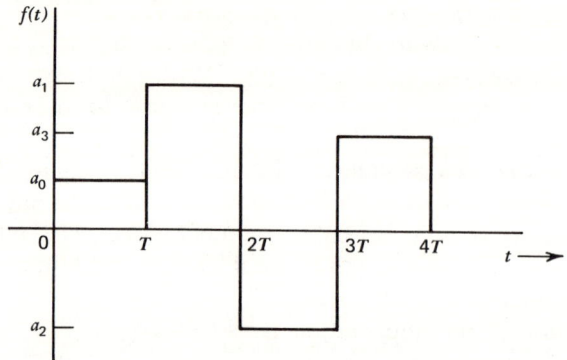

Figure 7-57. Train of N equal-time-duration video pulses.

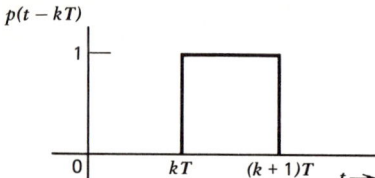

Figure 7-58. Rectangular pulse $p(t)$ delayed by kT.

where $P(s)$, the Laplace transform of $p(t)$, is

$$P(s) = \frac{1 - e^{-sT}}{s} \qquad (7.5\text{-}3)$$

The amplitude density spectrum $|F(j\omega)|$ is then

$$|F(j\omega)| = |P(j\omega)| \left| \sum_{k=0}^{N-1} a_k e^{-j\omega kT} \right| = T \left| \frac{\sin(\omega T/2)}{\omega T/2} \right| R(\omega) \qquad (7.5\text{-}4)$$

which is the product of the amplitude density spectrum of the basic pulse $p(t)$ and the shaping factor $R(\omega)$, given by

$$R(\omega) = \left[\left(\sum_{k=0}^{N-1} a_k \cos k\omega T \right)^2 + \left(\sum_{k=0}^{N-1} a_k \sin k\omega T \right)^2 \right]^{1/2} \qquad (7.5\text{-}5)$$

Thus $(\sin(\omega T/2))/(\omega T/2)$, the envelope of $|F(j\omega)|$, is aperiodic and depends on the basic pulse width, whereas $R(\omega)$ is periodic with period $\omega_P = 2\pi/T$ and is a function of the pulse amplitudes. The zeros of the envelope function enable us to estimate the maximum possible RF signal bandwidth as $1/2T$ hertz, independent of the a_k values. At $\omega = 0$,

$$|F(j0)| = T \left| \sum_{k=0}^{N-1} a_k \right| \qquad (7.5\text{-}6)$$

For the RF case, $f(t)$ is the amplitude modulating function of the carrier frequency ω_c, and if most of the energy in $f(t)$ lies in a bandwidth that is small compared to ω_c, the spectrum for the RF signal is obtained by shifting the LP spectrum to ω_c. This condition is often encountered in practice.

Phase-Coded Waveforms

For the Barker sequences $R(\omega)$ can be expressed as

$$R_B(\omega) = \left| N \pm \left(\frac{\sin N\omega T}{\sin \omega T} - 1 \right) \right|^{1/2} \quad (7.5\text{-}7)$$

where the positive sign is used for $N = 5$ and $N = 13$, and the negative sign is used for $N = 3, 7,$ and 11. At $\omega = 0$,

$$|F(j0)| = \begin{cases} T & N = 3, 7, 11 \\ 3T & N = 5 \\ 5T & N = 13 \end{cases} \quad (7.5\text{-}8)$$

Figures 7-59 and 7-60 show the Barker spectra and the envelope functions for $N = 11$ and $N = 13$, respectively. The spectrum of each is denormalized by multiplying the ordinate values by T. For example, the value at $\omega = 0$ for $N = 13$ is $5T$. Both spectra are essentially the same except in the vicinity of $\omega T = 0$ and $\omega T =$ odd multiples of π. These differences result from the differences in the basic structure of the two codes.

7.5.3 Sidelobe Reduction

The matched filter (Barker decoder) response to the Barker sequences, like that of the FM signal matched filter response, exhibits sidelobes that are undesirable in multiple target environments. These sidelobes can mask smaller adjacent targets, thus limiting the system dynamic range and/or range resolution. Similar to the FM signal situation, a suitable solution is to place an appropriately designed transversal filter in cascade with the matched filter. Efforts in this direction have yielded improved but nonoptimal responses; that is, the peak-to-sidelobe ratio (PSR) is not maximized for a given filter order [6, 10].

Here we describe a time-domain procedure for selecting the filter tap weights to optimize the PSR at the output of the transversal filter. Furthermore, the resulting equations can be manually solved to yield explicit solutions. However, for the higher-order sequences the arithmetic becomes tedious, and the solutions are obtained on a digital computer by a linear programming algorithm.

Figure 7-61 illustrates a scheme for processing the Barker codes. Two output terminals are included, permitting either the matched filter response or the response with the optimum PSR to be obtained. The Barker decoder output, which is the autocorrelation function of the input Barker sequence, is expressible as the sequence

$$\cdots \pm 1 \quad 0 \quad \pm 1 \quad 0 \quad N \quad 0 \quad \pm 1 \quad 0 \quad \pm 1 \cdots \quad (7.5\text{-}9)$$

where the plus or minus sign depends on the specific code, and the sequence terminates after N numbers on each side of N. These numbers are the function values at multiples of the subpulse duration T. For example, from Fig. 7-55e, the autocorrelation function for the fifth-order code is

$$0 \quad 1 \quad 0 \quad 1 \quad 0 \quad 5 \quad 0 \quad 1 \quad 0 \quad 1 \quad 0 \quad (7.5\text{-}10)$$

The transversal filter is assumed to be symmetric about the center tap, which means that its output sequence is also symmetric. We assume N nonzero tap weights (the code order), with the odd-order tap weights set to zero. This

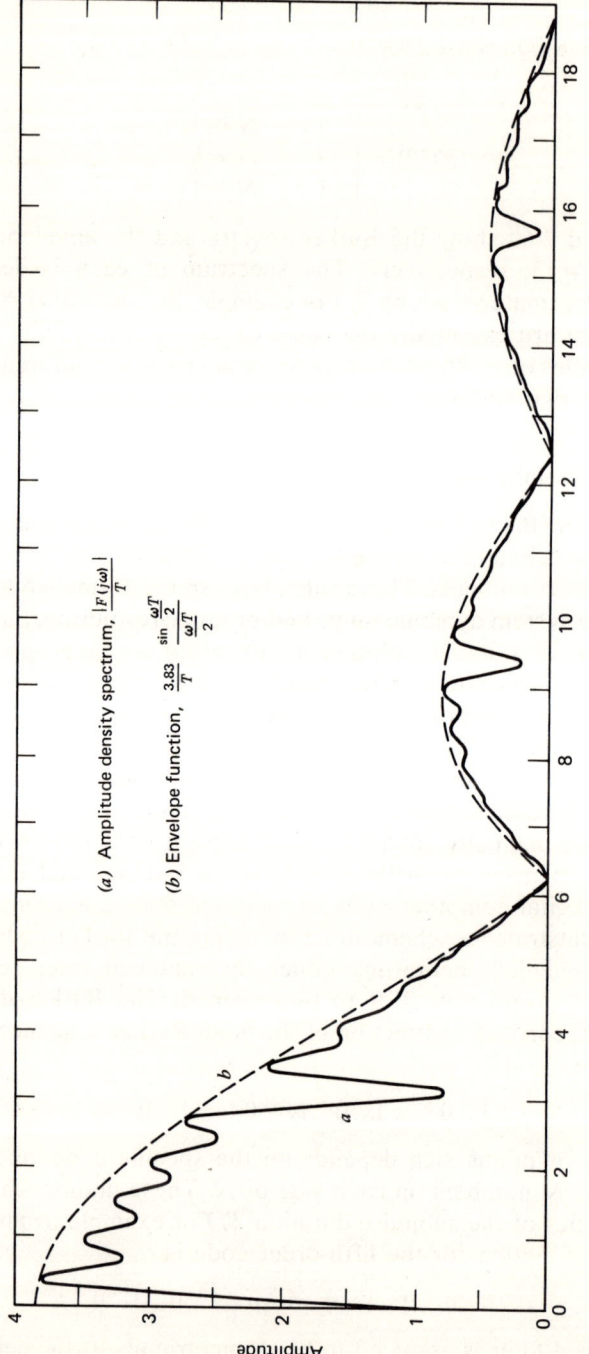

Figure 7-59. Amplitude density spectrum (*a*) and envelope function (*b*) for the $N = 11$ Barker sequence; T is subpulse duration.

(*a*) Amplitude density spectrum, $\dfrac{|F(j\omega)|}{T}$

(*b*) Envelope function, $\dfrac{3.83}{T} \dfrac{\sin\frac{\omega T}{2}}{\frac{\omega T}{2}}$

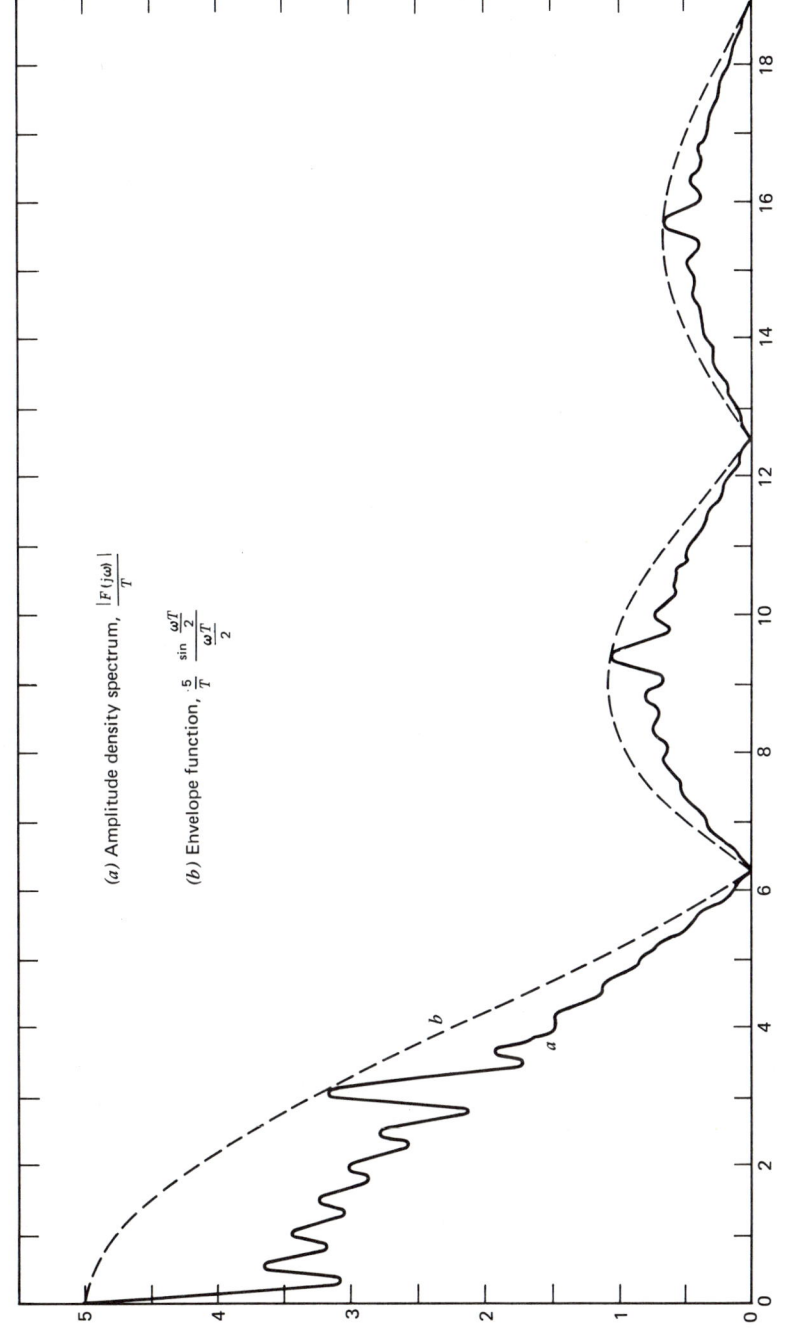

Figure 7-60. Amplitude density spectrum (*a*) and envelope function (*b*) for the $N = 13$ Barker sequence; T is subpulse duration.

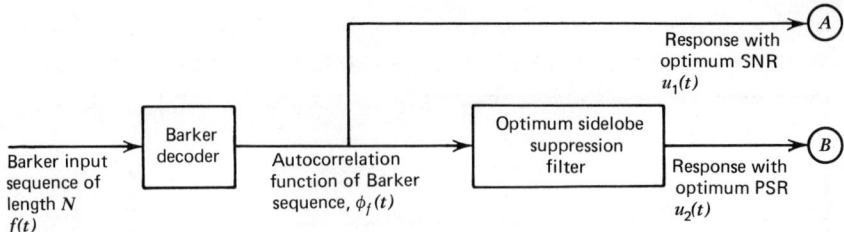

Figure 7-61. Barker code processor yielding either optimum SNR at A or optimum PSR at B.

sequence is then similar to the Barker sequence autocorrelation function and is given by

$$0 \quad a_{N-1} \quad \cdots \quad 0 \quad a_4 \quad 0 \quad a_2 \quad 0 \quad a_0 \quad 0 \quad a_2 \quad 0 \quad a_4 \quad 0 \quad \cdots \quad a_{N-1} \quad 0 \tag{7.5-11}$$

for the odd-ordered codes. The even-order codes ($N = 2, 4$) are not of practical interest.

The filter output, obtained by multiplying the sequences in (7.5-9) and (7.5-11), is illustrated for the fifth-order code.

								0	1	0	1	0	5	0	1	0	1	0	Output of Barker decoder
								a_4	0	a_2	0	a_0	0	a_2	0	a_4			Transversal filter impulse Response
×																			
								0	a_4 0	a_4 0	$5a_4$ 0	a_4 0	a_4 0						
						0	a_2 0	a_2 0	$5a_2$ 0	a_2 0	a_2 0								
				0	a_0 0	a_0 0	$5a_0$ 0	a_0 0											
		0	a_2 0	a_2 0	$5a_2$ 0	a_2 0	a_2 0												
0	a_4 0	a_4 0	$5a_4$ 0	a_4 0	a_4 0														
0	b_8 0	b_6 0	b_4 0	b_2 0	b_0 0	b_2 0	b_4 0	b_6 0	b_8 0										Transversal filter response

where

$$b_0 = 5a_0 + 2a_2 + 2a_4 \quad \text{Main response amplitude}$$

$$\left.\begin{array}{l} b_2 = a_0 + 6a_2 + a_4 \\ b_4 = a_0 + a_2 + 5a_4 \\ b_6 = a_2 + a_4 \\ b_8 = a_4 \end{array}\right\} \quad \text{Sidelobe amplitudes}$$

The design task is to select the tap weights in a way that maximizes the PSR; we wish to maximize the amplitude of the main lobe while constraining the amplitude of the sidelobes to be between -1 and $+1$. For the case of $N = 5$, the problem is to maximize the function

$$b_0 = 5a_0 + 2a_2 + 2a_4 \tag{7.5-12}$$

subject to the constraints

$$\begin{aligned} |b_2| &= |a_0 + 6a_2 + a_4| \leq 1 \\ |b_4| &= |a_0 + a_2 + 5a_4| \leq 1 \\ |b_6| &= |a_2 + a_4| \leq 1 \\ |b_8| &= |a_4| \leq 1 \end{aligned} \tag{7.5-13}$$

Phase-Coded Waveforms

This is a linear programming problem for which methods of solution are well established [3]. Furthermore the solution algorithm is especially well suited for computer implementation, and various such programs are available. The algorithm is now performed manually for the fifth-order case for explicit solution of the tap weights.

The function b_0 in (7.5-12) is called the object function. Eight nonnegative variables $(x_1, x_2, \ldots, x_8)$, known as slack variables, are introduced to convert the four inequalities in (7.5-13) to the following eight equalities:

$$\begin{aligned} a_4 + x_1 &= 1 \\ a_4 - x_2 &= -1 \\ a_2 + a_4 + x_3 &= 1 \\ a_2 + a_4 - x_4 &= -1 \\ a_0 + a_2 + 5a_4 + x_5 &= 1 \\ a_0 + a_2 + 5a_4 - x_6 &= -1 \\ a_0 + 6a_2 + a_4 + x_7 &= 1 \\ a_0 + 6a_2 + a_4 - x_8 &= -1 \end{aligned} \tag{7.5-14}$$

We now successively eliminate a_4, a_2, and a_0 to express b_0 as a function of the slack variables. Then b_0 is maximized subject to the resulting constraint equations in terms of the x's. From (7.5-14)

$$a_4 = 1 - x_1 \tag{7.5-15}$$

and the remaining seven equations and the object function become

$$\begin{aligned} x_1 + x_2 &= 2 \\ a_2 - x_1 + x_3 &= 0 \\ a_2 - x_1 - x_4 &= -2 \\ a_0 + a_2 - 5x_1 + x_5 &= -4 \\ a_0 + a_2 - 5x_1 - x_6 &= -6 \\ a_0 + 6a_2 - x_1 + x_7 &= 0 \\ a_0 + 6a_2 - x_1 - x_8 &= -2 \\ b_0 &= 5a_0 + 2a_2 - 2x_1 + 2 \end{aligned} \tag{7.5-16}$$

From (7.5-16)

$$a_2 = x_1 - x_3 \tag{7.5-17}$$

and it follows that

$$\begin{aligned} x_3 + x_4 &= 2 \\ a_0 - 4x_1 - x_3 + x_5 &= -4 \\ a_0 - 4x_1 - x_3 - x_6 &= -6 \\ a_0 + 5x_1 - 6x_3 + x_7 &= 0 \\ a_0 + 5x_1 - 6x_3 - x_8 &= -2 \\ b_0 &= 5a_0 - 2x_3 + 2 \end{aligned} \tag{7.5-18}$$

From (7.5-18)

$$a_0 = 4x_1 + x_3 - x_5 - 4 \tag{7.5-19}$$

and it follows that
$$x_5 + x_6 = 2$$
$$9x_1 - 5x_3 - x_5 + x_7 = 4$$
$$9x_1 - 5x_3 - x_5 - x_8 = 2 \quad (7.5\text{-}20)$$
$$b_0 = 20x_1 + 3x_3 - 5x_5 - 18$$

Subtracting the second and third equations of (7.5-20) yields the relationship
$$x_7 + x_8 = 2 \quad (7.5\text{-}21)$$

To introduce x_7 into the expression for b_0, we solve the second equation in (7.5-20) for x_1 as
$$x_1 = \frac{4 + 5x_3 + x_5 - x_7}{9} \quad (7.5\text{-}22)$$

and substitute it into the expression for b_0 to yield
$$b_0 = \frac{127}{9} x_3 - \frac{25}{9} x_5 - \frac{20}{9} x_7 - \frac{82}{9} \quad (7.5\text{-}23)$$

Therefore b_0 is maximized when $x_5 = x_7 = 0$, their smallest allowable nonnegative value, and $x_3 = 2$, its largest permissible value from the constraint in (7.5-18). Then from (7.5-23), (7.5-22), and the constraint equations,
$$b_0 = \frac{172}{9}, \quad x_1 = \frac{14}{9}, \quad x_2 = \frac{4}{9}, \quad x_3 = 2, \quad x_4 = 0, \quad x_5 = 0, \quad x_6 = 2, \quad x_7 = 0, \quad x_8 = 2 \quad (7.5\text{-}24)$$

The tap weights are then obtained from (7.5-15), (7.5-17), and (7.5-19) as
$$a_0 = \frac{38}{9}, \quad a_2 = -\frac{4}{9}, \quad a_4 = -\frac{5}{9} \quad (7.5\text{-}25)$$

The transversal filter output is the sequence
$$0 \quad -\frac{5}{9} \quad 0 \quad -1 \quad 0 \quad 1 \quad 0 \quad 1 \quad 0 \quad \frac{172}{9} \quad 0 \quad 1 \quad 0 \quad 1 \quad 0 \quad -1 \quad 0 \quad -\frac{5}{9} \quad 0 \quad (7.5\text{-}26)$$

shown in Fig. 7-62, normalized about $\tau = 0$. The PSR has increased from 5 (13.98 dB) for the matched condition to 172/9 (25.63 dB). The filter output is normalized to unity by dividing all tap weights by 172/9.

The cascade of the decoder and the transversal filter can be considered to be one filter whose impulse response is the convolution of the individual filter impulse responses. This is obtained by multiplying the tap weight sequences of the two filters as follows:

					$-\frac{5}{9}$	0	$-\frac{4}{9}$	0	$\frac{38}{9}$	0	$-\frac{4}{9}$	0	$-\frac{5}{9}$	Transversal filter impulse response
				$\times$					1	1	1	-1	1	Barker decoder impulse response
					$-\frac{5}{9}$	0	$-\frac{4}{9}$	0	$\frac{38}{9}$	0	$-\frac{4}{9}$	0	$-\frac{5}{9}$	
				$\frac{5}{9}$	0	$\frac{4}{9}$	0	$-\frac{38}{9}$	0	$\frac{4}{9}$	0	$\frac{5}{9}$		
			$-\frac{5}{9}$	0	$-\frac{4}{9}$	0	$\frac{38}{9}$	0	$-\frac{4}{9}$	0	$-\frac{5}{9}$			
		$-\frac{5}{9}$	0	$-\frac{4}{9}$	0	$\frac{38}{9}$	0	$-\frac{4}{9}$	0	$-\frac{5}{9}$				
$-\frac{5}{9}$	0	$-\frac{4}{9}$	0	$\frac{38}{9}$	0	$-\frac{4}{9}$	0	$-\frac{5}{9}$						
$-\frac{5}{9}$	$-\frac{5}{9}$	$-\frac{9}{9}$	$\frac{1}{9}$	$\frac{29}{9}$	$\frac{42}{9}$	$\frac{30}{9}$	$-\frac{42}{9}$	$\frac{29}{9}$	$-\frac{1}{9}$	$-\frac{9}{9}$	$\frac{5}{9}$	$-\frac{5}{9}$	Impulse response of cascaded filters (in reverse sequence)

Phase-Coded Waveforms

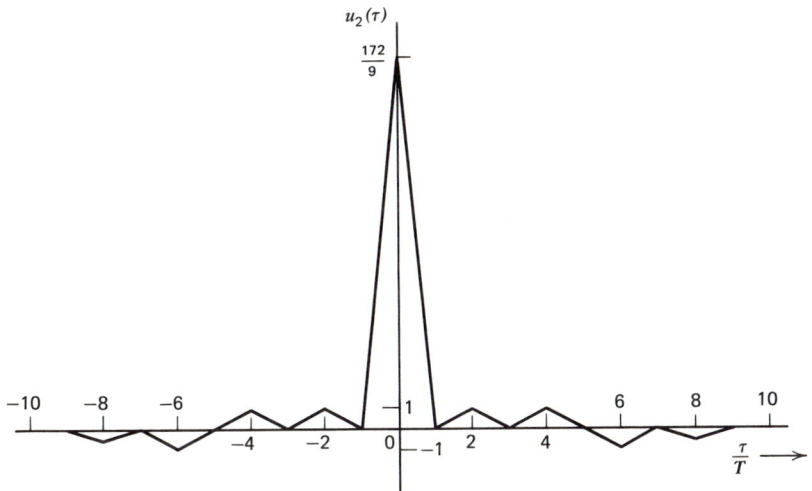

Figure 7-62. Response of sidelobe-reducing transversal filter for the $N = 5$ Barker sequence.

The response of this filter to the fifth-order Barker sequence yields the sequence of (7.5-26). To verify this, we convolve the Barker sequence with this filter's impulse response. First we scale the impulse response by 9 to remove fractions. After the computations, we shall divide the final sequence by 9. Then

					−5	5	−9	−1	29	−42| 30	42	29	1	−9	−5	−5	9 × overall impulse response
		×											1	1	1	−1	1	Barker sequence
					−5	5	−9	−1	29	−42| 30	42	29	1	−9	−5	−5	
			5	−5	9	1	−29	42	−30	−42	−29	−1	9	5	5			
		−5	5	−9	−1	29 |−42	30	42	29	1	−9	−5	−5				
	−5	5	−9	−1	29 |−42	30	42	29	1	−9	−5	−5					
−5	5	−9	−1	29 |−42	30	42	29	1	−9	−5	−5						
−5	0	−9	0	9	0	9	0	172	0	9	0	9	0	−9	0	−5		9 × output sequence

This sequence, divided by 9, is identical to (7.5-26).

The SNR at the output of a transversal filter for a white noise input is

$$\text{SNR} = \frac{(\text{peak output})^2}{\sum_i a_i^2} \qquad (7.5\text{-}27)$$

where the a's are the tap weight values, and the average output power for a white noise input is the sum of the squares of the tap weights. For the matched condition, all N tap weights are unity magnitude, the peak output is N, and

$$\text{SNR} = \frac{N^2}{N} = N \qquad (7.5\text{-}28)$$

which corresponds to

$$\text{SNR} = 10 \log N \text{ decibels} \qquad (7.5\text{-}29)$$

With the sidelobe filter in cascade, the tap weights corresponding to the cascaded combination are used in (7.5-27). For the case of $N = 5$,

$$\text{SNR} = \frac{(172/9)^2}{\left(\frac{1}{9}\right)^2 \{(30)^2 + 2[(5)^2 + (5)^2 + (9)^2 + 1 + (29)^2 + (42)^2]\}} = \frac{14792}{3187} = 4.641 \quad (7.5\text{-}30)$$

corresponding to 6.666 dB. From (7.5-29), for the matched condition,

$$\text{SNR} = 10 \log 5 \text{ dB} = 6.990 \text{ dB} \quad (7.5\text{-}31)$$

The sidelobe reducing filter therefore degrades the SNR by 0.324 dB.

The technique just described was applied to each odd-order Barker code, and the digital computer was then used to maximize the resulting object function subject to the inequality constraints. Table 7-4 gives the obtained tap weights of the sidelobe suppression filter, normalized for a peak output of unity. The odd tap weights are zero. Table 7-5 lists the PSR and SNR at the output of both the Barker decoder and the sidelobe suppression filter. For the $N = 13$ code, the sidelobes are suppressed to 34.8 dB with only a 0.11-dB SNR loss.

Note in Table 7-4 that for a given value of N, all tap weights except a_0 are approximately the same value. Realization of these small differences may be impractical for certain devices, such as acoustic wave filters or digital filters. If we assume that these tap weights are the same, the optimization process yields the results in Table 7-6. The PSR has slightly decreased for each code (except $N = 3$), but this decrease is offset by the increased ease of synthesis and manufacture. The

Table 7-4 **Optimum Tap Weights Normalized for Peak Output of Unity (Odd Tap Weights are Zero)**

Tap	$N = 3$	$N = 5$	$N = 7$	$N = 11$	$N = 13$
a_0	0.4	0.22093	0.163084	0.100177	7.97249×10^{-2}
a_2	0.1	-0.023256	0.026687	0.0119006	-2.391476×10^{-3}
a_4		-0.029069	0.023721	0.0111222	-2.67558×10^{-3}
a_6			0.020386	0.010267	-2.941092×10^{-3}
a_8				9.33984×10^{-3}	-3.186169×10^{-3}
a_{10}				8.348125×10^{-3}	-3.409138×10^{-3}
a_{12}					-3.60843×10^{-3}

Table 7-5 **PSR and SNR With and Without Sidelobe Suppression Filter**

	Peak-to-Sidelobe Ratio		Signal-to-Noise Ratio (dB)	
N	Matched Condition	Optimum Sidelobe Suppression	Matched Condition	Optimum Sidelobe Suppression
3	3 (9.54 dB)	10 (20 dB)	4.77	4.20
5	5 (13.98 dB)	19.11 (25.63 dB)	6.99	6.66
7	7 (16.90 dB)	14.13 (23 dB)	8.45	8.15
11	11 (20.83 dB)	19.62 (25.85 dB)	10.41	10.22
13	13 (22.28 dB)	54.91 (34.79 dB)	11.14	11.03

Matched Filter Technology

Table 7-6 Optimum PSR; All Tap Weights Equal, Except Center Tap; Normalized for Peak Output of Unity

N	a_0	a_2	$\left|\dfrac{a_0}{a_2}\right|$	PSR	SNR (dB)
3	2/5	1/10	4	10 (20 dB)	4.20
5	2/9	$-1/36$	8	18 (25.10 dB)	6.62
7	1/6	1/36	6	12 (21.58 dB)	7.55
11	4/39	1/78	8	15.6 (23.86 dB)	10.10
13	2/25	$-1/300$	24	50 (33.98 dB)	10.99

filter with unequal tap weights, however, remains as the standard against which other realizations are to be compared (assuming the same number of filter tap weights). If further suppression of the sidelobes is desired, the number of filter tap weights can be increased, and the procedure described here can then be applied.

7.6 MATCHED FILTER TECHNOLOGY

The complexity required to implement the matched filter often determines the technology used. The system designer must choose between a signal for which a known simple matched filter exists and signal that has specific useful characteristics even though system complexity is increased as a result of realizing the optimum filter.

Here we briefly discuss four processing technologies that can be applied to realize the matched filter—namely, optical, acoustic surface waves, digital, and charge transfer.

7.6.1 Optical Signal Processor

Signal processing techniques using optical components have been used with varying degrees of success. There are techniques for multiplication and integration, and these basic operations have been applied to compute Fourier and Laplace transforms, cross-correlation, autocorrelation, and convolution.

Optical processing has been most successful in situations requiring recognition of a fixed, known waveform. Under these conditions, a mask, which is usually film, is constructed to represent the waveform(s) to be detected. If necessary, the mask can be purposely distorted to represent the effects of a signal Doppler shift.

Ultrasonic delay lines can be used instead of film to form a mask when real-time reference signals are required. A real-time correlator uses one ultrasonic delay line for both the received signal and the reference signal, each propagating in opposite directions through the line, as in Fig. 7-63. The polarized light is applied normal to the surface of the bulk delay line and collected at the opposite side by a photosensor. The light polarization is changed by the ultrasonic waves, and after passing through both waves, the polarization is a good approximation to the product of the received signal and the reference signal. Integration is spatial along the length of the line, thus providing continuous convolution.

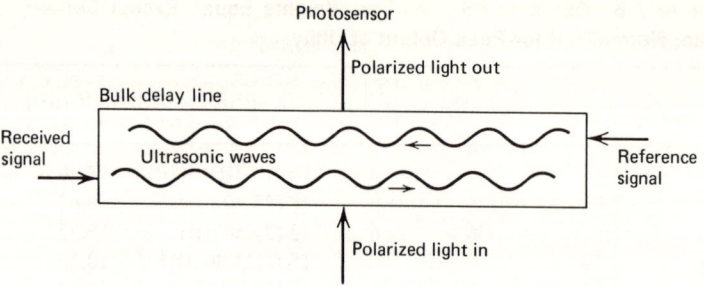

Figure 7-63. Optical correlation using a bulk delay line.

The major practical problem in optical systems is the change of the critical dimensions with temperature. These dimensions are necessary to achieve peak performance, thus operation over a wide temperature range is difficult. Also, special components, such as polarizing lenses, cannot operate over the required temperature range. There are also problems in packaging the large delay lines to survive in rugged environments.

7.6.2 Surface Acoustic Wave Device

Surface acoustic wave technology is based on the propagation of acoustic waves (Rayleigh waves) on the surface of the appropriate piezoelectric substrate, usually quartz crystal or lithium niobate. When an RF signal is applied to a surface wave interdigital transducer, the piezoelectric effect produces a surface stress that propagates along the surface toward a similar output transducer. The design of these transducers determines the response of the device. Surface wave devices can perform several useful signal-processing functions, such as simple delay (<1 cm long for delays of about 1 μsec), pulse compression, bandpass filtering, and matched filtering.

Surface wave devices can be considered to be tapped delay lines where the tap weights are simulated by the transducer metallic fingers. Consequently a surface wave device can be designed as a matched filter for binary phase-coded waveforms. The taps are matched to the particular input sequence ensuring that a negligible output occurs except when each signal sample coincides with the proper tap. Then a large output occurs.

Acoustic surface wave devices are passive, hence they consume no power. One problem associated with surface wave tapped delay lines, however, is the permanancy of the tap polarities. Thus a separate analog filter is necessary for each different code. This problem can be overcome by inserting code-changing switching circuits on both sides of each tap. The problem of Doppler frequency shifts is solved by using a bank of tapped delay lines.

7.6.3 Digital Filter

A digital filter is a computational process or algorithm by which a number sequence acting as an input is transformed into a second number sequence termed the output. The algorithm may be implemented as a special purpose computer or it may be

Matched Filter Technology

programmed on a general purpose computer. If the input signal $f(t)$ is not in digital form, it is first converted to a sequence of numbers by an analog-to-digital (A/D) converter (at point A in Fig. 7-64). The digital filter operates on this sequence and the filter output sequence at point B in Fig. 7-64 is then converted to an analog signal $u(t)$ by a digital-to-analog (D/A) converter.

Figure 7-64. Signal processing by digital filtering.

If the digital filter algorithm is designed so that the equivalent "impulse response" of the filter is the time-reversed samples of the input samples, the response is the desired matched filter response. This technique is usually used for matched filters when synchronization information is available to define the beginning and end of the signal. Because the operations required to implement this class of digital filter are identical to those required to compute the statistical correlation between the signal and the impulse response, this approach is sometimes called a correlation implementation of a matched filter. The digital filter is examined in detail in Chapter 9.

7.6.4 Charge Transfer Device

A charge transfer device (CTD) operates on the principle of signal charge transfer from one storage capacitance to the next. This group includes charge coupled devices (CCDs) and bucket brigade devices (BBDs). Many of the characteristics are similar, but basic operation differs among devices.

The CCD, first reported in 1970, is a metal-oxide-semiconductor (MOS) structure, as schematized in Fig. 7-65. It consists of a P-type silicon substrate on which 1000 to 3000 Å of silicon dioxide (SiO_2) is grown. Then metal electrodes are deposited on the SiO_2, where they form MOS capacitors. The electrodes are separated by about 0.1 mil, and every third electrode is connected to a common clock driver voltage; V_1, V_2, and V_3 are the three phases of the clock voltages in the "three-phase" arrangement. This configuration is the most commonly used, although there are also "two-phase" and "four-phase" systems.

Pulsing the CCD (an array of MOS capacitors) by the appropriate clock pulses transfers charge packets in discrete-time intervals between localized potential wells at the Si–SiO_2 interface. Between clock pulses, the minority charges (holes)

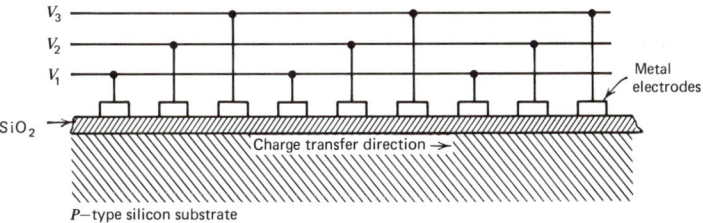

Figure 7-65. Basic structure of a charge coupled device.

are stored under the capacitors (metal electrodes) whose surface potential is most negative. The charge packets can then be detected at the output by capacitive coupling.

The BBD, first reported in 1958, is a series of MOS transistors connected so that the drain of one is common with the source of the next (Fig. 7-66). Charge is stored on the equivalent Miller capacitance between the gate and the drain and is transferred from one Miller capacitance to the next by alternately clocking V_1 and V_2.

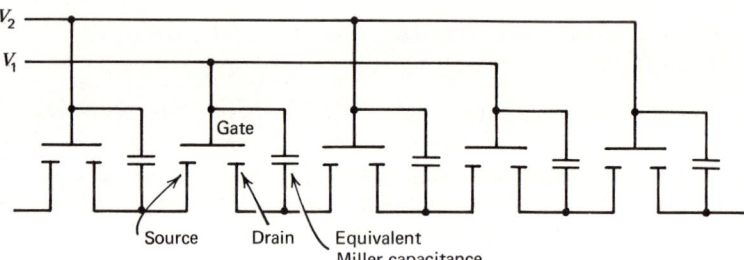

Figure 7-66. Basic structure of a bucket brigade device.

Charge transfer devices (CTDs) are basically analog shift registers (delay lines), and as such they are useful for implementing transversal filters and recursive filters. Then the CTD, with some additional circuitry, acts as a correlator or convolver of analog signals. Furthermore, since the tap weights of the transversal filter can be externally controlled, a programmable correlator can be realized. Charge transfer devices represent a major advance for analog signal processing, for the only successful analog time delay other than CTD is the surface wave device, which is limited to the UHF/VHF regions.

REFERENCES

1. Bucy, R. S., "Linear and Nonlinear Filtering," *Proc. IEEE*, Vol. 58, pp. 854–864, June 1970.
2. Cook, C. E., and M. Bernfeld, *Radar Signals, An Introduction to Theory and Application*, Academic Press, New York, 1967.
3. Dantzig, G. B., *Linear Programming and Extensions*, Princeton University Press, Princeton, N.J., 1963.
4. Kalman, R. E., "A New Approach to Linear Filtering and Prediction Problems," *Trans. ASME, J. Basic Eng.*, Vol. 82D, pp. 35–45, March 1960.
5. Kalman, R. E., and R. S. Bucy, "New Results in Linear Filtering and Prediction Theory," *Trans. ASME, J. Basic Eng.*, Vol. 83D, pp. 95–108, December 1961.
6. Key, E. L., E. N. Fowle, and R. D. Haggarty, "A Method of Sidelobe Suppression in Phase Coded Pulse Compression Systems," M.I.T. Lincoln Lab., Cambridge, Mass., Lexington Tech. Report 209, November 1959.
7. Lee, Y. W., *Statistical Theory of Communication*, Wiley, New York, 1963.
8. North, D. O., "An Analysis of the Factors Which Determine Signal/Noise Discrimination in Pulsed-Carrier Systems," *Proc. IEEE*, Vol. 51, pp. 1016–1027, July 1963.
9. Papoulis, A., *Probability, Random Variables and Stochastic Processes*, McGraw-Hill, New York, 1965.
10. Rihaczek, A. W., and R. M. Golden, "Range Sidelobe Suppression for Barker Codes," *IEEE Trans. Aerosp. Electron. Sys.*, Vol. AES-7, No. 6, pp. 1087–1092, November 1971.

Problems

11. Skolnik, M. I., *Radar Handbook*, McGraw-Hill, New York, 1970.
12. Sorenson, H. W., "Least-Squares Estimation: From Gauss to Kalman," *IEEE Spectrum*, Vol. 7, pp. 63–68, July 1970.
13. Kailath, T., "A View of Three Decades of Linear Filtering Theory," *IEEE Trans. Inform. Theory*, Vol. IT-20, No. 2, pp. 146–181, March 1974.

PROBLEMS

7-1. (a) What is the autocorrelation function of the pulse below?
(b) What is the total energy in $f(t)$?

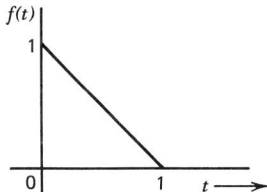

Figure P.7-1.

7-2. (a) What is the power density spectrum of the rectangular pulse below?
(b) What is the autocorrelation function of this pulse?

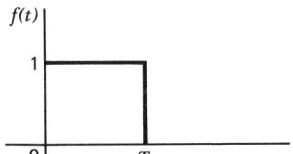

Figure P.7-2.

7-3. What is the output power density spectrum of an nth-order Butterworth LP filter with cutoff frequency ω_c when the input power density spectrum is N_0 watts per hertz?

7-4. What is the power density spectrum of the random function $f(t)$ whose autocorrelation function is

$$\phi_f(\tau) = e^{-a^2\tau^2} \qquad (a > 0)$$

7-5. What is the cross-correlation function between $u_{-1}(t)e^{-\alpha t}$ and the pulse in Problem 7-2?

7-6. For the pulse in Problem 7-1,
(a) What is the causal impulse response of the filter matched to this pulse?
(b) What is the transfer function of the matched filter?
(c) Synthesize this filter using adders, integrators, delays, and multipliers.

7-7. Verify the results in Tables 7-4 and 7-5 for the case of $N = 3$.

7-8. We now wish to further reduce the autocorrelation function sidelobes of the third-order Barker code, so we use a fifth-order sidelobe filter with tap weights a_4 0 a_2 0 a_0 0 a_2 0 a_4.
 (a) What is the optimum PSR?
 (b) What are then the values of a_0, a_2, and a_4 normalized for a peak output of unity?
 (c) What is the SNR for optimum PSR?

Chapter Eight
Time-Domain Operations

Throughout we have emphasized that the primary description of a filtering system is in the time domain, yet the majority of synthesis, analysis, and approximation procedures for linear time-invariant systems occur in the frequency domain. One reason for this is the historic development [28]. Network synthesis originated in the frequency domain and approximation developments remained there. In the earlier days of electronics, the magnitude response was considered to be the important network function; thus there was little reason to investigate the time-domain responses. Even the phase response received little attention, since it was well known that phase distortion did not affect speech transmission. It was not until the advent of television and other systems using pulses that it became apparent that the system impulse response was the more important system characterization.

Another reason is simply that many problems, such as synthesizing networks and approximating various frequency-domain functions, are more easily solved in the frequency domain than in the time domain. For example, no equivalent time-domain procedure is known to generate network element values given a specified impulse response. As a second example, if one wishes to obtain a maximally flat gain function, it is advantageous to remain in the frequency domain, where the error criterion is readily established. However, we must recognize that this frequency-domain requirement is specified to reduce the amplitude distortion encountered by time-domain signals whose frequency content lies within the constant magnitude region.

Time-domain approximations have not been totally neglected, but the desired time-domain responses are sometimes obtained by frequency-domain approximations. For example, the approximation to the Gaussian impulse response is obtained by approximating the Gaussian magnitude response, since the Fourier transform of a Gaussian function is another Gaussian function. Yet in many instances a specified time-domain response is desired, and for these cases it is advantageous to approximate in the time domain. An example of this is the rectangular impulse response approximation [20].

Although frequency-domain approximations were more extensively used than time-domain approximations, it does not necessarily follow that the resulting responses optimize system performance. It is now becoming apparent that the design of transmission systems and filters should essentially be carried out in the time domain for optimum performance. This, however, may be computationally difficult. A step in this direction was taken by Jess and Schüssler (Section 3.6.2 and Ref. 12), who initiated the design primarily in the time domain and obtained a filter with a minimum time-bandwidth product. This filter possesses equiripple transient responses and achieves the highest baseband digit rates for a given bandwidth. The surprising feature of these filters is the nonconstant group delays, some exhibiting a large peak in the stop band (Fig. 3-35). Before starting the design, the suspicion that the resulting group delay would be more constant might lead one to use this fact in formulating a frequency-domain error criterion. Thus he would miss this unorthodox group delay variation, which is achieved by the time-domain approach.

Here we first introduce two time-domain approximations—the method of time moments and the least-squares approximation by exponentials. The former has the advantage of being analytically tractable, while the latter, although yielding a better fit to the specified time function, results in a set of nonlinear simultaneous equations that must be solved on a computer. Our second consideration is the time-domain approach to the analysis of both time-variant and time-invariant linear systems excited by modulated waveforms. This approach allows the important attributes of the input signal to be explicitly retained in the response formulation and additionally provides insight into the system performance. The concept of an average system delay is then introduced by a least-squares formulation in the time domain. The average delay of three basic signals—the impulse function, the step function, and the rectangular pulse—is computed to show that the method yields results consistent with physical reasoning.

8.1 APPROXIMATIONS TO A PRESCRIBED FUNCTION

We now consider the problem of approximating a specified time-domain function $k(t)$ by the network impulse response $h_a(t)$. Because the network is assumed to be linear, lumped, and time-invariant, $h_a(t)$ can always be expressed as

$$h_a(t) = \sum_{i=1}^{N} (C_0^i + C_1^i t + \cdots + C_{N-1}^i t^{N-1}) A_i e^{-s_i t} u_{-1}(t) \qquad (8.1\text{-}1)$$

where the t polynomial reflects the presence of multiple poles in the network transfer function $H_a(s)$. We assume that $k(t)$ is real, zero for $t < 0$, piecewise continuous, and square integrable[†]; these conditions are satisfied in most practical problems.

There are many techniques in both the frequency domain and the time domain for achieving this goal (see Ref. 23 and the references there), but here we discuss only two time-domain techniques. The method of moments [8] yields a set of linear equations in the unknowns, which is easily solved by conventional means. The method of least squares results in a set of nonlinear equations that are solved by a search algorithm on a digital computer. The latter method yields a superior approximation at the expense of more tedious computations.

8.1.1 Method of Moments

The nth moment of an impulse response $h(t)$, m_n, is defined as

$$m_n = \int_0^\infty t^n h(t)\, dt \qquad n = 0, 1, 2, \ldots \qquad (8.1\text{-}2)$$

[†]$k(t)$ is square integrable if $\int_0^\infty k^2(t)\, dt < \infty$.

Approximations to a Prescribed Function

The zeroth moment m_0 is the area under the impulse response. If we assume that $h(t)$ is normalized so that $m_0 = 1$, then m_1 is interpreted as the "center of gravity" of $h(t)$ and m_2 as the "moment of inertia" about the line $t = 0$. The latter is a measure of the spread of the impulse response. The third moment is related to the skewness, that is, degree and direction of asymmetry, and the fourth moment is associated with the kurtosis, a measure of the relative peakedness of the impulse response.

The Laplace transform in (2.4-25) provides the relationship between the impulse response moments and the coefficients of the transfer function $H(s)$. Expand the exponential there into a power series and integrate term by term to give

$$H(s) = \int_0^\infty h(t)\left[1 - st + \frac{s^2 t^2}{2!} - \frac{s^3 t^3}{3!} + \cdots\right] dt$$

$$= \int_0^\infty h(t)\, dt - s\int_0^\infty t h(t)\, dt + \frac{s^2}{2}\int_0^\infty t^2 h(t)\, dt + \cdots \quad (8.1\text{-}3)$$

But from (8.1-2) each integral is a moment of $h(t)$, and $H(s)$ is then expressible as

$$H(s) = d_0 - d_1 s + d_2 s^2 - d_3 s^3 + \cdots \quad (8.1\text{-}4)$$

where

$$d_n = \frac{m_n}{n!} \quad (8.1\text{-}5)$$

Since s may be considered to be a differential operator, the expansion in (8.1-4) can be interpreted as the transform of the differential operator that relates the output function $u(t)$ and the input $f(t)$,

$$u(t) = \left[d_0 - d_1 \frac{d}{dt} + d_2 \frac{d^2}{dt^2} - \cdots\right] f(t) \quad (8.1\text{-}6)$$

The first term of the output function $d_0 f(t)$ is simply a scaled version of $f(t)$; the remaining terms, which are the successive derivatives of $f(t)$, can be considered distortion terms.

To approximate a specified time function $k(t)$ by the impulse response $h_a(t)$, we desire to match as many successive moments of $h_a(t)$ and $k(t)$ as possible. First we compute the moments of $k(t)$ and express $K(s)$, the Laplace transform of $k(t)$, as a power series similar to (8.1-4). Then we equate $K(s)$ to $H_a(s)$, the filter transfer function, and solve the resulting linear equations for the unknown coefficients in $H_a(s)$.

The method of moments gives a good approximation to the duration of $k(t)$, although the shape match may not be good, especially if $k(t)$ is discontinuous at some instants. Disadvantages of this technique are that the approximation error cannot be estimated in advance, and it is not always possible to match the moments with realizable networks. However this technique gives satisfactory approximations when $k(t)$ is continuous, has finite area, and is reasonably well behaved.

The method of moments has an entirely different, but familiar, interpretation in the frequency domain. It is equivalent to approximating $K(s)$ by $H_a(s)$ in the Taylor sense about the origin (see Section 3.2.1). Whereas the moments consider

$h(t)$ over its entire occurrence, the approximation in the frequency domain considers only the frequency $s = 0$. In the frequency domain, the moments need not be computed, for they are related to the coefficients of the power series expansion about $s = 0$, as shown in (8.1-4). Then $H_a(s)$ is expanded by long division about $s = 0$ and subtracted from $K(s)$ to form the error function

$$e(s) = K(s) - H_a(s) = e_0 + e_1 s + e_2 s^2 + \cdots \tag{8.1-7}$$

Setting successive coefficients of powers of s equal to zero yields the Taylor approximation about $s = 0$. Generally speaking, $H_a(s)$ approximates $K(s)$ very closely in the vicinity of the origin, but the approximation can be very poor at values of s far from $s = 0$.

We have just described equivalent time-domain and frequency-domain approximations, but the latter does not provide insights into the time function's attributes. Only the former—namely, the matching of the moments—reveals these characteristics. Usually, however, frequency-domain approximations do not have simple interpretations in the time domain, and for this reason a desired time function is best achieved by a time-domain approximation.

8.1.2 Approximation of the Rectangular Pulse by the Method of Moments

The rectangular pulse

$$k(t) = \begin{cases} 1 & 0 \leq t \leq 1 \\ 0 & \text{elsewhere} \end{cases} \tag{8.1-8}$$

is now approximated by the impulse response $h_a(t)$, whose transform is

$$H_a(s) = \frac{1}{a_0 s^2 + a_1 s + 1} \tag{8.1-9}$$

The moments of $k(t)$ from (8.1-2) are

$$m_n = \int_0^1 t^n(1) \, dt = \frac{1}{n+1} \tag{8.1-10}$$

from (8.1-5)

$$d_n = \frac{1}{(n+1)n!} = \frac{1}{(n+1)!} \tag{8.1-11}$$

and from (8.1-4)

$$K(s) = 1 - \frac{1}{2} s + \frac{1}{6} s^2 - \frac{1}{24} s^3 + \cdots \tag{8.1-12}$$

Then equate $K(s)$ to $H_a(s)$ to give

$$(1 + a_1 s + a_0 s^2)\left(1 - \frac{1}{2} s + \frac{1}{6} s^2 - \frac{1}{24} s^3 + \cdots\right) = 1 \tag{8.1-13}$$

or

$$1 + \left(a_1 - \frac{1}{2}\right) s + \left(a_0 + \frac{1}{6} - \frac{a_1}{2}\right) s^2 + \cdots = 1 \tag{8.1-14}$$

Equating corresponding powers of s, we find that $a_1 = \frac{1}{2}$ and $a_0 = \frac{1}{12}$. The approx-

Approximations to a Prescribed Function

imating transfer function is then

$$H_a(s) = \frac{12}{s^2 + 6s + 12} \tag{8.1-15}$$

and the approximating impulse response is

$$h_a(t) = 4\sqrt{3}\, e^{-3t} \sin \sqrt{3}\, t\, u_{-1}(t) \tag{8.1-16}$$

shown in Fig. 8-1 together with $k(t)$. The first three moments of each are identical. The duration of each response is similar, whereas the shapes are not. However, the approximation is good considering that $H_a(s)$ is only second order, $k(t)$ has a discontinuity at $t = 1$, and the arithmetic used is simple.

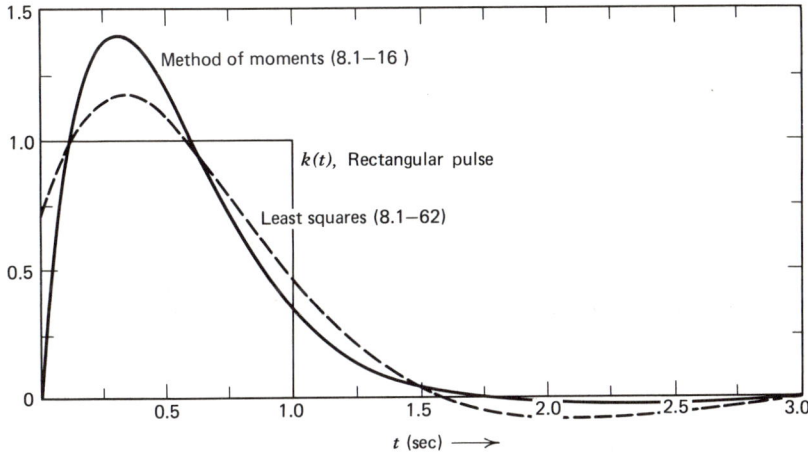

Figure 8-1. Approximations to the rectangular pulse by second-order system impulse responses.

To illustrate the same approximation in the frequency domain, we expand $H_a(s)$ in a power series about $s = 0$ by long division as

$$H_a(s) = 1 - a_1 s + (a_1^2 - a_0) s^2 + \cdots \tag{8.1-17}$$

Then form $e(s)$ in (8.1-7) from (8.1-12) and (8.1-17) as

$$e(s) = \left(a_1 - \frac{1}{2}\right)s + \left(\frac{1}{6} - a_1^2 + a_0\right)s^2 + \cdots \tag{8.1-18}$$

The Taylor approximation about $s = 0$ is accomplished when

$$a_1 - \tfrac{1}{2} = 0 \quad \text{and} \quad a_1^2 - a_0 = \tfrac{1}{6} \tag{8.1-19}$$

or $a_1 = \tfrac{1}{2}$, $a_0 = \tfrac{1}{12}$, the solutions previously obtained.

8.1.3 Approximation of the Triangular Function by the Method of Moments

To show the improvement in shape approximation when $k(t)$ is continuous at $t = 1$, we approximate the function

$$k(t) = \begin{cases} 2(1-t) & 0 \leq t \leq 1 \\ 0 & \text{elsewhere} \end{cases} \quad (8.1\text{-}20)$$

by the impulse response whose transform is

$$H_a(s) = \frac{b_0 s + 1}{a_0 s^2 + a_1 s + 1} \quad (8.1\text{-}21)$$

The additional numerator term reflects as a nonzero value for $h_a(0)$. The moments of $k(t)$ are

$$m_n = 2 \int_0^1 t^n (1-t)\, dt = \frac{2}{(n+2)(n+1)} \quad (8.1\text{-}22)$$

and from (8.1-5)

$$d_n = \frac{2}{(n+2)!} \quad (8.1\text{-}23)$$

Then equate $K(s)$ in the form of (8.1-4) to $H_a(s)$ as

$$(1 + a_1 s + a_0 s^2)\left(1 - \frac{1}{3}s + \frac{1}{12}s^2 - \frac{1}{60}s^3 + \cdots\right) = 1 + b_0 s \quad (8.1\text{-}24)$$

Equate the coefficients of corresponding powers of s to give the equations

$$\begin{aligned} a_1 - \frac{1}{3} &= b_0 \\ a_0 - \frac{1}{3}a_1 &= -\frac{1}{12} \\ \frac{1}{12}a_1 - \frac{1}{3}a_0 &= \frac{1}{60} \end{aligned} \quad (8.1\text{-}25)$$

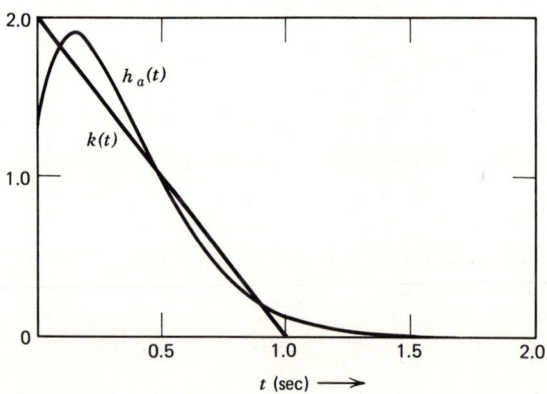

Figure 8-2. Approximation of $k(t)$ in (8.1-20) by $h_a(t)$ in (8.1-27).

Approximations to a Prescribed Function

Solutions to (8.1-25) are $a_1 = \frac{2}{5}$, $a_0 = \frac{1}{20}$, and $b_0 = \frac{1}{15}$. Then

$$H_a(s) = \frac{4}{3} \frac{s+15}{s^2 + 8s + 20} \qquad (8.1\text{-}26)$$

and the approximating impulse response is

$$h_a(t) = \frac{4}{3} e^{-4t} \left(\cos 2t + \frac{11}{2} \sin 2t \right) u_{-1}(t) \qquad (8.1\text{-}27)$$

which appears in Fig. 8-2 along with $k(t)$. In addition to achieving a good approximation to the duration of $k(t)$, we have closely fit its shape because $k(t)$ is continuous beyond $t = 0$. A network for realizing $h_a(t)$ is given in Fig. 8-3.

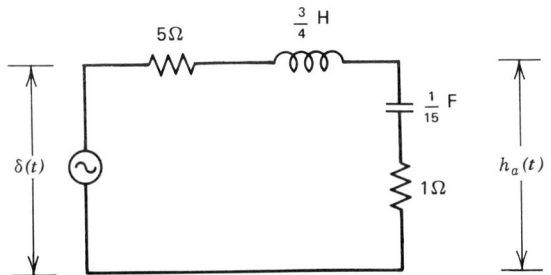

Figure 8-3. Network with the impulse response in (8.1-27).

8.1.4 Method of Least Squares

For convenience we assume that the approximating transfer-function poles are simple, however, the least-squares analysis is readily extended to treat multiple poles. Consequently all C's in (8.1-1) are zero except for C_0^i, which is unity. Then the unknowns are the s_i's, the complex frequencies that best characterize $k(t)$, and the A_i's, the amplitudes of the exponentials.

We wish to determine the $\{s_i\}$† and the $\{A_i\}$ so that

$$\varepsilon = \int_0^\infty \left[k(t) - \sum_{i=1}^N A_i e^{-s_i t} \right]\left[k(t) - \sum_{j=1}^N A_j^* e^{-s_j^* t} \right] dt \qquad (8.1\text{-}28)$$

the integrated square of the error

$$e(t) = k(t) - \sum_{i=1}^N A_i e^{-s_i t} \qquad (8.1\text{-}29)$$

is minimum. Since $k(t)$ is assumed to be real, complex values of A_i and s_i must occur in conjugate pairs. The asterisk in (8.1-28) indicates the complex conjugate (e.g., s_i^* is the complex conjugate of s_i).

Least-squares approximation of a specified time function by the sum of exponentials

$$h_a(t) = \sum_{i=1}^N A_i e^{-s_i t} \qquad (8.1\text{-}30)$$

†The symbol $\{\ \}$ indicates the "set of"; thus $\{s_i\}$ is "set of the s_i's".

has been extensively used in filter transient synthesis [14] and signal theory [9, 10]. Here we introduce the significant aspects of this subject and illustrate with some simple examples.

8.1.5 Least-Squares Approximation for Specified Exponentials

For a specified $\{s_i\}$, the $\{A_i\}$ that minimize ε must be a solution of the equations

$$\frac{\partial \varepsilon}{\partial A_j^*} = \int_0^\infty \left[k(t) - \sum_{i=1}^N A_i e^{-s_i t} \right] [-e^{-s_j^* t}] \, dt$$

$$= -\int_0^\infty k(t) e^{-s_j^* t} \, dt + \sum_{i=1}^N A_i \int_0^\infty e^{-(s_i + s_j^*) t} \, dt = 0, \qquad j = 1, 2, \ldots, N \quad (8.1\text{-}31)$$

But

$$\int_0^\infty k(t) e^{-s_j^* t} \, dt = K(s_j^*) \tag{8.1-32}$$

the Laplace transform of $k(t)$ evaluated at $s = s_j^*$, and

$$\int_0^\infty e^{-(s_i + s_j^*) t} \, dt = \frac{1}{s_i + s_j^*} \tag{8.1-33}$$

Then (8.1-31) reduces to

$$\sum_{i=1}^N \frac{A_i}{s_i + s_j^*} = K(s_j^*), \qquad j = 1, 2, \ldots, N \tag{8.1-34}$$

which is a set of linear equations in the A_i's.

In matrix notation, (8.1-34) is

$$\underline{S} \underline{A} = \underline{K} \tag{8.1-35}$$

where

$$\underline{S} = \begin{bmatrix} \frac{1}{s_1 + s_1^*} & \cdots & \frac{1}{s_N + s_1^*} \\ \vdots & & \vdots \\ \frac{1}{s_1 + s_N^*} & \cdots & \frac{1}{s_N + s_N^*} \end{bmatrix} \quad \begin{array}{l} \text{the Gram matrix} \\ \text{associated with} \\ \text{the exponential} \\ \text{set } \{e^{-s_i t}\} \end{array} \tag{8.1-36}$$

$$\underline{A} = (A_1, \ldots, A_N), \qquad \text{a column matrix} \tag{8.1-37}$$

$$\underline{K} = (K(s_1^*), \ldots, K(s_N^*)), \qquad \text{a column matrix} \tag{8.1-38}$$

and

$$\underline{A} = \underline{S}^{-1} \underline{K} \tag{8.1-39}$$

Miller[21] gives closed-form solutions for $\underline{S}^{-1}$ by use of which the inaccuracies in the $\{A_i\}$ arising from the conventional inversion of $\underline{S}$ are largely eliminated.

Approximations to a Prescribed Function

The minimum error number is obtained by expanding (8.1-28) as

$$\varepsilon = \int_0^\infty k^2(t)\,dt - \sum_{i=1}^N A_i \int_0^\infty k(t)e^{-s_i t}\,dt - \sum_{j=1}^N A_j^* \int_0^\infty k(t)e^{-s_j^* t}\,dt + \sum_{j=1}^N A_j^* \sum_{i=1}^N \frac{A_i}{s_i + s_j^*} \tag{8.1-40}$$

and substituting the result of (8.1-34) to obtain

$$\varepsilon = \int_0^\infty k^2(t)\,dt - \sum_{i=1}^N A_i K(s_i) \tag{8.1-41}$$

Square and integrate $h_a(t)$ in (8.1-30), and then use (8.1-34) to yield the equality

$$\int_0^\infty h_a^2(t)\,dt = \sum_{i=1}^N A_i K(s_i) \tag{8.1-42}$$

Thus the minimum error number in (8.1-41) is the area under the square of $k(t)$ minus the area under the square of $h_a(t)$. If the time functions are interpreted as voltages across a one-ohm resistor, the minimum value of ε is the energy in $k(t)$ minus the energy in the approximating impulse response $h_a(t)$.

8.1.6 Approximation of the Rectangular Pulse by Two Specified Exponentials

To illustrate the foregoing development we approximate the rectangular pulse in Section 8.1.2 by the two complex exponentials $e^{(-1+j)t}$ and $e^{(-1-j)t}$. The transform of the pulse is

$$K(s) = \int_0^1 1 \cdot e^{-st}\,dt = \frac{1 - e^{-s}}{s} \tag{8.1-43}$$

and from (8.1-34)

$$\frac{A_1}{s_1 + s_1^*} + \frac{A_2}{s_2 + s_1^*} = K(s_1^*)$$
$$\frac{A_1}{s_1 + s_2^*} + \frac{A_2}{s_2 + s_2^*} = K(s_2^*) \tag{8.1-44}$$

Since $A_1 = A_2^*$, we give A_2 as

$$A_2 = \frac{(s_1 + s_2^*)(s_2 + s_2^*)(s_2 + s_1^*)}{(s_2 - s_1)(s_1^* - s_2^*)}\left[\frac{s_1 + s_1^*}{s_1 + s_2^*}K(s_1^*) - K(s_2^*)\right] \tag{8.1-45}$$

Furthermore $s_2 = s_1^* = 1 + j$, and A_2 reduces to

$$A_2 = -4\left[\frac{K(1+j)}{1-j} - K(1-j)\right] = 4\left[\frac{1 - e^{-(1-j)}}{1-j} - \frac{1 - e^{-(1+j)}}{2}\right]$$

$$= \frac{2\sin 1}{e} + j2\left(1 - \frac{2\sin 1}{e} - \frac{\cos 1}{e}\right) = 0.61912 + j0.36423 \tag{8.1-46}$$

The approximating function is then the real function

$$h_a(t) = A_1 e^{-(1-j)t} + A_2 e^{-(1+j)t} = 2e^{-t}[0.61912 \cos t + 0.36423 \sin t]\,u_{-1}(t) \tag{8.1-47}$$

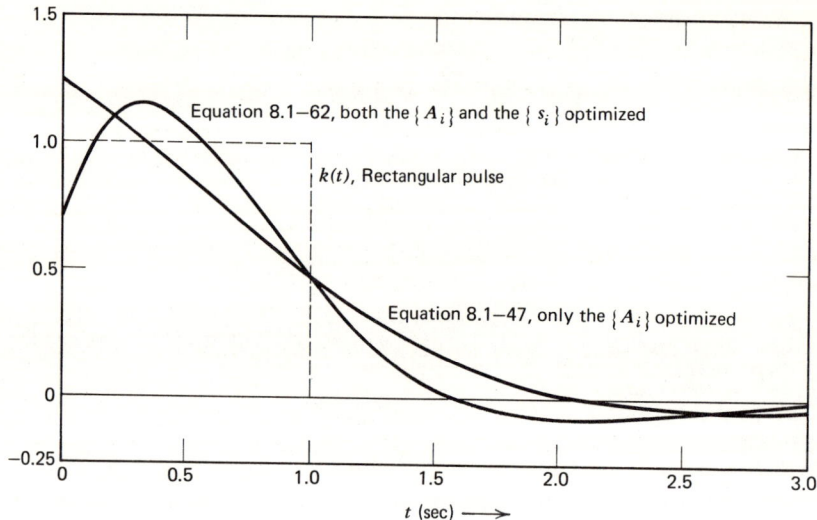

Figure 8-4. Approximation of the rectangular pulse by the least-squares technique.

in Fig. 8-4. The error number, from (8.1-41), is

$$\varepsilon = 1 - A_1 K(s_1) - A_2 K(s_2) = 1 - A_1 K(s_1) - A_1^* K(s_1^*) = 0.1332 \quad (8.1\text{-}48)$$

or 13.32% relative to the rectangular pulse energy.

8.1.7 Least-Squares Approximation for Optimum Amplitudes and Exponents

In the previous development we specified the exponents and optimized the amplitudes. Now we consider simultaneous minimization of the error ε over both the $\{A_i\}$ and the $\{s_i\}$. The $\{s_j^*\}$ that minimizes ε must be a solution of the equations

$$\frac{\partial \varepsilon}{\partial s_j^*} = \int_0^\infty \left[k(t) - \sum_{i=1}^N A_i e^{-s_i t} \right] A_j^* t e^{-s_j^* t} dt$$

$$= A_j^* \left[\int_0^\infty t k(t) e^{-s_j^* t} dt - \sum_{i=1}^N A_i \int_0^\infty t e^{-(s_i + s_j^*)t} dt \right] = 0 \quad j = 1, 2, \ldots, N \quad (8.1\text{-}49)$$

However

$$\int_0^\infty t e^{-(s_i + s_j^*)t} dt = \frac{1}{(s_i + s_j^*)^2} \quad (8.1\text{-}50)$$

and from the Laplace transform relationship between $k(t)$ and $K(s)$,

$$\int_0^\infty t k(t) e^{-s_j^* t} dt = -\frac{dK}{ds} \bigg|_{s=s_j^*} = -K'(s_j^*) \quad (8.1\text{-}51)$$

Approximations to a Prescribed Function

Then (8.1-49) becomes

$$\sum_{i=1}^{N} \frac{A_i}{(s_i + s_j^*)^2} = -K'(s_j^*) \qquad j = 1, 2, \ldots, N \qquad (8.1\text{-}52)$$

Equations 8.1-34 and 8.1-52, originally obtained by Aigrain and Williams [1], are nonlinear in the s_i's; hence an analytic solution for the optimum $\{A_i\}$ and $\{s_i\}$ is not possible, even if the functional form of $K(s)$ is known. Consequently a digital computer is used, and various algorithms for obtaining these values are given in Refs. 7, 19, 24, and the references listed there.

8.1.8 Approximation of the Rectangular Pulse by One Exponential

We now approximate the rectangular pulse by one real exponential to show the nonlinear relationship that arises even in this simple case. From (8.1-34)

$$\frac{A_1}{2s_1} = K(s_1) \qquad (8.1\text{-}53)$$

and from (8.1-52)

$$\frac{A_1}{4s_1^2} = -K'(s_1) \qquad (8.1\text{-}54)$$

Substituting A_1 from (8.1-53) into (8.1-54) yields

$$2s_1 K'(s_1) + K(s_1) = 0 \qquad (8.1\text{-}55)$$

where $K(s)$ is given by (8.1-43) and

$$K'(s) = \frac{(s+1)e^{-s} - 1}{s^2} \qquad (8.1\text{-}56)$$

Then (8.1-55) reduces to

$$2s_1 + 1 = e^{s_1} \qquad (8.1\text{-}57)$$

which is iteratively solved to give $s_1 = 1.25643$. From (8.1-53), $A_1 = 1.43066$, and the least-squares approximation with one exponential is

$$h_a(t) = 1.43066\, e^{-1.25643 t} u_{-1}(t) \qquad (8.1\text{-}58)$$

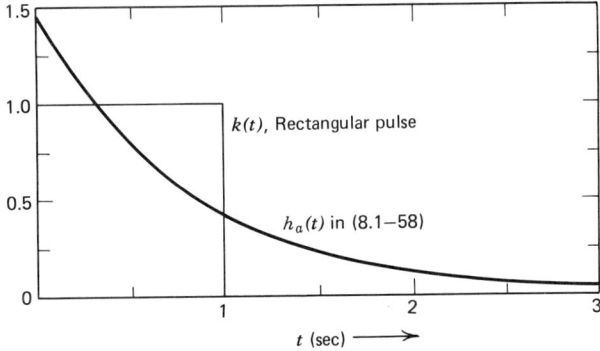

Figure 8-5. Optimum approximation of the rectangular pulse by one exponential.

shown in Fig. 8-5. The minimum error number, from (8.1-41), is

$$\varepsilon = 1 - 1.43066 \times \frac{1 - e^{-1.25643}}{1.25643} = 0.18547 \qquad (8.1\text{-}59)$$

or 18.55% relative to the rectangular pulse energy. The approximation in Section 8.1.6 by two exponentials is superior to the optimum approximation here even though the exponents earlier were not optimized. This is not always true, for we certainly can specify two exponentials that yield an error number greater than 0.18547.

8.1.9 Approximation of the Rectangular Pulse by Two Exponentials

We now use two exponentials to approximate the rectangular pulse, optimizing both the exponents and amplitudes. Remember that $A_2 = A_1^*$ and $s_2 = s_1^*$. Two equations are given by (8.1-44), and from (8.1-52)

$$\begin{aligned}\frac{A_1}{(s_1 + s_1^*)^2} + \frac{A_2}{(s_2 + s_1^*)^2} &= -K'(s_1^*) \\ \frac{A_1}{(s_1 + s_2^*)^2} + \frac{A_2}{(s_2 + s_2^*)^2} &= -K'(s_2^*)\end{aligned} \qquad (8.1\text{-}60)$$

Since these four simultaneous equations cannot be solved analytically, a computer is used to obtain [18]

$$\begin{aligned} s_1 &= 1.51094 + j1.82618 \\ s_2 &= 1.51094 - j1.82618 \end{aligned} \qquad (8.1\text{-}61)$$

leading to the approximation

$$h_a(t) = e^{-1.51094 t}[0.70673 \cos 1.82618t + 2.32457 \sin 1.82618t]\, u_{-1}(t) \qquad (8.1\text{-}62)$$

shown in Fig. 8-1. This approximation more closely approximates the pulse shape, for this aspect is considered in the approximation, whereas the method of moments emphasizes the function's area, spread, peakedness, and similar properties. Equation 8.1-62 also appears in Fig. 8-4, verifying that simultaneous minimization significantly improves the approximation. Also the error number is reduced from 0.1332 to 0.0860.

Least-squares approximation to the rectangular pulse has more than academic value, for a filter with a rectangular impulse response is the matched filter to a rectangular pulse, as discussed in Chapter 7. Meyer [20] uses the least-squares criterion to obtain filter impulse responses that approximate the rectangular pulse, with the restriction that finite transfer-function zeros occur only on the $j\omega$ axis. The resulting LP schematics are the same form as the filter networks in Fig. 3-37. We now have used the least-squares approximation in both the time and the frequency domains, and in Chapter 9 we use it again for the design of nonrecursive digital filters.

8.1.10 Orthogonal Filter

Once the appropriate exponential functions have been determined, they can be orthonormalized by a simple and elegant method that is described in Refs. 9, 10,

14, and 23. This procedure leads to a synthesized system known as an orthogonal filter, whose structure is similar to the transversal filter shown in Fig. 7-48. The orthogonal filter, however, is composed of elementary filter sections rather than the constant delay sections of the transversal filter. For example, the complete set of Laguerre functions is obtained if the frequencies (exponents) are all identical. The realization for this set is the Laguerre filter described in Ref. 16 in which the first section is a first-order low-pass filter and each succeeding section is the first-order all-pass filter of Section 5.2.1. Thus linear time-invariant systems, including the optimum filters in Chapter 7, can be synthesized by means of orthonormal functions, and this technique provides an alternative realization to the standard lattice and ladder (longitudinal) configurations.

This concludes our discussion of time-domain synthesis. We have attempted to introduce the topic but at the same time to present approximation methods that are useful in practice. Furthermore we now can view a filter as a pulse generator, for it can be designed to ensure that its impulse response approximates a desired pulse shape. Indeed, this is a common way of generating a pulse in practice.

8.2 RESPONSE OF LINEAR, TIME-INVARIANT SYSTEMS TO MODULATED WAVEFORMS

Almost every electronically transmitted signal, such as radar, radio, telegraph, television, and telephone, is generated by modulation. If one considers the switching on of a signal as the modulation of that signal by a step function, all transmitted signals fall into this category. At the receiver every waveform contains modulation, whether as a result of changes in the signal introduced at the transmitter, changes introduced by atmospheric conditions, or, as in a radar system, changes in the signal introduced by the target.

The response of the receiver to modulated signals must be determined before the overall system can be designed and analyzed. The receiver response, at least the linear portion, is essentially the response of the receiver filter. Two approaches are generally used to determine the response of filters excited by modulated inputs.

The first is the spectral, or Fourier, approach in which the driving function is decomposed into sinusoidal components, each of which is modified in magnitude and phase in passing through the filter. At the output these components are summed yielding the response function.

The second is the dynamic approach, in which the filter is regarded as a dynamic system whose response at any instant depends on the values of the input at that instant and prior instants. In this approach the input signal retains its form and need not be decomposed into more fundamental signals. This analysis is performed in the time domain, thereby supplying insights into the system behavior that are not obtainable by the spectral approach.

8.2.1 Spectral Approach

The spectral approach to the analysis of specific systems with specific input signals is often hampered by long and complex calculations. This is especially true when the input signals are modulated waveforms.† Many of the attributes of interest, such as instantaneous amplitude, instantaneous frequency, and other modulation details of the input signal, are lost in the complicated mathematical expressions.

Consider a filter with transfer function $H(s)$, which at $s = j\omega$ is expressible as

$$H(j\omega) = |H(j\omega)|e^{j\theta(\omega)} \qquad (8.2\text{-}1)$$

The filter response to $\sin \omega_0 t$ is then

$$u(t) = |H(j\omega_0)| \sin [\omega_0 t + \theta(\omega_0)] \qquad (8.2\text{-}2)$$

where $|H(j\omega_0)|$ is the magnitude of the sine wave response and $\theta(\omega_0)$ is its phase shift. We can also use the general input function $e^{j\omega_0 t}$, rather than $\sin \omega_0 t$ because it often leads to simpler mathematical analysis, but we usually imply the real part or the imaginary part. Consequently, by the principle of superposition, we then consider only the real or imaginary part of the filter response. For this input the response is

$$u(t) = |H(j\omega_0)|e^{j[\omega_0 t + \theta(\omega_0)]} \qquad (8.2\text{-}3)$$

which reduces to (8.2-2) if the input function is $\sin \omega_0 t$, the imaginary part of $e^{j\omega_0 t}$.

The response to any input function that can be decomposed into a sum of sinusoids can be obtained by considering each component as a separate input, determining its response from (8.2-3), and summing these component responses. For the input function

$$f(t) = \sum_k a_k e^{j\omega_k t} \qquad (8.2\text{-}4)$$

the filter response is

$$u(t) = \sum_k a_k |H(j\omega_k)| e^{j[\omega_k t + \theta(\omega_k)]} \qquad (8.2\text{-}5)$$

This is the spectral approach. For modulated inputs, the output function is so complicated that the important characteristics such as instantaneous frequency and instantaneous amplitude are very difficult to determine. Furthermore, decomposing the input function into the form of (8.2-4) may itself be a tedious task.

8.2.2 FM Waveform Analysis by the Spectral Approach

We now illustrate the spectral approach by determining the filter response to the FM waveform in (2.6-28), namely, frequency modulation by a sinusoid. The Fourier representation of this function, given in (2.6-29), is

$$f(t) = \sum_{k=-\infty}^{\infty} J_k(\mu) \cos(\omega_c + k\omega_m)t \qquad (8.2\text{-}6)$$

The response of the filter described by the transfer function $H(s)$ is

$$u(t) = \sum_{k=-\infty}^{\infty} J_k(\mu)|H(j\omega_c + jk\omega_m)| \cos[\omega_c t + k\omega_m t + \theta(\omega_c + k\omega_m)] \qquad (8.2\text{-}7)$$

We now know the amplitude and phase of each frequency component of the output wave, but what is the instantaneous frequency of the output? This is the

---

†A complete listing of the extensive literature on this topic is impractical. As a sample of the contributions to this area, see Refs. 5, 6, and 17.

Responses to Modulated Waveforms

function of interest, and it is hopelessly immersed in the complicated expression for $u(t)$. The price of obtaining the output by this relatively simple technique is failure to determine easily the important attributes of the signal.

As a side note, Carson's rule is an approximate simple formula for the necessary bandwidth about ω_c to reasonably reproduce the FM wave. He gives this bandwidth as

$$\text{bandwidth} \approx 2(\Delta f + f_m) = 2(1 + \mu)f_m \qquad (8.2\text{-}8)$$

For large modulation index ($\mu \gg 1$), (8.2-8) reduces to

$$\text{bandwidth} \approx 2\mu f_m = 2\Delta f \qquad (8.2\text{-}9)$$

That is, the necessary bandwidth is twice the maximum deviation from the carrier frequency.

8.2.3 Dynamic Approach

Attention turned to the "dynamic" viewpoint to overcome the computational problems inherent in the spectral approach to modulated waveform excitation [26]. In this approach the response, expressed as the sum of a quasi-stationary term plus a correction term, retains the identity of the instantaneous frequency and instantaneous amplitude of the input signal.

The value of the quasi-stationary term at any instant t is defined as the value of the input signal at that instant, multiplied by the value of the conventional sinusoidal steady-state transfer function in which the frequency variable ω is replaced by the value of the instantaneous frequency of the input wave $\omega(t)$ at the instant t.

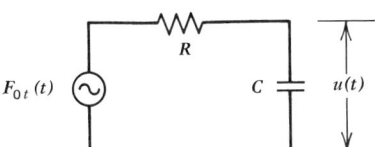

Figure 8-6. RC network in Example 8-1.

Example 8-1. As an example of the quasi-stationary response, consider the RC network in Fig. 8-6, excited by the frequency-modulated voltage

$$F_{0t}(t) = \exp\left[j \int_0^t \omega(\xi) \, d\xi\right] \qquad (8.2\text{-}10)$$

where the subscripts 0 and t on $F_{0t}(t)$ indicate the integral limits. The network transfer function at $s = j\omega$ is

$$H(j\omega) = \frac{1}{1 + j\omega RC} \qquad (8.2\text{-}11)$$

Then the quasi-stationary transfer function $H[j\omega(t)]$ is

$$H[j\omega(t)] = \frac{1}{1 + jRC\omega(t)} \qquad (8.2\text{-}12)$$

and the quasi-stationary response is $H[j\omega(t)]F_{0t}(t)$.

The quasi-stationary response neglects the transient response of the system to changes in the instantaneous frequency and instantaneous amplitude of the input signal. Thus the true response also includes a correction term that accounts for the inability of the system to respond as rapidly as the quasi-stationary term demands.

The quasi-stationary term approximates the system response if its magnitude is large compared to the correction term. This situation is satisfied in low-distortion transmission systems. Other applications for which the quasi-stationary term represents the desired system response are as follows:

1. The use of a linear FM sweep to display the steady-state magnitude characteristics of a filter.
2. The response of a comb filter to a burst of carrier for purposes of determining the frequency of the carrier.
3. The demodulation of a frequency-modulated wave by an FM discriminator.
4. The transmission of an AM wave through the RF and IF amplifiers of a conventional AM receiver.

Let us examine item 1 in more detail to see why the quasi-stationary response is the desired response for this situation. A typical circuit for measuring a filter magnitude response is presented in Fig. 8-7a. The filter is terminated properly at both ends, the excitation is a constant voltage source of variable frequency, and the response indicator is a high-impedance voltmeter connected across the load resistor. For each input frequency we obtain an output voltage, designated by a circle in Fig. 8.7b. A curve through those circles is then proportional to the filter magnitude response $|H(j\omega)|$. Instead of manually setting the signal generator to a new frequency for each measurement, suppose we electronically sweep the frequency linearly across the band of interest $(\omega_b - \omega_a)$, then return to ω_a and repeat the sweep, as indicated in Fig. 8-8. The input is no longer a sinusoid, but it is

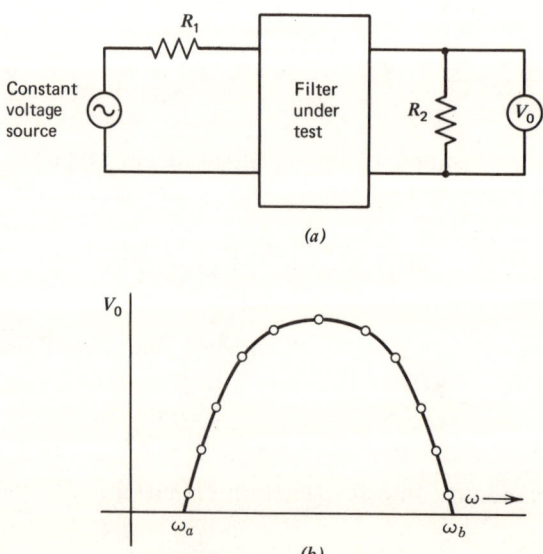

Figure 8-7. (a) Test circuit for measuring filter magnitude response and (b) the filter response.

Responses to Modulated Waveforms

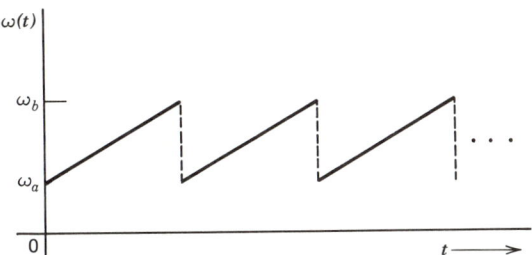

Figure 8-8. Linear FM sweep between ω_a and ω_b.

an FM signal with instantaneous frequency $\omega(t)$. Rather than use a voltmeter, we now display the output on an oscilloscope. Clearly, if the sweep rate is slow enough, the oscilloscope display is the curve in Fig. 8.7b; that is, the response is $|H[j\omega(t)]|$. What is "slow enough"? The sweep rate is "slow enough" when the filter response is essentially the quasi-stationary response $H[j\omega(t)] \exp[j \int_0^t \omega(\xi)\, d\xi]$. The absolute value of this response is the desired magnitude response. As the sweep rate is increased, the correction term is no longer negligible, and the total response no longer approximates the desired quasi-stationary response.

In 1937 Carson and Fry [3] introduced the concept of a quasi-stationary response and a correction term, in one of the earlier publications giving a mathematical analysis of the dynamic response of filters to frequency-modulated signals. They expressed the correction term as an infinite series in terms of the system admittance function and its derivatives with respect to frequency, evaluated at the frequency of the unmodulated carrier; however its convergence properties were obscured by mathematical complexity. Subsequent investigators (see bibliography of Ref. 26) failed to obtain this correction term in a closed form until Hupert [11] derived the desired result for the case of a frequency-modulated input. He determined the error term as a limit process by approximating the instantaneous frequency of the driving function by a series of small steps, then using the principle of superposition.

A major development in the evolution of the quasi-stationary approach was produced by Weiner and Leon [25, 26]. They used the simple mathematical process of integration by parts to demonstrate that for all stable, lumped, linear, time-invariant systems, the response to AM and FM inputs could readily be put in the desired form, with the exact correction term expressed as an integral. In addition, simple bounds on the correction term were given for several filters of interest. It is their work that we now describe.

8.2.4 Quasi-Stationary Analysis

Consider an Nth-order system with impulse response†

$$h(t) = u_{-1}(t) \sum_{k=1}^{N} A_k e^{-s_k t} \qquad (8.2\text{-}13)$$

---

†Here we consider transfer functions with only simple poles. The analysis to follow is readily extended to the multiple-pole case (see Ref. 26). It is noteworthy that previous investigators did not make use of the fact that the impulse response is expressible as a sum of exponentials.

excited by the combined amplitude- and frequency-modulated signal

$$f(t) = \begin{cases} B_1 \exp(j\omega_\nu t) & t < 0 \\ B(t) F_{0t}(t) & t \geq 0 \end{cases} \qquad (8.2\text{-}14)$$

where $F_{0t}(t)$ is given in (8.2-10) and $B(t)$ and $\omega(t)$ are both real. [We omit explicit designation of the real part of the complex-valued functions and write $\exp(j\omega_\nu t)$ instead of $\cos \omega_\nu t$]. For $t \geq 0$, the amplitude and the frequency are given at any instant by the functions $B(t)$ and $\omega(t)$, respectively, which are assumed to be continuous except for possible discontinuities at $t = 0$, where $B(0) = B_0$ and $\omega(0) = \omega_0$ (Fig. 8-9). If $B(t) = B_1$ for all t, we have frequency modulation, and if $\omega(t) = \omega_\nu$ for all t, we have amplitude modulation. For phase modulation $\phi(t)$, replace $F_{0t}(t)$ by $\exp[j\phi(t)]$. Thus (8.2-14) is a good representation for modulated signals used in communication, radar, and other electronic systems.

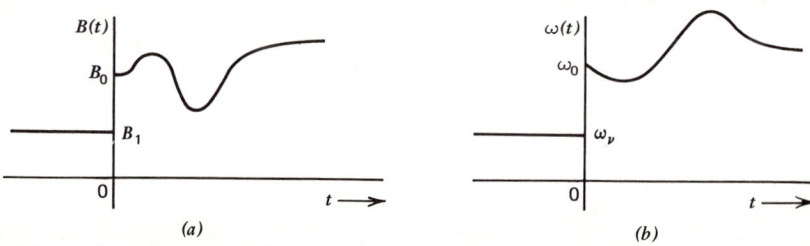

Figure 8-9. (a) Instantaneous amplitude and (b) instantaneous frequency of modulated input signal.

Substitution from (8.2-13) and (8.2-14) into the convolution integral (1.8.8) with $t_0 = -\infty$ (the system is initially at rest there) gives the system response for $t \geq 0$ as

$$u(t) = \sum_{k=1}^{N} Y_k(t) \int_{-\infty}^{t} e^{s_k \tau} f(\tau) \, d\tau; \qquad (Y_k(t) = A_k e^{-s_k t})$$

$$= \sum_{k=1}^{N} Y_k(t) \int_{-\infty}^{0_-} B_1 e^{(s_k + j\omega_\nu)\tau} \, d\tau + \sum_{k=1}^{N} Y_k(t) \int_{0}^{t} B(\tau) e^{s_k \tau} F_{0\tau}(\tau) \, d\tau \qquad (8.2\text{-}15)$$

where $t = 0_-$ is the value of time immediately prior to $t = 0$. The first integral is

$$\int_{-\infty}^{0_-} B_1 e^{(s_k + j\omega_\nu)\tau} \, d\tau = \frac{B_1}{s_k + j\omega_\nu} \qquad (8.2\text{-}16)$$

An integration by parts

$$\int_{0}^{t} u \, dv = uv \Big|_{0}^{t} - \int_{0}^{t} v \, du \qquad (8.2\text{-}17)$$

is performed on the second integral as follows:

$$u = \frac{B(\tau)}{s_k + j\omega(\tau)} \qquad v = e^{s_k \tau} F_{0\tau}(\tau)$$

$$\frac{du}{d\tau} = \frac{d}{d\tau} \left[\frac{B(\tau)}{s_k + j\omega(\tau)} \right] \qquad \frac{dv}{d\tau} = [s_k + j\omega(\tau)] e^{s_k \tau} F_{0\tau}(\tau) \qquad (8.2\text{-}18)$$

Responses to Modulated Waveforms

Substitution from (8.2-18) into (8.2-17) yields

$$\int_0^t B(\tau)e^{s_k\tau}F_{0\tau}(\tau)\,d\tau = \frac{B(\tau)e^{s_k\tau}F_{0\tau}(\tau)}{s_k+j\omega(\tau)}\bigg|_0^t - \int_0^t \frac{d}{d\tau}\left[\frac{B(\tau)}{s_k+j\omega(\tau)}\right]e^{s_k\tau}F_{0\tau}(\tau)\,d\tau$$

$$= \frac{B(t)e^{s_k t}F_{0t}(t)}{s_k+j\omega(t)} - \frac{B_0}{s_k+j\omega_0} - \int_0^t \frac{d}{d\tau}\left[\frac{B(\tau)}{s_k+j\omega(\tau)}\right]e^{s_k\tau}F_{0\tau}(\tau)\,d\tau$$

(8.2-19)

Now substitute from (8.2-16) and (8.2-19) into (8.2-15) to give

$$u(t) = B(t)F_{0t}(t)\sum_{k=1}^{N}\frac{A_k}{s_k+j\omega(t)} + \sum_{k=1}^{N}Y_k(t)\left[\frac{B_1}{s_k+j\omega_\nu} - \frac{B_0}{s_k+j\omega_0}\right]$$

$$-\sum_{k=1}^{N}Y_k(t)\int_0^t \frac{d}{d\tau}\left[\frac{B(\tau)}{s_k+j\omega(\tau)}\right]e^{s_k\tau}F_{0\tau}(\tau)\,d\tau \qquad (8.2\text{-}20)$$

However, the system transfer function $H(s)$ is the Laplace transform of $h(t)$, and from (8.2-13),

$$H(s) = \sum_{k=1}^{N}\frac{A_k}{s+s_k} \qquad (8.2\text{-}21)$$

therefore

$$\sum_{k=1}^{N}\frac{A_k}{s_k+j\omega(t)} = H[j\omega(t)] \qquad (8.2\text{-}22)$$

the quasi-stationary transfer function. The system response from (8.2-20) is now expressible as

$$u(t) = u_q(t) + u_b(t) + u_d(t) \qquad (8.2\text{-}23)$$

where

$$u_q(t) = H[j\omega(t)]B(t)F_{0t}(t)$$

$$u_b(t) = \sum_{k=1}^{N}Y_k(t)\left[\frac{B_1}{s_k+j\omega_\nu} - \frac{B_0}{s_k+j\omega_0}\right]$$

$$u_d(t) = -\sum_{k=1}^{N}Y_k(t)\int_0^t \frac{d}{d\tau}\left[\frac{B(\tau)}{s_k+j\omega(\tau)}\right]e^{s_k\tau}F_{0\tau}(\tau)\,d\tau$$

This is the desired result: $u(t)$ expressed as the sum of a quasi-stationary term $u_q(t)$ plus a closed-form correction term $u_b(t) + u_d(t)$.

Insight into the response characteristics is obtained by examination of the individual terms in (8.2-23). The first term $u_q(t)$ is the quasi-stationary term already discussed. The $u_b(t)$ is due to the discontinuity in the envelope and frequency functions at $t = 0$. These discontinuities create transients that for stable systems die away. At $t = 0$ the envelope is continuous if $B_1 = B_0$ and the frequency is continuous if $\omega_\nu = \omega_0$. When both these conditions are satisfied, $u_b(t)$ vanishes. The last term $u_d(t)$ is the most complicated of all, for it reflects the inability of the system to follow the variations of $B(t)$ and $\omega(t)$. The integrand contains the first derivatives of $B(t)$ and $\omega(t)$; hence for sufficiently slow variations of these quantities, the contribution of this term is negligible.

If the quasi-stationary term and the correction term are each considered to be rotating phasors, the system response can be interpreted as the interference between these two phasors. In this manner the pertinent features of linear system responses to modulated inputs are pictorially explained. For AM inputs, this approach allows us to interpret such quantities as envelope overshoots and undershoots and their time of occurrence, while for FM inputs this phasor approach explains such features of the output instantaneous frequency as overshoots, undershoots, and final value. An example of this interpretation for an FM case is given in Ref. 26.

Although the response is presented in a partitioned closed form, the computation of the correction term may be laborious. However a typical engineering problem is to determine the conditions for which $u_a(t)$ is indeed a good approximation to the system response. Upper bounds on the correction term are often sufficient to answer this question. The goal in bounding functions is to obtain tight bounds, for arbitrarily large bounds are not particularly useful. Therefore ingenuity and detailed knowledge of the specified functions are necessary to achieve the desired useful bound. An example of a simple bound (though not necessarily a tight bound) on $u_d(t)$ in (8.2-23) is

$$|u_d(t)| \leq \sum_{k=1}^{N} \left| Y_k(t) \int_0^t \frac{d}{d\tau} \left[\frac{B(\tau)}{s_k + j\omega(\tau)} \right] e^{s_k \tau} F_{0\tau}(\tau) \, d\tau \right| \qquad (8.2\text{-}24)$$

A larger bound is

$$|u_d(t)| \leq \sum_{k=1}^{N} |Y_k(t)| \int_0^t \left| \frac{d}{d\tau} \left[\frac{B(\tau)}{s_k + j\omega(\tau)} \right] \right| \left| e^{s_k \tau} \right| d\tau \qquad (8.2\text{-}25)$$

If $s_k = \alpha_k + j\beta_k$ and M is the maximum value of

$$\left| \frac{d}{d\tau} \left[\frac{B(\tau)}{s_k + j\omega(\tau)} \right] \right|$$

then $u_d(t)$ is further bounded by

$$|u_d(t)| \leq M \sum_{k=1}^{N} \frac{1}{\alpha_k} |A_k| (1 - e^{-\alpha_k t}) \qquad (8.2\text{-}26)$$

which approaches the asymptotic value

$$|u_d(t)| \leq |u_d(\infty)| = M \sum_{k=1}^{N} \frac{1}{\alpha_k} |A_k| \qquad (8.2\text{-}27)$$

Certainly if $|u_d(\infty)|$ is small compared with $u_a(t)$, no further examination is necessary. However a large value of $|u_d(\infty)|$ does not necessarily indicate a large correction term. It may mean only that the bound is large. The trend exhibited in (8.2-24) to (8.2-27) is often encountered in bounding functions. The tightness of the bound decreases as the computation is made easier.

We now examine (8.2-23) when $f(t)$ is an amplitude-modulated signal and when $f(t)$ is a frequency-modulated signal.

8.2.5 Response to an Amplitude-Modulated Signal

The instantaneous frequency $\omega(t)$ in (8.2-14) is the constant-valued frequency ω_ν, and the terms in (8.2-23) then become

$$u_q(t) = H(j\omega_\nu)B(t)e^{j\omega_\nu t}$$

$$u_b(t) = (B_1 - B_0)\sum_{k=1}^{N}\frac{A_k e^{-s_k t}}{s_k + j\omega_\nu}$$ (8.2-28)

$$u_d(t) = -\sum_{k=1}^{N}\frac{A_k e^{-s_k t}}{s_k + j\omega_\nu}\int_0^t \frac{dB(\tau)}{d\tau}e^{(s_k + j\omega_\nu)\tau}\,d\tau$$

The quasi-stationary term is the input signal multiplied by the transfer function evaluated at the carrier frequency, and $u_b(t)$ is proportional to the difference in the amplitude modulation at $t = 0$. If $B_1 = B_0$, this transient term vanishes. Note that the first derivative of $B(t)$ appears in the integrand of the expression for $u_d(t)$.

We now determine the response of a first-order LP filter with cutoff ω_c, whose transfer function is

$$H(s) = \frac{\omega_c}{s + \omega_c}$$ (8.2-29)

excited by the AM signal

$$f(t) = \begin{cases} (1+m)e^{j\omega_\nu t} & t < 0 \\ (1 + m\cos\omega_m t)e^{j\omega_\nu t} & t \geq 0 \end{cases}$$ (8.2-30)

From (8.2-14), $B(t) = 1 + m\cos\omega_m t$, and the quasi-stationary response from (8.2-28) and (8.2-29) is

$$u_q(t) = \frac{\omega_c}{\omega_c + j\omega_\nu}(1 + m\cos\omega_m t)e^{j\omega_\nu t}$$ (8.2-31)

We now see why the quasi-stationary response is the desired response of the RF and IF amplifiers of a conventional AM receiver. The envelope of $u_q(t)$ is $(\omega_c/\sqrt{\omega_c^2 + \omega_\nu^2})(1 + m\cos\omega_m t)$, which is proportional to the AM modulation, $1 + m\cos\omega_m t$, the desired information. This function is easily recovered by a standard envelope detector.

Since $f(t)$ is continuous at $t = 0$, $u_b(t)$ in (8.2-28) is zero. From (8.2-21) and (8.2-29), $N = 1$, $s_1 = \omega_c$, and $A_1 = \omega_c$. Also,

$$\frac{dB(t)}{dt} = -m\omega_m \sin\omega_m t$$ (8.2-32)

Then, from (8.2-28), the correction term is

$$u_d(t) = -\frac{\omega_c e^{-\omega_c t}}{\omega_c + j\omega_\nu}\int_0^t -m\omega_m \sin\omega_m \tau\, e^{(\omega_c + j\omega_\nu)\tau}\,d\tau$$

$$= m\omega_c\omega_m \frac{e^{j\omega_\nu t}[(\omega_c + j\omega_\nu)\sin\omega_m t - \omega_m \cos\omega_m t] + \omega_m e^{-\omega_c t}}{(\omega_c + j\omega_\nu)[(\omega_c + j\omega_\nu)^2 + \omega_m^2]}$$ (8.2-33)

With $p = \omega_m/\omega_c$ and $q = \omega_\nu/\omega_c$, the magnitude of the correction term is

$$|u_d(t)| = \frac{mp\sqrt{R^2(t) + I^2(t)}}{\sqrt{1 + q^2}\sqrt{(1 - q^2 + p^2)^2 + 4q^2}}$$ (8.2-34)

where

$$x = \omega_c t$$
$$R(t) = \cos qx [\sin px - p \cos px] - q \sin qx \sin px + pe^{-x}$$
$$I(t) = \sin qx [\sin px - p \cos px] + q \sin px \cos qx$$

and the magnitude of the quasi-stationary response is

$$|u_q(t)| = \frac{1 + m \cos px}{\sqrt{1 + q^2}} \tag{8.2-35}$$

To determine the relative magnitude of $|u_d(t)|$, we assume $m = 0.5$, $q = 0.6$ and then compute $|u_d(t)|/|u_q(t)|$ for $p = 0.1$, 0.3, and 0.4. The results appear in Figs. 8-10a, b, and c, respectively. The bounds on the envelope of the system response $u(t)$ are

$$|u_q(t)| \pm |u_d(t)| = |u_q(t)| \left[1 \pm \frac{|u_d(t)|}{|u_q(t)|} \right] \tag{8.2-36}$$

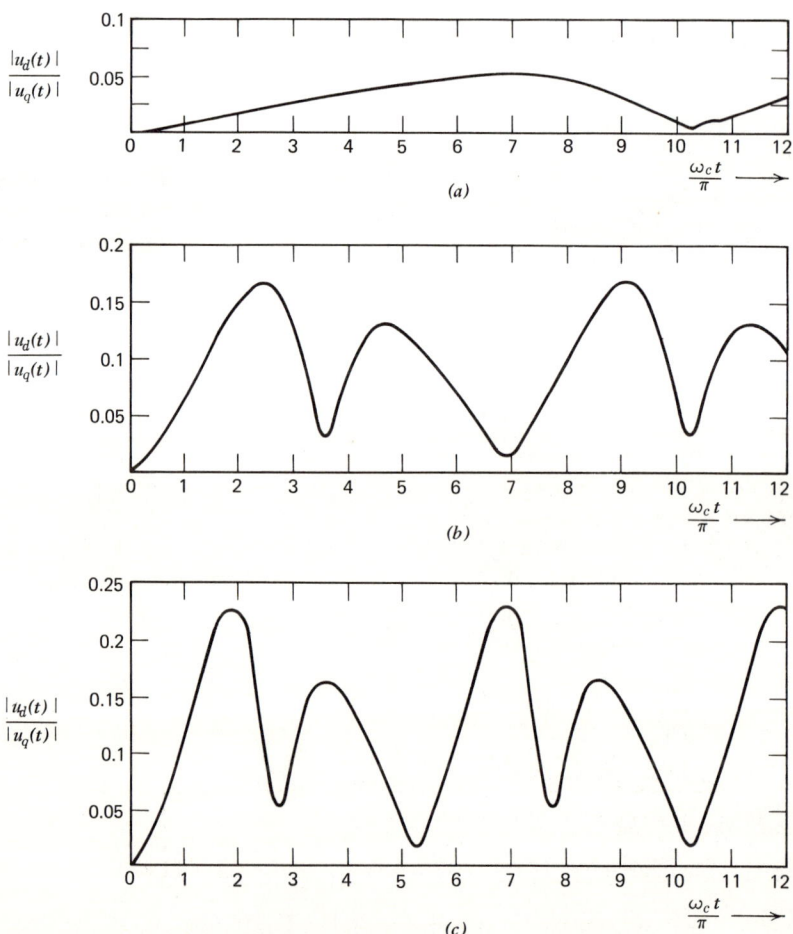

Figure 8-10. Values of correction-term envelope relative to quasi-stationary response envelope for the network in Section 8.2.5. (a) $p = 0.1$, (b) $p = 0.3$, (c) $p = 0.4$.

Responses to Modulated Waveforms

Therefore the quasi-stationary response, as an approximation to the system response, is in error by the values given in Fig. 8-10, and the maximum deviations from $|u_q(t)|$ are listed in Table 8-1. For $p = 0.1$, $u_q(t)$ is an excellent approximation to $u(t)$; but $p = 0.3$ results in a maximum error of 17% and for $p = 0.4$ the maximum error increases to 23%.

Table 8-1 **Maximum Deviations from $|u_q(t)|$**

p	$1 - \dfrac{|u_d(t)|}{|u_q(t)|}$	$1 + \dfrac{|u_d(t)|}{|u_q(t)|}$
0.1	0.95	1.05
0.3	0.83	1.17
0.4	0.77	1.23

These values may be better appreciated by referring to Fig. 8-11. The AM waveform can be decomposed into three sinusoids: a carrier at frequency ω_ν and two sidebands, at $\omega_\nu - \omega_m$ and $\omega_\nu + \omega_m$. Figure 8-11 shows their relative passband positions with $q = 0.6$. An increase in the value of p corresponds to an increased separation of carrier and sideband frequencies. For $p = 0.1$ all three frequencies essentially experience the same attenuation and phase in passing through the filter, and $u_q(t)$ is an excellent approximation to the system response. For larger values of p, each component is attenuated and phase shifted differently, resulting in a larger value for the correction term. With $p = 0.4$, for example, the lower sideband is attenuated 0.17 dB, the carrier is attenuated 1.34 dB, and the upper sideband is attenuated 3 dB.

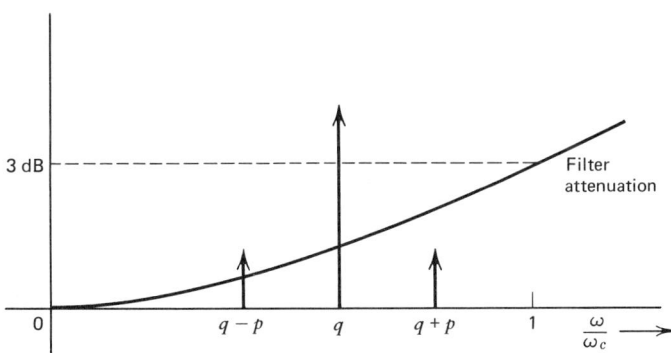

Figure 8-11. Relative positions of AM signal frequency components in filter passband with $q = 0.6$.

8.2.6 Response to a Frequency-Modulated Signal

The instantaneous envelope $B(t)$ in (8.2-14) is the constant-valued amplitude B_1, and the terms in (8.2-23) then become

$$u_q(t) = H[j\omega(t)]B_1 F_{0t}(t)$$

$$u_b(t) = jB_1(\omega_0 - \omega_\nu) \sum_{k=1}^{N} \frac{A_k e^{-s_k t}}{(s_k + j\omega_\nu)(s_k + j\omega_0)} \tag{8.2-37}$$

$$u_d(t) = jB_1 \sum_{k=1}^{N} A_k e^{-s_k t} \int_0^t \frac{d\omega(\tau)}{d\tau} \frac{e^{s_k \tau} F_{0\tau}(\tau)}{[s_k + j\omega(\tau)]^2} d\tau$$

The quasi-stationary term is the input signal multiplied by the transfer function evaluated at the instantaneous frequency $\omega(t)$. The $u_b(t)$ is proportional to the difference in the instantaneous frequency at $t = 0$. If $\omega_\nu = \omega_0$, this transient term vanishes. Note that no derivative of $\omega(t)$ higher than the first appears in $u_d(t)$. This is significant because the infinite series obtained by Carson and Fry [3] and others contained higher-order derivatives, which, as we now see, do not appear in $u_d(t)$.

The information of the input FM signal is contained in the instantaneous frequency $\omega(t)$, and it is this quantity that is preserved as $F_{0t}(t)$ passes through a low-distortion system. It is usually detected by a discriminator circuit followed by an LP filter. Unfortunately computation of the output instantaneous frequency is very laborious, but the phasor approach discussed earlier yields a simple interpretation of this function's genesis.

An important form of frequency modulation in communication applications is the abrupt change of a sinusoid's frequency from ω_ν to ω_0, known as frequency-shift keying (FSK). The instantaneous frequency of this signal is schematized in Fig. 8-12, and the quasi-stationary response is then the familiar steady-state response ($B_1 = 1$),

$$u_q(t) = H(j\omega_0)e^{j\omega_0 t} \tag{8.2-38}$$

The correction term in (8.2-37) is due solely to the discontinuity in the instantaneous frequency at $t = 0$ and is proportional to the frequency difference $(\omega_0 - \omega_\nu)$. The $u_d(t)$ is zero because $(d\omega(t))/dt = 0$ for $t \geq 0$. Oscillograms of the instantaneous frequency of a tuned circuit's response to this FM input are given in Ref. 26 along with an explanation of their behavior by way of the phasor interference approach.

It is also shown there that a general bound on the correction term of a

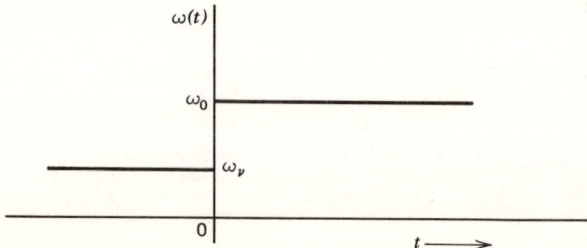

Figure 8-12. Instantaneous frequency of an FSK signal.

narrow-band BP filter response to an FM input whose instantaneous frequency is

$$\omega(t) = \omega_c + \Delta\omega b(t) \qquad (8.2\text{-}39)$$

is

$$|u_d(t)| \leq \frac{K\Delta\omega}{(\omega_2 - \omega_1)^2} \left|\frac{db(t)}{dt}\right|_{\max} \qquad (8.2\text{-}40)$$

where we assume $|b(t)| < 1$, thus the maximum frequency deviation is $\Delta\omega$, $(\omega_2 - \omega_1)$ is the radian 3-dB bandwidth of the BP filter, and K is a constant that depends on the BP filter response. The upper bound is proportional to the maximum frequency deviation and the maximum rate of change of $b(t)$, and inversely proportional to the filter bandwidth squared. These results agree with physical reasoning. A reduction in either numerator factor corresponds to a decrease in the signal bandwidth. Thus this change or an increase in the filter bandwidth allows the quasi-stationary response to better approximate the true filter response.

We have now shown that when the analysis is performed in the time domain, the system response can be expressed as a sum of terms, each corresponding to specific attributes of the signal and system. Furthermore, when the modulation parameters vary slowly enough, the quasi-stationary response is a good approximation to the system response. The partitioning of the response allows a quantitative analysis to determine what is "slow enough."

8.3 RESPONSE OF LINEAR, TIME-VARIANT SYSTEMS TO MODULATED WAVEFORMS

The analysis of linear time-variant systems to modulated inputs has received little attention because most responses of these systems can be obtained only by machine computation, and only for a specific input. A disadvantage of this approach is the loss of insight into the response characteristics that is often gained from analytic procedures.

Time-variant systems can be described by non-constant-coefficient differential equations, and as demonstrated in Chapter 1, the task of solving these equations can be reduced to the apparently simpler problem of solving the homogeneous portion. Unlike the constant-coefficient differential equation, whose homogeneous solution can always be determined as a sum of products of polynomials and exponentials, the non-constant-coefficient differential equation does not admit of a general form for its homogeneous solution, except for $n = 1$. The solution, if it can be computed analytically, is usually expressed as an infinite series whose convergence properties must be thoroughly examined. This form is usually undesirable for system analysis because many of the design parameters and system quantities of interest are not explicitly expressed.

To avoid the difficulties encountered in solving the equations exactly, investigators have obtained approximate solutions using simplifying assumptions; the most common of these are that the system is time-invariant with small perturbations or that the parameters vary slowly compared to the time scale of the system [27]. These results are useful when the assumptions are valid, but erroneous results are obtained if they are not valid. Another method also considers the time-variant system as a perturbation problem; however the unperturbed system is not a time-invariant system but the time-variant system for which an analytic solution is available [15]. This approach is advantageous for analyzing systems whose parameters have large variations about the stationary values.

The solution may also be computed for assigned values of system parameters to any desired degree of accuracy on a digital computer. An objection to this technique is that analytical expressions for the solutions in terms of the general-valued system parameters are not obtained. Furthermore one must still summarize and extrapolate these results in a meaningful manner.

However, the quasi-stationary response *is* characterized in terms of the system parameters, thus this approach to time-variant system analysis to modulated inputs has practical value.

8.3.1 Quasi-Stationary Response

The quasi-stationary response of a time-variant system is obtained at any arbitrary instant t by considering to be time invariant the system and the input signal generator in which the values of the signal and system parameters at that instant hold constant for all prior time. Thus at each instant for which the output signal value is desired, an appropriate time-invariant system is evaluated. Accordingly, for a frequency-modulated input signal, the system element values are replaced by the value of their time-varying counterparts at that instant t, and the frequency parameter ω is replaced by the value of the instantaneous frequency of the input waveform $\omega(t)$ at the same instant.

Example 8-2. As an example of the quasi-stationary response of a time-variant system, consider the time-varying network in Fig. 8-13 excited by $F_{0t}(t)$, the FM voltage in (8.2-10). If the network were time-invariant, the transfer function at $s = j\omega$ would be

$$H(j\omega) = \frac{1}{1 + j\omega CR} \tag{8.3-1}$$

Hence the quasi-stationary transfer function $T[j\omega(t), t]$, is

$$T[j\omega(t), t] = \frac{1}{1 + j\omega(t)C(t)R(t)} \tag{8.3-2}$$

and the quasi-stationary response is $T[j\omega(t), t]F_{0t}(t)$. The variable t is shown in the argument of T to remind us that T characterizes a time-varying system.

Even for a high-order time-varying system (with known parameters) we can determine $T[j\omega(t), t]$ relatively easily, for the primary task is then to determine the transfer function of an equivalent time-invariant system. The quasi-stationary

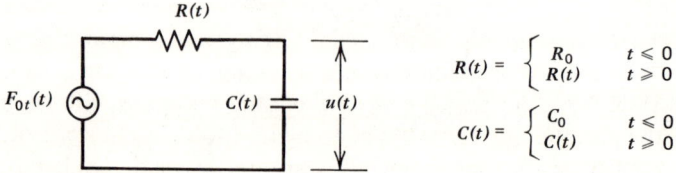

Figure 8-13. Time-varying RC network.

Response to Modulated Waveforms

response ignores the transient terms arising from the instantaneous variations in the frequency and amplitude of the input signal as well as in the system parameters. Therefore the quasi-stationary response is only an approximation to the true response. When the parameters of either the input signal or the system vary appreciably during the memory span of the system, the quasi-stationary response may be a poor approximation. Thus the true response is the sum of a correction term plus the quasi-stationary term.

As in the time-invariant case, the quasi-stationary term is often the desired response, the correction term representing distortion. Two such applications are the use of a time-variant network to track an FM signal and to minimize the acquisition time of a phase-lock loop.

8.3.2 Quasi-Stationary Analysis of a First-Order System

Weiner and Leon [25, 26] showed that for a first-order time-variant system, the response can be obtained as the sum of a quasi-stationary term plus a closed-form correction term, just as they had done for time-invariant systems. Furthermore the phasor interpretation is also valid for this first-order system. We now illustrate their technique by determining the response of the first-order system in Fig. 8-13 excited by the combined amplitude- and frequency-modulated voltage in (8.2-14).

The differential equation describing the RC network is

$$\frac{du(t)}{dt} + \left[\frac{1}{R(t)C(t)} + \frac{1}{C(t)}\frac{dC(t)}{dt}\right]u(t) = \frac{f(t)}{R(t)C(t)} \tag{8.3-3}$$

The solution to the homogeneous equation associated with (8.3-3) is

$$u_h(t) = \frac{1}{C(t)}\exp\left[-\int_a^t \frac{d\eta}{R(\eta)C(\eta)}\right] = \frac{1}{C(t)}G_{at}(t) \tag{8.3-4}$$

where a is an arbitrary constant. The subscripts a and t on $G_{at}(t)$ indicate the integral limits. Then, from (1.4-3), (1.4-20), and (1.8-3), the network impulse response is

$$h(t,\tau) = u_{-1}(t-\tau)\frac{G_{\tau t}(t,\tau)}{R(\tau)C(t)} \tag{8.3-5}$$

The variable τ is included in the argument of G to show that G is also a function of τ. From Fig. 8-13, $R(t)$ and $C(t)$ are constants for $t < 0$. Therefore $h(t,\tau)$ is

$$h(t,\tau) = u_{-1}(t-\tau)\begin{cases} \alpha_0 e^{-\alpha_0(t-\tau)} & t < 0 \\ \dfrac{1}{R_0 C(t)} e^{\alpha_0 \tau} G_{0t}(t) & \tau < 0, t \geq 0 \\ \dfrac{1}{R(\tau)C(t)} G_{\tau t}(t,\tau) & \tau \geq 0, t \geq 0 \end{cases} \tag{8.3-6}$$

where

$$\alpha(\eta) = \frac{1}{R(\eta)C(\eta)}; \quad \alpha_0 = \frac{1}{R_0 C_0}$$

and $R(0) = R_0$; $C(0) = C_0$; that is, the resistance and capacitance are continuous at $t = 0$.

Substituting $h(t, \tau)$ and $f(t)$ from (8.2-14) into the superposition integral (1.8-7), with $t_0 = -\infty$, we obtain the system response for $t \geq 0$ as

$$u(t) = \frac{B_1 G_{0t}(t)}{R_0 C(t)} \int_{-\infty}^{0_-} e^{\alpha_0 \tau} e^{j\omega_\nu \tau} \, d\tau + \frac{1}{C(t)} \int_0^t \frac{1}{R(\tau)} G_{\tau t}(t, \tau) B(\tau) F_{0\tau}(\tau) \, d\tau$$

$$= \frac{B_1 G_{0t}(t)}{R_0 C(t)(\alpha_0 + j\omega_\nu)} + \frac{1}{C(t)} \int_0^t \frac{B(\tau)}{R(\tau)} \exp\left[-\int_\tau^t \alpha(\eta) \, d\eta\right] \exp\left[j \int_0^\tau \omega(\xi) \, d\xi\right] d\tau$$
(8.3-7)

The integration by parts of (8.2-17) is now performed on the remaining integral with the following identification:

$$u = \frac{B(\tau)}{R(\tau)[\alpha(\tau) + j\omega(\tau)]} = C(\tau) B(\tau) T[j\omega(\tau), \tau]$$
(8.3-8)

$$v = G_{\tau t}(t, \tau) F_{0\tau}(\tau) = \exp\left[-\int_\tau^t \alpha(\eta) \, d\eta + j \int_0^\tau \omega(\xi) \, d\xi\right]$$

where

$$T[j\omega(\tau), \tau] = \frac{1}{1 + j\omega(\tau) C(\tau) R(\tau)} = \text{quasi-stationary transfer function [see (8.3-2)]}$$

Then $u(t)$ is expressible as

$$u(t) = u_q(t) + u_b(t) + u_d(t) \tag{8.3-9}$$

where

$$u_q(t) = B(t) F_{0t}(t) T[j\omega(t), t]$$

$$u_b(t) = \frac{C_0}{C(t)} [B_1 T(j\omega_\nu, 0) - B_0 T(j\omega_0, 0)] G_{0t}(t)$$

$$u_d(t) = -\frac{1}{C(t)} \int_0^t \frac{d}{d\tau} \{B(\tau) C(\tau) T[j\omega(\tau), \tau]\} G_{\tau t}(t, \tau) F_{0\tau}(\tau) \, d\tau$$

The response of the time-variant RC network is now expressed as the sum of a quasi-stationary term $u_q(t)$ plus a correction term $u_b(t) + u_d(t)$. As in the time-invariant case, $u_b(t)$ results from discontinuities in the envelope and instantaneous frequency at $t = 0$, while $u_d(t)$ now reflects the inability of the system to respond to variations in $B(t)$ and $\omega(t)$ as well as to variations in $C(t)$ and $R(t)$.

When the input frequency and amplitude are constant for all time, $B(t) = B_1$, $\omega(t) = \omega_\nu$, $u_b(t) = 0$, and $u(t)$ is

$$u(t) = B_1 T[j\omega_\nu, t] e^{j\omega_\nu t} - \frac{B_1}{C(t)} \int_0^t \frac{d}{d\tau} \{C(\tau) T[j\omega_\nu, \tau]\} G_{\tau t}(t, \tau) e^{j\omega_\nu \tau} \, d\tau \quad (8.3\text{-}10)$$

For the time-invariant solution, the second term $u_d(t)$ vanishes.

The first-order system response can be expressed as a quasi-stationary term plus a closed-form correction term because the system impulse response is expressible as an exponential function. Consequently the integration by parts employed for time-invariant system analysis is then applicable. The impulse response of higher-order systems, however, is not obtainable in this form, thus this analysis technique is not applicable. Furthermore, there are no simple ways of obtaining the system impulse response in terms of the general-valued system parameters regardless of its form.

8.3.3 Separable Systems

Because the basic difficulty concerning the analysis of time-variant systems is computational rather than theoretical, it seems reasonable to direct this computational effort toward general rather than specific results. Such an approach was taken by Blinchikoff and Huggins [2], who initially obtain a representation of the system impulse response $h(t,\tau)$ that is compatible with the objective of expressing the response to modulated inputs as a quasi-stationary response plus a correction term. This representation $h_s(t,\tau)$ is an approximation of $h(t,\tau)$, and the discrepancy between the two functions can be made sufficiently small to satisfy most engineering applications. They assume that $h(t,\tau)$ is known, for if we are ever able to analyze systems for which $h(t,\tau)$ is not explicitly known, we certainly must be able to analyze systems when we have an analytic expression for $h(t,\tau)$.

Blinchikoff and Huggins approximate the time-variant system by a *separable* system. Once the input and output quantities are identified, a separable system is characterized in the frequency domain by a time-dependent system function of the general form.

$$H(s,t) = \sum_{k=1}^{N} A_k(t)\Phi_k(s) \tag{8.3-11}$$

that is, the time-varying memoryless elements $A_k(t)$ are separated from the time-invariant dynamic elements $\Phi_k(s)$.

Zadeh [27] first introduced the time-varying system function and defined it as

$$H(s,t) = e^{-st} \int_{-\infty}^{t} h(t,\tau)e^{s\tau}\, d\tau \tag{8.3-12}$$

where $h(t,\tau)$ is the system impulse response. This integral is the superposition integral [see (1.8-7)], so $H(s,t)e^{st}$ is then the response of a time-variant system when the input is an exponential function. The $H(s,t)$ is a generalization of the time-invariant transfer function $H(s)$, which describes the complex envelope of the system response to an exponential input signal. Finding $H(s,t)$, however, is generally no simpler than finding $h(t,\tau)$.

The change of variable $\xi = t - \tau$ allows (8.3-12) to be written as

$$H(s,t) = \int_{0}^{\infty} h(t, t-\xi)e^{-s\xi}\, d\xi \tag{8.3-13}$$

where $H(s,t)$ is now the Laplace transform of $h(t, t-\xi)$ and ξ is the age variable. When determining the transform, t is treated as a constant. This is analogous to the time-invariant situation in which $H(s)$ is the Laplace transform of $h(t)$.

Conversely, the impulse response is the inverse Laplace transform of $H(s, t)$,

$$h(t, t - \xi) = \frac{1}{2\pi j} \int_c H(s, t) e^{s\xi} \, ds \qquad (8.3\text{-}14)$$

where c is a suitable contour in the s-plane.

The impulse response of the separable system is then expressed in terms of the age variable ξ by substituting from (8.3-11) into (8.3-14) to give

$$h(t, t - \xi) = \sum_{k=1}^{N} A_k(t) \frac{1}{2\pi j} \int_c \Phi_k(s) e^{s\xi} \, ds = u_{-1}(\xi) \sum_{k=1}^{N} A_k(t) \phi_k(\xi) \qquad (8.3\text{-}15)$$

When $\Phi_k(s)$ is rational in s, its inverse transform $\phi_k(\xi)$ is necessarily a sum of products of polynomials and exponentials. Then $H(s, t)$ in (8.3-11) takes the form

$$H(s, t) = \sum_{k=1}^{P} \sum_{m=0}^{Q_k} \frac{A_{km}(t)}{(s + s_k)^{m+1}} \qquad (8.3\text{-}16)$$

Note that the $\{s_k\}$, the poles of $H(s, t)$, are time-invariant. Kaplan [13], among others, has shown that the necessary and sufficient condition for a system function $H(s, t)$ that is rational in s to characterize a linear system described by an ordinary linear differential equation is that all the s-plane poles be time invariant. Thus time-varying poles cannot be associated with a system described by an ordinary linear differential equation. Time-variable poles may be useful in system analysis, but they do not occur in the synthesis of linear differential systems.

When $H(s, t)$ has simple poles, Q_k in (8.3-16) is zero, and the impulse response from (8.3-15) is

$$h(t, t - \xi) = u_{-1}(\xi) \sum_{k=1}^{N} A_k(t) e^{-s_k \xi} \qquad (8.3\text{-}17)$$

where $A_k(t)$ is the time-varying "residue" at the pole $s = -s_k$. This impulse response is used to characterize the time-variant system in subsequent analyses because it allows the method of integration by parts introduced in Section 8.2.4 to be applied.

A separable system offers the advantage that its response $u(t)$ to an input $f(t)$ whose Laplace transform is $F(s)$, may be found by familiar Laplace transform methods. The transform of the response is the product of $H(s, t)$ and $F(s)$; thus

$$u(t) = \mathcal{L}^{-1}[H(s, t) F(s)] = \sum_{k=1}^{N} A_k(t) \mathcal{L}^{-1}[\Phi_k(s) F(s)] \qquad (8.3\text{-}18)$$

Example 8-3. The first-order time-variable system, characterized by the system function

$$H(s, t) = \frac{A_1(t)}{s + 1}, \qquad \left(\Phi_1(s) = \frac{1}{s + 1} \right) \qquad (8.3\text{-}19)$$

is excited by the step function

$$f(t) = u_{-1}(t) \qquad (8.3\text{-}20)$$

From (8.3-18), the response is

$$u(t) = A_1(t) \mathcal{L}^{-1}\left[\frac{1}{s(s + 1)}\right] = A_1(t) \mathcal{L}^{-1}\left[\frac{1}{s} - \frac{1}{s + 1}\right] = A_1(t)[1 - e^{-t}] u_{-1}(t) \qquad (8.3\text{-}21)$$

Response to Modulated Waveforms

This simple example shows that even though the system is time variant, its response is determined by time-invariant methods. This is one of the major advantages of separable systems.

The quantity $\mathscr{L}^{-1}[\Phi_k(s)F(s)]$ in (8.3-18) represents the response of the time-invariant system with transfer function $\Phi_k(s)$ to the input $f(t)$. This leads to the simple realization presented in Fig. 8-14, a parallel connection of N time-invariant systems, each followed by a time-varying scalar $A_k(t)$. Each time-invariant system possesses a memory, but the time-varying scalars are memoryless. This separation of the dynamic and time-varying aspects aids in understanding the system behavior. For the impulse response of (8.3-17), each $\Phi_k(s)$ is of the form

$$\Phi_k(s) = \frac{1}{s + s_k} \tag{8.3-22}$$

where s_k may be a complex number.

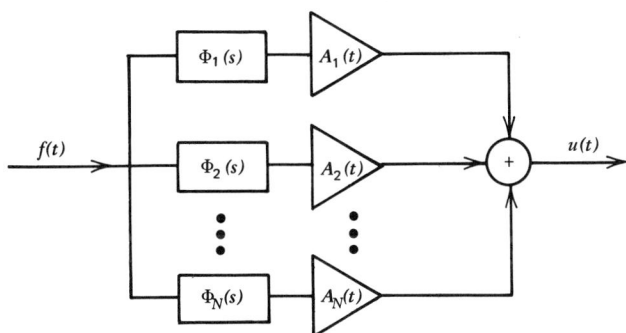

Figure 8-14. Realization of a separable system.

8.3.4 Quasi-Stationary Analysis of Separable Systems

We now determine the response of a separable system whose impulse response is given by (8.3-17) excited by the combined amplitude- and frequency-modulated signal, following the technique in Ref. 2. Substitute from (8.2-14) and (8.3-17) into (1.8-7), with $t_0 = -\infty$, to yield the output expressed by (8.2-15). For this time-varying condition, however, $Y_k(t)$ is defined as

$$Y_k(t) = A_k(t)e^{-s_k t} \tag{8.3-23}$$

Thus the only change from the time-invariant system analysis is the change of the residue at the pole $-s_k$ from a constant value to a time-varying scalar. Because the form of $h(t, \tau)$ in (8.3-17) is similar to the time-invariant system impulse response, the method of integration by parts and analysis introduced in Section 8.2.4 for time-invariant systems is likewise applicable here. The response, expressed as the sum of a quasi-stationary response $u_q(t)$ plus other terms is

$$\begin{aligned} u(t) &= u_q(t) + [u_a(t) - u_q(t)] + u_b(t) + u_d(t) \\ &= u_q(t) + u_{aq}(t) + u_b(t) + u_d(t) \end{aligned} \tag{8.3-24}$$

where

$$u_q(t) = T[j\omega(t), t]B(t)F_{0t}(t)$$
$$u_a(t) = H[j\omega(t), t]B(t)F_{0t}(t)$$
$$u_b(t) = \sum_{k=1}^{N} A_k(t)e^{-s_k t}\left[\frac{B_1}{s_k + j\omega_\nu} - \frac{B_0}{s_k + j\omega_0}\right]$$
$$u_d(t) = -\sum_{k=1}^{N} A_k(t)e^{-s_k t}\int_0^t \frac{d}{d\tau}\left[\frac{B(\tau)}{s_k + j\omega(\tau)}\right] e^{s_k \tau} F_{0\tau}(\tau)\, d\tau$$

Note that the quasi-stationary term is introduced by subtracting it from $u_a(t)$ to form $u_{aq}(t)$ and adding to it $u_{aq}(t)$ so that the right-hand side of (8.3-24) is unchanged.

The first term is the quasi-stationary response already discussed. The second term $u_{aq}(t)$ is associated with the system's time variation. For a time-invariant system, Zadeh's system function reduces to the system transfer function [$H(j\omega) = T(j\omega)$] and this term is then zero. But if the system parameters change rapidly, this term can be appreciable. The remaining two terms $u_b(t)$ and $u_d(t)$ have the same interpretation as their corresponding terms in the time-invariant system response in (8.2-23).

8.3.5 Approximation of a Nonseparable System by a Separable System

The remaining task is to obtain a suitable approximation of an arbitrary (nonseparable) $h(t, \tau)$ by the impulse response in (8.3-17), designated $h_s(t, \tau)$. Minimization of the integrated squared error

$$e(t) = \int_0^\infty [h(t, \tau) - h_s(t, \tau)]^2\, d\tau \tag{8.3-25}$$

is used in Ref. 2 with the result that the $A_k(t)$'s are the solution of the linear set of equations

$$\sum_{j=1}^{N} \frac{A_j(t)}{s_j + s_k^*} = H(s_k^*, t) \qquad k = 1, \ldots, N \tag{8.3-26}$$

The derivation of (8.3-26) follows that described in Section 8.1.5. Note the similarity between (8.1-34) and (8.3-26).

An attempt to find the $\{s_k\}$ by the technique described in Section 8.1.7 results in a solution in which each s_k is a function of the instant t at which the minimization is evaluated. Each s_k, however, must be constant; thus this approach is invalid. Instead the $\{s_k\}$ is chosen to minimize the *weighted* average of the integrated squared error $e(t)$ over the interval $(0 \leq t \leq T)$; given by

$$\mathscr{E} = \frac{1}{T}\int_0^T w(t)\, e(t)\, dt \tag{8.3-27}$$

The weighting factor enables the accuracy to be concentrated into the region of greatest interest. The simultaneous solution of (8.3-26) and (8.3-27) yields the optimum complex $\{A_k(t)\}$ and $\{s_k\}$. As with time-invariant systems, an analytic solution is not possible, and this task is performed on a computer by a search algorithm.

8.3.6 Summary

The messy problem of expressing the response of a time-variant system excited by a modulated waveform in terms of the signal and system parameters has been treated here. Approximating the system impulse response by the impulse response of a separable system allowed us to apply the integration by parts suggested in Ref. 2 and thereby express the response as the sum of a quasi-stationary term plus a correction term. The accuracy of the approximation can be improved as desired by increasing the order of the separable system. The correction term is partitioned to ensure that each term corresponds to a specific system behavior, thereby providing insights into the response characteristics. The extent to which the quasi-stationary response approximates the true system response can be determined by examining the bounds on the correction term. When this approximation is valid, the response is explicitly expressed in terms of the system and signal parameters.

The obtained results are an extension of those presented in Section 8.2.4 for time-invariant systems. A time-invariant system is a separable system in which each $A_k(t)$ is constant (the residue at the pole $s = -s_k$). Then the quasi-stationary transfer function is identical to Zadeh's time-varying system functions $[H(j\omega) = T(j\omega)]$ and (8.3-24) reduces to (8.2-23).

8.4 AVERAGE DELAY THROUGH A TIME-INVARIANT SYSTEM

In many instances the system group delay function is nonconstant over the signal bandwidth, and for these cases group delay is not meaningful for describing the delay between the input and output signals. Even if the group delay is reasonably constant, the magnitude function may be nonconstant, further complicating this delay-time determination. It is then useful to introduce the concept of an average time delay, which is determined from time-domain considerations and is valid for both deterministic and random signals whose autocorrelation functions are known [22].

8.4.1 Definition of Average Delay

Let $f(t)$ be the input signal and $u(t)$ be the output signal of a linear time-invariant system. We first form the error function

$$e(t, \tau) = u(t) - f(t - \tau) \qquad (8.4\text{-}1)$$

and then form the integral of this error squared as

$$\varepsilon(\tau) = \lim_{T \to \infty} \frac{1}{2T} \int_{-T}^{T} e^2(t, \tau) \, dt \qquad (8.4\text{-}2)$$

where τ is the delay of the input signal. The average delay between the input and output signals is that value of τ which minimizes $\varepsilon(\tau)$. Expansion of (8.4-2) and the definitions in Tables 7-1 and 7-2 give

$$\varepsilon(\tau) = \lim_{T\to\infty} \frac{1}{2T} \int_{-T}^{T} u^2(t)\, dt + \lim_{T\to\infty} \frac{1}{2T} \int_{-T}^{T} f^2(t-\tau)\, dt - 2 \lim_{T\to\infty} \frac{1}{2T} \int_{-T}^{T} u(t) f(t-\tau)\, dt$$

$$= \phi_u(0) + \phi_f(0) - 2\phi_{uf}(-\tau) \tag{8.4-3}$$

When the integrand function contains finite area, the limit operation is removed and the integrals are evaluated from $-\infty$ to $+\infty$.

Both $\phi_u(0)$ and $\phi_f(0)$ are positive numbers that are the autocorrelation functions of $u(t)$ and $f(t)$, respectively, evaluated at zero shift. Consequently $\varepsilon(\tau)$ is minimized when $\phi_{uf}(-\tau)$ attains its maximum positive value. From (7.1-40), $\phi_{uf}(-\tau) = \phi_{fu}(\tau)$, the cross-correlation function between $f(t)$ and $u(t)$, and, from (7.1-45), $\phi_{fu}(\tau)$ can equivalently be expressed in terms of the input autocorrelation function as

$$\phi_{fu}(\tau) = \phi_{uf}(-\tau) = \int_{-\infty}^{\infty} h(-t) \phi_f(\tau - t)\, dt \tag{8.4-4}$$

where $h(t)$ is the system impulse response. Thus knowledge of the input signal autocorrelation function and $h(t)$ is sufficient to determine the average delay.

In many cases the minimization of $\varepsilon(\tau)$ becomes computationally difficult. However simple results are obtained when the input signal is the sinusoid, the impulse function, the step function, and the rectangular pulse.

8.4.2 Average Delay of a Sinusoid

We now show that for the input signal $\sin \omega_0 t$, the average signal delay is the system phase delay at ω_0. The response of the filter with transfer function $H(j\omega) = A(\omega) e^{j\theta(\omega)}$ to this sinusoid, is

$$u(t) = A_0 \sin(\omega_0 t + \theta_0) \tag{8.4-5}$$

where $A_0 = A(\omega_0)$ and $\theta_0 = \theta(\omega_0)$. Then, from (8.4-3),

$$\phi_{uf}(-\tau) = \lim_{T\to\infty} \frac{1}{2T} \int_{-T}^{T} A_0 \sin(\omega_0 t + \theta_0) \sin \omega_0(t - \tau)\, dt$$

$$= \frac{A_0}{2} \lim_{T\to\infty} \frac{1}{2T} \left[\int_{-T}^{T} \cos(\omega_0 \tau + \theta_0)\, dt - \int_{-T}^{T} \cos(2\omega_0 t - \omega_0 \tau + \theta_0)\, dt \right] \tag{8.4-6}$$

As $T \to \infty$ the contribution of the second integral vanishes. Then

$$\phi_{uf}(-\tau) = \frac{A_0}{2} \cos(\omega_0 \tau + \theta_0) \lim_{T\to\infty} \frac{2T}{2T} = \frac{A_0}{2} \cos(\omega_0 \tau + \theta_0) \tag{8.4-7}$$

The maximum value of $\phi_{uf}(-\tau)$, which is $A_0/2$, occurs when

$$\tau = -\frac{\theta_0}{\omega_0} \tag{8.4-8}$$

Average Delay Through a Time-Invariant System

the system phase delay at ω_0. For a sinusoidal input the average delay is the exact delay.

8.4.3 Average Delay of the Impulse Function

When the system input is the impulse function $\delta(t)$, $u(t) = h(t)$, the system impulse response, and $\phi_{uf}(-\tau)$, from (8.4-3), is

$$\phi_{uf}(-\tau) = \int_{-\infty}^{\infty} h(t)\delta(t - \tau)\,dt = h(\tau) \qquad (8.4\text{-}9)$$

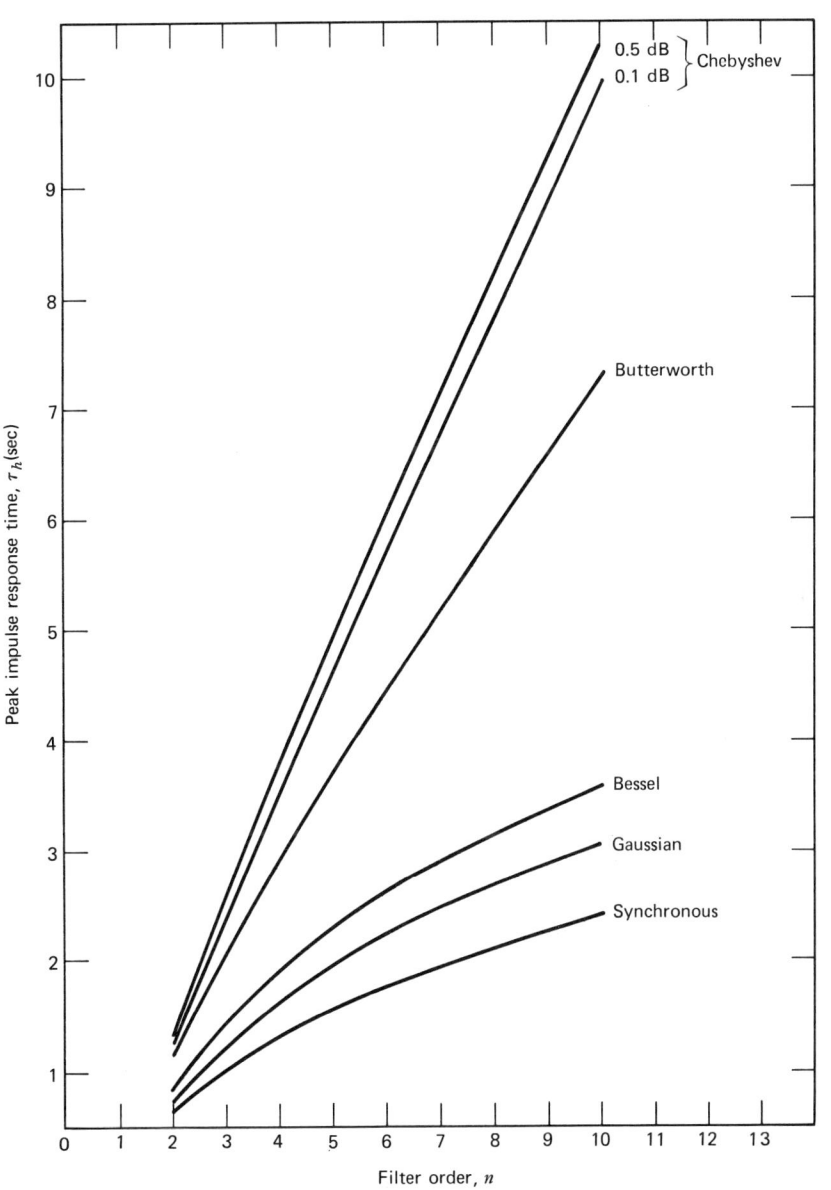

Figure 8-15. Peak impulse response time (τ_h) for various LP filter types ($\omega_c = 1$).

For the impulse function input, $\phi_f(0)$ is infinite, but we still consider that ε is minimized when ϕ_{uf} is maximized. This occurs when τ takes on the value for which $h(\tau)$ is maximum, designated τ_h. This value is consistent with previously obtained results. The impulse response of the rectangular magnitude LP filter with linear phase is shown in Fig. 3-6. There the impulse response peaks at $t = n\pi/2\omega_c$, the value of the filter group delay. For this case τ_h as computed from (8.4-9) is the exact value for the system delay.

Figure 8-15 gives the value of τ_h for various LP filter types, all normalized to unity-radian bandwidth. The delay of the constant-delay type filters is less than the delay of filters approximating the reactangular frequency response. These values were determined on a digital computer, but the synchronous filter delay is given analytically by

$$\tau_h = (n-1)\sqrt{2^{1/n} - 1} \tag{8.4-10}$$

8.4.4 Average Delay of the Step Function

For the step function input $u_{-1}(t)$ (8.4-2) becomes

$$\varepsilon(\tau) = \int_0^\infty [g(t) - u_{-1}(t-\tau)]^2 dt \tag{8.4-11}$$

where $g(t)$ is the system step response. The delayed input superimposed over $g(t)$ appears in Fig. 8-16 for an arbitrary value of τ. The value of τ_s, the average system delay, is now obtained by directly minimizing $\varepsilon(\tau)$.

Because $u_{-1}(t - \tau)$ is zero for $t < \tau$, $\varepsilon(\tau)$ can be rewritten as

$$\varepsilon(\tau) = \int_0^\tau g^2(t)\, dt + \int_\tau^\infty [g(t) - 1]^2\, dt \tag{8.4-12}$$

A necessary condition for minimum ε is

$$\frac{d\varepsilon}{d\tau} = 0 \tag{8.4-13}$$

which, using Leibnitz's rule for differentiating integrals [13], yields

$$g^2(\tau_s) - [g(\tau_s) - 1]^2 = 0 \tag{8.4-14}$$

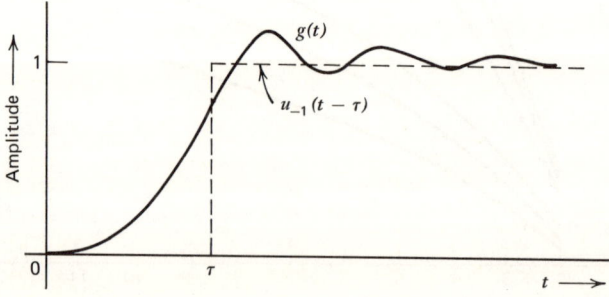

Figure 8-16. The step response and delayed step function for arbitrary value of delay τ.

Average Delay Through a Time-Invariant System

and

$$g(\tau_s) = \frac{1}{2} \qquad (8.4\text{-}15)$$

Thus ε is minimized when $\tau = \tau_s$, the time at which the step response equals one-half, as in Fig. 8-17. This value of τ_s is very near that given by t_d in (3.8-3), as computed by Su [23]. In fact, at t_d the value of the standard step-response delay in Fig. 3-42 is one-half. Again this definition of average delay yields a value consistent with previous results.

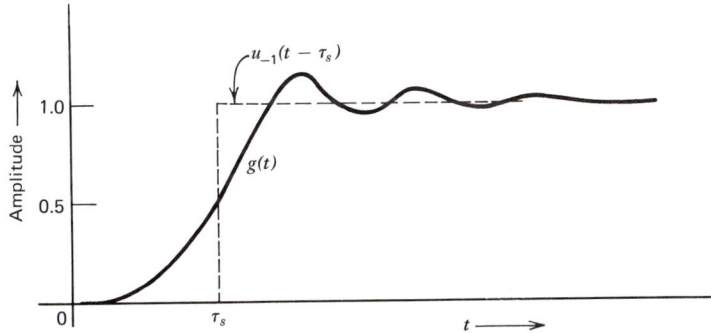

Figure 8-17. Step function position for minimum error.

Example 8-4. The value of τ_s is now analytically determined for the first-order system with transfer function

$$H(s) = \frac{\omega_c}{s + \omega_c} \qquad (8.4\text{-}16)$$

The step response (Fig. 8-18) is

$$g(t) = (1 - e^{-\omega_c t})u_{-1}(t) \qquad (8.4\text{-}17)$$

and (8.4-15) then requires that

$$e^{-\omega_c \tau_s} = \frac{1}{2} \qquad (8.4\text{-}18)$$

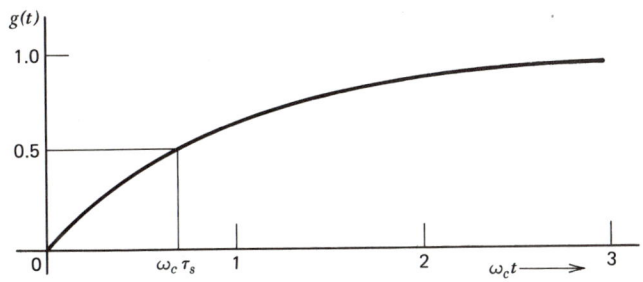

Figure 8-18. Step response of a first-order system. Average delay is τ_s.

Thus the average delay of the step function through this system is

$$\tau_s = \frac{1}{\omega_c} \ln 2 = \frac{0.693}{\omega_c} \tag{8.4-19}$$

as indicated in Fig. 8-18.

8.4.5 Average Delay of the Rectangular Pulse

The system response to the unit-amplitude rectangular pulse of duration τ_0 is designated $p(t)$. The delayed input and $p(t)$ are plotted in Fig. 8-19 for an

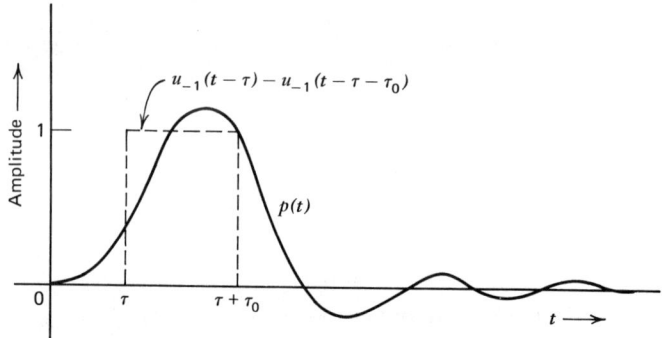

Figure 8-19. The pulse response and delayed input pulse for an arbitrary value of delay τ.

arbitrary value of delay τ. Similar to the development in Section 8.4.4, we express the error as

$$\varepsilon(\tau) = \int_0^\tau p^2(t)\,dt + \int_\tau^{\tau+\tau_0} [p(t)-1]^2\,dt + \int_{\tau+\tau_0}^\infty p^2(t)\,dt \tag{8.4-20}$$

From (8.4-13), the value of delay for minimum error τ_p satisfies

$$p^2(\tau_p) + [p(\tau_p + \tau_0) - 1]^2 - [p(\tau_p) - 1]^2 - p^2(\tau_p + \tau_0) = 0 \tag{8.4-21}$$

yielding the relationship

$$p(\tau_p) = p(\tau_p + \tau_0) \tag{8.4-22}$$

At the delay time τ_p the output pulse value is the same as the output pulse value τ_0 seconds later. Figure 8-20 illustrates this condition for the response of a fifth-order Butterworth LP filter ($\omega_c = 1$) excited by a rectangular pulse with $\tau_0 = 7$ sec. The input pulse is positioned so that (8.4-22) is satisfied, that is,

$$p(3.4) = p(10.4) \tag{8.4-23}$$

Thus $\tau_p = 3.4$ sec, consistent with the values in Table 8-2 for $n = 5$. As the input pulse narrows, $p(t)$ approaches the filter impulse response and τ_p approaches the value at which this response is maximum, namely τ_h given in Fig. 8-15.

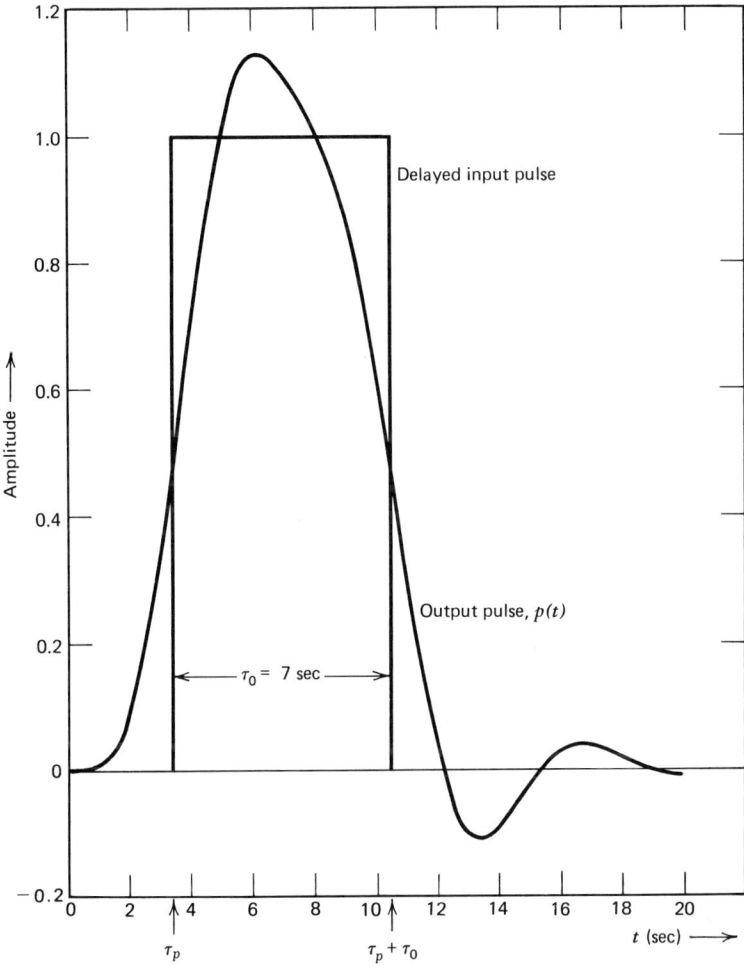

Figure 8-20. Pulse response of $n = 5$ Butterworth LP filter ($\omega_c = 1$) with pulse positioned for minimum integrated squared error. Average delay is τ_p.

Table 8-2 **Delay Values from Various Definitions, for Butterworth Filters ($\omega_c = 1$)**

Order (n)	$D_L(0)$	D_{av}	τ_h	τ_s	τ_d
2	1.41	1.57	1.11	1.43	1.49
3	2.00	2.36	2.05	2.14	2.14
4	2.61	3.14	2.89	2.82	2.80
5	3.24	3.93	3.68	3.50	3.45
6	3.86	4.71	4.43	4.17	4.11
7	4.49	5.50	5.17	4.83	4.76
8	5.12	6.28	5.89	5.49	5.41

8.4.6 Summary of Delay Definitions

We have now considered several measures of system delay, some depending only on the system characteristics and others on both the system and input signal characteristics. To obtain an appreciation of these values, we list the various delays in Table 8-2 for the unity-radian bandwidth Butterworth LP filters. The $D_L(0)$ is the group delay at $\omega = 0$ and D_{av} is the average passband group delay,

$$D_{av} = \int_0^1 D_L(\omega)\, d\omega = -\int_0^1 \frac{d\theta_L(\omega)}{d\omega}\, d\omega = \theta_L(0) - \theta_L(1) = -\theta_L(1) = \frac{n\pi}{4} \quad (8.4\text{-}24)$$

The average delay is just the negative of the phase function evaluated at the cutoff frequency $\omega_c = 1$. We have already discussed τ_h and τ_s and τ_d is the delay computed by Su [23] from (3.8-3).

For a specified order $n > 2$, $D_L(0)$, τ_s, and τ_d are all within 14% of τ_h. The D_{av} is larger than the others because the Butterworth delay increases in the vicinity of the cutoff frequency. For the constant-delay type filters, all five definitions yield approximately the same value for a specified order. However other systems may not have such well-behaved magnitude and phase characteristics, and these definitions of delay may then give values that differ appreciably. For these cases, the most meaningful value is the average system delay for the particular input signal.

In summary, we have obtained a value of system delay by minimizing the integrated squared error between the input signal and the appropriately delayed output signal. Most important, this theoretically obtained value is consistent with previous results and our physical reasoning. The various examples bear this out.

The average delay time can also be denormalized for the BP equivalents of the previously discussed signals applied to a BP filter by multiplying τ by $1/\pi \Delta f$, where Δf is the BP 3-dB bandwidth in hertz and the LP prototype has unity-radian cutoff frequency. The LP response then corresponds to the envelope of the BP response.

REFERENCES

1. Aigrain, P. R., and E. M. Williams, "Synthesis of n-Reactance Networks for Desired Transient Response," *J. Appl. Phys.*, Vol. 20, pp. 597–600, June 1949.
2. Blinchikoff, H. J., and W. H. Huggins, "Quasi-Stationary Analysis of Linear Time-varying Systems Excited by Modulated Signals", *J. Franklin Inst.*, Vol. 292, pp. 369–385, November 1971.
3. Carson, J. R., and T. C. Fry, "Variable-Frequency Electric Circuit Theory," *Bell. Syst. Tech. J.*, Vol. 16, pp. 513–540, October 1937.
4. Chohan, V. C., "On the Usefulness of Group Delay and/or Phase Delay Parameters for Low Distortion Transmission," *IEEE Trans. Commun.*, Vol. COM-22, pp. 1147–1148, August 1974.
5. Clavier, A. G., "Application of Fourier Transforms to Variable-Frequency Circuit Analysis," *Proc. IRE*, Vol. 37, pp. 1287–1290, November 1949.
6. Gold, B., "The Solution of Steady-State Problems in FM," *Proc. IRE*, Vol. 37, pp. 1264–1269, November 1949.
7. Harman, R. K., and F. W. Fairman, "Exponential Approximation Via a Closed-Form Gauss-Newton Method," *IEEE Trans. Circuit Theory*, Vol. CT-20, pp. 361–369, July 1973.

8. Huggins, W. H., *Network Approximation in the Time Domain*, Report E5048A, Air Force Cambridge Research Laboratories, Cambridge, Mass., October 1949.
9. Huggins, W. H., "Signal Theory," *IRE Trans. Circuit Theory*, Vol. CT-3, pp. 210–216, December 1956.
10. Huggins, W. H., "Representation and Analysis of Signals, Part I, The Use of Orthogonalized Exponentials" (Report), The Johns Hopkins University, Department of Electrical Engineering, Baltimore, Md., September 1957.
11. Hupert, J. J., "Transient Response of Narrow-Band Networks to Angle-Modulated Signals," *Proc. NEC*, Vol. 18, pp. 458–468, October 1962.
12. Jess, J., and H. W. Schüssler, "On the Design of Pulse-Forming Networks," *IEEE Trans. Circuit Theory*, Vol. CT-12, pp. 393–400, September 1965.
13. Kaplan, W., *Operational Methods for Linear Systems*, Addison-Wesley, Reading, Mass., 1962.
14. Kautz, W. H., "Transient Synthesis in the Time Domain," *IRE Trans. Circuit Theory*, Vol. CT-1, pp. 29–39, September 1954.
15. Kinariwala, B. K., "Analysis of Time-Varying Networks," *IRE Int. Conv. Rec.*, Part 4, pp. 268–276, 1961.
16. Lee, Y. W., *Statistical Theory of Communication*, Wiley, New York, 1963.
17. McCoy, R. E., "FM Transient Response of Band-Pass Circuits," *Proc. IRE*, Vol. 42, pp. 574–579, March 1954.
18. McDonough, R. N., "Representation and Analysis of Signals; Part 15, Matched Exponents for the Representation of Signals" (Report), Department of Electrical Engineering, The Johns Hopkins University, Baltimore, Md., April 1963.
19. McDonough, R. N., and W. H. Huggins, "Best Least-Squares Representation of Signals by Exponentials," *IEEE Trans. Autom. Control*, Vol. AC-13, pp. 408–412, August 1968.
20. Meyer, P. A., "Filters with an Approximately Rectangular Impulse Response," *NTZ-CJ*, No. 3, pp. 108–115, March 1966.
21. Miller, G., "Closed-Form Inversion of the Gram Matrix Arising in Certain Least-Squares Problems," *IEEE Trans. Circuit Theory*, Vol. CT-16, pp. 237–240, May 1969.
22. Simpson, R. S., and R. C. Houts, "A Definition of Average Time Delay for a Linear System," *Proc. IEEE*, Vol. 55, pp. 1733–1734, October 1967.
23. Su, K. L., *Time-Domain Synthesis of Linear Networks*, Prentice-Hall, Englewood Cliffs, N.J., 1971.
24. Svensson, T., "An Approximation Method for Time Domain Synthesis of Linear Networks," *IEEE Trans. Circuit Theory*, Vol. CT-20, pp. 142–144, March 1973.
25. Weiner, D. D. and B. J. Leon, "Analysis of Time-Varying Systems: An Analogy with Time-Invariant Systems," Proc. First Allerton Conference on Circuit and System Theory, University of Illinois, pp. 81–99, November 1963.
26. Weiner, D. D., and B. J. Leon, "The Quasi-Stationary Response of Linear Systems to Modulated Waveforms," *Proc. IEEE*, Vol. 53, pp. 564–575, June 1965.
27. Zadeh, L. A., "Frequency Analysis of Variable Networks," *Proc. IRE*, Vol. 38, pp. 291–299, March 1950.
28. Zverev, A. I., "The Golden Anniversary of Electric Wave Filters," *IEEE Spectrum*, Vol. 3, pp. 129–131, March 1966.

PROBLEMS

8-1. Use the method of time moments to approximate the impulse response below by the impulse response corresponding to the transfer function

$$H_a(s) = \frac{b_0 s^2 + b_1 s + 1}{a_0 s^3 + a_1 s^2 + a_2 s + 1}$$

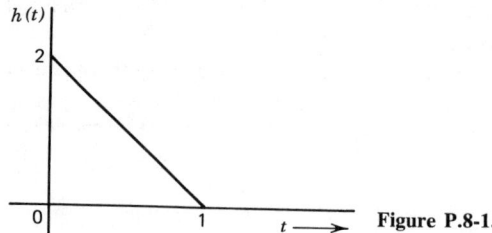

Figure P.8-1.

That is, determine the a's and b's. Then determine $h_a(t)$ and see how closely it approximates $h(t)$. One pole of $H_a(s)$ is $s = -5.66$.

8-2. (a) What is the transfer function of the filter below such that its impulse response approximates the triangular pulse, using the method of time moments?

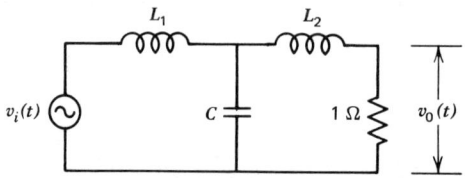
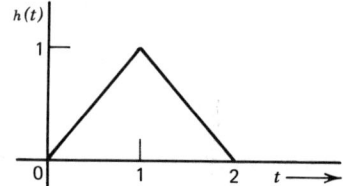

Figure P.8-2.

(b) What are the element values corresponding to this transfer function? Equate the filter transfer function to the approximating transfer function and equate coefficients of like powers of s.

8-3. Determine A_1 and A_2 such that $f(t) = A_1 e^{-t} + A_2 e^{-2t}$ approximates $k(t) = e^{-(3t/2)} u_{-1}(t)$ in the least-squares sense.

8-4. For the $k(t)$ and $f(t)$ in Problem 8-3, determine A_1 and A_2 so that $k(0) = f(0)$ and $k'(0) = f'(0)$. We have not discussed this approximation in this chapter, but it is related to the maximally flat approximation discussed in Chapter 3. Compare this approximation with the one obtained in Problem 8-3. Note the improvement in the vicinity of $t = 0$.

8-5. What is the quasi-stationary transfer function of the third-order Butterworth LP filter ($\omega_c = 1$) excited by the FM input signal given by (8.2-10)?

8-6. The synchronous filter, whose transfer function is given by (3.7-1), is excited by the AM waveform

$$f(t) = (\cos \omega_a t + \cos \omega_b t) \cos \omega_\nu t \qquad (\omega_\nu \gg \omega_a \text{ and } \omega_b)$$

What is the quasi-stationary response of this filter?

8-7. Determine the response of the separable system with impulse response given by (8.3-17), excited by the signal $f(t) = u_{-1}(t) e^{-\alpha t}$.

8-8. Consider the first-order time-variant system characterized by $h(t, \tau)$ in (8.3-6), with $\tau \geq 0$. What functional form of $\alpha(t)$ will ensure that this system is then separable in the sense of (8.3-17)?

Problems

8-9. What is the average delay of a rectangular pulse of duration T in passing through the first-order filter characterized by (8.4-16)?

8-10. What is the average delay of the exponential $u_{-1}(t)e^{-at}$ in passing through the first-order filter characterized by (8.4-16)?

Chapter Nine
Digital Filtering

We have taken the viewpoint that a filtering device is a signal processor, capable of signal discrimination on the basis of either frequency-domain or time-domain characteristics. In the past, electronic systems predominantly used continuous-time (analog) signals, but the last decade has seen these signals replaced by discrete-time (sampled) signals in many communication, radar, seismic, and biomedical applications. A sampled signal that is additionally quantized in amplitude becomes a sequence of numbers and is called a digital signal. The filter that processes these numbers is known as a digital filter, but "numerical filter" might be a more accurate name.

The aims of digital filtering are the same as those of continuous filtering, but digital hardware is used as the basic building block rather than the more conventional analog components. The increased use of digital circuitry results from the advances in integrated circuitry. As large-scale integration (LSI) progresses and circuit chips become readily available and inexpensive, the potential of digital filtering increases. Then more functions on a single chip will be possible, resulting in digital filters occupying less space and possessing greater accuracies than conventional analog filters.

Digital filters, in addition to incorporating the benefits of integrated circuitry, offer the following advantages over analog filters.

Finite-duration impulse responses are achievable.

Time-varying filters are realized without any special components by simply programming a different set of numbers into the filter.

Linear-phase filters are realizable.

Certain realization problems, such as negative element values, and practical problems, such as inconveniently large components at low frequencies, do not arise.

Greater accuracy is achieved.

Time sharing is possible, thus allowing more than one input to be processed.

Environmental conditions, such as temperature, humidity, and pressure, have little effect on the filter response.

The digital filter, however, is not a cureall and it does have certain limitations. For example, it is an active system and therefore requires power. Furthermore, digital filter performance is limited by

The quantization of the input samples and filter coefficients, and the use of finite arithmetic in computations. Thus input signals with a wide range of values cannot be accommodated if the number of bits for quantization is insufficient.

The Uniform Sampling Theorem

Present logic circuitry speed, which prohibits high-frequency application above approximately 50 MHz.

In spite of the foregoing differences between analog and digital filtering there are many similarities. Linear analog filters are described by *linear differential* equations, and linear digital filters are described by *linear difference* equations. The basic mathematical tool for analyzing and synthesizing analog filters is the Laplace transform, whereas the z-transform is useful in the study of digital filtering. Discrete-time convolution is analogous to the convolution integral, and the analog filter impulse response is equivalent to the digital filter response when the input is a single unit sample. These analogies, summarized in Table 9-1, allow us to use much of the analog filter theory and data to understand, design, and synthesize digital filters.

Table 9-1 **Analogy Between Analog and Digital Domains**

Analog	Digital
Continuous time	Discrete time
Differential equation	Difference equation
Convolution integral	Discrete-time convolution
Impulse function	Unit sample
Laplace transform	z-Transform
s-Plane	z-Plane

The value of the sampling interval T has both theoretical and practical significance. The sampling theorem establishes an upper bound for T that permits a band-limited signal to be unambiguously recovered from the sampled signal. In practice the lower limit for T is dictated by the logic circuitry speed.

We show that the digital filter can be considered to be the processor of the more general sampled data filter. The iterative nature of the difference equation characterizing the digital filter is described, and this calculation procedure suggests an implementation of the digital filter on a general purpose computer.

We introduce and review the important properties of the z-transform and show how this transform is used to solve difference equations. From this development we obtain the system function and the discrete-time convolution, both useful for design, analysis, and synthesis. Three design methods are discussed—two for recursive filters and one for nonrecursive filters. We then examine sources of error in filter responses, introduce the fast Fourier transform, and discuss its application to digital filtering.

9.1 THE UNIFORM SAMPLING THEOREM

The uniform sampling theorem was introduced by Shannon in 1948 [13], although Nyquist had disclosed the idea in 1928. This theorem is fundamental in the study of sampled systems for it establishes the theoretical maximum sampling interval for complete signal reconstruction. Although the theorem applies to physically unrealizable bandlimited signals, such signals are often good approximations to many signals encountered in practice.

9.1.1 Theorem and Proof

If $F(\omega)$, the Fourier transform of $f(t)$, is bandlimited as in Fig. 9-1, that is,

$$F(\omega) = 0 \quad \text{for} \quad |\omega| \geq \omega_c \quad (\omega_c = 2\pi f_c) \tag{9.1-1}$$

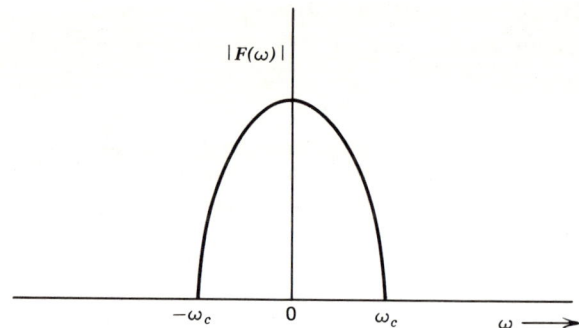

Figure 9-1. The Fourier spectrum of a band-limited function $f(t)$.

then $f(t)$ is uniquely determined by its values at uniform time intervals that are $1/2f_c$ seconds apart; namely

$$f_n = f\left(\frac{n}{2f_c}\right) = f\left(\frac{n\pi}{\omega_c}\right) \tag{9.1-2}$$

Thus we can reconstruct $f(t)$ from the f_n's by first multiplying each sample by a cardinal function centered at the sample time and then summing the resulting terms as

$$f(t) = \sum_{n=-\infty}^{\infty} f_n \frac{\sin(\omega_c t - n\pi)}{\omega_c t - n\pi} \tag{9.1-3}$$

The interval $1/2f_c$ is called the Nyquist interval and $2f_c$ is known as the Nyquist frequency.

The proof of this theorem is now given. Since $F(\omega)$ contains no frequencies higher than ω_c rad/sec, the inversion formula of (2.1-6) takes the form

$$f(t) = \frac{1}{2\pi} \int_{-\omega_c}^{\omega_c} F(\omega) e^{j\omega t} \, d\omega \tag{9.1-4}$$

and f_n is then given by $f(t)$ evaluated at $t = n\pi/\omega_c$,

$$f_n = f\left(\frac{n\pi}{\omega_c}\right) = \frac{1}{2\pi} \int_{-\omega_c}^{\omega_c} F(\omega) e^{j(n\pi\omega/\omega_c)} \, d\omega \tag{9.1-5}$$

The band-limitedness of $f(t)$ allows us to expand $F(\omega)$ into a Fourier series $F_p(\omega)$

The Uniform Sampling Theorem

whose fundamental period is $2\omega_c$. In this interval,

$$F(\omega) = F_p(\omega) = \sum_{n=-\infty}^{\infty} c_n e^{-j(n\pi\omega/\omega_c)} \qquad -\omega_c \leq \omega \leq \omega_c \qquad (9.1\text{-}6)$$

where the c_n's are the Fourier coefficients of the exponential series defined as

$$c_n = \frac{1}{2\omega_c} \int_{-\omega_c}^{\omega_c} F(\omega) e^{j(n\pi\omega/\omega_c)} d\omega \qquad (9.1\text{-}7)$$

Comparing (9.1-5) and (9.1-7) we see that

$$c_n = \frac{\pi}{\omega_c} f_n \qquad (9.1\text{-}8)$$

Equation 9.1-6 shows that $F(\omega)$ is completely described if the c_n's are known, and (9.1-8) reveals that c_n is expressible in terms of the time-function samples f_n at intervals $1/2f_c$ seconds apart. Since $F(\omega)$ is completely described, $f(t)$ is also completely described and the theorem is proved.

9.1.2 Reconstruction of Time Function

In addition to the desired spectrum, the Fourier series representation produces additional spectra (dashed lines in Fig. 9-2). These phenomena are of no

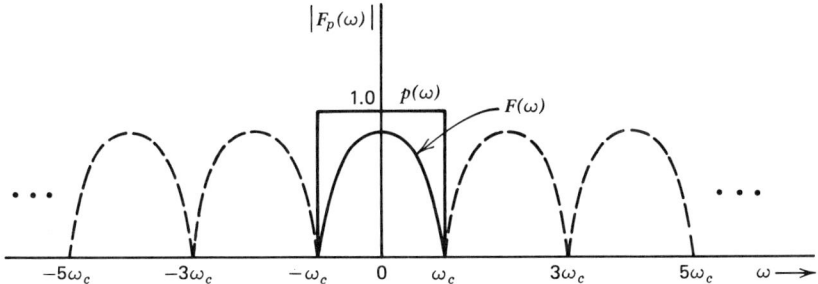

Figure 9-2. Representation of $F(\omega)$ by the Fourier series $F_p(\omega)$.

consequence because the representation is valid only over the frequency range from $-\omega_c$ to ω_c. Mathematically, $F(\omega)$ can be expressed as the product

$$F(\omega) = p(\omega) F_p(\omega) = p(\omega) \sum_{n=-\infty}^{\infty} \frac{\pi}{\omega_c} f_n e^{-j(n\pi\omega/\omega_c)} \qquad (9.1\text{-}9)$$

where $p(\omega)$ is the rectangular function of unit amplitude in Fig. 9-2. The time function $f(t)$, from (2.1-6), is

$$f(t) = \frac{\pi}{\omega_c} \sum_{n=-\infty}^{\infty} f_n \frac{1}{2\pi} \int_{-\omega_c}^{\omega_c} \exp\left[j\omega\left(t - \frac{n\pi}{\omega_c}\right)\right] d\omega \qquad (9.1\text{-}10)$$

yielding the result in (9.1-3).

If the uniform sampling level is less than $1/2f_c$, $f(t)$ can still be exactly reconstructed. To show this, we follow the proof in Section 9.1.1. We assume the sampling interval is $1/(2f_c + f_b)$ and expand $F(\omega)$ as a Fourier series whose fundamental period is $2\omega_c + \omega_b$, as in Fig. 9-3. The Fourier coefficients are proportional to the sampled values that are $1/(2f_c + f_b)$ second apart and $F(\omega)$ is completely described.

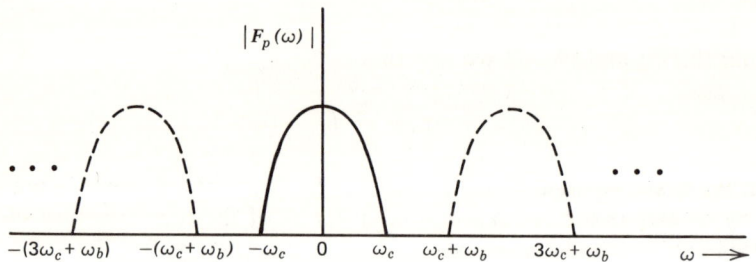

Figure 9-3. Representation of $F(\omega)$ by the Fourier series with the period $(2\omega_c + \omega_b)$.

We can now state that any uniform sampling interval less than or equal to $1/2f_c$ allows the band-limited function $f(t)$ to be exactly reconstructed. Note that if the sampling interval is greater than $1/2f_c$, the spectra in Fig. 9-3 overlap and recovery of $F(\omega)$ is impossible. This condition, known as aliasing or foldover, is discussed in Section 9.4.3.

9.1.3 Physical Interpretation of the Sampling Theorem

The rectangular function $p(\omega)$ in Fig. 9-2 corresponds to an ideal LP filter with radian cutoff frequency ω_c, which separates the signal spectrum from the infinitely repeating spectra of Fig. 9-2. Thus the filter output spectrum is the original spectrum in Fig. 9-1, which means that the continuous signal has been reconstructed from the sampled signal.

Additional insight into the reconstruction of $f(t)$ from the sampled values f_n is obtained by examining this process in the time domain. The ideal LP filter impulse response,† from (3.1-4), is

$$h(t) = \frac{\omega_c}{\pi} \frac{\sin \omega_c t}{\omega_c t} \qquad (9.1\text{-}11)$$

We convert the sample f_1 into an impulse function of area $(\pi/\omega_c) f_1$ (π/ω_c is the sampling interval $T = 2\pi/\omega_s$ with $\omega_s = 2\omega_c$), which then excites the LP filter. The response $h_1(t)$ is then $h(t)$ of (9.1-11) weighted by $(\pi/\omega_c) f_1$ and delayed by π/ω_c,

$$h_1(t) = f_1 \frac{\sin (\omega_c t - \pi)}{\omega_c t - \pi} \qquad (9.1\text{-}12)$$

shown in Fig. 9-4. The sample f_2 is then converted to an impulse function of

---

†For simplicity, the phase shift is assumed to be zero. The inclusion of the linear phase shift only causes a delay in the reconstructed signal.

The Uniform Sampling Theorem

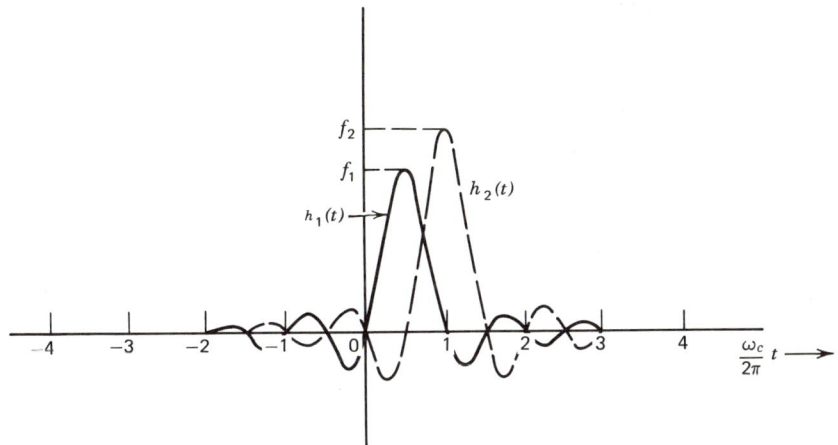

Figure 9-4. Impulse response of the ideal LP filter due to samples f_1 and f_2.

weight $(\pi/\omega_c)f_2$, which then excites the LP filter yielding the weighted impulse response delayed by twice the sampling interval,

$$h_2(t) = f_2 \frac{\sin(\omega_c t - 2\pi)}{\omega_c t - 2\pi} \tag{9.1-13}$$

(dashed line in Fig. 9-4). Continuing this process yields the impulse response for the nth sample as

$$h_n(t) = f_n \frac{\sin(\omega_c t - n\pi)}{\omega_c t - n\pi} \tag{9.1-14}$$

The total response is then the superposition of these individually weighted impulse responses, and this is expressed mathematically by (9.1-3).

Note from Fig. 9-4 that the value of $f(t)$ at $t = \pi/\omega_c$ is determined only by the impulse response due to the sample f_1 because the impulse responses due to other samples are all zero there. Likewise the value of $f(t)$ at $t = 2\pi/\omega_c$ is determined only by the impulse response due to the sample f_2. In general, $h_n(t) = f_n$ at $t = n\pi/\omega_c$ but is zero at all other sampling times. The cardinal function can thus be considered to be an interpolation function, since the weighted sum of these functions determines the values of $f(t)$ between samples.

It is very surprising, to put it mildly, that values of $f(t)$ between sample points can be uniquely determined when we know nothing about the behavior there. However we must remember that we are dealing with band-limited signals and filters having ideal response characteristics. Band-limitedness implies a nonrealizable signal that has been on for all time. An infinite number of samples prior to the application of the input are then available allowing complete reconstruction. Yet real-life signals are time-limited, rather than band-limited, because they were "turned on" at some previous instant. Also, as shown in Section 3.1.1, the ideal filter used for reconstruction possesses an impulse response that is physically unrealizable (anticipatory), and the use of such filters yields unrealizable responses. It is the use of these nonrealizable models for signal and filter that permit unambiguous reconstruction.

We can never exactly reconstruct the time signal from its sampled values. However the spectra of some signals are insignificant beyond a certain frequency and can be neglected. When this is not true, the resulting error should be examined [14].

9.1.4 Interpolation Functions

The conversion of equally spaced sample values f_n to a continuous function requires interpolation between samples. Linear interpolation is achieved by the triangular component function $\Delta_l(t)$ in Fig. 9-5, where T is the time between

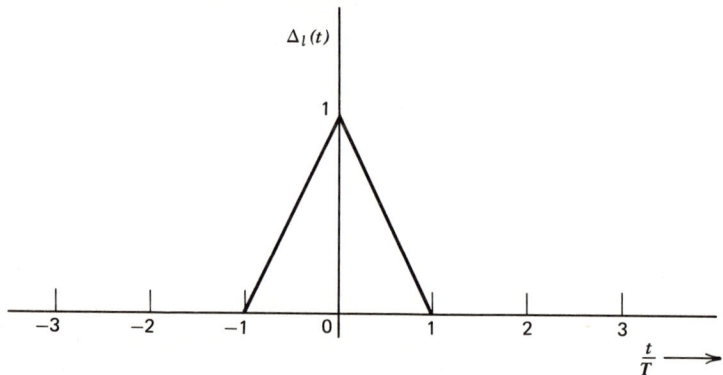

Figure 9-5. The triangular component function $\Delta_l(t)$.

samples. Unfortunately this function is incapable of reproducing a smooth curve because the resulting curve is straight lines from one sample point to the next. The reconstructed function $f_1(t)$ is then

$$f_1(t) = \sum_{n=-\infty}^{\infty} f_n \Delta_l(t - nT) \qquad (9.1\text{-}15)$$

Note that $\Delta_l(t)$ has two desirable properties of a good interpolation function: unity at the sampling instant of interest and zero at all other sampling times.

There are many other acceptable interpolation functions but we now consider the cardinal function because it is the interpolation function for band-limited functions. Furthermore its derivative exists for all values of t; hence it is capable of reproducing smooth curves. Since we initially know nothing about the behavior between samples, we can assume that the sampling frequency is twice the highest significant spectrum frequency. Therefore we can reconstruct the band-limited function $f_2(t)$ that has the sample values f_n. From the sampling theorem, this function is

$$f_2(t) = \sum_{n=-\infty}^{\infty} f_n \frac{\sin(t/T - n)\pi}{(t/T - n)\pi} \qquad (9.1\text{-}16)$$

We say that $f_2(t)$ offers a band-limited interpolation between the samples f_n.

9.2 DEFINITION OF A DIGITAL FILTER

A digital filter is a computational process or algorithm by which a number sequence acting as an input is transformed into a second number sequence termed the output. The algorithm may be implemented in software as a computer subroutine for a general-purpose machine or in hardware as a special-purpose computer. The term "digital filter" is then applied to the subroutine or to the hardware [15].

The digital filter performs the same filtering function as the analog filter except the former operates on numbers and the latter operates on functions. Real-time filtering implies that the filter computations are performed in a time less than the sampling interval T, that is, before the next sample arrives.

The hardware digital filter can equivalently be defined as a special-purpose digital computer that numerically convolves the input number sequence and the filter impulse response. Thus the filter's output number sequence is the solution of the linear difference equation that describes the filter behavior.

Numerical convolution requires an approximation between samples, but the approximating functions are usually hidden in the design techniques because many techniques (we discuss some in Section 9.6) are defined in the frequency domain. However another design approach is to remain in the time domain and explicitly approximate analog convolution by discrete-time convolution [7, 11].

Since an analog signal cannot be applied directly to a digital filter, additional circuitry is necessary to convert the continuous-time signal to numbers. An intermediate step in this scheme is the conversion of the input signal to discrete-time data, which are also suitable as input to a sampled data filter (SDF), a close relative of the digital filter (DF). The DF, however, requires its input quantized in amplitude as well. In fact, the DF may be called the processor of the SDF. Let us examine this aspect more closely by considering the signal processing model in Fig. 9-6.

The continuous input $f(t)$ at point A is quantized in time by the sampler there every $T = 2\pi/\omega_s$ seconds, where ω_s is the sampling frequency in radians per second. In real-life each sample is a narrow amplitude-modulated pulse, but for convenience we consider the sample width to be zero. The sampled signal at point

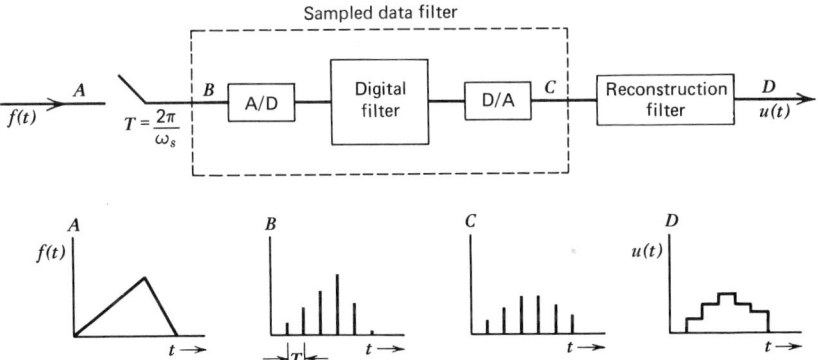

Figure 9-6. Model of discrete-time system for processing continuous-time signals.

B is then operated on by the SDF, either directly (analog means) or by a DF internal to the SDF (thus the notion that the DF is the processor of the SDF). For the latter case the sampled signal is further quantized in amplitude by the analog-to-digital (A/D) converter preceding the digital filter. The input to the DF is a number usually expressed as a series of binary digits (bits) representing the amplitude of $f(t)$ at the sampling instant. The accuracy of this representation depends on the number of bits. The DF operates on these numbers, producing an output number sequence, also represented by a series of bits. A digital-to-analog (D/A) converter then transforms this sequence to a sampled signal that is quantized in time only (point C). The continuous function $u(t)$ at point D is created by the reconstruction filter.

The impulse response of the reconstruction filter is the interpolating function between samples, and we have already discussed the triangular function (linear interpolation) and the cardinal function (band-limited interpolation). The hold function is often used for reconstruction, that is, the sample value is "held" until the next sample occurs. This is illustrated in Fig. 9-6, where the waveform at point D is the output of a holding circuit (reconstruction filter).

The signal processor in Fig. 9-6 is simplified for various system characteristics. If the input signal at point A is already digitized, the sampler and the A/D converter are not necessary. In an all-digital system only the digital filter is necessary, for the input and output are already in the desired digital form. For this case the SDF and DF are one and the same. The closer the A/D and D/A converters are located to the input and output, respectively, of the actual system, the more of the system can be realized with digital circuitry [12].

Digital filters are usually classified as being recursive or nonrecursive. These terms are recommended as descriptions of the filter realization, a subject discussed in Section 9.7.1. The present output sample of a recursive filter is a weighted sum of the present and the past input samples, as well as past output samples. The present output sample of a nonrecursive filter, however, is a weighted sum of present and past input samples only and is independent of past output samples.

9.3 THE DIFFERENCE EQUATION

The operation of a digital filter is often clarified by first examining the iterative nature of the difference equation describing the filter. Consider a digital filter having input samples $f(nT)$ and output samples $u(nT)$, where T is the sampling interval. The operation of this filter is described by a difference equation that relates $u(nT)$ as a function of the present input sample $f(nT)$ and any number of past input and output samples. This system description is analogous to a differential equation that relates the input and output functions of a continuous-time system. An mth-order linear difference equation with constant coefficients L_i and K_i may be written as

$$u(nT) = \sum_{i=0}^{q} L_i f(nT - iT) - \sum_{i=1}^{m} K_i u(nT - iT) \qquad (9.3\text{-}1)$$

This equation is a recursion formula, for the new output $u(nT)$ can be computed

The Difference Equation

given the m previous values of the output and the $q + 1$ most recent values of the input.

9.3.1 Example of a Difference Equation Computation

We now illustrate the recursive nature of the first-order difference equation

$$u(nT) = 3f(nT) - 2f(nT - T) + 6f(nT - 2T) + 2f(nT - 3T) - u(nT - T) \quad (9.3\text{-}2)$$

by computing the output at successive sampling instants. The assumed input sampled function $f_S(t)$ appears in Fig. 9-7a. Comparing (9.3-2) with (9.3-1),

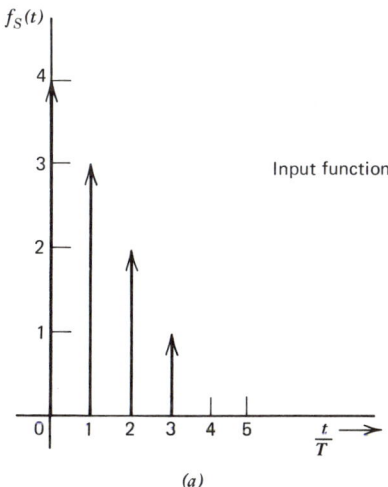

(a)

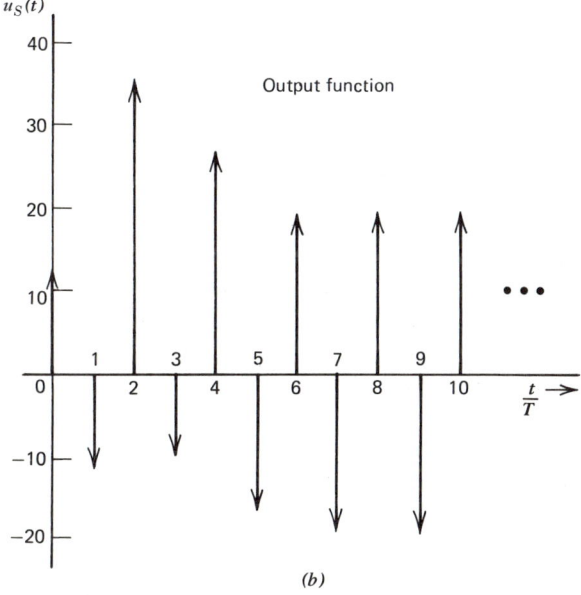

(b)

Figure 9-7. Solution of (9.3-2) for input function $f_S(t)$.

$m = 1, q = 3, K_1 = 1, L_0 = 3, L_1 = -2, L_2 = 6$, and $L_3 = 2$. The system described by (9.3-2) is assumed to be causal; thus the sampled output function $u_S(t)$ is zero for $t < 0$, since $f_S(t)$ is zero for $t < 0$. Successive computations of $u(nT)$ are given in Table 9-2 and these values are shown in Fig. 9-7b. For $n > 7, u(nT) = (-1)^n 18$.

The recursiveness of the difference equation suggests that a general-purpose digital computer can easily be programmed to perform these arithmetic operations. This is true, but the result is a digital filter that is larger and has greater capabilities than necessary. Therefore, if we build a special-purpose computer to simulate (9.3-2), we will have a practical digital filter. Thus one method of digital filter design is to directly simulate the difference equation. The filter constants are then the coefficients of the difference equation. Furthermore, the filter output is the same as the convolution of the input samples and the equivalent impulse response of the filter. Simulation of the difference equation may require much computation time; thus the fast Fourier transform (FFT) method (Section 9.7) can be considered a competitive scheme.

9.3.2 Digital Filter Simulation of a First-Order Difference Equation

The three basic components of a digital filter are introduced by examining the digital filter in Fig. 9-8, which is described by the first-order difference equation

$$u(nT) = K_1 u(nT - T) + L_0 f(nT) \tag{9.3-3}$$

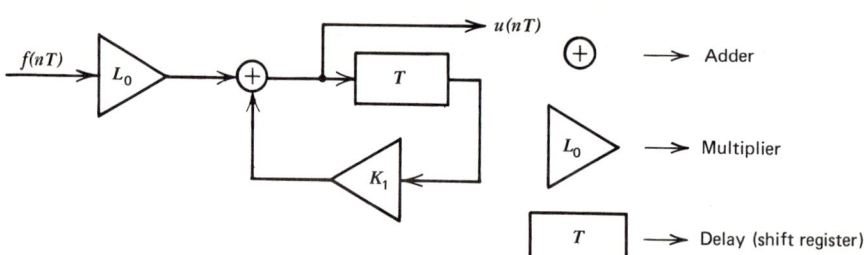

Figure 9-8. Realization of a first-order recursive digital filter.

The multiplier is indicated by a triangle with the multiplying constant inside it. The delay is represented by a rectangle with the sampling interval T inside it, but after the z-transform is later introduced, the symbol z^{-1} is sometimes used instead of T. A plus sign inside a circle designates the adder.

The computation scheme of the digital filter in Fig. 9-8 is a direct mechanization of the difference equation. The sample $f(nT)$ is multiplied by L_0 and is added to the previous output multiplied by K_1 to yield an output sample $u(nT)$.

9.4 THE z-TRANSFORM

The z-transform is useful in the study of discrete-time systems just as the Laplace transform helps in the study of continuous-time systems. Whereas the Laplace transform reduces constant-coefficient linear differential equations to linear

Table 9-2 Successive Computations of $u(nT)$ from Difference Equation 9.3-2

n	$f(nT)$	$3f(nT) - 2f(nT-T) + 6f(nT-2T) + 2f(nT-3T) - u(nT-T)$	$u(nT)$
0	4	$3f(0) \quad -2f(-T) + 6f(-2T) + 2f(-3T) - u(-T) = 3 \times 4$	12
1	3	$3f(T) \quad -2f(0) + 6f(-T) + 2f(-2T) - u(0) = (3 \times 3) - (2 \times 4) - 12$	-11
2	2	$3f(2T) \quad -2f(T) + 6f(0) + 2f(-T) - u(T) = (3 \times 2) - (2 \times 3) + (6 \times 4) - (-11)$	35
3	1	$3f(3T) \quad -2f(2T) + 6f(T) + 2f(0) - u(2T) = (3 \times 1) - (2 \times 2) + (6 \times 3) + (2 \times 4) - 35$	-10
4	0	$3f(4T) \quad -2f(3T) + 6f(2T) + 2f(T) - u(3T) = -(2 \times 1) + (6 \times 2) + (2 \times 3) - (-10)$	-26
5	0	$3f(5T) \quad -2f(4T) + 6f(3T) + 2f(2T) - u(4T) = (6 \times 1) + (2 \times 2) - 26$	-16
6	0	$3f(6T) \quad -2f(5T) + 6f(4T) + 2f(3T) - u(5T) = (2 \times 1) - (-16)$	18
7	0	$3f(7T) \quad -2f(6T) + 6f(5T) + 2f(4T) - u(6T) = -(18)$	-18
$n > 7$	0	$-u(nT-T)$	$(-1)^n 18$

algebraic equations, the z-transform reduces constant-coefficient linear difference equations to linear algebraic equations. The z-transform is the last of the four major transforms that we need in the analysis, synthesis, and understanding of linear time-invariant systems, either continuous or discrete time. The other three are the Fourier, Laplace, and Hilbert transforms.

9.4.1 Definitions and Properties

The z-transform is defined through the concept of the ideal sampler (Fig. 9-9). The input to the sampler is the continuous function $f(t)$, and $f_S(t)$ is the output of the sampler. The switch closes every T seconds. As is customary, the Laplace transform of the time function is indicated by a capital letter. Thus $F_S(s)$ is the Laplace transform of $f_S(t)$.

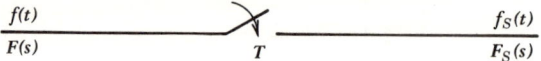

Figure 9-9. The ideal sampler circuit.

Assume that $f(t)$ is zero for $t < 0$ because we are anticipating the application of this mathematical treatment to real-life causal signals. Then Fig. 9-10 shows the continuous input $f(t)$ and the output $f_S(t)$. Sampled values of $f(t)$, indicated by dots, occur every T seconds. In practice $f_S(t)$ is a narrow-amplitude-modulated pulse, but for convenience we assume a zero pulse width.

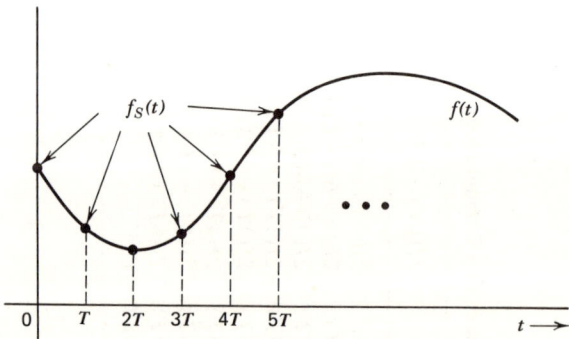

Figure 9-10. Input and output of the ideal sampler.

It is often helpful to relate the sampling process to familiar ideas such as signal spectra, modulation, and filtering processes. Therefore we consider $f(t)$ as modulating $d(t)$, a train of periodic impulse functions of area T extending over all time, as illustrated in Fig. 9-11. With this understanding, $f_S(t)$ is then expressed as

$$f_S(t) = f(t)\,d(t) = T \sum_{n=0}^{\infty} f(nT)\delta(t - nT) \tag{9.4-1}$$

where $f(nT)$ is the value of $f(t)$ at $t = nT$ and $\delta(t - nT)$ is the impulse function

The z-Transform

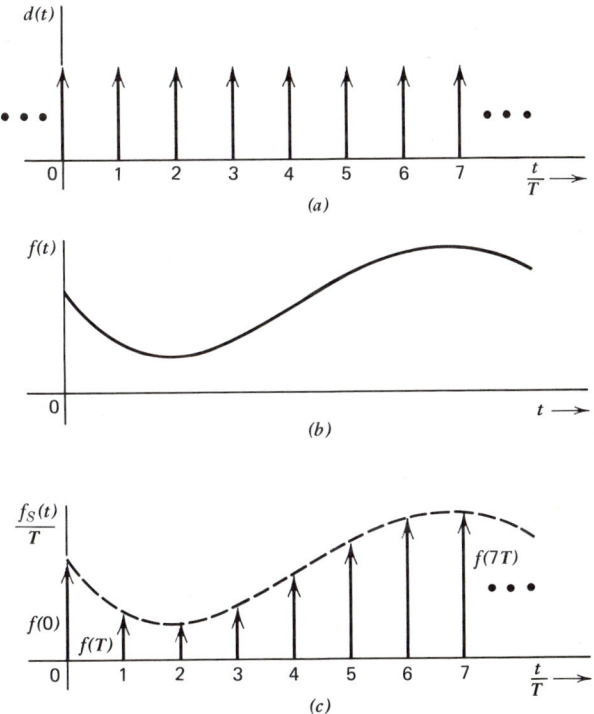

Figure 9-11. The sampling process regarded as a modulation process. (*a*) Train of periodic impulse functions of area T (train extends from $-\infty$ to $+\infty$), (*b*) continuous function, (*c*) sampled function.

occurring at $t = nT$.† Information therefore exists at the sampler output only at the instants $t = 0, T, 2T, \ldots$, and we may consider $f(nT)$ to be the weighting factor of the impulse function $\delta(t - nT)$, as indicated by the values in Fig. 9-11c.

The Laplace transform of both sides of (9.4-1) (neglecting T), from Table 2-1, entry 1, is

$$F_S(s) = \sum_{n=0}^{\infty} f(nT) e^{-nTs} \qquad (9.4\text{-}2)$$

The Laplace transform is useful in analog system analysis because many of the commonly-encountered time functions convert to rational functions in s. Unfortunately $F_S(s)$ contains exponentials, hence is nonalgebraic in s. However the single substitution

$$z = e^{sT} \qquad \left(\text{or} \quad s = \frac{1}{T} \ln z\right) \qquad (9.4\text{-}3)$$

reduces (9.4-2) to an algebraic expression in z.

---

†The constant T in (9.4-1) is usually neglected, ensuring that the introduction of the variable z results in the standard z-transform of $f_S(t)$. Accordingly we also neglect T. By doing this we simplify the notation and eliminate a proportionality constant that is unnecessary for succeeding developments. Remember, however, that the analysis in Section 9.1.3 required the introduction of T.

We define $F(z)$, the z-transform of $f_S(t)$, as

$$\mathscr{Z}[f_S(t)] = F(z) = F_S(s)|_{s=(1/T)\ln z} = \sum_{n=0}^{\infty} f(nT)z^{-n} \qquad (9.4\text{-}4)$$

where the symbol $\mathscr{Z}$ is spoken "the z-transform of."

Thus z^{-1} can be interpreted as the unit-delay variable, hence the coefficient of z^{-n} is the nth sample of $f(t)$. The z-transform does not immediately appear attractive because it may contain an infinite number of terms. However, if $F(s)$ is a rational function in s, then $F(z)$ can be obtained as a rational function in z.

Example 9-1. The Laplace transform of $f(t) = e^{-3t}$ is rational in s; hence the z-transform of $f_S(t)$ is obtainable as a rational function in z. Substitute e^{-3nT} into (9.4-4) as

$$F(z) = \sum_{n=0}^{\infty} e^{-3nT} z^{-n} \qquad (9.4\text{-}5)$$

and rearrange as

$$F(z) = \sum_{n=0}^{\infty} (e^{-3T} z^{-1})^n \qquad (9.4\text{-}6)$$

or

$$F(z) = 1 + (e^{-3T} z^{-1}) + (e^{-3T} z^{-1})^2 + \cdots \qquad (9.4\text{-}7)$$

This, however, is the same form as the series $1 + x + x^2 + x^3 + \cdots$, for which the closed-form expression is $1/(1-x)$ for $|x| < 1$. Identify $e^{-3T} z^{-1}$ with x to yield

$$F(z) = \frac{1}{1 - e^{-3T} z^{-1}} = \frac{z}{z - e^{-3T}}, \qquad |z| > e^{-3T} \qquad (9.4\text{-}8)$$

a rational function in z.

The method outlined in Example 9-1 is impractical when $F(z)$ is a complicated function, for it is then difficult to determine the closed-form expression. In one systematic technique for obtaining this expression, the time function $f(t)$ whose Laplace transform is a rational function in s with simple poles s_i, has the form

$$f(t) = \sum_{i=1}^{n} A_i e^{s_i t} \qquad (9.4\text{-}9)$$

Then, from the result of Example 9-1, the z-transform of $f_S(t)$ is

$$F(z) = \sum_{i=1}^{n} A_i \frac{z}{z - e^{s_i T}} \qquad (9.4\text{-}10)$$

and we observe that $F(z)$ is indeed expressible as a rational function in z. Furthermore A_i is the residue of $F(s)$ for the pole at $s = s_i$; hence $F(z)$ may alternately be expressed as

$$F(z) = \frac{1}{2\pi j} \int_c \frac{F(s)}{1 - e^{sT} z^{-1}} ds \qquad (9.4\text{-}11)$$

where the contour c encloses the poles of $F(s)$ only and z is considered to be a constant for integration purposes. The residue of $F(s)$ for each pole s_i is obtained

The z-Transform

using complex variable theory, and the solution of (9.4-11) is

$$F(z) = \sum_{i=1}^{n} \left[\frac{(s - s_i) F(s)}{1 - e^{sT} z^{-1}} \right]_{s = s_i} \quad (9.4\text{-}12)$$

If $F(s)$ includes multiple poles,

$$F(z) = \sum_{i=1}^{n} \frac{1}{(m_i - 1)!} \frac{d^{m_i - 1}}{ds^{m_i - 1}} \left[\frac{(s - s_i)^{m_i} F(s)}{1 - e^{sT} z^{-1}} \right]_{s = s_i} \quad (9.4\text{-}13)$$

where $s = s_i$ is the ith pole of $F(s)$; $i = 1, \ldots, n$; m_i is the order of multiplicity of the pole s_i, and n is the total number of different poles. We encountered a computationally similar expression in (2.3-33) for finding inverse Laplace transforms, and as there, $F(s)$ must be a proper fraction before (9.4-12) and (9.4-13) are applicable.

Example 9-2. We now determine the z-transform of the sampled exponential function in Example 9-1, using (9.4-12). Since $F(s) = 1/(s + 3)$,

$$F(z) = \frac{s + 3}{1 - e^{sT} z^{-1}} \times \frac{1}{s + 3} \bigg|_{s = -3} = \frac{1}{1 - e^{-3T} z^{-1}} \quad (9.4\text{-}14)$$

which is identical to the result in (9.4-8).

Example 9-3. Consider the function $f(t) = t u_{-1}(t)$, which from Table 2-1, entry 9, has the Laplace transform $F(s) = 1/s^2$. Now $F(s)$ has a double root at $s = 0$, and (9.4-13) is used to find the z-transform of $f_S(t)$. The total number of different poles n is unity, $s_i = 0$, and $m_i = 2$. Then

$$F(z) = \frac{d}{ds} \left[\frac{s^2}{1 - e^{sT} z^{-1}} \times \frac{1}{s^2} \right]\bigg|_{s=0} = \frac{T e^{sT} z^{-1}}{(1 - e^{sT} z^{-1})^2}\bigg|_{s=0} = \frac{T z^{-1}}{(1 - z^{-1})^2} \quad (9.4\text{-}15)$$

The expansion of (9.4-15) by long division yields the power series given by (9.4-4),

$$F(z) = T z^{-1} + 2T z^{-2} + 3T z^{-3} + \cdots \quad (9.4\text{-}16)$$

It is not obvious from (9.4-16) that the closed-form expression for $F(z)$ is given by (9.4-15). Use of (9.4-13) is advantageous here.

Example 9-4. The z-transform of the time sequence $f_S(t)$ delayed by i units of T is obtained by substituting $f(nT - iT)$ into (9.4-4) as

$$\mathscr{Z}[f_S(t - iT)] = \sum_{n=0}^{\infty} f(nT - iT) z^{-n} \quad (9.4\text{-}17)$$

Introduce the new variable $N = n - i$,

$$\mathscr{Z}[f_S(t - iT)] = \sum_{N=-i}^{\infty} f(NT) z^{-(N+i)} \quad (9.4\text{-}18)$$

But we have assumed that $f(NT) = 0$ for $N < 0$; thus

$$\mathscr{Z}[f_S(t - iT)] = z^{-i} \sum_{N=0}^{\infty} f(NT) z^{-N} = z^{-i} F(z) \quad (9.4\text{-}19)$$

The transform is just z^{-i} times the transform of the undelayed sequence. This is quite reasonable, since delaying the sequence by i units of T is equivalent to changing z^{-n} in (9.4-4) to $z^{-(n+i)}$ and the nth sample of the undelayed sequence then becomes the $(n+i)$th sample of the delayed sequence. Delaying the sampled exponential function in Example 9-1 by $4T$ introduces the delay variable z^{-4} into (9.4-8) to yield the z-transform as

$$F(z) = \frac{z^{-4}}{1 - e^{-3T}z^{-1}} \tag{9.4-20}$$

Example 9-5. The discrete-time convolution of $f_S(t)$ and $h_S(t)$, very important in the study of discrete-time systems, is discussed in Section 9.5.3. Its z-transform, after substitution from (9.5-11) into (9.4-4) is

$$\mathscr{Z}[f_S * h_S] = \sum_{n=0}^{\infty} \left[\sum_{k=0}^{\infty} f(kT)h(nT - kT)\right] z^{-n} \tag{9.4-21}$$

The summation limit is legitimately changed to ∞ because $h(nT - kT) = 0$ for $k > n$. Interchanging the order of summation yields

$$\mathscr{Z}[f_S * h_S] = \sum_{k=0}^{\infty} f(kT) \left[\sum_{n=0}^{\infty} h(nT - kT)z^{-n}\right] \tag{9.4-22}$$

and from (9.4-19) the bracketed term is $z^{-k}H(z)$. Then

$$\mathscr{Z}[f_S * h_S] = H(z) \sum_{k=0}^{\infty} f(kT)z^{-k} = H(z)F(z) \tag{9.4-23}$$

Thus the z-transform of the convolution of $f_S(t)$ and $h_S(t)$ is the product of their transforms, a result analogous to that obtained for continuous-time functions.

As with Laplace transforms, tables of z-transforms have been developed (see, e.g., Refs. 1 and 9), and the transform pairs of some elementary functions are given in Table 9-3. The linearity property of z-transforms and the transforms in Table 9-3 allow us to obtain the transforms of more complicated functions.

Example 9-6. Let us now determine the z-transform of $f_S(t)$, given the Laplace transform of $f(t)$ as

$$F(s) = \frac{1}{s(s+a)(s+b)} \tag{9.4-24}$$

Although $F(z)$ can be evaluated from (9.4-12), alternatively we can first find $f(t)$ and then refer to Table 9-3 for the z-transform of $f_S(t)$. Proceeding in this manner we expand $F(s)$ into the sum of partial fractions

$$F(s) = \frac{1/ab}{s} + \frac{1/[a(a-b)]}{s+a} + \frac{-1/[b(a-b)]}{s+b}$$

$$= \frac{1}{a-b}\left[\frac{a-b}{ab}\frac{1}{s} + \frac{1}{a(s+a)} - \frac{1}{b(s+b)}\right] \tag{9.4-25}$$

From Table 9-3, the corresponding time function is

$$f(t) = \frac{1}{a-b}\left[\frac{a-b}{ab} + \frac{1}{a}e^{-at} - \frac{1}{b}e^{-bt}\right]u_{-1}(t) \tag{9.4-26}$$

and the z-transform is obtained by identifying the first term with entry 1 of Table 9-3, the next two terms with entry 3 of Table 9-3, and remembering that the

The z-Transform

Table 9-3 z-Transform Pairs

$f(t)$ is uniformly sampled at $0, T, 2T, \ldots$

	$f(t)$	$F(s)$	$f(nT)$	$F(z)$
1.	1	$\dfrac{1}{s}$	1	$\dfrac{z}{z-1}$
2.	t	s	nT	$\dfrac{zT}{(z-1)^2}$
3.	$e^{-\alpha t}$	$\dfrac{1}{s+\alpha}$	$e^{-\alpha nT}$	$\dfrac{z}{z-e^{-\alpha T}}$
4.	$e^{-\alpha t}\sin\beta t$	$\dfrac{\beta}{(s+\alpha)^2+\beta^2}$	$e^{-\alpha nT}\sin\beta nT$	$\dfrac{ze^{-\alpha T}\sin\beta T}{z^2-2ze^{-\alpha T}\cos\beta T+e^{-2\alpha T}}$
5.	$e^{-\alpha t}\cos\beta t$	$\dfrac{s+\alpha}{(s+\alpha)^2+\beta^2}$	$e^{-\alpha nT}\cos\beta nT$	$\dfrac{z^2-ze^{-\alpha T}\cos\beta T}{z^2-2ze^{-\alpha T}\cos\beta T+e^{-2\alpha T}}$
6.	$\sin\beta t$	$\dfrac{\beta}{s^2+\beta^2}$	$\sin\beta nT$	$\dfrac{z\sin\beta T}{z^2-2z\cos\beta T+1}$
7.	$\cos\beta t$	$\dfrac{s}{s^2+\beta^2}$	$\cos\beta nT$	$\dfrac{z^2-z\cos\beta T}{z^2-2z\cos\beta T+1}$
8.	$u_{-1}(t-iT)f(t-iT)$	$e^{-sit}F(s)$	$u_{-1}(nT-iT)f(nT-iT)$	$z^{-i}F(z)$
9.	$\displaystyle\int_0^\infty f_1(\tau)f_2(t-\tau)\,d\tau$	$F_1(s)F_2(s)$	$\displaystyle\sum_{k=0}^\infty f_1(kT)f_2(nT-kT)$	$F_1(z)F_2(z)$

z-transform is a linear transformation. Then

$$F(z) = \frac{1}{a-b}\left[\frac{a-b}{ab}\frac{z}{z-1}+\frac{1}{a}\frac{z}{z-e^{-aT}}-\frac{1}{b}\frac{z}{z-e^{-bT}}\right] \quad (9.4\text{-}27)$$

which can be rearranged as a rational function in z.

9.4.2 Mapping the s-Plane into the z-Plane

We now examine the effects of mapping the complex s-plane into the complex z-plane by the transformation of (9.4-3). All points on the $j\omega$ axis in the s-plane transform to points on the unit circle in the z-plane, because for $s = j\omega$

$$z = e^{j\omega T} \quad (9.4\text{-}28)$$

a complex function whose magnitude is unity and whose phase angle is ωT. This is a periodic function of period 2π. The significance of this fact will be more apparent when we discuss the spectrum of $f_S(t)$.

For points in the left half of the s-plane, $s = -\alpha + j\beta$, $(\alpha > 0)$, and

$$z = e^{(-\alpha+j\beta)T} = e^{-\alpha T}e^{j\beta T} \quad (9.4\text{-}29)$$

This is a complex number whose magnitude is less than unity, thus these points lie interior to the unit circle. Similarly, points in the right half of the s-plane map into points exterior to the unit circle in the z-plane. These important regions are illustrated in Fig. 9-12. This simple mapping enables us to predict the general

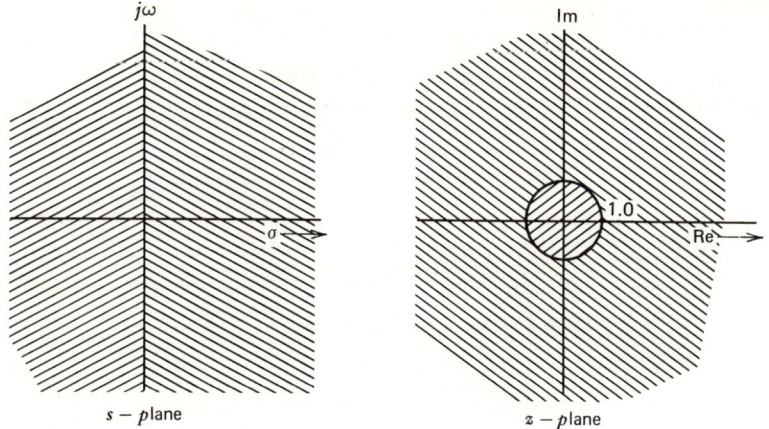

Figure 9-12. *s*-plane and *z*-plane relationships.

behavior of $f_S(t)$ by observing the *z*-plane poles, since we know corresponding behavior in the *s*-plane. Poles interior to the unit circle characterize sampled time functions that exponentially decrease with increasing time, while poles exterior to the unit circle characterize sampled time functions that exponentially increase with increasing time. For example, the exponential in Example 9-1 has a *z*-transform pole at $z = e^{-3T}$, which is less than unity. Therefore $f_S(t)$ exponentially decreases with time, and for this case this is precisely the behavior of $f_S(t)$. Poles on the unit circle correspond to oscillating sampled time functions with the exception of poles at $z = 1$, which correspond to constant or increasing functions depending on the multiplicity of the pole (see Table 9-3, entries 1 and 2).

9.4.3 Frequency-Domain Characteristics

The Laplace transform of $f_S(t)$, given by (9.4-2), is important as an intermediate step in deriving the *z*-transform of $f_S(t)$, but it does not reveal the significant frequency-domain characteristic of $f_S(t)$. To obtain this result, we recall the representation of $f_S(t)$ given in (9.4-1),

$$f_S(t) = f(t)\, d(t) \tag{9.4-30}$$

Since $d(t)$ is periodic, it can be expressed as the Fourier series,

$$d(t) = \sum_{n=-\infty}^{\infty} k_n e^{jn\omega_s t} \quad \left(\omega_s = \frac{2\pi}{T} \text{ is the sampling frequency}\right) \tag{9.4-31}$$

where, analogous to (9.1-7),

$$k_n = \frac{1}{T} \int_{-T/2}^{T/2} f(t) e^{jn\omega_s t}\, dt = \frac{1}{T} \int_{-T/2}^{T/2} \delta(t) e^{jn\omega_s t}\, dt \tag{9.4-32}$$

The value of the integral is unity (see (1.6-3); thus all coefficients are equal, independent of n, and given by

$$k_n = \frac{1}{T} \tag{9.4-33}$$

The z-Transform

The sampled signal is then written as

$$f_S(t) = \frac{1}{T}\sum_{n=-\infty}^{\infty} f(t)e^{jn\omega_s t} \tag{9.4-34}$$

and its Laplace transform, after interchanging integral and summation signs, is

$$F_S(s) = \frac{1}{T}\sum_{n=-\infty}^{\infty}\int_0^{\infty} f(t)e^{-(s-jn\omega_s)t}\,dt = \frac{1}{T}\sum_{n=-\infty}^{\infty} F(s-jn\omega_s) + \frac{1}{2}f(0_+) \tag{9.4-35}$$

The constant term in (9.4-35) appears when $f(0_+) \neq 0$ because the impulse at the origin is split in half in taking the Laplace transform with its lower limit at $t = 0_+$. At $s = j\omega$,

$$F_S(j\omega) = \frac{1}{T}\sum_{n=-\infty}^{\infty} F[j(\omega - n\omega_s)] + \frac{1}{2}f(0_+) \tag{9.4-36}$$

exposing the most significant property of the sampled function's spectrum. The effect of sampling is to reproduce the spectrum of the original signal about each harmonic of ω_s, as in Fig. 9-13, which illustrates a function that is not band-limited; hence complete recovery of $f(t)$ is impossible because the repetitive spectra overlap the primary spectrum. This overlap is known as aliasing or foldover. The composite spectrum is given by the dashed curve.

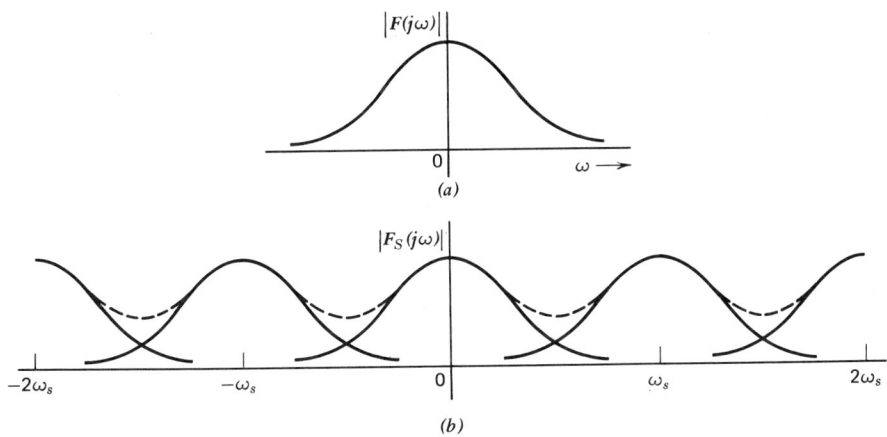

Figure 9-13. Spectrum change due to the sampling process. (a) Spectrum of $f(t)$, (b) spectrum of $f_S(t)$.

If $f(t)$ is band-limited, as in Fig. 9-14a, two situations can occur. We can either sample at a frequency greater than or equal to twice the highest frequency present (ω_c) (Fig. 9-14b) or less than twice the highest frequency (Fig. 9-14c). For the former we can unambiguously recover $f(t)$ as explained in Section 9.1. For the latter condition, $\omega_s < 2\omega_c$, and spectra overlap again does not allow recovery of $f(t)$. The discussion here and in Section 9.1 should have made it clear that for a function $f(t)$ to be reasonably recovered from the sampled function $f_S(t)$, the sampling frequency should be greater than twice the highest significant frequency in $f(t)$.

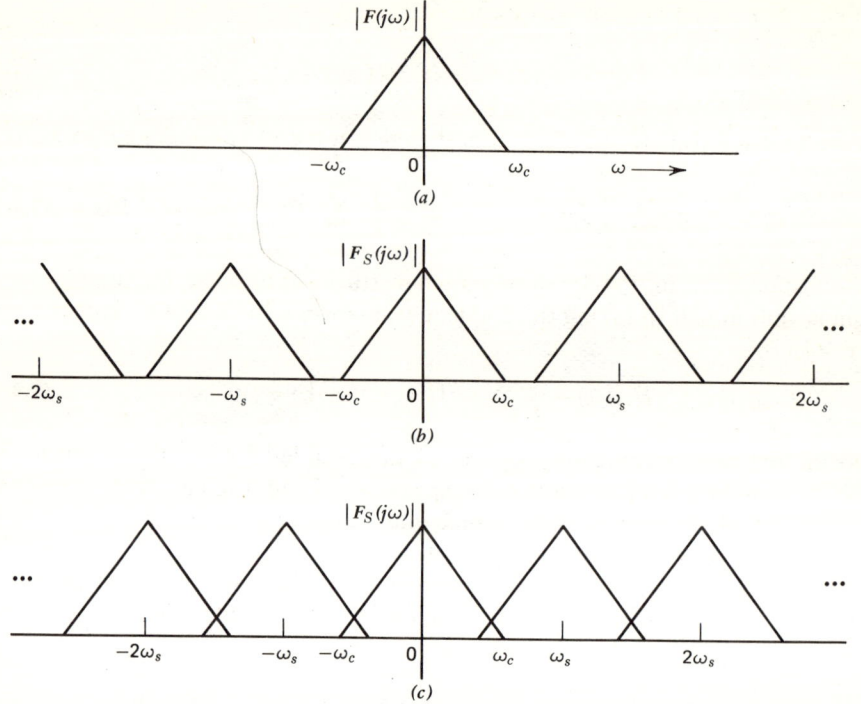

Figure 9-14. (a) Spectrum of band-limited function. Two possible spectra by sampling $f(t)$. (b) Complete recovery of $f(t)$ is possible ($\omega_s > 2\omega_c$), (c) complete recovery of $f(t)$ is impossible because of aliasing ($\omega_s < 2\omega_c$).

9.4.4 The Inverse z-Transform

We now consider methods for finding the sampled value $f(nT)$ given the transform $F(z)$, that is, we wish to find $\mathscr{Z}^{-1}[F(z)]$. The symbol $\mathscr{Z}^{-1}$ is spoken "the inverse z-transform of." The inversion formula is expressed as the contour integral

$$\mathscr{Z}^{-1}[F(z)] = f(nT) = \frac{1}{2\pi j} \int_c F(z) z^{n-1} \, dz \qquad (9.4\text{-}37)$$

where c is a suitable contour in the complex z-plane [9]. Evaluation of (9.4-37) requires complex variable theory, but assuming that $F(z)$ contains k simple poles z_i, the explicit solution is

$$f(nT) = \sum_{i=1}^{k} [(z - z_i) z^{n-1} F(z)]_{z=z_i} \qquad (9.4\text{-}38)$$

Example 9-7. Let us apply (9.4-38) to recover $f(nT)$ from the z-transform in Example 9-1,

$$F(z) = \frac{z}{z - e^{-3T}} \qquad (9.4\text{-}39)$$

Here $k = 1$, and $z_1 = e^{-3T}$. Then

$$f(nT) = (z - e^{-3T}) z^{n-1} \frac{z}{z - e^{-3T}} \bigg|_{z = e^{-3T}} = e^{-3nT} \qquad (9.4\text{-}40)$$

which is the correct sampled value.

The z-Transform

Another technique for recovering the sampled time function corresponding to a given z-transform is simply to expand the z-transform into a power series by ordinary long division. This obviates the needs to find roots of polynomials, evaluate residues, and perform contour integration as the finding of inverse Laplace transforms requires. Thus the inverse transform is obtained by a simple arithmetic operation, hence is easily adaptable to a digital computer. The z-transform given in (9.4-4) expresses $F(z)$ as this series, with the sampled values given by the coefficients of z^{-n}.

Example 9-8. Consider the z-transform in Example 9-3,

$$F(z) = T\frac{z}{(z-1)^2} = T\frac{z^{-1}}{1 - 2z^{-1} + z^{-2}} \qquad (9.4\text{-}41)$$

By long division,

$$\begin{array}{r}
z^{-1} + 2z^{-2} + 3z^{-3} + \cdots \\
1 - 2z^{-1} + z^{-2} \overline{\smash{\big)}\, z^{-1}} \\
\underline{z^{-1} - 2z^{-2} + z^{-3}} \\
2z^{-2} - z^{-3} \\
\underline{2z^{-2} - 4z^{-3} + 2z^{-4}} \\
3z^{-3} - 2z^{-4} \\
\underline{3z^{-3} - 6z^{-4} + 3z^{-5}} \\
4z^{-4} - 3z^{-5}
\end{array}$$

and

$$F(z) = Tz^{-1} + 2Tz^{-2} + 3Tz^{-3} + \cdots = \sum_{n=0}^{\infty} nTz^{-n} \qquad (9.4\text{-}42)$$

Hence the time-function samples are the coefficients of z^{-n}; namely,

$$f(nT) = nT; \qquad n = 0, 1, 2, \ldots \qquad (9.4\text{-}43)$$

A third method of finding the inverse z-transform is to first expand $(F(z))/z$ into a sum of partial fractions, multiply both sides by z, and then find the inverse of each partial fraction from a table of z-transforms. Finally, replace the variable t by nT. This technique is similar to that described in Section 2.3 for finding Laplace transform inverses in that the method is systematic, it avoids integration in the complex z-plane, and it permits the user to obtain closed-form solutions for $f_S(t)$.

Example 9-9. We now determine the time sequence associated with the z-transform

$$F(z) = \frac{z(e^{-T} - e^{-2T})}{z^2 - (e^{-T} + e^{-2T})z + e^{-3T}} \qquad (9.4\text{-}44)$$

by the three previously discussed methods. The poles of $F(z)$ are e^{-T} and e^{-2T}; and $F(z)$ is rewritten as

$$F(z) = \frac{z(e^{-T} - e^{-2T})}{(z - e^{-T})(z - e^{-2T})} \qquad (9.4\text{-}45)$$

Method 1. From (9.4-38), $k = 2$ and

$$f(nT) = (z - e^{-T})z^{n-1} \frac{z(e^{-T} - e^{-2T})}{(z - e^{-T})(z - e^{-2T})}\bigg|_{z=e^{-T}}$$

$$+ (z - e^{-2T})z^{n-1} \frac{z(e^{-T} - e^{-2T})}{(z - e^{-T})(z - e^{-2T})}\bigg|_{z=e^{-2T}}$$

$$= e^{-nT} - e^{-2nT} \tag{9.4-46}$$

Method 2. Expand $F(z)$ in (9.4-44) by long division as

$$z^2 - (e^{-T} + e^{-2T})z + e^{-3T} \overline{\big) \begin{array}{l} (e^{-T} - e^{-2T})z^{-1} + (e^{-2T} - e^{-4T})z^{-2} + \cdots \\ \underline{(e^{-T} - e^{-2T})z} \\ (e^{-T} - e^{-2T})z - (e^{-2T} - e^{-4T}) + (e^{-4T} - e^{-5T})z^{-1} \\ \overline{(e^{-2T} - e^{-4T}) - (e^{-4T} - e^{-5T})z^{-1}} \end{array}}$$

noting that the coefficients of powers of z^{-n} are the sampled values given by (9.4-46).

Method 3. First expand $(F(z))/z$ as a sum of partial functions,

$$\frac{F(z)}{z} = \frac{e^{-T} - e^{-2T}}{z^2 - (e^{-T} + e^{-2T})z + e^{-3T}} = \frac{1}{z - e^{-T}} - \frac{1}{z - e^{-2T}} \tag{9.4-47}$$

Then

$$F(z) = \frac{z}{z - e^{-T}} - \frac{z}{z - e^{-2T}} \tag{9.4-48}$$

and each fraction, from Table 9-3, entry 3, is the transform of a sampled exponential. Therefore

$$f(nT) = e^{-nT} - e^{-2nT} \tag{9.4-49}$$

which is the correct sampled value.

9.5 APPLICATION OF z-TRANSFORMS TO DIFFERENCE EQUATIONS

By applying the z-transform method to difference equations, we now derive discrete-time convolution, the system function $H(z)$, and the unit-sample response, the discrete-time response analogous to the impulse response of a continuous-time system.

9.5.1 Classical Solution of Difference Equations

Before using the z-transform to solve (9.3-1) explicitly, we emphasize that there is a classical method for solving difference equations just as there is a classical method for solving differential equations. Furthermore much of the terminology is common to both equations. For example, if all the L_i in (9.3-1) equal zero, the resulting equation

$$\sum_{i=0}^{m} K_i u(nT - iT) = 0 \tag{9.5-1}$$

Application of z-Transforms to Difference Equations

is known as the homogeneous equation. To solve (9.5-1) we assume the solution

$$u(nT) = c^{nT} \tag{9.5-2}$$

where c is a constant, possibly complex, and substitute it into (9.5-1), yielding

$$\sum_{i=0}^{m} K_i c^{(n-i)T} = c^{nT} \sum_{i=0}^{m} K_i c^{-iT} = 0 \tag{9.5-3}$$

This is an mth-order polynomial in c^{-T}, and its solutions are used to form a system response analogous to the impulse response of a continuous-time system. This response then serves to form the discrete-time convolution, which is analogous to the convolution integral. The procedure is reminiscent of the continuous-time procedure used in Chapter 1, where the assumed homogeneous differential equation solution is e^{mt}.

9.5.2 Solution by z-Transforms

To solve (9.3-1), we take the z-transform of both sides, yielding

$$U(z) = \sum_{i=0}^{q} L_i \mathscr{Z}[f(nT-iT)] - \sum_{i=1}^{m} K_i \mathscr{Z}[u(nT-iT)] \tag{9.5-4}$$

and from Table 9-3, entry 8,

$$U(z) = \sum_{i=0}^{q} L_i z^{-i} F(z) - \sum_{i=1}^{m} K_i z^{-i} U(z) \tag{9.5-5}$$

All initial conditions are assumed to be zero, but nonzero initial conditions can be included in a manner similar to their inclusion in the theory of Laplace transforms. Equation 9.5-5 is rewritten as

$$U(z) = F(z) \frac{\sum_{i=0}^{q} L_i z^{-i}}{1 + \sum_{i=1}^{m} K_i z^{-i}} \tag{9.5-6}$$

whose inverse transform yields the explicit solution to the difference equation.

Example 9-10. We now solve the difference equation in (9.3-2) using z-transforms. The z-transform of $f_S(t)$ in Fig. 9-7a is

$$F(z) = 4 + 3z^{-1} + 2z^{-2} + z^{-3} \tag{9.5-7}$$

and from (9.5-6)

$$U(z) = F(z) \frac{3 - 2z^{-1} + 6z^{-2} + 2z^{-3}}{1 + z^{-1}}$$

$$= \frac{12 + z^{-1} + 24z^{-2} + 25z^{-3} + 16z^{-4} + 10z^{-5} + 2z^{-6}}{1 + z^{-1}} \tag{9.5-8}$$

Ordinary long division yields the power series in z^{-1} whose coefficients are the values of $u_S(t)$ listed in Table 9-2.

9.5.3 System Function and Unit-Sample Response

Dividing both sides of (9.5-6) by $F(z)$ reveals that the right side of the resulting equation is independent of the input and output functions. It depends only on the difference equation coefficients, which are functions of the system parameters. The function designated $H(z)$ is defined as the system function, analogous to the transfer function $H(s)$ associated with continuous-time systems. Thus

$$H(z) = \frac{U(z)}{F(z)} = \frac{\sum_{i=0}^{q} L_i z^{-1}}{1 + \sum_{i=1}^{m} K_i z^{-1}} \qquad (9.5\text{-}9)$$

and for constant-coefficient linear difference equations, $H(z)$ is always a rational function in z. Furthermore the z-transform of the output sequence is equal to the z-transform of the input sequence multiplied by the system function $H(z)$. If all K_i's are zero, $H(z)$ characterizes a nonrecursive filter, but if all K_i's are not zero, $H(z)$ characterizes a recursive filter.

Expand $U(z)$ in (9.5-9), where $H(z)$ is the z-transform of $h_S(t)$ and $F(z)$ is the z-transform of $f_S(t)$, as

$$\begin{aligned}
U(z) &= F(z)H(z) \\
&= [f(0) + f(T)z^{-1} + f(2T)z^{-2} + \cdots][h(0) + h(T)z^{-1} + h(2T)z^{-2} + \cdots] \\
&= h(0)f(0) + [h(0)f(T) + h(T)f(0)]z^{-1} \\
&\quad + [h(0)f(2T) + h(T)f(T) + h(2T)f(0)]z^{-2} + \cdots \\
&= u(0) + u(T)z^{-1} + u(2T)z^{-2} + \cdots \qquad (9.5\text{-}10)
\end{aligned}$$

Equating coefficients of like powers of z^{-1} enables us to explicitly express $u(nT)$ as

$$u(nT) = \sum_{m=0}^{n} f(mT)h(nT - mT) \qquad (9.5\text{-}11)$$

This equation is the discrete-time convolution, analogous to the convolution integral of continuous-time systems. The $h_S(t)$ is similar to the impulse response, for the unit sample

$$f_1(nT) = \begin{cases} 1 & \text{for} \quad n = 0 \\ 0 & \text{for} \quad n \neq 0 \end{cases} \qquad (9.5\text{-}12)$$

plays the same role in discrete-time systems as the impulse function plays in continuous-time systems.

To show this we substitute the z-transform of $f_1(nt)$, $F_1(z) = 1$ into (9.5-9) and determine that $U(z) = H(z)$. Therefore $u(nT)$, the inverse transform of $U(z)$, is also $h(nT)$, the inverse transform of $H(z)$, that is,

$$h(nT) = \mathcal{Z}^{-1}[H(z)] \qquad (9.5\text{-}13)$$

For discrete-time systems, (9.5-13) is the relationship that links the time domain and the frequency domain. Henceforth $h_S(t)$, whose sampled values are $h(nT)$, is referred to as the unit-sample response. The graphical interpretation of discrete convolution is the same as that for the continuous-time convolution described in Section 1.5, except instead of integrating functions, we now sum numbers.

Application of z-Transforms to Difference Equations

9.5.4 Example of System Response Calculation

We now determine the system function and the unit-sample response for the system described by (9.3-2). From (9.5-9)

$$H(z) = \frac{3 - 2z^{-1} + 6z^{-2} + 2z^{-3}}{1 + z^{-1}} \qquad (9.5\text{-}14)$$

and by long division

$$H(z) = 3 - 5z^{-1} + 11z^{-2} - 9z^{-3} + 9z^{-4} + \cdots + (-1)^k 9z^{-k} + \cdots \qquad (9.5\text{-}15)$$

Thus the unit-sample response samples $h(nT)$, given by the coefficients of z^{-n}, are

$$\begin{aligned} h(0) &= 3 \\ h(T) &= -5 \\ h(2T) &= 11 \\ h(3T) &= -9 \\ &\vdots \\ h(kT) &= (-1)^k 9 \end{aligned} \qquad (9.5\text{-}16)$$

The system response to the input in Fig. 9-7a has already been obtained as the coefficients of the power series $H(z)F(z)$ in Example 9-10. These values are now explicitly generated from (9.5-11). For reference, $f(0) = 4$, $f(T) = 3$, $f(2T) = 2$, and $f(3T) = 1$.

$$u(0) = f(0)h(0) = 4 \times 3 = 12$$

$$u(T) = \sum_{m=0}^{1} f(mT)h(T - mT) = f(0)h(T) + f(T)h(0)$$

$$= (4)(-5) + (3)(3) = -11 \qquad (9.5\text{-}17)$$

$$u(2T) = \sum_{m=0}^{2} f(mT)h(2T - mT) = f(0)h(2T) + f(T)h(T) + f(2T)h(0)$$

$$= (4)(11) + (3)(-5) + (2)(3) = 35$$

Continued use of the discrete-time convolution yields the values in Table 9-2.

Let us now determine $u_S(t)$ graphically to show the similarity to continuous-time convolution. First $h_S(t)$ is reversed in time, and this function's values are indicated by crosses in Fig. 9-15. Then $h_S(-t)$ slides across $f_S(t)$, whose values are indicated by dots, and the output at each sampling time is the sum of the products of these two functions. The relative positions of these functions for $t = -T$ appear in Fig. 9-15a, and since no overlap occurs, the output is zero. Similarly the output is zero for all $t < 0$. The relative positions at $t = 0$ are plotted in Fig. 9-15b. Overlap of the samples occurs only at $t = 0$, and the product of these sample values is 12. Therefore $u(0) = 12$. At $t = T$ (Fig. 9-15c) the sum of products is $u(T) = (4)(-5) + (3)(3) = -11$. As $h_S(-t)$ is successively advanced by T seconds, the computed sequence is that in Fig. 9-7b.

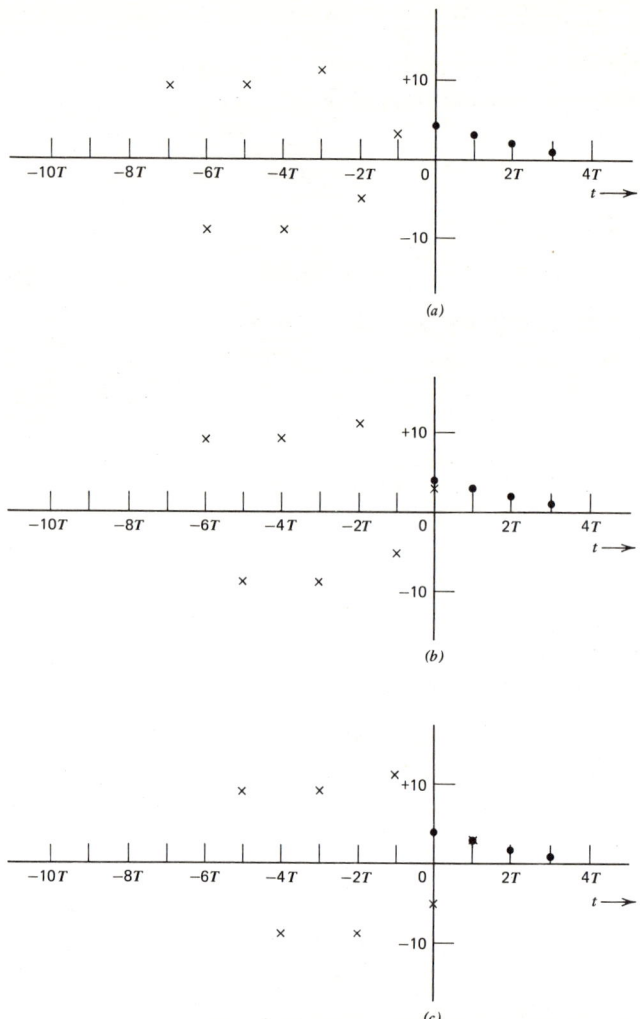

Figure 9-15. Graphical interpretation of the discrete-time convolution in (9.5-17). (a) $t = -T$ ($n = -1$), (b) $t = 0$ ($n = 0$), (c) $t = T$ ($n = 1$).

9.5.5 Frequency-Domain Functions

A useful feature of $H(z)$ is its interpretation as a frequency function. Evaluation of $H(s)$ at $s = j\omega$ is equivalent to evaluating $H(z)$ at $z = e^{j\omega t}$. This substitution allows us to determine the frequency characteristics of a sampled signal in the same manner used to determine these characteristics for a continuous-time signal. However the sampled signal characteristics are periodic at the sampling rate.

Consider a digital filter with system function $H(z)$ in (9.5-9) excited by the sampled sine wave of frequency ω_1 (Fig. 9-16a). The filter output sample values are given by

$$u(nT) = H(e^{j\omega_1 T}) \sin n\omega_1 T \qquad (9.5\text{-}18)$$

which are also samples of the sine wave of frequency ω_1. However the output waveform is scaled in magnitude and shifted in phase relative to the input

Application of z-Transforms to Difference Equations

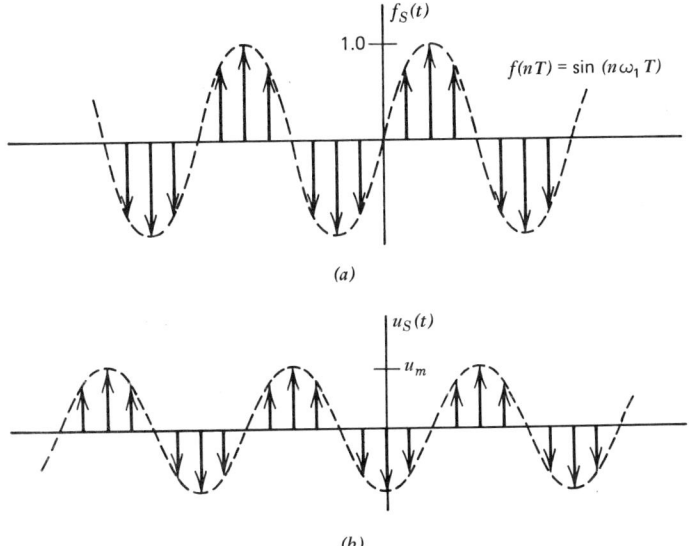

Figure 9-16. (a) Filter input sampled sine wave, (b) filter response to sampled sine wave.

waveform, as in Fig. 9-16b. The magnitude and phase of $u_S(t)$ are computed from the real and imaginary parts of $H(e^{j\omega_1 T})$ in the same manner we employed to treat the continuous-time frequency functions. The maximum possible value of $u_S(t)$ is $u_m = |H(e^{j\omega_1 T})|$. Since ω_1 is arbitrary, the unit subscript can be dropped, yielding $H(e^{j\omega T})$ as the filter frequency response.

The frequency response can be obtained geometrically from the normalized z-plane poles and zeros in the same manner that served to give us the frequency response of analog filters from the s-plane poles and zeros (see Section 2.5.4). However, for discrete-time systems, frequency is measured along the unit circle in the z-plane, whereas for continuous-time systems, frequency is measured along the $j\omega$ axis.

Example 9-11. We now determine the system function, frequency responses, and unit-sample response of the first-order recursive digital filter characterized by the difference equation

$$u(nT) = \alpha f(nT) + e^{-\alpha T} u(nT - T) \qquad (9.5\text{-}19)$$

Comparing (9.5-19) with (9.3-1) we find that $L_0 = \alpha$, $K_1 = -e^{-\alpha T}$, and from (9.5-9), the system function is

$$H(z) = \frac{\alpha}{1 - e^{-\alpha T} z^{-1}} \qquad (9.5\text{-}20)$$

The frequency response is

$$H(e^{j\omega T}) = \frac{\alpha}{1 - e^{-\alpha T} e^{-j\omega T}} = \frac{\alpha}{1 - e^{-\alpha T}(\cos \omega T - j \sin \omega T)} \qquad (9.5\text{-}21)$$

the magnitude response is

$$|H(e^{j\omega T})| = \frac{\alpha}{\sqrt{(1 - e^{-\alpha T} \cos \omega T)^2 + e^{-2\alpha T} \sin^2 \omega T}} = \frac{\alpha}{\sqrt{1 + e^{-2\alpha T} - 2e^{-\alpha T} \cos \omega T}} \qquad (9.5\text{-}22)$$

and the phase response is

$$\theta(\omega) = \tan^{-1} \frac{\sin \omega T}{\cos \omega T - e^{\alpha T}} \qquad (9.5\text{-}23)$$

Note that $|H(e^{j\omega T})|$ and $\theta(\omega)$ are each periodic every $\omega = 2\pi/T$. The unit-sample response is the inverse transform of $H(z)$ in (9.5-20), and from Table 9-3, entry 3, its sample values are

$$h(nT) = \alpha e^{-\alpha n T} u_{-1}(t) \qquad (9.5\text{-}24)$$

The impulse response of a first-order RC network is

$$h(t) = \alpha e^{-\alpha t} u_{-1}(t) \qquad (9.5\text{-}25)$$

where $\alpha = 1/RC$. Comparison of (9.5-24) and (9.5-25) indicates that both responses have the same values at the sampling instants. This feature is useful for design and is further explored in Section 9.6.1.

9.6 INTRODUCTION TO DESIGN TECHNIQUES

The digital filter design problem is the determination of the appropriate filter unit-sample response $h_S(t)$. However, knowledge of $h_S(t)$ is equivalent to knowing the system function $H(z)$ or the difference equation characterizing the system. Thus the design problem is reduced to finding the constants K_i and L_i of the system function in (9.5-9). These constants can be determined by suitable approximation techniques in either the time domain or in the frequency domain.

An example of the time-domain approach is the approximation of a specified convolution integral by the discrete-time convolution [7]. In the frequency domain, the approximation is obtained by specifying an error criterion and then determining the poles and zeros of $H(z)$ until this criterion is achieved. Again, the maximally flat, equiripple, and least-squares criteria are the most popular, just as they were for analog filter approximations. The Fourier series is equivalent to a least-squares approximation, as we showed in Section 3.2.3, and this approach is used to design nonrecursive filters.

It seems reasonable to try to use the vast amount of design data already compiled for analog filters, such as catalogued in Ref. 19, to reduce the amount of work in designing digital filters. This has been done. We now present two recursive filter design techniques that use the analog filter data to realize digital filters whose characteristics retain the important attributes of the analog filter responses. The first is the impulse-invariant method and the second is the bilinear transformation.

9.6.1 The Impulse-Invariant Method

Recursive filters designed by this method have a unit-sample response equal to the sampled impulse response of a given analog filter. Figure 9-17a shows $h(t)$, the impulse response of an RC filter; the unit-sample responses obtained by this design procedure for $T = 1, 2,$ and 3 appear in Fig. 9-17b, c, and d, respectively. Note that independent of the value of T, the sampled values are equal to the RC filter impulse response at that instant. The digital filter system function having this impulse-invariant property is now derived.

Introduction to Design Techniques

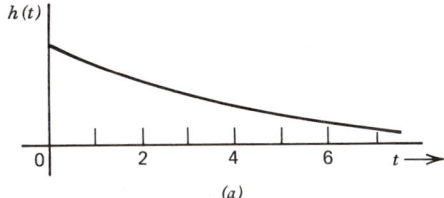
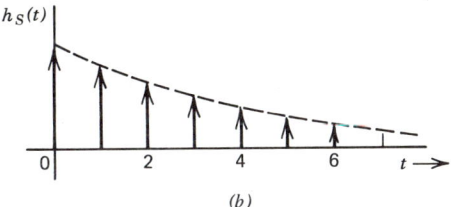
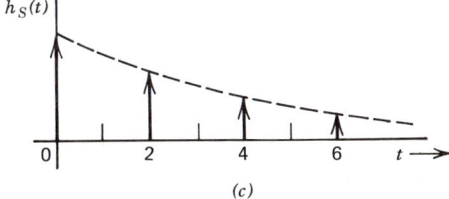
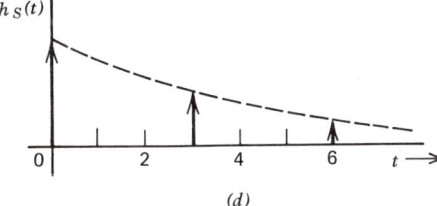

Figure 9-17. Unit-sample responses obtained by impulse-invariant method. (a) RC filter impulse response, (b) $T = 1$ sec, (c) $T = 2$ sec, (d) $T = 3$ sec.

Consider the analog filter transfer function $H_c(s)$ with m simple poles,

$$H_c(s) = \sum_{i=1}^{m} \frac{A_i}{s + s_i} \qquad (9.6\text{-}1)$$

and the corresponding impulse response

$$h_c(t) = u_{-1}(t) \sum_{i=1}^{m} A_i e^{-s_i t} \qquad (9.6\text{-}2)$$

We desire that the digital filter unit-sample response $h_S(t)$ equal $h_c(t)$ at each sample time, that is,

$$h(nT) = h_c(t), \qquad t = 0, T, 2T, \ldots \qquad (9.6\text{-}3)$$

or

$$h(nT) = u_{-1}(t) \sum_{i=1}^{m} A_i e^{-s_i nT} \qquad (9.6\text{-}4)$$

Then take the z-transform of both sides of (9.6-4) to obtain the digital filter system function $H(z)$,†

$$H(z) = \sum_{n=0}^{\infty} h(nT)z^{-n} = \sum_{i=1}^{m} A_i \sum_{n=0}^{\infty} e^{-s_i nT} z^{-n} = \sum_{i=1}^{m} \frac{A_i}{1 - e^{-s_i T} z^{-1}} \quad (9.6\text{-}5)$$

To obtain the digital filter system function we express the analog filter transfer function as a sum of partial fractions and then use the correspondence given by (9.6-5). But this is equivalent to taking the z-transform of the analog filter impulse response with t replaced by nT, as in (9.6-4).

The form of (9.6-5) has one disadvantage. If s_i is a complex number, so also is A_i, and additional algebraic manipulations are necessary to reduce $H(z)$ to a real function of z. By initially partitioning $H(s)$ into terms corresponding to real time-functions, we eliminate this extra algebra. This technique is illustrated in Section 9.6.2.

The impulse-invariant method has two interesting and useful properties. First, it is equivalent to approximating the convolution integral by a sum of rectangles, where sample values of the integrand are taken to the right of each rectangle [7, 11]. Second, for the class of band-limited input signals, this method yields the following optimal approximation in the time domain. If $u_d(t)$ is the response of the digital filter simulating the analog filter whose response is $u_a(t)$, the error between those responses at the sampling times is $e(nT) = u_a(nT) - u_d(nT)$. The maximum value of $|e(nT)|$ is minimized when the digital filter unit-sample response is obtained by the impulse-invariant method [3].

A drawback of this design method is that the frequency response of the digital filter may differ markedly from that of the corresponding analog filter if the latter has significant frequency response above one-half the sampling frequency. This foldover effect can be reduced by increasing the sampling rate, but the upper limit is dictated by practical considerations such as the A/D converters. Furthermore it can be shown that the gain of a narrow-band digital resonator is proportional to the sampling frequency [6]. Thus as this frequency is increased to reduce the aliasing error, the gain may become so large that overflow in the computer system is a possibility.

Another scheme often used in practice to reduce aliasing error is to "band-limit" the incoming signal and then choose the sampling frequency in accordance with the sampling theorem. "Band-limiting" is achieved by a highly selective LP filter, usually a Butterworth or Chebyshev type, known as a guard filter. Its passband encompasses the frequencies over which the digital filter system function $H(z)$ is to approximate $H(s)$, the analog filter transfer function. Of course its phase should also be linear, and this may require all-pass filters for phase equalization.

Another limiting feature of the standard z-transformation is the failure of the moments of $h(t)$ to remain invariant, that is, in general

$$\sum_{q=0}^{\infty} (qT)^n h(qT) \neq \int_0^{\infty} t^n h(t) \, dt \quad (9.6\text{-}6)$$

---

†To be completely general in (9.6-1) we should include terms of the form $A_i/(s + s_i)^q$. Each pole of multiplicity q then corresponds to a term in $H(z)$ equal to

$$\left[A_i \frac{(-1)^{q-1}}{(q-1)!} \frac{\partial^{q-1}}{\partial a^{q-1}} \frac{1}{1 - e^{-aT} z^{-1}} \right]\bigg|_{a=s_i}$$

Introduction to Design Techniques

However, when necessary, the numerator coefficients of $H(z)$ may usually be varied slightly to satisfy (9.6-6) for the first few moments without seriously affecting the overall frequency characteristics of $H(z)$ [8].

9.6.2 Example Illustrating Impulse-Invariant Method

We now use the impulse-invariant method to obtain a unit-sample response equal to the third-order Butterworth impulse response at the sampling instants. We partition $H(s)$ to ensure that the corresponding time function for each term is real. For the present, let the poles be the general values

$$s_1 = -\alpha_0$$
$$s_{2,3} = -\alpha \pm j\beta \quad (9.6\text{-}7)$$

Then the third-order transfer function is

$$H(s) = \frac{\alpha_0(\alpha^2 + \beta^2)}{(s + \alpha_0)[(s + \alpha)^2 + \beta^2]} = \frac{\alpha_0(\alpha^2 + \beta^2)}{(\alpha - \alpha_0)^2 + \beta^2} \left[\frac{1}{s + \alpha_0} - \frac{s + (2\alpha - \alpha_0)}{(s + \alpha)^2 + \beta^2} \right] \quad (9.6\text{-}8)$$

which is rewritten as

$$H(s) = \frac{\alpha_0(\alpha^2 + \beta^2)}{(\alpha - \alpha_0)^2 + \beta^2} \left[\frac{1}{s + \alpha_0} - \frac{s + \alpha}{(s + \alpha)^2 + \beta^2} - \frac{\alpha - \alpha_0}{(s + \alpha)^2 + \beta^2} \right] \quad (9.6\text{-}9)$$

In this form the inverse transform of $H(s)$, using entries 3, 4, and 5 in Table 2-1, is the sum of real time-functions, expressed as

$$h(t) = \frac{\alpha_0(\alpha^2 + \beta^2)}{(\alpha - \alpha_0)^2 + \beta^2} \left[e^{-\alpha_0 t} - e^{-\alpha t} \cos \beta t - \frac{(\alpha - \alpha_0)}{\beta} e^{-\alpha t} \sin \beta t \right] u_{-1}(t) \quad (9.6\text{-}10)$$

and at $t = nT$, we have

$$h(nT) = \frac{\alpha_0(\alpha^2 + \beta^2)}{(\alpha - \alpha_0)^2 + \beta^2} \left[e^{-\alpha_0 nT} - e^{-\alpha nT} \cos \beta nT - \frac{(\alpha - \alpha_0)}{\beta} e^{-\alpha nT} \sin \beta nT \right] \quad (9.6\text{-}11)$$

The digital filter system function, obtained by taking the z-transform of both sides of (9.6-11), is

$$H(z) = \frac{\alpha_0(\alpha^2 + \beta^2)}{(\alpha - \alpha_0)^2 + \beta^2} \left[\frac{z}{z - e^{-\alpha_0 T}} - \frac{z^2 - (e^{-\alpha T} \cos \beta T)z}{z^2 - (2e^{-\alpha T} \cos \beta T)z + e^{-2\alpha T}} - \frac{\alpha - \alpha_0}{\beta} \frac{(e^{-\alpha T} \sin \beta T)z}{z^2 - (2e^{-\alpha T} \cos \beta T)z + e^{-2\alpha T}} \right] \quad (9.6\text{-}12)$$

For the Butterworth response with unity-radian cutoff, $\alpha_0 = 1$, $\alpha = 1/2$, $\beta = \sqrt{3}/2$. Then

$$H(z) = \frac{z}{z - e^{-T}} - \frac{z^2 - \left(e^{-T/2} \cos \frac{\sqrt{3}}{2} T\right)z}{z^2 - \left(2e^{-T/2} \cos \frac{\sqrt{3}}{2} T\right)z + e^{-T}}$$

$$+ \frac{\sqrt{3}}{3} \frac{\left(e^{-T/2} \sin \frac{\sqrt{3}}{2} T\right)z}{z^2 - \left(2e^{-T/2} \cos \frac{\sqrt{3}}{2} T\right)z + e^{-T}} \quad (9.6\text{-}13)$$

which can be rewritten as the rational function in z^{-1},

$$H(z) = \frac{b_1 z^{-1} + b_2 z^{-2}}{1 + a_1 z^{-1} + a_2 z^{-2} + a_3 z^{-3}} \qquad (9.6\text{-}14)$$

where

$$a_1 = -\left(2e^{-T/2} \cos \frac{\sqrt{3}}{2} T + e^{-T}\right)$$

$$a_2 = e^{-T} + 2e^{-3T/2} \cos \frac{\sqrt{3}}{2} T$$

$$a_3 = -e^{-2T}$$

$$b_1 = e^{-T} + e^{-T/2} \left(\frac{\sqrt{3}}{3} \sin \frac{\sqrt{3}}{2} T - \cos \frac{\sqrt{3}}{2} T\right)$$

$$b_2 = e^{-T} - e^{-3T/2} \left(\frac{\sqrt{3}}{3} \sin \frac{\sqrt{3}}{2} T + \cos \frac{\sqrt{3}}{2} T\right)$$

Expressing $H(z)$ as the power series

$$H(z) = b_1 z^{-1} + (b_2 - a_1 b_1) z^{-2} + \cdots \qquad (9.6\text{-}15)$$

yields the first three sample values of $h_s(t)$ as

$$h(0) = 0$$

$$h(T) = b_1 = e^{-T} + e^{-T/2} \left(\frac{\sqrt{3}}{3} \sin \frac{\sqrt{3}}{2} T - \cos \frac{\sqrt{3}}{2} T\right) \qquad (9.6\text{-}16)$$

$$h(2T) = b_2 - a_1 b_1 = e^{-2T} + e^{-T} \left(\frac{\sqrt{3}}{3} \sin \sqrt{3} T - \cos \sqrt{3} T\right)$$

which are the values of the analog filter impulse response of (9.6-10) evaluated at $t = 0$, T, and $2T$, respectively, with $\alpha_0 = 1$, $\alpha = 1/2$, and $\beta = \sqrt{3}/2$. Thus the design procedure is verified.

The frequency response, however, bears investigation. For unity-radian cutoff ($\omega_c = 1$), the highest significant frequency is about $3\omega_c$; thus sampling at twice this frequency ($T = \pi/3\omega_c = 1.05$) should be sufficient for signal reconstruction. The magnitude response, computed from $|H(e^{j\omega T})|$ in (9.6-14), is shown in Fig. 9-18 for $T = 0.1$, 1, and 2 (f_s is the sampling frequency in hertz). For $T = 1$, the approximation to the Butterworth response is excellent from $\omega = 0$ to $\omega = 3$, as expected, and it then deteriorates because the response is repetitive every $\omega = 2\pi$. When the sampling rate is too slow, as the $T = 2$ curve demonstrates, the foldover error is large and the resulting response does not even resemble the Butterworth response. With $T = 0.1$ the response repeats every $\omega = 20\pi$, and the digital filter response is an excellent approximation to the theoretical response from $\omega = 0$ to $\omega = 10\pi$. Thus we see that even though the analog filter impulse response is preserved at the sampling instant, the frequency responses can be distorted, depending on the sampling frequency.

Introduction to Design Techniques

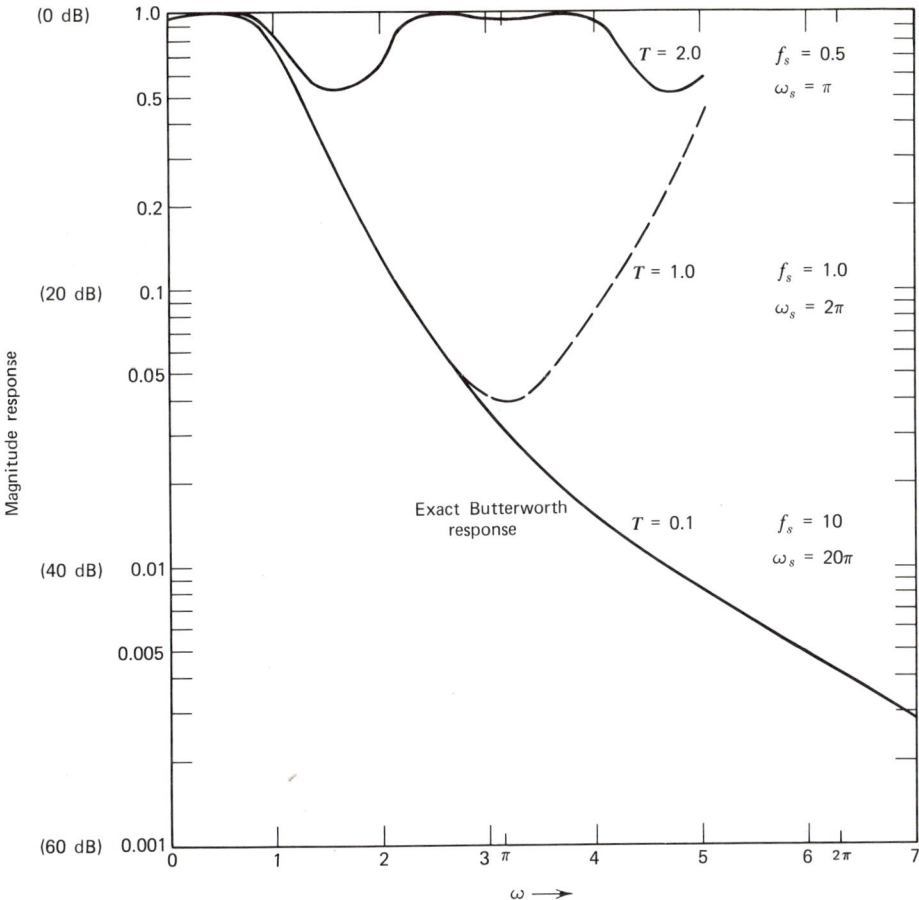

Figure 9-18. Magnitude response of a third-order Butterworth digital filter designed to be impulse invariant, with $T = 0.1, 1, 2$ (Section 9.6.2).

9.6.3 The Bilinear z-Transform

The standard z-transform, resulting in impulse invariance, is useful on continuous functions that are suitably band-limited, such as Butterworth, Chebyshev, and Legendre (we assume that no guard filter is used). However, non-band-limited functions such as high-pass and bandstop functions may also be realized by digital filters by use of the bilinear z-transform, defined as

$$s = \frac{2}{T} \tanh\left(\frac{s_1 T}{2}\right) = \frac{2}{T} \frac{1 - e^{-s_1 T}}{1 + e^{-s_1 T}} \qquad (9.6\text{-}17)$$

which, after substituting $z^{-1} = e^{-s_1 T}$, becomes

$$s = \frac{2}{T} \frac{1 - z^{-1}}{1 + z^{-1}} \qquad (9.6\text{-}18)$$

The transformation in (9.6-17) eliminates the foldover problem by mapping the entire complex s-plane into a horizontal strip in the s_1-plane, bounded by the lines

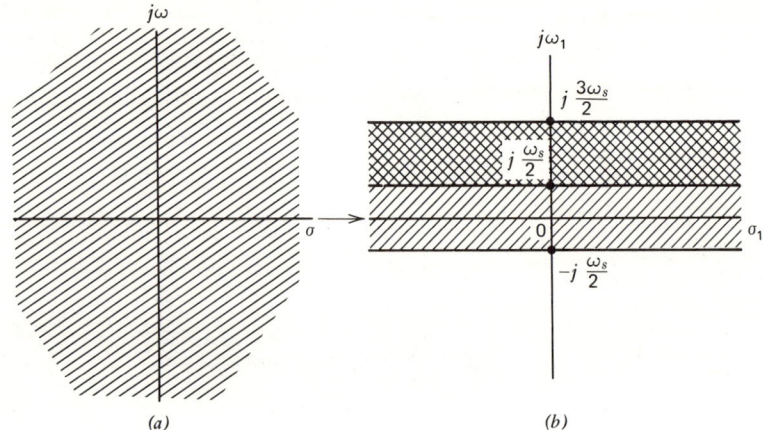

Figure 9-19. Equation 9.6-17 maps entire s-plane (a) into horizontal strips in the s_1-plane (b).

$s_1 = -j(\omega_s/2)$ and $s_1 = +j(\omega_s/2)$, as in Fig. 9-19. Thus any magnitude response becomes band-limited in the s_1-plane. However (9.6-18) also causes $H(s)$ to be mapped identically in each of the horizontal strips bounded by the lines $s_1 = j(n - \tfrac{1}{2})\omega_s$ and $s_1 = j(n + \tfrac{1}{2})\omega_s$, where n is an integer and ω_s is the radian sampling frequency. The horizontal strip for $n = 1$ also appears in Fig. 9-19b as the cross-hatched area.

Equation 9.6-18 retains the desirable property of uniquely mapping the left half of the s-plane into the interior of the unit circle in the z-plane, the $j\omega$ axis again transforming to the unit circle in the z-plane. Foldover errors are eliminated because the magnitude function is band-limited by the transformation. Thus we may consider the bilinear z-transform to be a band-limiting transformation. The digital filter system function is obtained by replacing the variable s in $H(s)$ by the function of z in (9.6-18) as

$$H(z) = H(s) \qquad (9.6\text{-}19)$$

evaluated at

$$s = \frac{2}{T}\left(\frac{1 - z^{-1}}{1 + z^{-1}}\right)$$

and $H(z)$ is again a rational function in z^{-1}.

Although there is a one-to-one correspondence between values of $H(s)$ evaluated on the $j\omega$ axis and values of $H(z)$ evaluated on the unit circle, the two functions are not the same because the frequency scales are distorted relative to one another. This is the price paid for the band-limiting property of the transformation. This nonlinear warping of the frequency scale is obtained by first substituting $s_1 = j\omega_1$ and $s = j\omega$ into (9.6-17) to obtain

$$\frac{\omega T}{2} = \tan \frac{\omega_1 T}{2} \qquad (9.6\text{-}20)$$

or

$$fT = \frac{1}{\pi} \tan \pi f_1 T \qquad (9.6\text{-}21)$$

Introduction to Design Techniques

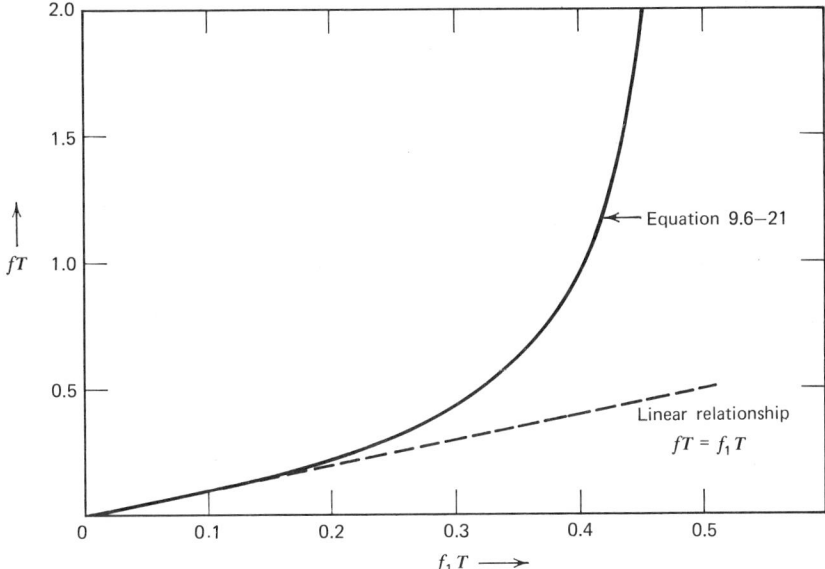

Figure 9-20. Frequency scale distortion introduced by bilinear transformation.

This relationship is illustrated in Fig. 9-20 together with the linear relationship (indicated by the dashed curve). If significant frequencies of the analog filter response are greater than about one-fourth the sampling frequency ($fT > 0.25$), this frequency warping may be intolerable; but we can compensate for the warping by prewarping the analog filter design to ensure that applying the transformation shifts the critical frequencies back to the desired values.

This transformation has two primary advantages. First, the magnitude characteristic of the analog filter is carried directly over to the digital filter except for a warping of the frequency scale. Thus an analog filter with equiripple magnitude response transforms to an equiripple digital filter, with only the positions of the maxima and minima being shifted in frequency by a calculable and compensatable amount. Second, this transform is algebraic in form and can be applied equally well to the rational transfer function in either polynomial or factored form.

A disadvantage of the transformation is that the digital frequency and time responses may be considerably different from the desired analog filter responses. This is especially true of the time-domain responses.

9.6.4 Example of Bilinear z-Transform Design

We now apply the bilinear z-transform with compensation for frequency warping to digitize the third-order Butterworth LP filter; then we compare the magnitude response with the response in Section 9.6.2, obtained by the standard z-transform. Assume that the sampling time T is unity. Then, to ensure that the radian cutoff frequency of the digital filter (ω_1) is unity, we prewarp the analog filter cutoff ω_c according to (9.6-20). Then

$$\omega_c = 2 \tan \tfrac{1}{2} = 1.0926 \qquad (9.6\text{-}22)$$

the poles of the analog filter transfer function with $\omega_c = 1.0926$ are then

$$-1 \times 1.0926 = -1.0926$$

$$\left(-\frac{1}{2} \pm j\frac{\sqrt{3}}{2}\right) \times 1.0926 = -0.5463 \pm j0.94622 \qquad (9.6\text{-}23)$$

and the prewarped transfer function is

$$H(s) = \frac{1.30432}{s^3 + 2.1852s^2 + 2.38755s + 1.30432} \qquad (9.6\text{-}24)$$

The digital filter system function, obtained from (9.6-24) by replacing the variable s by $2[(1-z^{-1})/(1+z^{-1})]$, is

$$H(z) = \frac{1.30432(1+z^{-1})^3}{22.82022 - 24.05274z^{-1} + 14.39706z^{-2} - 2.72998z^{-3}} \qquad (9.6\text{-}25)$$

The magnitude response $|H(e^{j\omega T})|$ appears in Fig. 9-21 together with the analog filter response and the digital filter response of Section 9.6.2 using the standard z-transform. The Butterworth response has been compressed into the frequency interval from $-\omega_s/2$ to $+\omega_s/2$, accompanied by the frequency scale warping. However the cutoff frequency is unity because of the prewarping employed.

9.6.5 Nonrecursive Filters

If the present output sample of a digital filter does not depend on past output samples, this filter is referred to as either a nonrecursive or a transversal filter. It is closely related to the analog tapped delay line filter, as we note in discussing this digital filter's realization.

For a nonrecursive filter, all the K_i's in the difference equation of (9.3-1) are zero and the system function, from (9.5-9), reduces to the polynomial in z^{-1};

$$H(z) = \sum_{n=0}^{m} L_n z^{-n} \qquad (9.6\text{-}26)$$

Comparison of (9.6-26) with (9.4-4) shows that $H(z)$ is the z-transform of the time sequence $L_0, L_1, \ldots, L_m$. But since we know that $H(z)$ is the z-transform of the unit-sample response $h_S(t)$, the following relationship is established:

$$h(nT) = \begin{cases} L_n & n = 0, 1, 2, \ldots, m \\ 0 & n > m \end{cases} \qquad (9.6\text{-}27)$$

Then the system function can be written as

$$H(z) = \sum_{n=0}^{m} h(nT) z^{-n} \qquad (9.6\text{-}28)$$

and by inspection one can determine the unit-sample response from the system function, eliminating use of the inversion techniques described in Section 9.4.4.

Equation 9.6-27 reveals that a nonrecursive filter unit-sample response has finite duration (assuming a finite number of taps); hence the filter has finite memory. The unit-sample response of the recursive filter usually encountered in practice is infinite in length, thus has infinite memory. However finite unit-sample responses

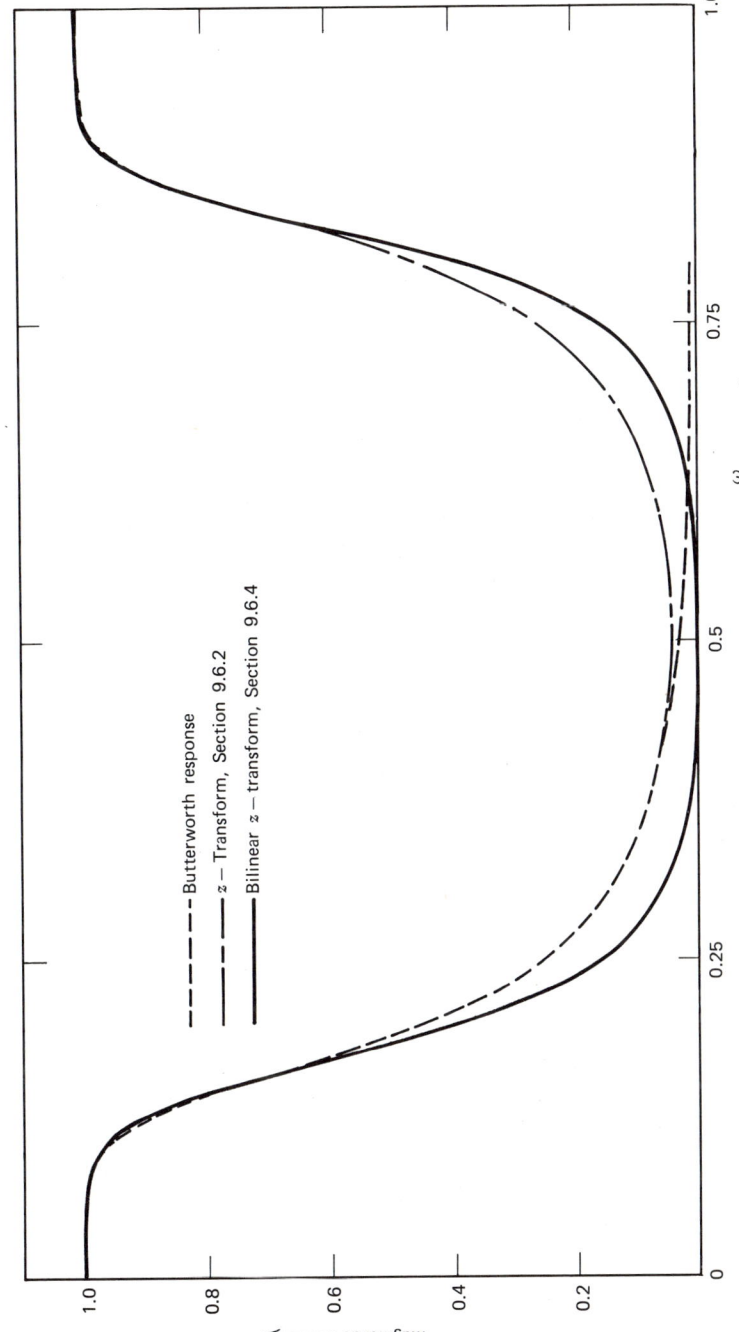

Figure 9-21. Digital filter magnitude approximations to the third-order Butterworth response ($\omega_{3dB} = 1$, $T = 2\pi/\omega_s = 1$).

can be realized recursively. We later show that a linear phase nonrecursive filter is realizable whereas linear phase is not realizable by the design procedures in Sections 9.6.1 and 9.6.3.

The polynomial form of the system function reflects in a realization containing no feedback paths. The canonic form of a general nonrecursive digital filter (Fig. 9-22), clearly exposes the filter computational scheme. The first sample $f(0)$ is multiplied by $h(0)$ and sent to the summer; $f(0)$ is also delayed by T seconds, multiplied by $h(T)$ and sent to the summer; $f(0)$ is then delayed by $2T$ seconds, multiplied by $h(2T)$, and sent to the summer, and so on. The filter processes succeeding input samples in the same way. Then all samples with the same delay are added to yield the output sequence.

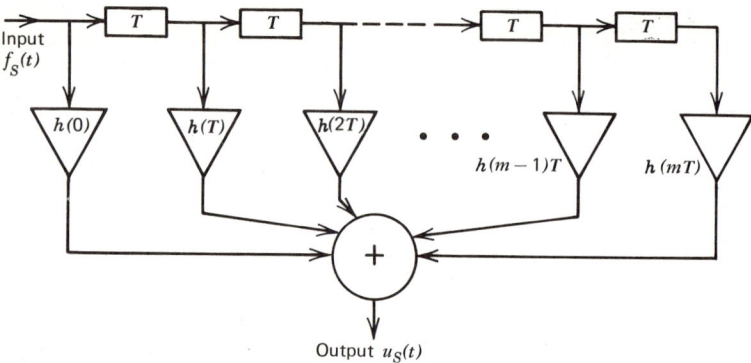

Figure 9-22. Canonic form of a nonrecursive digital filter.

The nonrecursive filter output sequence is the weighted sum of the delayed outputs, and we now see why this filter is often associated with the analog tapped delay line. In a digital filter mechanization the delay T is realized by a shift register, whereas an analog filter realizes T by a delay line. The weighting coefficients are the real constants $h(0), h(T), \ldots$, the filter unit-sample response. Table 9-4 compares the characteristics of recursive and nonrecursive filters.

Nonrecursive filters, with their finite-duration unit-sample responses, are especially useful for realizing matched filters. Thus for the sampled input $f_S(t)$ in Fig. 9-23a, the matched filter unit-sample response is $f_S(t)$ reversed in time and delayed by $3T$ (Fig. 9-23b). These sample values are just the tap weights in the canonical realization. Whereas the determination of the matched analog filter generally requires solving both the approximation and realization problems, the nonrecursive digital matched filter is easily obtained.

9.6.6 Example of Nonrecursive Filter Computation

Let the unit-sample response of an $m = 2$ nonrecursive filter be $h(0) = 0.8$, $h(T) = 0.6$, and $h(2T) = 0.2$, and the input sequence be $f(0) = 2$, $f(T) = 0.7$, and $f(2T) = 0.5$, as in Figs. 9-24b and 9-24a, respectively. With the delay variable z^{-1}

Introduction to Design Techniques

Table 9-4 Comparison of Recursive and Nonrecursive Filter Characterizations

	Recursive	Nonrecursive
Difference equation	$u(nT) = \sum_{i=0}^{q} L_i f(nT - iT) - \sum_{i=1}^{m} K_i u(nT - iT)$	$u(nT) = \sum_{i=0}^{q} L_i f(nT - iT)$
System function	$H(z) = \dfrac{\sum_{i=0}^{q} L_i z^{-i}}{1 + \sum_{i=1}^{m} K_i z^{-i}}$	$H(z) = \sum_{i=0}^{q} L_i z^{-i}$
Unit-sample response	$h_S(t) = \mathcal{Z}^{-1}[H(z)]$ (infinite duration)	$h_S(t) = \sum_{i=0}^{q} L_i \delta(t - iT)$ (finite duration)
Realization	Contains feedback paths	Does not contain feedback paths

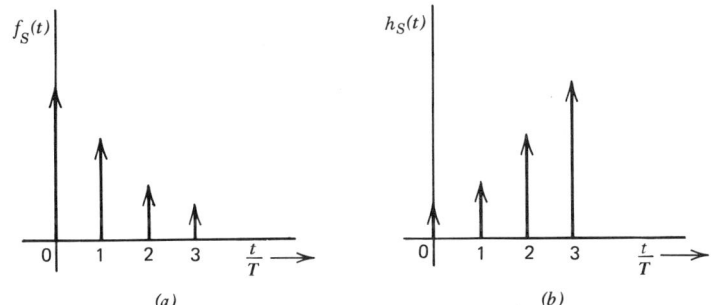

Figure 9-23. (a) Input samples and (b) unit-sample response matched to $f_S(t)$.

introduced, the transform of the response to $f(0) = 2$, from Fig. 9-22, is

$$f(0)h(0) + f(0)h(T)z^{-1} + f(0)h(2T)z^{-2} = 2(0.8 + 0.6z^{-1} + 0.2z^{-2})$$
$$= 1.6 + 1.2z^{-1} + 0.4z^{-2} \qquad (9.6\text{-}29)$$

The response transform for the input sample $f(T)$ is

$$0.7z^{-1}(0.8 + 0.6z^{-1} + 0.2z^{-2}) = 0.56z^{-1} + 0.42z^{-2} + 0.14z^{-3} \qquad (9.6\text{-}30)$$

and the response transform to $f(2T)$ is

$$0.5z^{-2}(0.8 + 0.6z^{-1} + 0.2z^{-2}) = 0.4z^{-2} + 0.3z^{-3} + 0.1z^{-4} \qquad (9.6\text{-}31)$$

Summing the response transforms to each sample input yields the output transform

$$U(z) = 1.6 + 1.76z^{-1} + 1.22z^{-2} + 0.44z^{-3} + 0.1z^{-4} \qquad (9.6\text{-}32)$$

The coefficients are the output sample values shown in Fig. 9-24c. Thus each computational step in determining the system response is clearly exposed by the filter realization in Fig. 9-22.

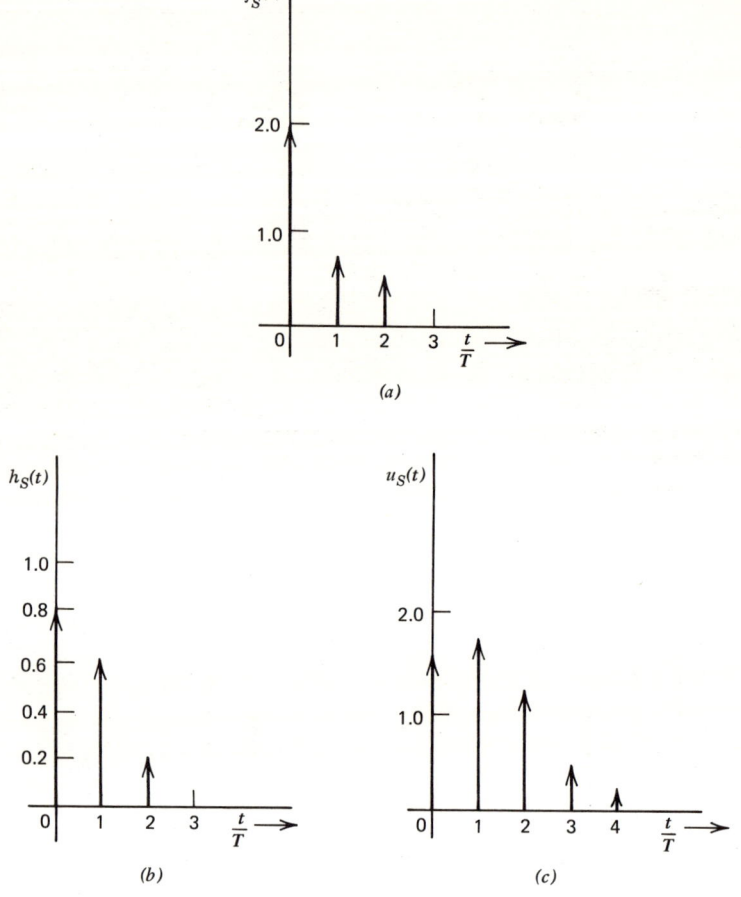

Figure 9-24. (a) Input $f_S(t)$, (b) unit-sample response $h_S(t)$, and (c) output $u_S(t)$ for nonrecursive filter of Section 9.6.6.

9.6.7 Fourier Series Approach to Nonrecursive Filtering

The frequency-domain characteristics of the transversal filter are obtained by evaluating $H(z)$ in (9.6-28) at $z = e^{j\omega T}$ as

$$H(e^{j\omega T}) = \sum_{n=0}^{m} h(nT) e^{-jn\omega T} = \sum_{n=0}^{m} h(nT) \cos n\omega T - j \sum_{n=0}^{m} h(nT) \sin n\omega T \quad (9.6\text{-}33)$$

This form reveals that the frequency response is the superposition of harmonically related sinusoids, each weighted by an appropriate unit-sample-response coefficient. The variable here is the radian frequency ω and, as with all digital filters, the magnitude and phase response are periodic with period $\omega_s = 2\pi/T$. This periodicity and the form of (9.6-33) suggest the use of a truncated Fourier series to approximate the desired frequency response. Let us further explore this possibility which, unlike the previous two design procedures, does not make use of the stockpile of analog filter data. Starting with a specified frequency characteristic, we propose to simultaneously solve the approximation and realization problems

Introduction to Design Techniques

by this technique. Once the coefficients $h(nT)$ in (9.6-33) are obtained, the tap weights in Fig. 9-22 are determined and the realization problem is solved.

Consider the representation of the periodic frequency function $G(j\omega)$ over the interval $-\omega_s/2$ to $\omega_s/2$ by the Fourier series

$$G(j\omega) = \sum_{k=-\infty}^{\infty} a_k e^{jkT\omega} \qquad (9.6\text{-}34)$$

where the coefficient a_k, given by

$$a_k = \frac{1}{\omega_s} \int_{-\omega_s/2}^{\omega_s/2} G(j\omega) e^{-jkT\omega}\, d\omega \qquad (9.6\text{-}35)$$

may be a complex number. The z-domain representation is then†

$$G(z) = \sum_{k=-\infty}^{\infty} a_k z^k \qquad (9.6\text{-}36)$$

Thus the digital filter to realize $G(z)$ requires an infinite number of taps, since each coefficient a_k corresponds to a tap weight. However, in practice the filter is limited to a finite number of taps (see Fig. 9-22) and it is unlikely that $H(z)$, the digital filter system function, will duplicate $G(z)$. A reasonable solution is to truncate $G(z)$ so that the tap weights in the finite series (9.6-28) are equal to the corresponding coefficients in the infinite series (9.6-36), that is,

$$\begin{aligned} h(0) &= a_0 \\ h(T) &= a_{-1} \\ &\vdots \\ h(mT) &= a_{-m} \end{aligned} \qquad (9.6\text{-}37)$$

However, this is an asymmetric truncation and does not permit the realization of a real or an imaginary frequency function. These functions are realized by a symmetric truncation.

The symmetric truncation of $G(z)$ is

$$G_T(z) = \sum_{k=-m}^{m} a_k z^k = a_m z^m + \cdots + a_1 z + a_0 + a_{-1} z^{-1} + \cdots + a_{-m} z^{-m} \qquad (9.6\text{-}38)$$

Although now containing a finite number of terms, $G_T(z)$ is not a realizable system function because the terms with $k > 0$ correspond to advance or prediction. Excitation at $t = 0$ results in output samples prior to $t = 0$, thus the system output anticipates the input. To associate $G_T(z)$ with a causal system, we introduce a constant delay of mT seconds, and the realizable system function $H(z)$ is then

$$H(z) = z^{-m} G_T(z) = z^{-m} \sum_{k=-m}^{m} a_k z^k = \sum_{k=-m}^{m} a_k z^{(k-m)} \qquad (9.6\text{-}39)$$

The introduction of the term z^{-m} does not change the magnitude response of $G_T(z)$ because $|e^{-jm\omega T}|$ is unity, but an arbitrary phase response is no longer realizable

†Here we shorten the notation and represent $G(1/T \ln z)$ by $G(z)$.

because the linear phase response $\theta(\omega) = -m\omega T$ is always present. However a linear phase response introduces no group delay distortion and is usually not objectionable.

The relationship between the digital filter tap weights $h(nT)$, and the Fourier series coefficients a_k is obtained by equating coefficients of like power of z^{-1} in (9.6-28) and (9.6-39) to give

$$\left.\begin{aligned} h(0) &= a_m \\ h(T) &= a_{m-1} \\ &\vdots \\ h(mT) &= a_0 \\ &\vdots \\ h(2mT - T) &= a_{-m+1} \\ h(2mT) &= a_{-m} \end{aligned}\right\} (2m+1) \text{ coefficients} \qquad (9.6\text{-}40)$$

Thus the mth-order approximation requires $(2m + 1)$ filter taps rather than the $m + 1$ tap weights of (9.6-28).

When $G(j\omega)$ is a real and even function, such as a magnitude function, we obtain real tap weights that are symmetric about the center tap weight [see (9.6-35) and (9.6-40)]. Then, from (9.6-39), $H(e^{j\omega T})$ is expressible as

$$H(e^{j\omega T}) = e^{-jm\omega T}\left[a_0 + 2\sum_{k=1}^{m} a_k \cos k\omega T\right] \qquad (9.6\text{-}41)$$

The magnitude response is given by the bracketed terms, since $|e^{-jm\omega T}| = 1$. The significant characteristic of this system function is the linear phase shift, which is not realizable with an analog filter or the usual recursive filters, hence constitutes one of the advantages of the nonrecursive design. This characteristic should not be surprising, for we showed in Section 2.1.3 that a symmetric impulse response corresponds to a linear phase response. Remember that the tap weights in (9.6-40) are the output sample values when the input is the unit sample, the digital equivalent of the impulse function.

9.6.8 Example of the Fourier Series Design Technique

We now realize the digital filter whose magnitude response approximates the rectangular magnitude response in Fig. 9-25 in the least-squares sense. (Note that

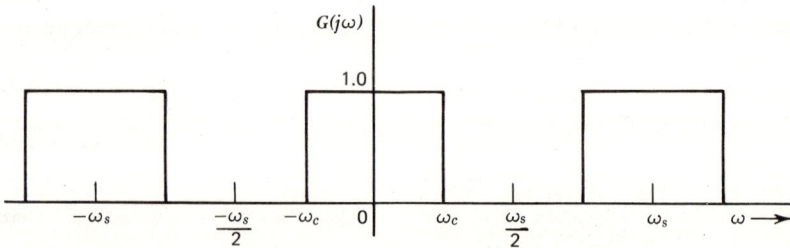

Figure 9-25. Rectangular magnitude response of a digital filter.

Introduction to Design Techniques

the representation by the truncated Fourier series minimizes the integrated-squared error.) For this example $G(j\omega)$ is the magnitude function, hence a real and even function. Then a_k, from (9.6-35), is

$$a_k = \frac{1}{\omega_s} \int_{-\omega_c}^{\omega_c} 1 \times \cos kT\omega \, d\omega = \frac{2}{\omega_s} \int_0^{\omega_c} \cos kT\omega \, d\omega = \frac{2 \sin kT\omega_c}{kT\omega_s} \quad (9.6\text{-}42)$$

But $T = 2\pi/\omega_s$, and if we let $r = \omega_c/\omega_s$, a_k becomes

$$a_k = 2r \frac{\sin 2\pi kr}{2\pi kr} \quad (9.6\text{-}43)$$

For this example we set $\omega_c = 1$ and $\omega_s = 4$. Then $r = \frac{1}{4}$ and

$$a_k = \frac{1}{2} \frac{\sin k(\pi/2)}{k(\pi/2)} \quad (9.6\text{-}44)$$

We choose $m = 5$ yielding the 11 filter coefficients from (9.6-40) as

$$h(0) = h(10) = a_5 = \frac{1}{5\pi}$$
$$h(1) = h(9) = a_4 = 0$$
$$h(2) = h(8) = a_3 = -\frac{1}{3\pi} \quad (9.6\text{-}45)$$
$$h(3) = h(7) = a_2 = 0$$
$$h(4) = h(6) = a_1 = \frac{1}{\pi}$$
$$h(5) = a_0 = \frac{1}{2}$$

and the digital filter system function, from (9.6-39), is

$$H(z) = \frac{1}{5\pi} - \frac{1}{3\pi} z^{-2} + \frac{1}{\pi} z^{-4} + \frac{1}{2} z^{-5} + \frac{1}{\pi} z^{-6} - \frac{1}{3\pi} z^{-8} + \frac{1}{5\pi} z^{-10} \quad (9.6\text{-}46)$$

The frequency response $H(e^{j\omega T})$ is obtained by factoring out z^{-5}, replacing z by $e^{j\omega T}$, and rearranging as follows:

$$H(e^{j\omega T}) = e^{-j5\omega T} \left[\frac{1}{5\pi} (e^{j5\omega T} + e^{-j5\omega T}) - \frac{1}{3\pi} (e^{j3\omega T} + e^{-j3\omega T}) + \frac{1}{\pi} (e^{j\omega T} + e^{-j\omega T}) + \frac{1}{2} \right]$$
$$= e^{-j5\omega T} \left[\frac{1}{2} + \frac{2}{\pi} \cos \omega T - \frac{2}{3\pi} \cos 3\omega T + \frac{2}{5\pi} \cos 5\omega T \right] \quad (9.6\text{-}47)$$

Now $H(e^{j\omega T})$ is in the form of (9.6-41). The first factor is unity magnitude, linear phase, and introduces a constant delay of $5T$ seconds. Because the phase is linear, the impulse response is symmetric as in Fig. 9-26. These sample values are the coefficients in (9.6-46), which are the samples of the truncated cardinal function.

The magnitude response, from (9.6-47), is

$$|H(e^{j\omega T})| = \left| \frac{1}{2} + \frac{2}{\pi} \cos \omega T - \frac{2}{3\pi} \cos 3\omega T + \frac{2}{5\pi} \cos 5\omega T \right| \quad (9.6\text{-}48)$$

shown in Fig. 9-27 with $T = 2\pi/\omega_s = \pi/2$. As more terms are included in the

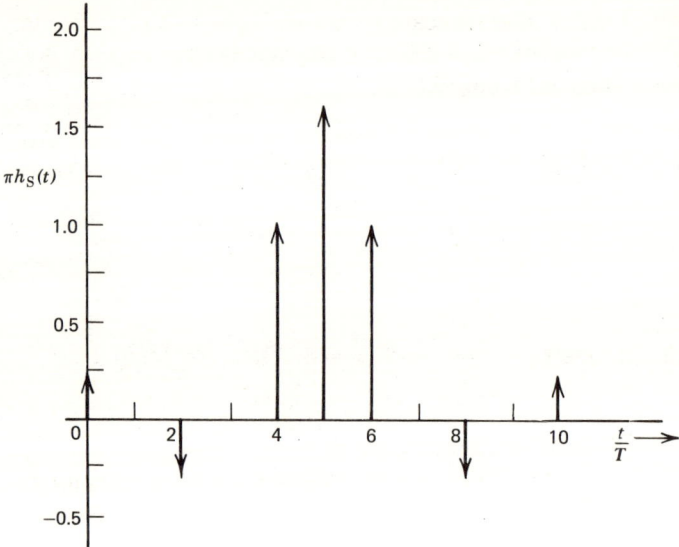

Figure 9-26. Digital filter unit-sample response of (9.6-45).

Fourier representation, the approximation in the least-squares sense improves; but overshoots, the familiar Gibbs phenomenon, occur in the vicinity of the discontinuity. The maximum deviation remains at about 9% of the constant passband response, independent of the number of additional included terms. Its width, however, decreases and approaches zero as the number of terms approaches infinity. Figure 9-27 also shows the approximation to the rectangular response with $m = 25$ to illustrate the ever-present Gibbs phenomenon.

9.6.9 Example of Nonrecursive Filter Realizations

The classical realization for the filter in Section 9.6.8 is presented in Fig. 9-28, and because the even magnitude-function results in a symmetric impulse response, three of the multipliers are duplicated. This symmetry allows us to realize this response with only four multipliers by folding the filter about its center, as in Fig. 9-29 [11]. This realization has practical importance because multiplication is a costly and slow operation. The $2T$ delays in Figs. 9-28 and 9-29 occur because some of the tap weights are zero.

9.6.10 Window (Weighting) Functions

In Section 9.6.7 we solved the approximation problem by the Fourier series technique and thereby obtained the filter system function. This in turn allowed us to realize the filter by the schematics in Figs. 9-28 and 9-29. This design process is useful for approximating well-behaved functions, but for functions containing discontinuities, the Gibbs phenomenon, already mentioned, appears. We attribute the Gibbs phenomenon to the manner in which the time function is truncated. These truncating functions are known as "window" or "weighting"

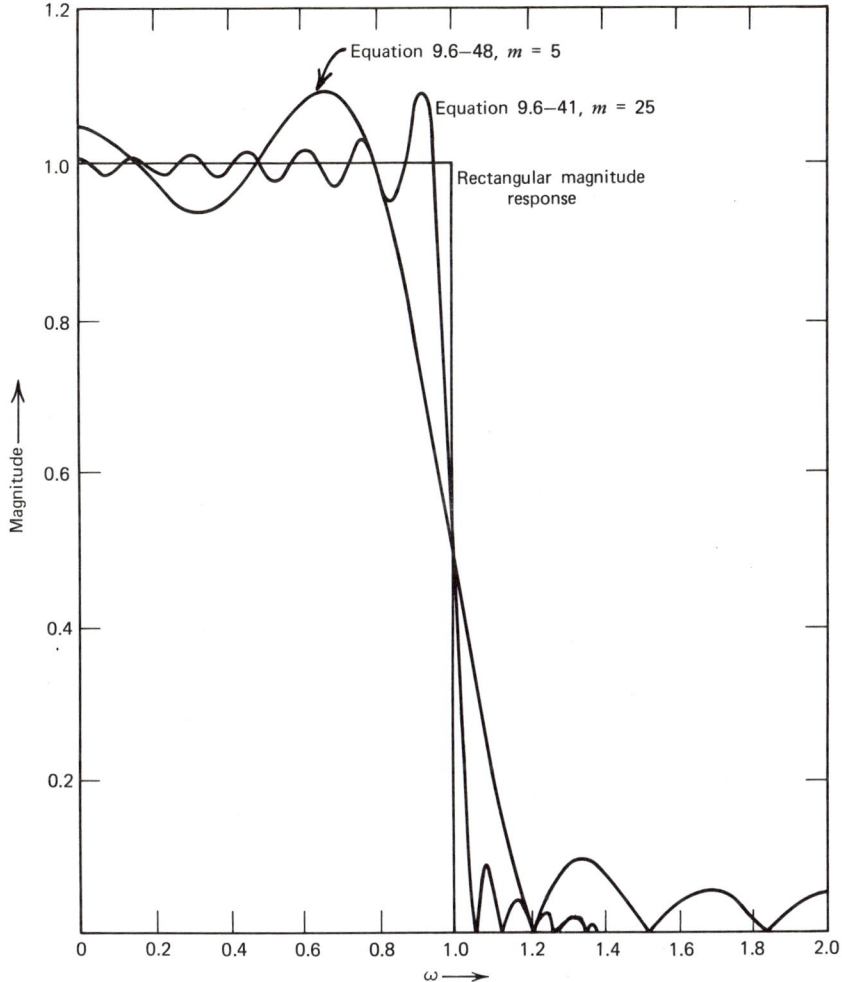

Figure 9-27. Fourier series approximation to the rectangular magnitude response.

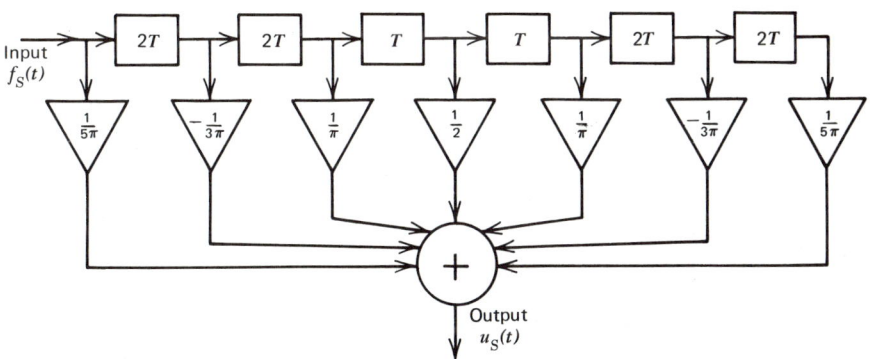

Figure 9-28. Classical realization of nonrecursive filter characterized by (9.6-45).

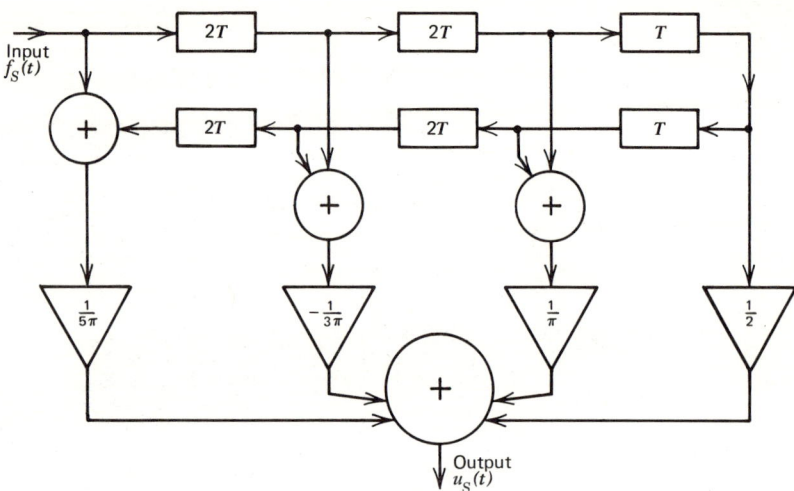

Figure 9-29. Equivalent realization of filter in Section 9.6.8 with only four multipliers.

functions. The unit-sample response in Fig. 9-26 is the result of time limiting the sampled cardinal function by a rectangular window function, as in Fig. 9-30.

The magnitude response overshoots are due to the abrupt truncation of the time response. If the time response were truncated differently, the magnitude characteristic would be smoother. Much work has been devoted to developing suitable window functions, and we do not dwell on this subject.† Rather, we mention that this technique exists for spectrum smoothing and we leave the details for the more interested readers [10, 11].

The more commonly encountered window functions are the Hamming window (a single cycle of a raised cosine function), the Kaiser window (a nearly optimum window function employing Bessel functions), the Hanning window, the Dolph–Chebyshev window (minimum transition bandwidth at a spectrum discontinuity for a specified ripple), and the Taylor window (a realizable approximation to the Dolph–Chebyshev window). Because multiplication in the time domain, as shown in Fig. 9-30, is equivalent to convolution in the frequency domain, the geometric procedure of shifting and sliding the spectra often helps in understanding why some window functions reduce the spectrum ripples and others do not.

9.6.11 Example of MTI Filter Design

A useful signal-processing scheme for separating a radar target return from clutter is to pass the composite signal through a nonrecursive filter that attenuates the clutter and does not appreciably affect the signal from the moving target [18]. Such filters are known as moving-target-indicator (MTI) filters, and a digital realization with three tap weights, known as a three-pulse canceler, is schematized in Fig. 9-31.

The design of this filter is the determination of a_0 and a_{-1}. We normalize a_1 to unity and then require that the response at $\omega = 0$ be zero, for the primary clutter

---

†This technique is similar to the sidelobe suppression procedure in Section 7.4.4 except that time-domain overshoots are considered there.

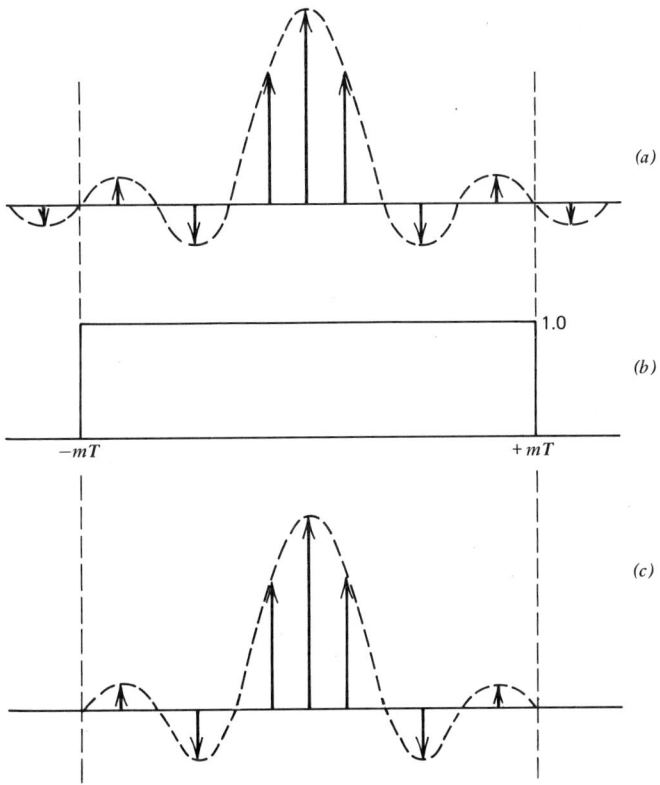

Figure 9-30. Time limiting the sampled cardinal function (*a*) by the rectangular window function (*b*) generates (*c*) unit-sample response of Fig. 9-26.

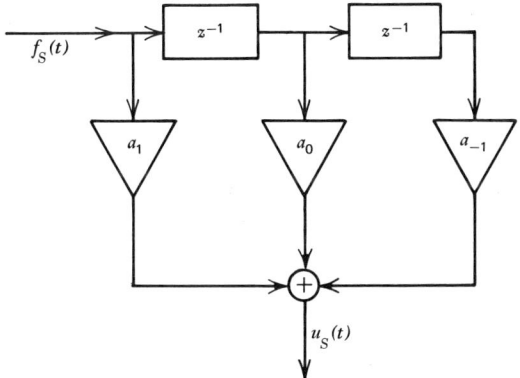

Figure 9-31. Three-pulse canceler.

energy is located there (since no Doppler shift is associated with clutter returns). To eliminate phase distortion we set $a_{-1} = a_1$ for a linear phase response. The filter system function, from (9.6-39), is then

$$H(z) = 1 + a_0 z^{-1} + z^{-2} \qquad (9.6\text{-}49)$$

and the frequency response is

$$H(e^{j\omega T}) = 1 + a_0 e^{-j\omega T} + e^{-j2\omega T} \qquad (9.6\text{-}50)$$

For zero output at $\omega = 0$, $1 + a_0 + 1 = 0$, and $a_0 = -2$. Then

$$H(e^{j\omega T}) = (1 - e^{-j\omega T})^2 = e^{-j\omega T}(e^{j(\omega T/2)} - e^{-j(\omega T/2)})^2$$

$$= -4 \sin^2 \frac{\omega T}{2} e^{-j\omega T} \qquad (9.6\text{-}51)$$

The phase is linear and the magnitude response (Fig. 9-32) is

$$|H(e^{j\omega T})| = 4 \sin^2 \frac{\omega T}{2} \qquad (9.6\text{-}52)$$

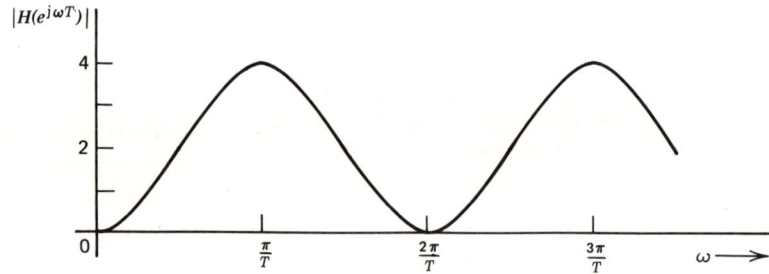

Figure 9-32. Magnitude response of the three-pulse canceler.

MTI filters are also realized with recursive filters, and sophisticated error criteria have been proposed for optimizing the MTI response. This example was presented only to show the application of a nonrecursive digital filter to a problem of practical importance.

9.7 DIGITAL NETWORKS, ERROR SOURCES, AND THE FFT

The following practical aspects of digital filtering are now considered.

The digital networks for realizing the specified filtering function.

Errors due to practical hardware.

The Fast Fourier transform and its application to digital filtering.

The survey-oriented discussion of these specialities is intended to introduce topics often obscured in the literature by information intended for designers only. We hope that this section provides insight into these subjects for system-oriented engineers who specify and evaluate the designs.

9.7.1 Digital Networks

We now consider the realization of the digital filter computation algorithm using the three basic digital components: adders, multipliers, and shift registers (delays). With digital hardware we are not faced with the tedious and time-consuming calculations encountered in synthesizing analog filters, for digital filter network values are easily obtained, often by inspection. This is in sharp contrast to the analog filtering case, where a change in one transfer-function coefficient requires the complete synthesis procedure to be repeated.

Two forms of the recursive filter are known as the direct forms (Fig. 9-33), where the multiplier coefficients are related to the coefficients of the difference

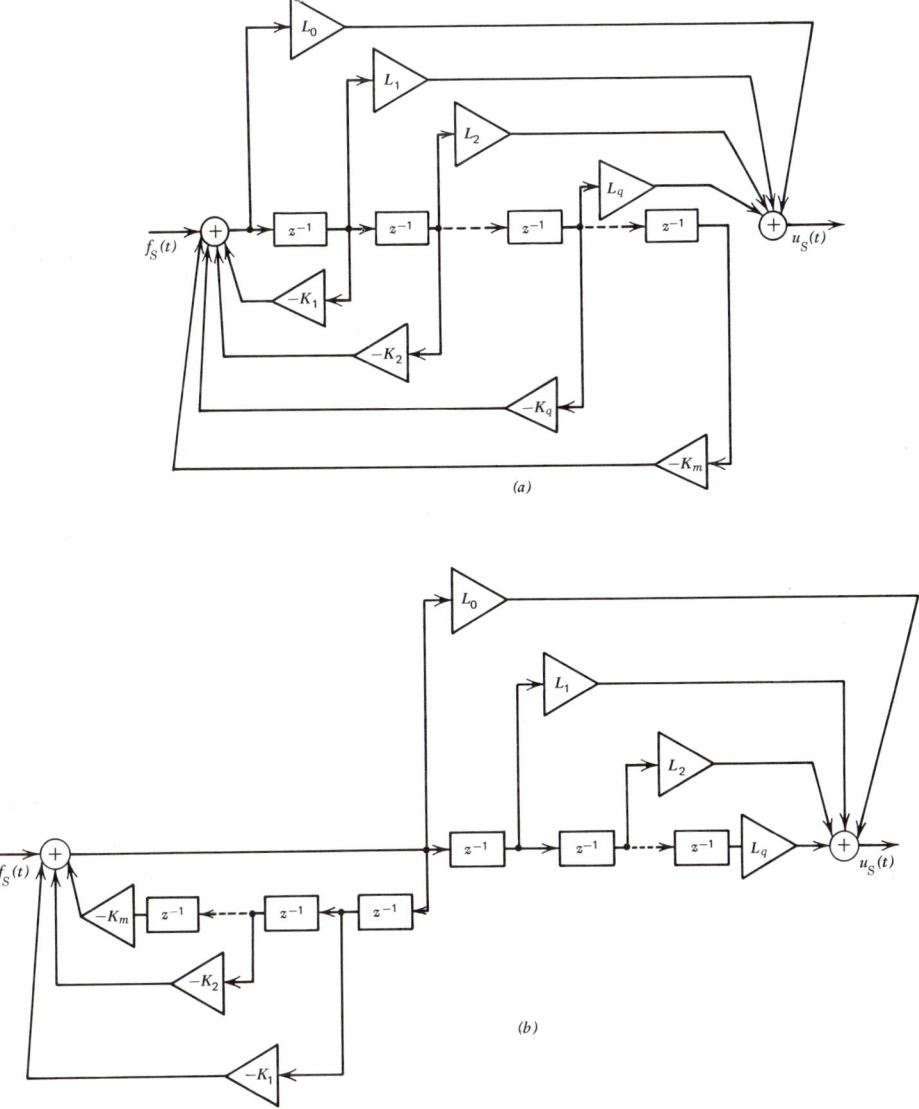

Figure 9-33. Direct forms of recursive filters. (a) Canonic, (b) noncanonic.

equation in (9.3-1), which reflect as the system function coefficients in (9.5-9). With respect to the latter, the numerator polynomial coefficients are the multiplier values in the feed-forward paths and the negative of the denominator polynomial coefficients are the multiplier values in the feedback paths. Given $H(z)$ as a ratio of polynomials in z^{-1}, one obtains the network values directly from the given function by inspection; hence the origin of the network nomenclature.

The direct form in Fig. 9-33a is called a canonic network; that is, an mth-order system is realized with m delays. Realization with fewer delays is impossible, and the network is no longer canonic if more than m delays are used. In the canonic form, the feed-forward and feedback paths share the same delays. For example, the network in Fig. 9-33b is not canonic.

If all the K's are zero, the network in Fig. 9-33a reduces to the nonrecursive form in Fig. 9-22. Furthermore, when the tap weights are symmetric about the center, almost one-half the multipliers can be eliminated by folding the filter about its center, as in Fig. 9-29. Once $H(z)$ of the nonrecursive filter is known, its realization is straightforward (see Section 9.6.5), and the two networks just discussed are the ones selected in practice. The remainder of this section is therefore devoted to recursive network realizations.

The direct forms in Fig. 9-33 should rarely be used in practice, except as first- and second-order realizations because the poles and zeros of the recursive filter system function are sensitive functions of the difference equation coefficients [6, 10]. The degree of sensitivity increases as the filter order increases. Consequently, the multiplier coefficient accuracy requirement is least when the filter order is smallest [10].

For this reason the second-order filter is used as a basic building block, since any stable system whose system function contains real coefficients can be realized by cascading or paralleling first- and second-order filters (Fig. 9-34). Figure 9-34 does not imply that the same subsystems are used in both the cascade and the parallel networks. Rather it shows the connection of the basic building blocks.

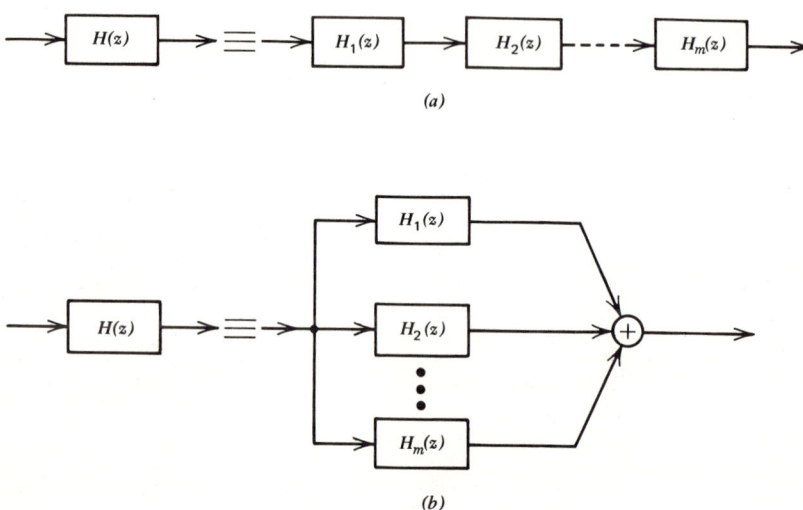

Figure 9-34. (a) Cascade form and (b) parallel form of digital filter realization using first- and second-order networks.

Digital Networks, Error Sources, and the FFT

With digital networks we do not encounter the interaction problems associated with the cascade or parallel connection of analog filters. We just connect the individual digital networks in the appropriate arrangement to realize the desired overall transfer function. It is not clear that one form is always more advantageous than the other, since the realization of the individual blocks in Fig. 9-34 enters into this decision. However, the parallel form is more widely used.

Figure 9-35 illustrates a canonic realization of the second-order system function with two zeros,

$$H(z) = \frac{L_0 + L_1 z^{-1} + L_2 z^{-2}}{1 + K_1 z^{-1} + K_2 z^{-2}} \qquad (9.7\text{-}1)$$

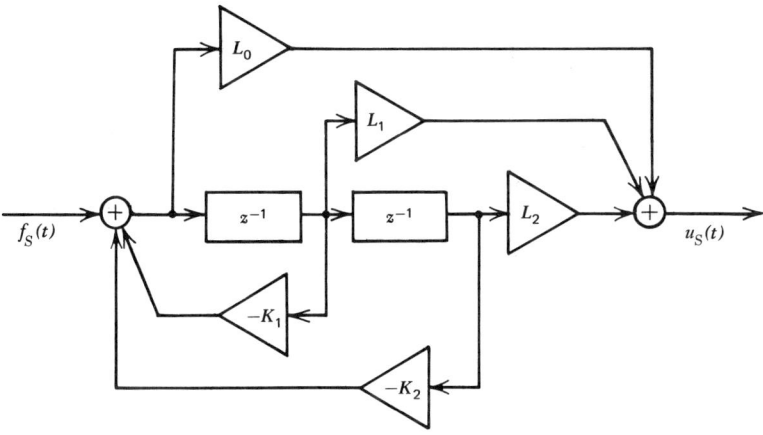

Figure 9-35. Canonic realization of a second-order system function with two zeros.

If only one zero is specified, L_2 is zero; hence this multiplier is removed. The first-order system function with one real zero,

$$H(z) = \frac{L_0 + L_1 z^{-1}}{1 + K_1 z^{-1}} \qquad (9.7\text{-}2)$$

is obtained from (9.7-1) by setting $L_2 = K_2 = 0$. Then the canonic realization of $H(z)$ is that in Fig. 9-36.

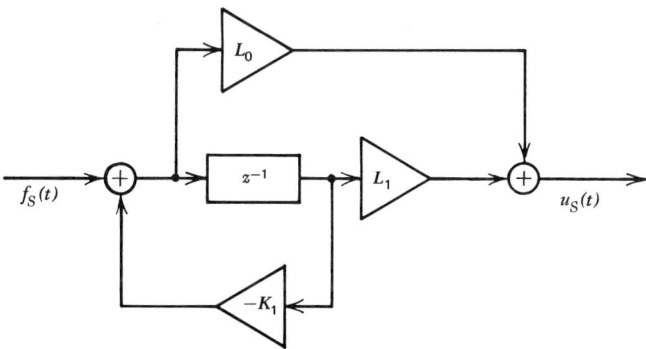

Figure 9-36. Canonic realization of a first-order system function with one real zero.

The network constants for the parallel form are obtained by first expressing the desired $H(s)$ as a partial fraction expansion and then applying the standard z-transform or the bilinear z-transform to each term. The terms with complex-conjugate poles can be combined to yield a second-order system function.

Care must be taken when realizing the cascade form. First, the desired $H(s)$ is expressed in factored form and then separated into first- and second-order transfer functions. The bilinear z-transform can then be applied to each factor because this transformation has the property that the transform of a product is equal to the product of the transforms, that is,

$$\mathscr{J}[G_1(s)G_2(s)] = \mathscr{J}[G_1(s)]\mathscr{J}[G_2(s)] \tag{9.7-3}$$

Since the standard z-transform (impulse-invariant method) does not possess this property, we must determine $H(z)$ from $H(s)$ and arrange $H(z)$ in factored form to realize the cascade arrangement.

9.7.2 Example of Filter Realization in Parallel and Cascade Forms

In Section 9.6.2 we designed a third-order Butterworth digital filter by the impulse-invariant method. Let us now realize this filter in both the parallel and cascade form.

Parallel Form. Combine the last two terms of (9.6-13) and rewrite $H(z)$ as

$$H(z) = \frac{1}{1 - e^{-T}z^{-1}} + \frac{-1 + e^{-T/2}\left(\cos\frac{\sqrt{3}}{2}T + \frac{\sqrt{3}}{3}\sin\frac{\sqrt{3}}{2}T\right)z^{-1}}{1 - \left(2e^{-T/2}\cos\frac{\sqrt{3}}{2}T\right)z^{-1} + e^{-T}z^{-2}} \tag{9.7-4}$$

Now $H(z)$ is expressed as the sum of a first-order system function and a second-order system function. Identify the first function with (9.7-2) and Fig. 9-36, where $L_0 = 1$, $L_1 = 0$, and $K_1 = -e^{-T}$. Identify the second term of (9.7-4) with (9.7-1) and Fig. 9-35, where $L_0 = -1$, $L_1 = e^{-T/2}(\cos(\sqrt{3}/2)T + (\sqrt{3}/3)\sin(\sqrt{3}/2)T)$, $L_2 = 0$, $K_1 = -2e^{-T/2}\cos(\sqrt{3}/2)T$, and $K_2 = e^{-T}$. The parallel form of Fig. 9-34b is then realized as shown in Fig. 9-37, where the dashed lines enclose the basic building blocks. Since only three delays are used, the network is canonic.

Cascade Form. Rearrange $H(z)$ in (9.6-13) in factored form as

$$H(z) = \frac{z^{-1}(L_0 + L_1 z^{-1})}{(1 - e^{-T}z^{-1})\left[1 - \left(2e^{-T/2}\cos\frac{\sqrt{3}}{2}T\right)z^{-1} + e^{-T}z^{-2}\right]} \tag{9.7-5}$$

where

$$L_0 = e^{-T} + e^{-T/2}\left(\frac{\sqrt{3}}{3}\sin\frac{\sqrt{3}}{2}T - \cos\frac{\sqrt{3}}{2}T\right)$$

$$L_1 = e^{-T} - e^{-3T/2}\left(\frac{\sqrt{3}}{3}\sin\frac{\sqrt{3}}{2}T + \cos\frac{\sqrt{3}}{2}T\right)$$

The arrangement of the cascaded form depends on the placement of the numerator

Digital Networks, Error Sources, and the FFT

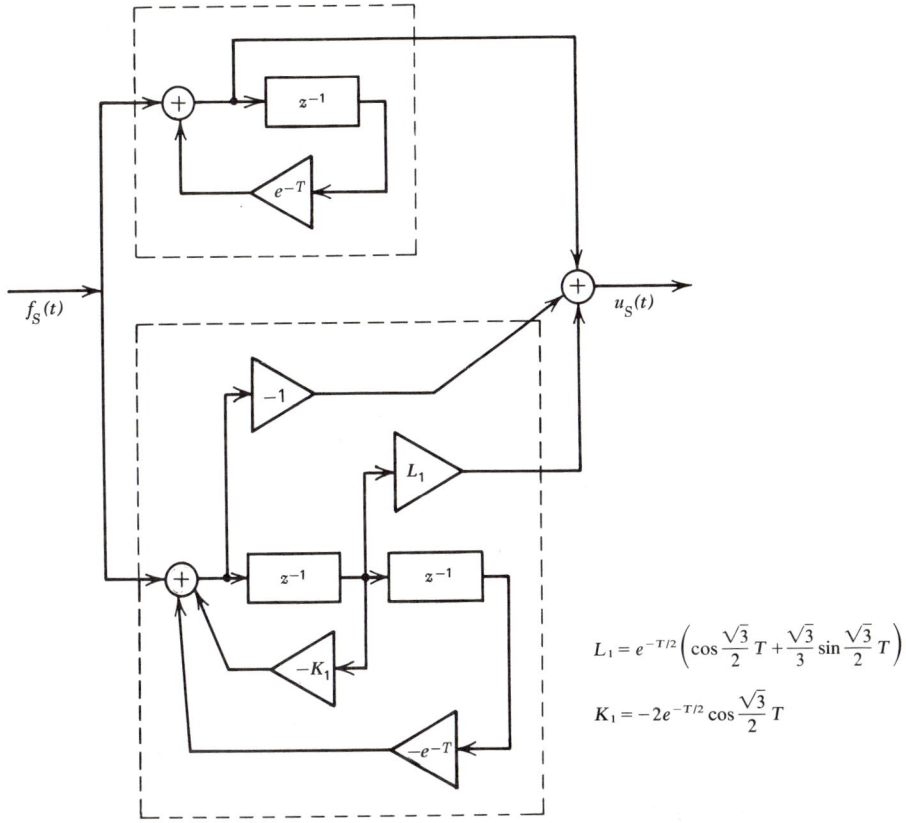

Figure 9-37. Third-order Butterworth filter in Section 9.6.2 realized in parallel form.

factors. We choose to partition $H(z)$ as

$$H(z) = \left(\frac{L_0 + L_1 z^{-1}}{1 - e^{-T} z^{-1}}\right) \left(\frac{z^{-1}}{1 - \left(2e^{-T/2} \cos \frac{\sqrt{3}}{2} T\right) z^{-1} + e^{-T} z^{-2}}\right) \qquad (9.7\text{-}6)$$

and then realize the first factor by the network in Fig. 9-36 and the second factor by the network in Fig. 9-35. The canonic cascade form of Fig. 9-34a for this realization is given in Fig. 9-38. Note that we did not first arrange $H(s)$ in factored form and then apply the z-transform to each factor. This is an illegitimate step and yields an incorrect digital system function.

9.7.3 Errors Due to Practical Hardware

The difference between the actual digital filter response and the theoretical response depends on the accuracy of the sampling frequency, the quantization error resulting from the A/D converter, the number of significant bits used to define the multiplier constants, and the round-off error caused by filter computations with finite word lengths. Many detailed theoretical studies of these errors have been made (see Refs. 6, 10, 11, and the references listed there), but here we only briefly describe these errors and their consequences.

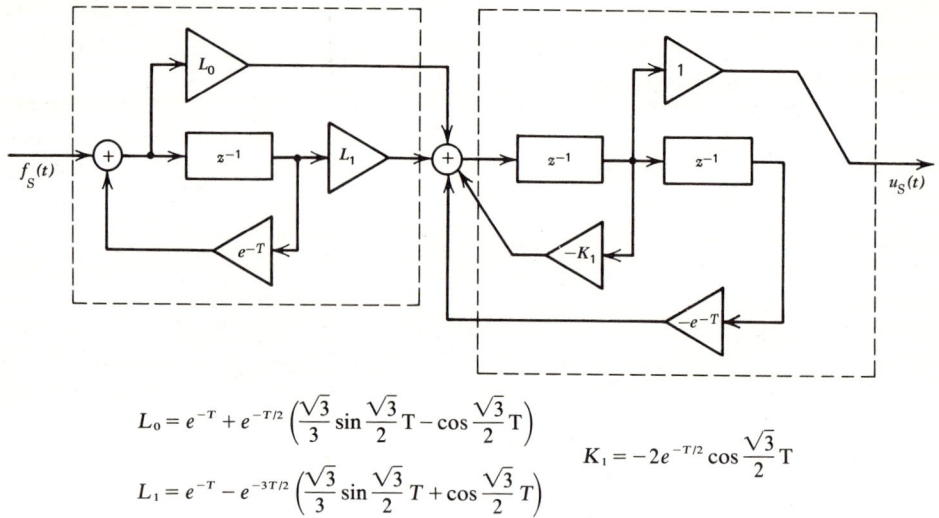

$$L_0 = e^{-T} + e^{-T/2}\left(\frac{\sqrt{3}}{3}\sin\frac{\sqrt{3}}{2}T - \cos\frac{\sqrt{3}}{2}T\right)$$

$$L_1 = e^{-T} - e^{-3T/2}\left(\frac{\sqrt{3}}{3}\sin\frac{\sqrt{3}}{2}T + \cos\frac{\sqrt{3}}{2}T\right)$$

$$K_1 = -2e^{-T/2}\cos\frac{\sqrt{3}}{2}T$$

Figure 9-38. Third-order Butterworth filter in Section 9.6.2 realized in cascade form.

The first source of error is negligible compared to the others because the sampling frequency is usually controlled by a stable, accurate quartz crystal or, more recently, by a more accurate cesium standard.

Quantization of the input samples means that each filter input sample value is not the actual value of the analog sample; rather, it is an approximation generated by the A/D converter (Fig. 9-6). The accuracy of this representation depends on the number of bits used in the conversion.

Example 9-12. Suppose the input analog sample is $\sqrt{2}$ V and the A/D converter is a 6-bit system, that is, $\sqrt{2}$ is to be approximated by the linear combination of

$$\begin{aligned} 2^0 &= 1 & 2^{-3} &= 0.125 \\ 2^{-1} &= 0.5 & 2^{-4} &= 0.0625 \\ 2^{-2} &= 0.25 & 2^{-5} &= 0.03125 \end{aligned} \quad (9.7\text{-}7)$$

The best approximation is $2^0 + 2^{-2} + 2^{-3} + 2^{-5} = 1.40625$, or in binary form 1.01101. This represents a percentage error of

$$\frac{\sqrt{2} - 1.40625}{\sqrt{2}} \times 100\% = 0.563\% \quad (9.7\text{-}8)$$

For a 6-bit system the error is less than ±1%, and for a 10-bit system the least significant bit has a value of only 1/1024 of the full dynamic range. The effect of quantization by the A/D converter is often considered as a noise source with a variance $E_0^2/12$, where E_0 is the fundamental quantizing step [6].

Multiplier coefficient quantization is equivalent to changing the coefficients of the difference equation describing the filter or changing the poles of the filter system function. This is a one-time change, and its consequences are discussed in Refs. 6 and 10.

Round-off error caused by filter computations with finite word lengths is perhaps the most important of all, but unfortunately it is the hardest to calculate.

Furthermore this error cannot be avoided, for an n-bit data sample multiplied by an n-bit coefficient results in a product that is $2n$ bits long, the next multiplication yields a $3n$-bit word, and so on. In a recursive filter the number of bits will increase indefinitely if we do not quantize the result of the various arithmetic operations. Various theoretical studies of this problem have been made, and one important finding is that the direct form is inferior to the cascade and parallel forms.

In spite of the many investigations of the error sources in digital hardware, it is impractical to use these results to accurately calculate filter response degradation when all the above-mentioned quantizations are present. As an alternative to the previous approaches, we suggest that the digital filter be simulated on a general-purpose digital computer. In this manner, the number of necessary bits for an acceptable response can be determined and even optimized. After all, the proof of the pudding is in the eating.

9.7.4 The Fast Fourier Transform

Whereas the Fourier integral transform is useful for the analysis of continuous functions, the discrete Fourier transform (DFT) is more appropriate for the analysis of sampled functions. The digital computer requires that input data be in sampled form. Thus the DFT has immediate application for time-series analysis, power spectrum analysis, and filter simulation on these machines.

The DFT of the sequence $\{f(0), f(1), \ldots, f(N-1)\}$ is defined as

$$F(k) = \sum_{n=0}^{N-1} f(n) \exp\left[-j\left(\frac{2\pi}{N}\right)nk\right] \tag{9.7-9}$$

yielding another sequence $\{F(0), F(1), \ldots, F(N-1)\}$. The original sequence is recovered from the $F(k)$'s by the inverse of the DFT, denoted by IDFT, as

$$f(n) = \frac{1}{N} \sum_{n=0}^{N-1} F(k) \exp\left[j\left(\frac{2\pi}{N}\right)nk\right] \tag{9.7-10}$$

Note that the DFT and IDFT are obtained by the same arithmetic operations and each operates on the data in batches of N samples.

Equations 9.7-9 and 9.7-10 are a mathematical transform pair and describe many physical variables. Here we associate the sequence $\{F(k)\}$ in (9.7-9) with frequency and the sequence $\{f(n)\}$ in (9.7-10) with time. The $f(n)$'s and $F(k)$'s may be complex numbers, and both are periodic with period N. Thus N values of $f(n)$ yield N values of $F(k)$ before repetition begins, and vice versa. This should not be surprising, since the spectrum of a discrete-time function has been shown to be periodic.

The DFT is a transform in its own right, but it has many mathematical properties that are analogous to those of the Fourier transform. However, the DFT theory, unlike the Fourier transform theory, assumes that the time samples belong to a band-limited function. If this is not true, the familiar aliasing error is present and the correct spectrum is not obtained. Therefore the sampling rate should be at least twice the highest significant frequency present in the waveform. Additional pitfalls are leakage and the picket-fence effect[2].

In 1965 Cooley and Tukey [5] reported an efficient procedure for computing the

DFT of a time series. This method, now widely known as the fast Fourier transform (FFT), has achieved immense popularity because of the considerable savings in computer time and the reduction in round-off errors associated with these computations. It is not an overstatement to say that the FFT has revolutionized the processing of digital waveforms.

The algorithm achieves efficiency by eliminating redundant computations and by computing the coefficients of the DFT iteratively, using previous values as stepping stones in subsequent calculations [2, 4, 6]. If N, the number of samples in the batch to be processed, is a power of 2, the FFT requires about $2N \log_2 N$ multiplications to evaluate all N coefficients associated with the DFT, whereas, from (9.7-9), the DFT requires N^2 multiplications. Table 9-5 compares the number of calculations by each technique. If $N = 1024$, the FFT reduces the computational effort by about 50.

Table 9-5 **Comparison of Required Multiplication Operations Using the FFT and the Direct DFT**

N	N^2 (direct DFT)	$2N \log_2 N$ (FFT)
64	4,096	768
128	16,384	1,792
256	65,536	4,096
512	262,144	9,216
1024	1,048,576	20,480

Filtering using the FFT results in less computation time than recursive or nonrecursive realizations when the number of multiplications for the former is less than the number of multiplications for the latter. The number of multiplications for each depends on the structure of the respective algorithms, such as the number of feed-forward and feedback paths in the filter and the number of batch samples N needed to approximate a similar filter by the FFT. This is why one approach is not always more efficient than the other.

Figure 9-39 illustrates frequency-domain processing of the input time samples using the FFT [17]. These samples are transformed to the frequency sampled function $F_S(\omega)$ by the FFT and then multiplied by samples of the stored filter system function $H_S(j\omega)$. The inverse FFT of this product, $U_S(\omega)$, yields the

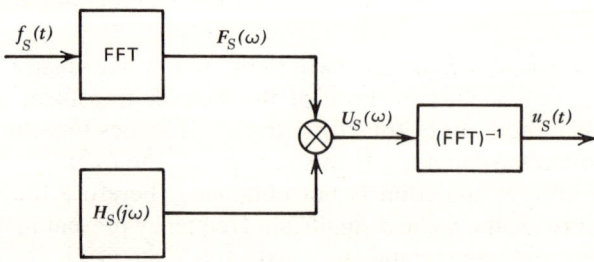

Figure 9-39. Frequency-domain processing using FFT.

sampled output $u_S(t)$. Mathematically,

$$u_S(t) = (\text{FFT})^{-1}[H_S(j\omega)F_S(\omega)] \qquad (9.7\text{-}11)$$

where

$$F_S(\omega) = \text{FFT}[f_S(t)] \qquad (9.7\text{-}12)$$

The FFT offers much flexibility. For example, if in Fig. 9-39, $H_S(j\omega) = G_S^*(\omega)$, where $G_S(\omega)$ is the transform of the transmitted time sequence, the processor becomes a matched filter for that sequence. The digital processor in Fig. 9-40

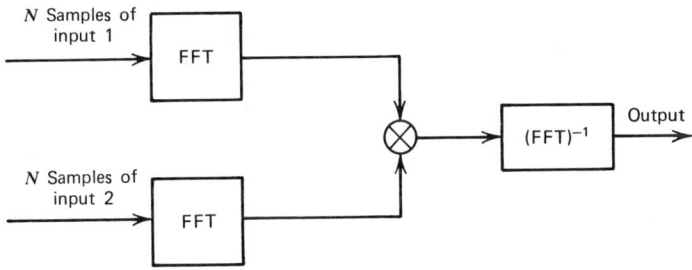

Figure 9-40. Digital convolver.

yields the convolution of the two inputs. Each input is transformed by the FFT to a sequence in the frequency domain and then the sequences are multiplied together. The inverse FFT of this product is the convolution of the two inputs, since multiplication in the frequency domain corresponds to convolution in the time domain. Since convolution and correlation are related by a sign change in the time variable, this digital convolver is easily converted to a digital correlator. Finally, the FFT itself is equivalent to a spectrum analyzer (Fig. 9-41). The FFT of N samples spaced T seconds apart is a series of N samples in the frequency domain spaced $1/NT$ hertz apart. Thus the FFT box is equivalent to a bank of N narrow BP filters spaced $1/NT$ hertz apart.

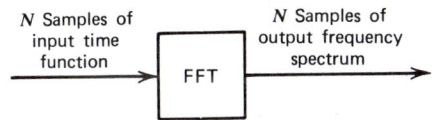

Figure 9-41. A digital spectrum analyzer.

9.7.5 Summary

In Section 9.7 we have introduced some of the practical considerations of digital filtering. The appropriate network realization, the minimum number of bits employed, the resulting computation error, and other variables, depend on the specific case under consideration and on the experience of the designer. There is still no "best" approach to digital filter design. Contributions in this area have come and continue to come from the areas of circuit theory, control theory, statistics, and numerical analysis, among others.

REFERENCES

1. Aseltine, J. A., *Transform Method in Linear System Analysis*, McGraw-Hill, New York, 1958.
2. Bergland, G. D., "A Guided Tour of the Fast Fourier Transform," *IEEE Spectrum*, Vol. 6, pp. 41–52, July 1969.
3. Caprihan, A., "An Optimal Property of the Impulse Invariant Method," *IEEE Trans. Circuit Theory*, Vol. CT-20, pp. 173–174, March 1973.
4. Cochran, W. T., et al., "What is the Fast Fourier Transform?" *Proc. IEEE*, Vol. 55, pp. 1664–1674, October 1967.
5. Cooley, J. W., and J. W. Tukey, "An Algorithm for the Machine Calculation of Complex Fourier Series," *Math. Comput.*, Vol. 19, pp. 297–301, April 1965.
6. Gold, B., and C. M. Rader, *Digital Processing of Signals*, McGraw-Hill, New York, 1969.
7. Harrison, S. R., and B. J. Lear, "Digital Filters for Approximating Continuous Convolutions," *IEEE Trans.* Circuit Theory, Vol. CT-18, pp. 743–745, November 1971.
8. Kaiser, J. F., "Design Methods for Sampled Data Filters," Proc. First Allerton Conference on Circuit and System Theory, University of Illinois, pp. 221–236, November 1963.
9. Kaplan, W., *Operational Methods for Linear Systems*, Addison-Wesley, Reading, Mass., 1962, Chap. 6.
10. Kuo, F. F., and J. F. Kaiser, *System Analysis by Digital Computer*, Wiley, New York, 1966, Chap. 7.
11. Leon, B. J., and S. C. Bass, "Designers' Guide to: Digital Filters," Electrical Design News (EDN), 6 parts, Jan. 20, 1974, pp. 30–36, Feb. 20, 1974, pp. 65–72, March 20, 1974, pp. 51–59, April 20, 1974, pp. 57–62, May 20, 1974, pp. 61–68, June 20, 1974, pp. 69–75.
12. Moschytz, G. S., "Inductorless Filters: A Survey," *Proc. 20th Electron. Components Conf.*, 1970, pp. 243–268.
13. Oliver, B. M., J. R. Pierce, and C. E. Shannon, "The Philosophy of Pulse-Code Modulation," *Proc. IRE*, Vol. 36, pp. 1324–1331, November 1948.
14. Papoulis, A., "Error Analysis in Sampling Theory," *Proc. IEEE*, Vol. 54, pp. 947–955, July 1966.
15. Rabiner, L. R., et al., "Terminology in Digital Signal Processing," *IEEE Trans. Audio Electroacoust.*, Vol. AU-20, pp. 322–337, December 1972.
16. Thiran, J. P., "Recursive Digital Filters with Maximally Flat Group Delay," *IEEE Trans. Circuit Theory*, Vol. CT-18, pp. 659–664, November 1971.
17. Thompson, B. J. and B. J. Goldstone, "Digital Signal Processing," *Comput. Des.*, pp. 44–49, May 1969.
18. Zverev, A. I., "Digital MTI Radar Filters," *IEEE Trans. Audio Electroacoust.*, Vol. AU-16, pp. 422–432, September 1968.
19. Zverev, A. I., *Handbook of Filter Synthesis*, Wiley, New York, 1967.

PROBLEMS

9-1. Find the z-transform of the sampled values of
(a) $\sin \omega_0 t\, u_{-1}(t)$
(b) $u_{-1}(t)$

9-2. Find the time sequence associated with the z-transform

$$F(z) = \frac{z(e^{-T} - e^{-2T})}{z^2 - (e^{-T} + e^{-2T})z + e^{-3T}}$$

Problems

9-3. A filter with the noncausal impulse response

$$h(t) = \frac{\sin 4\pi t}{\pi t}$$

is excited by the input signal $f(t) = u_{-1}(t)$. We sample the output signal every T seconds. What is the maximum value of T that permits the analog output signal to be exactly reconstructed from the sampled output signal?

9-4. The z-transform of a sampled time function $h_S(t)$ is

$$H(z) = \frac{2z^4 + 5z^3 + 6z^2 + 5z + 4}{z^5 + z^4 + 10z^3 + 7z^2 + 6z + 3}$$

What is the value of $h_S(t)$ at $t = 2T$?

9-5. What is the z-transform of the impulse response of the second-order Butterworth LP filter ($\omega_c = 1$) sampled every T seconds?

9-6. Find the time sequence associated with the z-transform

$$F(z) = \frac{z^2}{z^2 - (2 + e^{-T})z + 2e^{-T}}$$

9-7. A filter has the impulse response shown below.

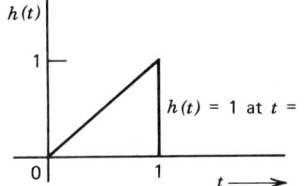

Figure P.9-7a.

(a) What is the transfer function for this filter?
(b) Is this filter realizable with a finite number of lumped elements? Why?
(c) A digital filter is then designed whose unit-sample response is the sampled values of $h(t)$. With $T = 0.2$ second, what is the digital filter system function as a rational function in z^{-1}?
(d) Is this digital filter recursive or nonrecursive? Why?
(e) Draw this filter's digital schematic.
(f) The input to this filter is sampled values of the pulse below ($T = 0.2$).

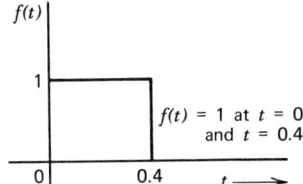

Figure P.9-7b.

What are the sampled values $u(nT)$ of the output signal?

9-8. A digital filter, described by the difference equation

$$u(nT) = e^{-T}u(nT - T) + f(nT)$$

is excited by the input sequence $f_S(t) = 1, 0, 0, 0, \ldots$
- (a) Compute the output sequence $u_S(t)$ at $t = 0, T, 2T, 3T, 4T$, and $5T$.
- (b) How does this response compare with the impulse response of an $RC\,LP$ network with $\omega_c = 1$?
- (c) Determine the digital filter's system function.
- (d) Determine the digital filter's magnitude and phase responses.

9-9. Apply the impulse-invariant method to the second-order Butterworth response to obtain the digital filter system function.

9-10. Draw the canonic (direct) schematic for the system function of Problem 9-4.

9-11. Consider the digital filter system function

$$H(z) = \frac{a_z - 1}{z - a} \qquad a < 1$$

- (a) Show that $H(z)$ is an all-pass response, that is, $|H(e^{j\omega T})| = 1$.
- (b) What is the phase function?
- (c) What is the group delay function?
- (d) How is the pole of $H(z)$ related to the zero?

Answers to Problems

CHAPTER 1

1-1. (a) L.I.
(b) L.I.
(c) L.I.
(d) L.D.
(e) L.I.
(f) L.D.
(g) L.I.
(h) L.I.

1-2. (a) $u_h(t) = c_1 e^{-4t} + c_2 e^{4t}$
(b) $u_h(t) = c_1 e^t + c_2 t e^t + c_3 t^2 e^t$
(c) $u_h(t) = c_1 + c_2 t + c_3 t^2 + c_4 t^3$
(d) $u_h(t) = c_1 e^{2t} + c_2 e^{-t} \cos \sqrt{3} t + c_3 e^{-t} \sin \sqrt{3} t$
(e) $u_h(t) = c_1 e^{-3t} \sin 4t + c_2 e^{-3t} \cos 4t$

1-3. (a) $a \ne b$, $\quad k(t-\tau) = \dfrac{1}{b-a}[e^{-a(t-\tau)} - e^{-b(t-\tau)}]$

$a = b$, $\quad k(t-\tau) = (t-\tau)e^{-b(t-\tau)}$

(b) From (1.4-4), $u(t) = \dfrac{1}{ab}\left[1 - \dfrac{be^{-at} - ae^{-bt}}{b-a}\right] u_{-1}(t)$

1-4. From (1.4-1), $v_1(\tau) = \dfrac{-u_2(\tau)}{a_0(\tau)W(\tau)}$, $v_2(\tau) = \dfrac{u_1(\tau)}{a_0(\tau)W(\tau)}$

1-5. (a) From Problem 1-4, $k(t,\tau) = \dfrac{\tau+5}{(t+5)(\tau+4)^2} - \dfrac{e^{-(t-\tau)}}{(\tau+4)^2}$

(b) $u(t) = \displaystyle\int_0^t k(t,\tau) f(\tau)\, d\tau + \dfrac{7}{8} e^{-t} + \dfrac{25}{8(t+5)}$

(c) From (1.8-3), $u(t) = h(t,2) = \dfrac{1}{36}\left[\dfrac{7}{t+5} - e^{-(t-2)}\right] u_{-1}(t-2)$

1-6. From (1.6-3)
(a) e^{-48}
(b) e^{π}
(c) 0

1-7.

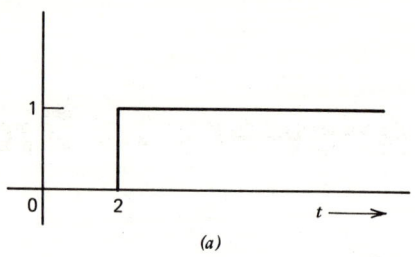

(a)

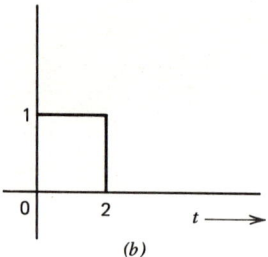

(b)

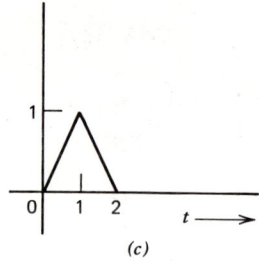

(c)

Figure A.1-7.

1-8. (a) $k(t-\tau) = \dfrac{R}{L} e^{-(R/L)(t-\tau)}$; $h(t-\tau) = u_{-1}(t-\tau)k(t-\tau)$

(b)

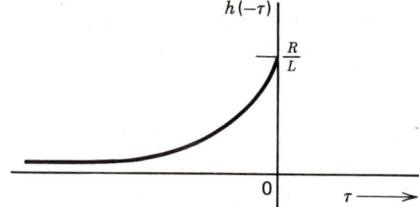

Figure A.1-8.

(c) $u(t) = \begin{cases} 1 - e^{-(R/L)t} & 0 \le t \le T \\ e^{-(R/L)t}(e^{(R/L)T} - 1) & t > T \end{cases}$

General shape is shown in Fig. 1-5.

(d) (i) Response is that of (c) with $T = \infty$
$$u(t) = (1 - e^{-(R/L)t})u_{-1}(t)$$

(ii) Impulse response ($\tau = 0$)
$$u(t) = \frac{R}{L} e^{-(R/L)t} u_{-1}(t)$$

(iii) Impulse response ($\tau = 5$)
$$u(t) = \frac{R}{L} e^{-(R/L)(t-5)} u_{-1}(t-5)$$

1-9. (a) $\alpha = 2$, $\beta = \pm 3$

(b) $k(t-\tau) = \dfrac{13}{3} e^{-2(t-\tau)} \sin 3(t-\tau)$

Answers to Problems

CHAPTER 2

2-1. (a) From (2.1-1), $F(\omega) = \dfrac{\sin \omega}{\omega} + j\dfrac{\cos \omega - 1}{\omega}$. From (2.1-4), $A(\omega) = \left|\dfrac{\sin(\omega/2)}{\omega/2}\right|$

(b) $F(\omega) = A(\omega) = \dfrac{2a}{a^2 + \omega^2}$

2-2. (a) $F(\omega) = \dfrac{R(1 - e^{-j\omega T})}{j\omega(R + j\omega L)}$

(b) $A(\omega) = \dfrac{RT}{\sqrt{R^2 + \omega^2 L^2}} \left|\dfrac{\sin(\omega T/2)}{\omega T/2}\right|$

(c) From (2.1-32) and (2.1-33), $E = \dfrac{T}{R} + \dfrac{L}{R^2}(e^{-(RT/L)} - 1)$

2-3. (a) From (2.5-46), $H(s) = \dfrac{15}{s^3 + 6s^2 + 15s + 15}$

(b) $\dfrac{d^3 u}{dt^3} + 6\dfrac{d^2 u}{dt^2} + 15\dfrac{du}{dt} + 15u = 15f(t)$

(c) From (2.6-9)
$$D(\omega) = \dfrac{225 + 45\omega^2 + 6\omega^4}{225 + 45\omega^2 + 6\omega^4 + \omega^6}$$

(d) $h(t) = \mathscr{L}^{-1}[H(s)] =$ impulse response
$= \{4.53 e^{-2.32t} + e^{-1.84t}[1.25 \sin 1.75t - 4.53 \cos 1.75t]\} u_{-1}(t)$

(e) $e^{-2.32t}$, $e^{-1.84t} \sin 1.75t$, $e^{-1.84t} \cos 1.75 t$

2-4. (a) $L = \sqrt{2}\text{H}$, $C = \dfrac{\sqrt{2}}{2}\text{F}$

(b) $|H(j\omega)| = \dfrac{1}{\sqrt{1 + \omega^4}}$

(c) $\omega = 1$

2-5. (a) $H(s) = \dfrac{s^2}{s^2 + (13/2)s + 25/16}$

(b) poles: $-1/4, -25/4$; zeros: $0, 0$

(c) $h(t) = \left[\delta(t) + \dfrac{1}{96} e^{-t/4} - \dfrac{625}{96} e^{-(25/4)t}\right] u_{-1}(t)$

(d) From (1.8-12), $g(t) = \dfrac{1}{24}[25 e^{-(25/4)t} - e^{-t/4}] u_{-1}(t)$

2-6. $u(t) = A e^{-t} u_{-1}(t)$

2-7. $\theta(\omega) = \dfrac{T_0}{4}(\omega - 3)(\omega - 1) + \theta_0$

2-8. From (2.6-21), $T = 2$ seconds; from (2.6-22), $\theta_0 = 6$ rads

2-9. (a) From Table 2-1, $H(s) = \dfrac{s}{s+a}$

(b) From (2.6-10), $D(\omega) = \dfrac{a}{a^2+\omega^2}$

(c) $g(t) = \mathcal{L}^{-1}\left[\dfrac{H(s)}{s}\right] = e^{-at}u_{-1}(t)$

(d) $u(t) = \dfrac{1}{a-\alpha}[ae^{-at} - \alpha e^{-\alpha t}]u_{-1}(t)$

(e) $|H(j\omega)| = \dfrac{\omega}{\sqrt{a^2+\omega^2}}$

2-10. From (1.8-12), $g(t) = \begin{cases} \frac{1}{2}t^2 & 0 \leq t \leq 1 \\ 2t - \frac{1}{2}t^2 - 1 & 1 \leq t \leq 2 \\ 1 & t \geq 2 \end{cases}$

CHAPTER 3

3-1. (a) $L = \dfrac{\sqrt{2}}{2}$ H, $C = \sqrt{2}$ F

(b) $L = 0.45388$ H, $C = 1.36165$ F

3-2. $\alpha/\beta = \sqrt{3}$

3-3. From (3.2-40), $a_1 = \dfrac{36\sqrt{2}}{35}$, $a_2 = \dfrac{-2\sqrt{2}}{7}$

3-4. From (3.4-2), $n = \dfrac{\log[10^{0.1A_s} - 1]}{2\log\omega_s}$

3-5. (a) From (3.2-34) and (3.4-17), $A = 40.414$ dB
(b) From (3.4-27), $\omega_{3dB} = 1.0739$

3-6. From Table 3-1, $\omega_{3dB} = 1.38899$. At $\omega = 2.77798$, $A = 21.48$ dB

3-7. $H(s) = \dfrac{945}{s^5 + 15s^4 + 105s^3 + 420s^2 + 945s + 945}$

3-8. (a) $L = \frac{1}{2}\sqrt{\sqrt{2}-1} = 0.3218$ H, $C = 4L = 1.2872$ F
(b) Phase delay = 1.245 sec

3-12. From (3.8-12), $\omega_N = \sqrt{\dfrac{\pi}{4\ln 2}}\,\omega_c = 1.0645\,\omega_c$

CHAPTER 4

4-1. $C_1 = C_5 = 983.6$ pF, $L_2 = L_4 = 2.575$ MH, $C_3 = 3183.1$ pF

Answers to Problems

4-2. (a) $n = 4$

(b) $h_{peak} = 0.382 \, \omega_c = 7200$ V, time of peak $= \dfrac{2.9}{\omega_c} = 153.8 \, \mu\text{sec}$

4-3. $t_r = \dfrac{2.15}{\omega_c} = 3.422 \, \mu\text{sec}$

4-4. (a) 10 kHz

(b) 79.6 μsec

4-5.

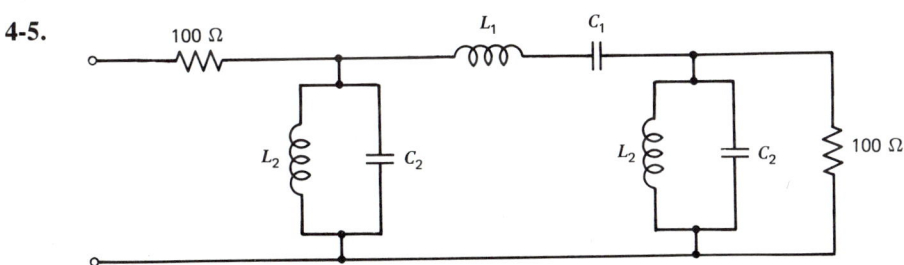

Figure A.4-5.

$L_1 = 0.636 \, \mu\text{H}$ $C_1 = 11.05$ pF
$L_2 = 0.221 \, \mu\text{H}$ $C_2 = 31.83$ pF

f (MHz)	Delay (nsec)
40	25.86
50	17.49
60	12.73
70	11.98
80	13.22
90	11.49

4-6.

f(MHz)	59	59.2	59.5	60	60.5	60.8	61
Delay (μsec)	0.401	0.442	0.376	0.318	0.369	0.430	0.395

From Fig. 3-26
  envelope max $= 0.4(\omega_2 - \omega_1) = 5.03 \times 10^6$ V
  time of peak $= \dfrac{2}{\omega_2 - \omega_1} \times 2 = 0.3183 \, \mu\text{sec}$

4-7. (a) From Fig. 3-24 and (4.4-11), $f_2 - f_1 = \dfrac{30 \times 10^6}{4} = 7.5$ MHz

(b) From (4.4-12), $f_0 = 25.98$ MHz

(c) From (4.4-21), $D_b(\omega_0) = 137.3$ nsec

4-8.

n	q_1	k_{12}	k_{23}	k_{34}	k_{45}	q_n
2	1.414	0.707				1.414
3	1.000	0.707	0.707			1.000
4	0.765	0.841	0.541	0.841		0.765
5	0.618	1.000	0.556	0.556	1.000	0.618

4-9. $L_I = L_{II} = L_{III} = 4$ mH, $C_{12} = C_{23} = \dfrac{\sqrt{2}}{100} \mu$F

$C_I = C_{III} = 254.5$ pF, $C_{II} = 259.2$ pF

$D_b(\omega_0) = 160$ μsec

4-10. $\dfrac{Z}{R_1} = \left(\dfrac{k_{23}}{k_{12}}\right)^2 \dfrac{q_3}{q_1}$. From (4.5-32) and (4.5-34), $\dfrac{Z}{R_1} = \dfrac{1}{R_s}$

4-11. (a)

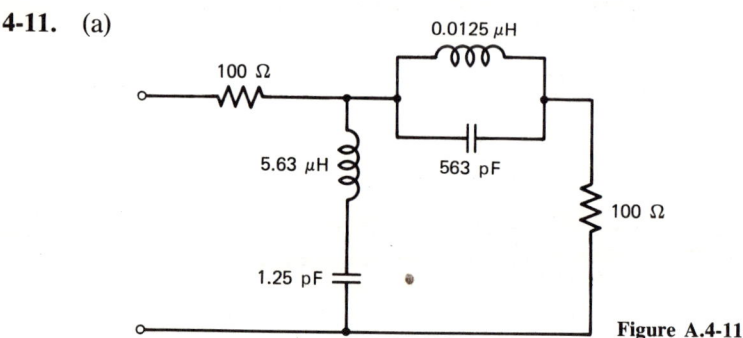

Figure A.4-11.

(b) $\Omega = 2$, attenuation $= 12.3$ dB
(c) $\Omega = \infty$, attenuation $= \infty$ dB

CHAPTER 5

5-1. (a) $D(\omega) = \dfrac{6\omega^2 + 12}{\omega^4 + 5\omega^2 + 4}$

(b) $h(t) = \delta(t) + [6e^{-t} - 12e^{-2t}]u_{-1}(t)$

5-2. $\gamma = 1$ or $\gamma = 0.2956$

5-3. $\gamma = 5000$

5-4. $\mu_p = 0.8$, $\bar{\omega}_0 = 7.755 \times 10^6$ rads/sec, $k = 1.14473$
$L = 14.76$ μH $L_1 = 11.26$ μH
$C = 1126.4$ pF $C_1 = 1476.1$ pF

5-5. $g_2(t) = (1 - 4\alpha t e^{-\alpha t})u_{-1}(t)$, $(\alpha = \bar{\omega}_0, \beta = 0)$

5-6. (a) π
(b) 2π

Answers to Problems

5-7. (a) From (5.2-19), $\alpha = \frac{3}{2}$, $\beta = \frac{1}{2}\sqrt{55}$. From (5.2-22), $k = \frac{3}{4}$, $\bar{\omega}_0 = 4$
From Fig. 5-20b, $L = \frac{3}{16}$ H, $C = \frac{1}{3}$ F, $C_1 = \frac{3}{16}$ F, $L_1 = \frac{1}{3}$ H
(b) From (5.2-24), $D_2(0) = \frac{3}{8}$ sec
(c) From (5.2-33)
$$h_2(t) = \delta(t) - 6e^{-(3/2)t}\left[\cos\frac{\sqrt{55}}{2}t - \frac{3}{\sqrt{55}}\sin\frac{\sqrt{55}}{2}t\right]u_{-1}(t)$$

5-10. (a) From (5.2-25), $k = 0.5$; from (5.2-26), $\bar{\omega}_0 = 10^6$
(b) From Figs. 5-20b and 5-23a,
$$2L = 5 \text{ MH}, \quad C = 400 \text{ pF}, \quad \frac{L_1}{2} = 5 \text{ MH}, \quad C_A = 266.67 \text{ pF}$$

CHAPTER 6

6-1. (a) $H(s + \frac{1}{4}) = \dfrac{1}{s^3 + (11/4)s^2 + (51/16)s + 105/64}$
(b) From (6.3-15), $l_d = 4.343$ dB; from (a), $l_d = 4.3$ dB

6-2. From (6.3-15) or Fig. 6-19, $q = 11.24$. From Fig. 6-9, lossless bandwidth = $5.32 = 10^4$ rads/sec; and from (6.2-20), $Q = 211$

6-3. (a) $Q = 300$, $q = 6$, $l_d = 4.69$ dB
(b) From Table 6-5, $q_{\min} = 3.236$, $Q_{\min} = 161.8$
(c) From Fig. 6-9, lossy bandwidth = $0.89 \times 2 \times 10^5 = 1.78 \times 10^5$ rads/sec

6-4. (a) $l_0 = 20 \log \dfrac{3}{\sqrt{5}}$ dB $= 2.553$ dB
(b) From (6.3-7), $l_m = 2.553$ dB
(c) $l_d = 0$ dB

6-5. (a) $l_0 = 6.005$ dB
(b) From (6.3-7), $l_m = 2.553$ dB
(c) From (6.3-10), $l_d = 3.452$ dB

6-6. (a) $l_0 = 20 \log \dfrac{186}{125} = 3.452$ dB
(b) $l_m = 0$ dB
(c) From (6.3-10), $l_d = 3.452$ dB

6-7. From Fig. 6-13, $f_c = \dfrac{100}{0.905} = 110.5$ kHz

6-8. (a) $-0.132683 \pm j0.923880$
$-0.673880 \pm j0.382683$
(b) From (6.4-5), $l_p = 10.49$ dB

6-9. From (6.4-5), $l_p = 6.91$ dB

6-10. (a) $L = \dfrac{q}{\sqrt{2}q - 2} H$, $C = \dfrac{q(\sqrt{2}q - 2)}{q^2 - \sqrt{2}q + 1} F$

(b) $\dfrac{q^2 - \sqrt{2}q + 1}{q^2}$

CHAPTER 7

7-1. (a) From (7.1-18), $\phi_f(\tau) = \begin{cases} \frac{1}{3} - \frac{1}{2}|\tau| + \frac{1}{6}|\tau|^3 & -1 \leq \tau \leq 1 \\ 0 & \text{elsewhere} \end{cases}$

(b) From (7.1-22), $\phi_f(0) = \dfrac{1}{3}$

7-2. (a) From (7.1-20), $\Phi_f(\omega) = T^2 \left(\dfrac{\sin(\omega T/2)}{\omega T/2} \right)^2$

(b) From (7.1-18), $\phi_f(\tau) = \begin{cases} T - |\tau| & -T \leq \tau \leq T \\ 0 & \text{elsewhere} \end{cases}$

7-3. From (7.1-49), $\Phi_u(\omega) = \dfrac{N_0}{1 + (\omega/\omega_c)^{2n}}$

7-4. From (7.1-1), $\Phi_f(\omega) = \dfrac{\sqrt{\pi}}{a} e^{-(\omega/2a)^2}$

7-5. From Table 7-2 (aperiodic functions)

$$\phi_{fg}(\tau) = \begin{cases} \dfrac{1 - e^{-\alpha(\tau + T)}}{\alpha} & -T \leq \tau \leq 0 \\ \dfrac{e^{-\alpha\tau}(1 - e^{-\alpha T})}{\alpha} & \tau \geq 0 \\ 0 & \tau \leq -T \end{cases}$$

7-6. (a)

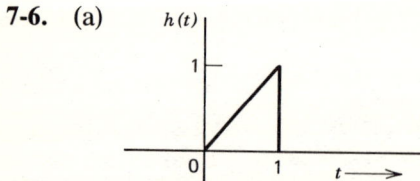

Figure A.7-6a.

(b) $H(s) = \dfrac{1 - (s + 1)e^{-s}}{s^2}$

Answers to Problems

(c)

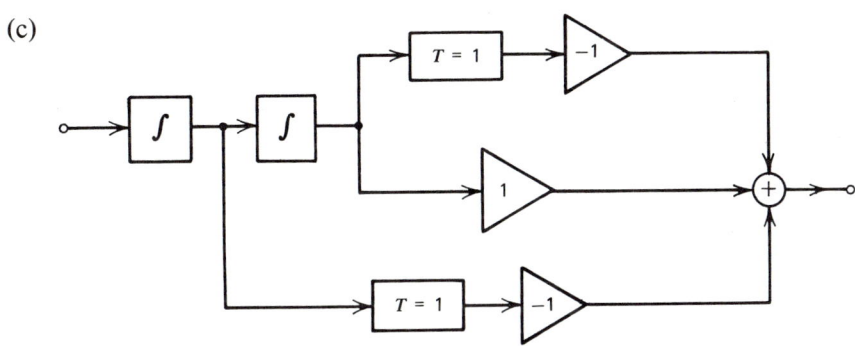

Figure A.7-6c.

7-8. (a) 28

(b) $a_0 = \frac{3}{7}$, $a_2 = \frac{1}{7}$, $a_4 = \frac{1}{28}$

(c) From (7.5-27), SNR $= \frac{784}{326} \to 3.81$ dB

CHAPTER 8

8-1. $b_1 = \frac{2}{21}$, $b_0 = \frac{1}{84}$, $a_2 = \frac{3}{7}$, $a_1 = \frac{1}{14}$, $a_0 = \frac{1}{210}$

$h_a(t) = [10.8e^{-5.66t} - 8.38e^{-4.67t} \cos(3.93t + 0.152)]u_{-1}(t)$

8-2. (a) $H(s) = \dfrac{12}{s^3 + 5s^2 + 12s + 12}$,

$h(t) = \left\{3e^{-2t} - 3e^{-(3/2)t}\left[\cos\dfrac{\sqrt{15}}{2}t - \dfrac{1}{\sqrt{15}}\sin\dfrac{\sqrt{15}}{2}t\right]\right\}u_{-1}(t)$

(b) $L_1 = \frac{4}{5}$ H, $L_2 = \frac{1}{5}$ H, $C = \frac{25}{48}$ F

8-3. From (8.1-34), $A_1 = \frac{12}{35}$, $A_2 = \frac{24}{35}$

8-4. $A_1 = A_2 = \frac{1}{2}$

8-5. $H[j\omega(t)] = \dfrac{1}{1 - 2\omega^2(t) + j\omega(t)[2 - \omega^2(t)]}$

8-6. From (8.2-28)

$$u_q(t) = \frac{a^n}{(a + j\omega_v)^n}(\cos \omega_a t + \cos \omega_b t)\cos \omega_v t$$

8-7. $u(t) = u_{-1}(t) \sum_{k=1}^{N} \dfrac{A_k(t)}{S_k - \alpha} (e^{-\alpha t} - e^{-s_k t})$

8-8. $\alpha(t) = $ a constant

8-9. From (8.4-22), $\tau_p = \dfrac{1}{\omega_c} \ln(2 - e^{-\omega_c T})$

8-10. $\tau = \dfrac{1}{a - \omega_c} \ln\left[\dfrac{a + \omega_c}{2\omega_c}\right]$

CHAPTER 9

9-1. (a) $F(z) = \dfrac{z \sin \omega_0 T}{z^2 - 2z \cos \omega_0 T + 1}$

(b) $F(z) = \dfrac{z}{z - 1}$

9-2. From (9.4-38), $f(nT) = e^{-nT} - e^{-2nT}$

9-3. $T_{\max} = 0.25$ sec

9-4. By long division, $h(2T) = 3$

9-5. $h(t) = \sqrt{2} e^{-(\sqrt{2}/2)t} \sin \dfrac{\sqrt{2}}{2} t$. From Table 9-3,

$$H(z) = \sqrt{2} \dfrac{\left(e^{-(\sqrt{2}/2)T} \sin \dfrac{\sqrt{2}}{2} T\right) z}{z^2 - \left(2e^{-(\sqrt{2}/2)T} \cos \dfrac{\sqrt{2}}{2} T\right) z + e^{-\sqrt{2}T}}$$

9-6. From (9.4-38)

$$f(nT) = \dfrac{1}{2 - e^{-T}} [2^{n+1} - e^{-(n+1)T}]$$

9-7. (a) $H(s) = \dfrac{1 - (s+1)e^{-s}}{s^2}$

(b) No. $H(s)$ is not a rational function.

(c) $H(z) = 0.2z^{-1} + 0.4z^{-2} + 0.6z^{-3} + 0.8z^{-4} + z^{-5}$

(d) Nonrecursive. Unit-sample response is finite.

Answers to Problems

(e)

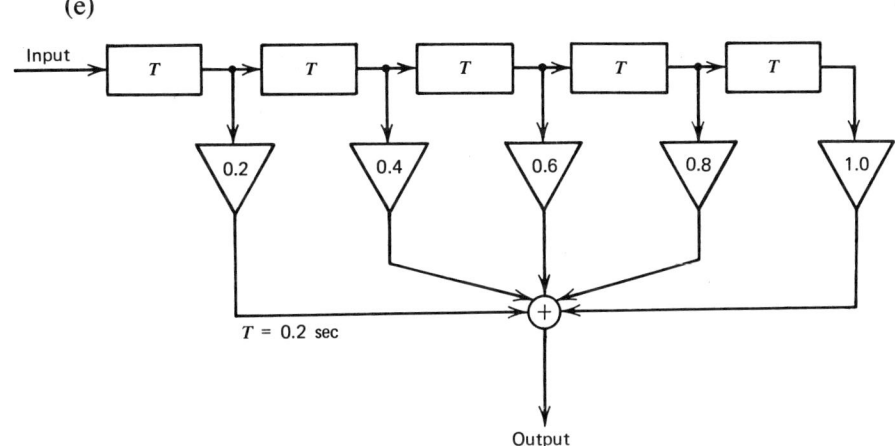

Figure A.9-7.

(f) $U(z) = F(z)H(z)$
$$= 0.2z^{-1} + 0.6z^{-2} + 1.2z^{-3} + 1.8z^{-4} + 2.4z^{-5} + 1.8z^{-6} + z^{-7}$$

9-8. (a) $u(0) = 1$, $u(T) = e^{-T}$, $u(2T) = e^{-2T}$, $u(3T) = e^{-3T}$, $u(4T) = e^{-4T}$, $u(5T) = e^{-5T}$
(b) It is the sampled values of the RC network's impulse response.
(c) $H(z) = \dfrac{1}{1 - e^{-T}z^{-1}}$

(d) $|H(e^{j\omega T})| = \dfrac{1}{\sqrt{1 + e^{-2T} - 2e^{-T}\cos \omega T}}$

$\theta(\omega) = \tan^{-1} \dfrac{\sin \omega T}{\cos \omega T - e^T}$

9-9. $H(z) = \sqrt{2} \dfrac{ze^{-(\sqrt{2}/2)T} \sin \dfrac{\sqrt{2}}{2} T}{z^2 - 2ze^{-(\sqrt{2}/2)T} \cos \dfrac{\sqrt{2}}{2} T + e^{-\sqrt{2}T}}$

9-10. See Fig. 9-33a and (9.5-9)
$L_0 = 0$, $L_1 = 2$, $L_2 = 5$, $L_3 = 6$, $L_4 = 5$, $L_5 = 4$
$K_1 = 1$, $K_2 = 10$, $K_3 = 7$, $K_4 = 6$, $K_5 = 3$

9-11. (b) $\theta(\omega) = \tan^{-1} \dfrac{(1 - a^2) \sin \omega T}{2a - (1 + a^2) \cos \omega T}$

(c) Group delay $= \dfrac{(1 - a^2)T(1 + a^2 - 2a \cos \omega T)}{[2a - (1 + a^2) \cos \omega T]^2 + (1 - a^2)^2 \sin^2 \omega T}$

(d) They are reciprocals. This is a necessary condition for an all-pass $H(z)$.

Index

Adder, 424
Age variable, 20, 399
Aigrain, P. R., 381
Aliasing, 433, 444
All-pass filter, applications of, 212-214
   for equalizing the guard filter, 444
   as a matched filter for linear FM, 349
   response, 81
   transfer function, 215
All-pass function, 90
   approximation techniques for, 214
   properties of, first-order, 215-217
   second order, 220-225
All-pass networks, 237-246
   constant-resistance property of, 215
   effects of losses in second-order, 242-244
All-pole networks, definition of, 107-108
Amplitude density spectrum, 28
Analog-to-digital (A/D) converter, 422
Anticipative system, definition of, 3
   relationship to differential equation, 5, 19
Aperiodic function analysis, 297-300
Approximation, Chebyshev, 100-103
   Fourier series, 417, 456-458
   least-squares, 103-106, 377-383
   by a separable system, 402-403
   Taylor, 96-100
   in time domain, 372
Arithmetic, center frequency, 178
   symmetry, 195, 196
Attenuation, in decibels, 60
   in nepers, 60
Attenuation characteristics, of Butterworth digital filter, 447, 451, 459
   of Butterworth filters, 111
   of equal-element filter, 282-283
   of lossy Butterworth filters, 252-265
   of lossy Chebyshev filters, 268-269
   of lossy second-order all-pass filter, 242
   of wide-band constant delay filters, 200
Autocorrelation function, of an aperiodic function, 297
   definition of, 293
   Fourier transform of, 293
   of linear system output, 307-308
   of a periodic function, 294
   properties of, 304
   of a random function, 301
Average delay, definition of, 403-404
   of the impulse function, 405-406
   of the rectangular pulse, 408-409
   of a sinusoid, 404-405

   of the step function, 406-408

Band-limiting, implication of, 88, 415-420, 433-434
   transformation, 448
Bandpass filter, 167-205
   conventional transformation, 167-177
   narrow-band transformation, 178-195
   response definition, 80
   wide-band constant delay, 196-205
Bandstop filter, 205-209
   conventional transformation, 205-207
   narrow-band transformation, 207-209
   response definition, 80
Bandwidth, relative, 168
Bandwidth change, of lossy Butterworth filters, 266
   of lossy Chebyshev filters, 271
   of lossy Gaussian filters, 273
Barker codes, 353-355
   sidelobe reduction, 357-365
Bessel filter, 124-127
   impulse response overshoots of, 131
Bilinear z-transform, 447-450
   example of, 449-450
Bits, 422, 467
Blinchikoff, H. J., 399
Bounded variation, definition of, 29
Bridged-Tee network, 91-92, 241
Bridge network, 91
   first-order all-pass, 240
   second-order all-pass, 240-241
Bucy, R. S., 309, 311
Bulk-wave delay line, 349
Butterworth, approximation, 109-117
   polynomials, 110
Butterworth delay equalization, with a first-order all-pass function, 217-219
   with a second-order all-pass function, 255-228, 235-237
Butterworth filter, analysis with losses, 251-270
   summary of delay definitions of, 410
Butterworth function, 97
   linear phase analysis of, 230-232
Butterworth-Thomson filters, 135

Canonic digital network, 463-464
Cardinal function, 82
   interpolation using the, 420
Carson, J. R., 387, 394
Carson's rule, 385
Cascade form of digital network, 464

example of, 466-467
Cauer filter, 137-139
    3-dB frequency of, 138-139
Cauer, W., 137
Causal system, *see* Nonanticipative system
Cavity filter, 187
Center frequency, arithmetic, 178
    geometric, 169
Characteristic equation, 7
Characteristic polynomial, 7, 50
Charge transfer device, 349, 367-368
Chebyshev, approximation, 100-103
    delay, 128, 214
    filter analysis with losses, 270
    3-dB frequency, 119
    magnitude approximation, 117-123
    polynomials, 102
Chebyshev response, relationship to Butterworth response, 119
Class L filter, 123
Cohn, S. B., 281
Complex frequency, 38
    imaginary part of, 56
    real part of, 56
Convolution, using the FFT, 471
    relationship to cross-correlation, 305-307
Convolution integral, 12-15
    definition of, 12
    graphical interpretation, 13-14
    physical interpretation, 21
    as the system output, 21
Cooley, J. W., 469
Correlation, using the FFT, 471
    by matched filter, 326-329
Cosine-power weighting, 346
Coupled resonator filters, 188-195
    features of, 194-195
    mesh design of, 189-190, 192-194
    nodal design of, 188-189, 191-192
Cross-correlation, function, 304-305
    practical uses of, 329
    relationship to convolution, 305-307
    relationship to matched filtering, 326-329
Current transfer function, 53
Cutoff frequency, 79-80

Decibel, definition of, 60
Delay, to AM signals, 73-74
    definitions for Butterworth filters, 410
    to FM signals, 74-75
    symbol for, 424
    system, 70-72
    through a time-invariant system, *see* Average delay
    see also Group delay and Phase delay
Delay equalization, 212
    of Butterworth filters, 217-219, 225-228, 235-237

for digital systems, 213
    effect on transient responses, 228-230
    example of bandpass, 244-246
    using least-squares criterion, 233-237
    of narrow-band filters, 232
Delta function, 15
Denormalization procedure, *see* Frequency transformation
Deterministic function, 293
Difference equation, 422-424
    classical solution, 436-437
    solution by z-transforms, 437
Differential bridge, 92
Differential equation, 3-12
    definition of, 3-5
    existence theorem for, 5
    solution by Laplace transforms, 49
Digital computer, as a digital filter, 424
Digital domain, analogy with analog domain, 415
Digital filter, 366-367, 414-471
    advantages of, 414
    definition of, 414, 421-422
    design techniques for, 442-462
    limitations of, 414-415
    simulation, 424
Digital networks, 463-467
Digital signal, 414
Digital system, response calculation of, 439-440
    system function of, 438
    unit-sample response of, 438
Digital-to-analog (D/A) converter, 422
Dirac delta function, 15
Dirac, P., 15
Direct form, of digital network, 463-464
Discrete Fourier transform, 469
Discrete-time convolution, 438
    approximating analog convolution, 421
Dishal, M., 186
Dissipation, *see* Loss
Distributed capacitance, 190, 240
Distribution theory, 29, 34, 37
Dolph-Chebyshev weighting, 346, 460
Dynamic approach to modulated signal analysis, 385-387

Element values, of Butterworth filters, 110, 115
    of Chebyshev filters, 119, 122
    of equal-element filter, 284
    of equiripple impulse response filter, 133
    of wide-band constant delay filter, 203-204
Elliptic filter, *see* Cauer filter
Energy, of an aperiodic function, 32
    of an exponential, 34
    of high-pass step response, 158
    of low-pass impulse response, 149, 158
Energy density function, of an aperiodic function, 294, 297

Index

relationship to amplitude density function, 32
Envelope delay, *see* Group delay
Envelope response, of bandpass filters, 185-186
  of bandstop filters, 208-209
Equal-element filter, 281
Equiripple approximation, *see also* Chebyshev, of transient response overshoots, 132-134
Error, criterion, 96
  function, 95
Errors, in digital hardware, 467-469
Estimation, linear mean-square, 309-312
Euler's identity, 8
Even function, 29

False alarm, 328
False-alarm probability, relationship to SNR, 333
Fast Fourier transform (FFT), 424, 469-471
Foldover, 433
Fourier, approach to modulated signal analysis, 383-385
  integral, 28
  spectrum, 28
Fourier series, analysis, 294-297
  approach to nonrecursive filtering, 454-456
Fourier transform, of causal time function, 30-31
  definition of, 27
  discrete, 469
  of the exponential, 32-34
  of the impulse function, 34
  inversion formula for, 28
  of real time-functions, 29-30
  relationship to Laplace transform, 59-60
  of sgn t, 35-36
  of the sinusoid, 35
  of the step function, 36-37
  of symmetric time functions, 31-32
Frequency-domain, characteristics of a sampled function, 432-434
Frequency modulation, definition of, 385
  linear, 335-338
Frequency response, of analog system, 58-60
  of a digital system, 440-442
Frequency transformation, bilinear, 447-450
  effect on group delay, 154
  low-pass to, bandpass, 167-178
    bandstop, 205-209
    denormalized low-pass, 154-157
    high-pass, 157-167
    narrow band, 178
Fry, T. C., 387, 394
Fubini, E. G., 287
Fukada, M., 123

Gain-bandwidth product, 108
Gain constant, 56-57
Gaussian-Chebyshev filter, 135
Gaussian filter, analysis with losses, 270-271
  envelope response to linear FM, 346
  impulse response overshoots of, 131
Gaussian response, 84-85
  approximation, 130-132
  definition of, 84
Geometric symmetry, of bandpass filters, 169
  discussion of, 194-195
Gibbs phenomenon, 458
Gram matrix, 378
Gram-Schmidt procedure, 106
Group delay, 66-75
  of all-pass filter, 215-216, 220-222
  of bandpass filter, 171
  definition of, 66
  derivative at band center, 171
  of high-pass filter, 158
  of low-pass filter, 154
  using narrow-band transformation, 179-183
  of normalized low-pass filter, 108
  relationship to envelope delay, 66, 74
  transformed, 154
  wide-band constant, 196
  at zero frequency, 108-109
Group delay characteristics of, Butterworth filters, 112
  equalized Butterworth filters, 219, 227, 236
  of filter with equiripple transient overshoots, 134, 135
  of first-order all-pass filter, 216
  of lossy Butterworth filter, 266
  of second-order all-pass filter, 222
  of wide-band constant delay filters, 201
Group delay equalization, *see* Delay equalization
Guard filter, 213, 444
Guillemin, E. A., 287

Hamming function, 346, 460
Hanning function, 460
Helical filter, 187
High-pass filter, conventional transformation, 157-165
  preservation of low-pass transients, 165-167
  response definition, 79
  step response energy, 158
Hilbert transform, 75-76, 93-94
Hold function, 422
Homogeneous difference equation, 436-437
Homogeneous differential equation, 6
  constant coefficient, 7
  solution of, 5-8
Huggins, W. H., 127, 399
Humpherys, D. S., 87, 103, 214

Impedance, denormalization, 153

generalized, 54
Improper fraction, 43
Impulse function, 15-16
Impulse-invariant method, 442-447
Impulse response, of anticipative system, 19
   definition of, 18
   example of causal system, 22-23
   of Gaussian filter, 85
   of lossy filter, 250
   of matched filter, 315
   of nonanticipative system, 18-19
   of rectangular magnitude response filter, 82
   relationship to transfer function, 52
   synthesis of, rectangular, 322-323
      trapezoidal, 323-326
   of time-variant system, 400
   of Wiener filter, 310
Impulse response characteristics, of Butterworth filters, 113
   of equalized Butterworth filter, 229
   of first-order all-pass filter, 217
   of second-order all-pass filter, 224
Impulse response energy, of low-pass filter, 149, 158
Impulse response overshoots, of Gaussian and Bessel filters, 131
   of linear phase Butterworth filters, 231
Impulse response peak, 83, 84, 86, 88
Impulse response pulse width, of Gaussian response, 86
   of rectangular magnitude response, 83
Insertion loss, comparison of, 282-286
   definition of, 274-275
   filter with minimum, 281
   mismatch, 275-276
   of predistorted filter, 287
   relationship to group delay, 278-279
   resistive, 276
Insertion loss characteristics, of Bessel filters, 280
   of Butterworth filters, 277
   of Chebyshev filters, 278
   of equal-element filters, 285
   of Gaussian filters, 279
Instantaneous frequency, 387-388, 390, 394-395
   of linear FM, 335
Integral, Fourier, 28
   Fourier inversion, 28
Integration, by parts, 41, 388, 398
   numerical, 233-234
Interpolation functions, 420, 422
Inverse, Fourier transform, 28
   Laplace transform, 43-49
   z-transform, 434-436

Jess, J., 132, 371

k and q values, 187
Kaiser window, 460
Kalman, R. E., 293, 309, 311
Kalman filtering, 311-312
Kaplan, W., 10, 400
Kernel function, 5, 9, 10, 19
   modified, 11
Kolmogoroff, 293, 309
Kronecker delta, 105

Ladder network, 91
Laguerre filter, 383
Laplace transform, 37-43
   definition of, 37
   inverse, 43-49
   region of convergence of, 39
   pairs, 39
Lattice network, 91
   first-order all-pass, 238
   second-order all-pass, 238-239
Least-pth approximation, 106
Least-squares approximation, 103-106
   applied to delay equalization, 233-237
   of delay, 128-129
   of magnitude, 123-124
   in nonrecursive filtering, 456-458
   in the time domain, 377-383
   to wide-band delay, 196
Legendre filter, 123
Leon, B. J., 387, 397
Linear FM signal, 335-338
Linear independence, definition of, 6
Linear independent solutions, of differential equation, 6-7
Linear phase response, 32
   Chebyshev approximation to, 128
   of minimum-phase network, 94
Linear programming algorithm, 361
Linear system, definition of, 2
Loss, Butterworth filter analysis with, 251-270
   Chebyshev filter analysis with, 270
   effect on, all-pass responses, 242-244
      s-plane locations, 249-250
      transfer function, 249-250
      transient responses, 250-251
   Gaussian filter analysis with, 270-271
   insertion, 274-286
Low-pass filter, delay at zero frequency, 108-109
   impulse response energy of, 158
   see also Frequency transformation
Low-pass response, characterization, 106-139
   definition of, 79
   Gaussian magnitude and linear phase, 84-85
   normalized, 107
   rectangular magnitude and linear phase, 81-82

Magnitude response, computation of, 60-61

definition of, 59
  relationship to transfer function, 59, 61
  from s-plane geometry, 64
  in terms of poles and zeros, 63
Matched filter, 293, 312-335
  decoding properties of, 327
  FFT, 471
  frequency-domain characterization, 319-321
  for linear FM, 338-341
  for nonoptimal conditions, 329-333
  for nonwhite noise, 333-335
  output envelope function for linear FM, 340
  realizations, 348-350
  for rectangular pulse, 316-319
  relationship to cross-correlation, 326-329
  for the RF case, 319
  sidelobe reduction, 345-346
  synthesis, 321-322
  technology, 365-368
  for white noise, 313-316
Maximally-flat, approximation, 98
  delay, 105, 124-127
  magnitude, 109-117
Mesh networks, see Coupled resonator filters
Message, 292
Meyer, P., 382
Microwave filtering, 281
Minimum insertion-loss filter, 281
Minimum-phase function, definition of, 89
Minimum-phase network, with linear phase response, 94
  with rectangular magnitude response, 93-94
Minimum quality factor, 286
Mismatch loss, 275-276
Modulated signal, amplitude and frequency of, 388
  analysis by, dynamic approach, 385-387
    spectral approach, 383-385
  delay, 73-75
  response of linear systems to, 383-403
Moments, method of, 372-377
  of unit-sample response, 444
Moving-target-indicator (MTI) filter, 460-462
Multiplier, symbol for, 424

Narrow-band filter, 178-196
  using the FFT, 471
Neper, definition of, 60
Network, all-pole, 107
  narrow-band, 188
  realizations, 92, 239-241
Network transformation, lattice to bridged-Tee, 92, 241
  low-pass to bandpass, 174
    bandstop, 206
    high-pass, 163
  for transformers, 190, 191
Nodal networks, see Coupled resonator filters

Noise, 292
Noise bandwidth, 146-148
  relationship to impulse response energy, 149
  of various filters, 148
Nonanticipative system, definition of, 3
  relationship to differential equation, 5, 19, 20
Noncausal system, see Anticipative system
Nondeterministic function, see Random function
Nonhomogeneous differential equation, solution of, 8-12
Nonminimum-phase function, definition of, 90
Nonrecursive filter, 450-462
  comparison with recursive filter, 453
  definition of, 422
  Fourier series design of, 454-456
  realizations, 458
  system function of, 450
  unit-sample response of, 450
Normalized filter parameters, 153-154
Normalized frequency, 153
  for bandpass filter, 168
  for bandstop filter, 205
  for high-pass filter, 157
  for low-pass filter, 154
Normalized low-pass filter, 153
North, D. O., 312
Nyquist, H., 415
Nyquist, frequency, 416
  interval, 416

Odd function, 29
Optical signal processor, 365-366
Orthogonal filter, 382-383
Orthogonal functions, approximation by, 104-106
  definition of, 104-105
Overshoots, of Bessel impulse response, 131
  of Bessel step response, 127
  of Butterworth impulse response, 231
  of Butterworth step response, 127
  of constant-delay filters, 129
  of equalized Butterworth impulse response, 231
  of Gaussian impulse response, 131

Paley-Wiener condition, 88-89
Papoulis, A., 87, 123
Parallel form of digital network, 464
  example of, 466-467
Parseval's theorem, 32
  for periodic functions, 296
Partial fractions, method of, 43, 44-47
Particular solution, of differential equation, 5
Periodic function analysis, 294-297
Phase characteristics of all-pass filter, first-order, 216
  second-order, 221

Phase-coded waveforms, 350-365
Phase delay, definition of, 66
  negative, 67-68
Phase intercept distortion, 72-73
Phase-lag angle, 60, 91
Phase response, computation of, 60-61
  definition of, 59
  from poles and zeros, 63
  from s-plane geometry, 64-66
Phase-splitting network, 213-214
Physically realizable system, 3
Physically unrealizable system, 3
Poles, of Butterworth delay equalizers, 218, 227, 236
  of Butterworth function, 110
  of Chebyshev function, 119
  definition of, 56
  of equal-element filter, 284
  of least-squares delay function, 129
  pictorial representation, 57
  of transfer function, 56-58
Power density function, of linear system output, 308-309
  of periodic function, 295
  of random function, 294
Predistortion, 286-289
Preselector, 248
Prewhitening filter, 334
Proper fraction, 43
Prototype low-pass filter, 153
Pulse compression, linear delay for, 213
  using linear FM, 335-350
  mechanism, 341-344
  by phase coding, 350
  ratio, 341
Pulse sequences, spectrum of, 355-357
Pulse width of impulse response, 83, 84, 86, 87

Quality factor (Q), all-pass responses due to finite, 242-244
  for insignificant response change, 251
  invariance under frequency scaling, 248-249
  minimum required, 286-287
  normalized, 187
  requirement for bandpass filters, 272
  see also Loss
Quantization, 467
Quasi-stationary analysis, of first-order time-variant system, 397-399
  of separable system, 401-402
  of time-invariant system, 387-395
Quasi-stationary response, applications of, 386, 397
  definition for, time-invariant system, 385
  time-variant system, 396-397
Quasi-stationary transfer function, for time-invariant system, 385
  for time-variant system, 396

Ramp function, 322
Random function, 293
  analysis, 301-304
Range, of radar target, 328
Rational function, 43
Reconstruction filter, 422
Rectangular impulse response, approximation of, 374-375, 379-382
  synthesis of, 322-323
Rectangular magnitude response, 81-82, 93-94, 456-458
  approximations, 109-124
Recursive filter, comparison with nonrecursive filter, 453
  definition of, 422
  design techniques, 442-450
  networks, 463-467
Reflection coefficient, 137
  zeros, 108
Region of convergence, 39, 60
Relative bandwidth, definition of, 168
Resistive loss, 276
Return loss, 137
Rise time, 84, 86-88, 139-146
  of Butterworth filters with linear phase, 231
  maximum slope definition, 143
  normalized to $\omega_{40 \text{ dB}}=1$, 145
  Su's definition, 144
  10 to 90% definition, 141
Round-off error, 468

Sampled data filter, 421
Sampled function, 426
  frequency-domain characteristics of, 432-434
Sampling interval, 415
Sampling theorem, 415-420
Schüssler, H. W., 132, 371
Schwarz inequality, 314
Semi-invariants, 127
Separable system, 399-401
  approximation by a nonseparable system, 402-403
  quasi-stationary analysis of, 401-402
Shannon, C. E., 415
Sidelobe reduction, of Barker waveform, 357-365
  filter, 345, 347
  of linear FM matched filter output, 345-346
Signal, detection, 292
  distortion, 70-72
  extraction, 292
  processing, 292
  processing technologies, 365-368
Signal-to-noise ratio (SNR), 312
  relationship to false-alarm probability, 333
Singly terminated filter, 147
Spectral approach to modulated signal analysis, 383-385

Index

Spectrum, energy density, 294
   Fourier, 26
   power density, 294
s-plane, definition of, 57
   geometry, 64-66
   mapping into z-plane, 431-432
   regions of, 57
   parameter shift due to loss, 249-250
   parameter shift for predistortion, 286
Steady state responses, 58-66
Step function, definition of, 17
   relationship to impulse function, 17
   for synthesis of pulses, 17-18
Step response, comparison of Butterworth and Bessel characteristics, 127
   energy of high-pass filter, 158
   example of causal system, 22-23
   of Gaussian filter, 86-87
   of lossy filter, 251
   monotonic with minimum rise time, 134-135
   of nonanticipative system, 21-22
   of rectangular magnitude response filter, 83-84
   relationship to transfer function, 52
Step response characteristics, of Butterworth filters, 114
   of equalized Butterworth (n=4), 229
   of first-order all-pass filter, 217
   of high-pass Bessel filters, 162
   of high-pass Butterworth filters, 159
   of high-pass Chebyshev filters, 160, 161
   of second-order all-pass filter, 225
Storch, L., 125
Stripline filter, 187
Su, K. L., 142, 410
Superposition integral, 10, 399
   relationship to convolution integral, 13
Superposition property, 2
Surface acoustic wave device, 349, 366
Symmetry, arithmetic, 195, 196
   geometric, 169, 194-195
Synchronous response, 136
System delay, 70-72
System function, of a digital system, 438
   interpretation as a frequency function, 440-442
   time-variant, 399

Tapped delay line, 324, 347-348
Tap weight, 324
Taylor, approximation, 96-100, 109, 373
   weighting, 346, 460
Thomson, W. E., 124
Thomson filter, 125
Threshold device, 328
Time-bandwidth product, 337
   minimization of, 132
Time domain, analysis, 1-24
   approximation, 129-135, 372-383

Time-invariant system, average delay of, 403-410
   definition of, 3
   response to modulated waveform, 383-395
Time-scale ordering, 3, 5, 18, 19
Time-variant system, definition of, 3
   impulse response of, 400
   response to modulated waveform, 395-403
Time-variant system function, 399, 402
   inverse Laplace transform of, 400
Time-varying poles, 400
Transfer admittance, 53
Transfer function, definition of, 50
   derivation, 53-56
   relationship to impulse and step response, 52
   zeros of, 92
Transfer impedance, 53
Transformations, see Frequency transformation and Network transformation
Transformer equivalent, of π network, 190
   of Tee network, 191
Transient response, equiripple approximation of, 132-134
   improvement due to delay equalization, 228-230
   of narrow-band filters, 184-186
   relationship between low-pass and high-pass, 158, 163
Transitional responses, 135-136
Transversal filter, 324, 347
Trapezoidal impulse response, 323-326
Tukey, J. W., 469

Unit impulse function, see Impulse function
Unit-sample response, 438

Voltage standing wave ratio (VSWR), 138
Voltage transfer function, 53, 55

Waveguide filter, 187
Weighting factor, 324
Weighting function, 21
   for digital filtering, 458-460
   for sidelobe reduction, 346
Weiner, D. D., 387, 397
Wiener, N., 293, 309
Wiener-Hopf equation, 310
Wiener-Khintchine Theorem, 294
Wiener-Kolmogoroff filtering, 309-311
White noise, 146, 308, 312, 333
Williams, E. M., 381
Window function, for digital filtering, 458-460
Wronskian determinant, definition of, 6

Zadeh, L. A., 399

Zero initial conditions, definition of, 5
Zeros, definition of, 56
   pictorial representation of, 57
   of transfer function, 56-58
z-plane, mapping the s-plane into the, 431-432

z-transform, 424-436
   application to difference equations, 436-442
   definitions and properties, 426-431
   inverse, 434-436
   pairs, 431
Zverev, A. I., 106, 287

SOUTHEASTERN MASSACHUSETTS UNIVERSITY
TK7872.F5 B56
Filtering in the time and frequency doma

3 2922 00124 428 1